T0137412

NEUROMETHODS

Series Editor
Wolfgang Walz
University of Saskatchewan,
Saskatoon, SK, Canada

For further volumes:
http://www.springer.com/series/7657

Neuromethods publishes cutting-edge methods and protocols in all areas of neuroscience as well as translational neurological and mental research. Each volume in the series offers tested laboratory protocols, step-by-step methods for reproducible lab experiments and addresses methodological controversies and pitfalls in order to aid neuroscientists in experimentation. *Neuromethods* focuses on traditional and emerging topics with wide-ranging implications to brain function, such as electrophysiology, neuroimaging, behavioral analysis, genomics, neurodegeneration, translational research and clinical trials. *Neuromethods* provides investigators and trainees with highly useful compendiums of key strategies and approaches for successful research in animal and human brain function including translational "bench to bedside" approaches to mental and neurological diseases.

Experimental Neurotoxicology Methods

Edited by

Jordi Llorens

Departament de Ciències Fisiològiques, Institut de Neurociènces, Universitat de Barcelona, Hospitalet de Llobregat, Catalunya, Spain; Institut d'Investigació Biomèdica de Bellvitge, IDIBELL, Hospitalet de Llobregat, Catalunya, Spain

Marta Barenys

Departament de Farmacologia, Toxicologia i Química Terapèutica, Universitat de Barcelona, Barcelona, Catalunya, Spain

 Humana Press

Editors
Jordi Llorens
Departament de Ciències Fisiològiques,
Institut de Neurociències
Universitat de Barcelona, Hospitalet de
Llobregat
Catalunya, Spain

Marta Barenys
Departament de Farmacologia, Toxicologia i Química
Terapèutica
Universitat de Barcelona
Barcelona, Catalunya, Spain

Institut d'Investigació Biomèdica de Bellvitge
IDIBELL, Hospitalet de Llobregat
Catalunya, Spain

ISSN 0893-2336   ISSN 1940-6045   (electronic)
Neuromethods
ISBN 978-1-0716-1639-0   ISBN 978-1-0716-1637-6   (eBook)
https://doi.org/10.1007/978-1-0716-1637-6

Cover Caption: This image is courtesy of Marta Barenys.

This Humana imprint is published by the registered company Springer Science+Business Media, LLC, part of Springer
Nature.
The registered company address is: 1 New York Plaza, New York, NY 10004, U.S.A.

Preface to the Series

Experimental life sciences have two basic foundations: concepts and tools. The *Neuromethods* series focuses on the tools and techniques unique to the investigation of the nervous system and excitable cells. It will not, however, short-change the concept side of things as care has been taken to integrate these tools within the context of the concepts and questions under investigation. In this way, the series is unique in that it not only collects protocols but also includes theoretical background information and critiques which led to the methods and their development. Thus, it gives the reader a better understanding of the origin of the techniques and their potential future development. The *Neuromethods* publishing program strikes a balance between recent and exciting developments like those concerning new animal models of disease, imaging, *in vivo* methods, and more established techniques, including, for example, immunocytochemistry and electrophysiological technologies. New trainees in neurosciences still need a sound footing in these older methods in order to apply a critical approach to their results.

Under the guidance of its founders, Alan Boulton and Glen Baker, the *Neuromethods* series has been a success since its first volume published through Humana Press in 1985. The series continues to flourish through many changes over the years. It is now published under the umbrella of Springer Protocols. While methods involving brain research have changed a lot since the series started, the publishing environment and technology have changed even more radically. *Neuromethods* has the distinct layout and style of the Springer Protocols program, designed specifically for readability and ease of reference in a laboratory setting.

The careful application of methods is potentially the most important step in the process of scientific inquiry. In the past, new methodologies led the way in developing new disciplines in the biological and medical sciences. For example, Physiology emerged out of Anatomy in the nineteenth century by harnessing new methods based on the newly discovered phenomenon of electricity. Nowadays, the relationships between disciplines and methods are more complex. Methods are now widely shared between disciplines and research areas. New developments in electronic publishing make it possible for scientists that encounter new methods to quickly find sources of information electronically. The design of individual volumes and chapters in this series takes this new access technology into account. *Springer Protocols* makes it possible to download single protocols separately. In addition, Springer makes its print-on-demand technology available globally. A print copy can therefore be acquired quickly and for a competitive price anywhere in the world.

Preface

Neurotoxicology is devoted to study the adverse effects of chemical and physical agents on the structure or function of the nervous system. At the halfway between toxicology and neuroscience, it has developed its own identity, approaches, and methods. These methods are sometimes adaptations of methods first developed in other areas of neuroscience. In other cases, however, they are original methods developed to respond to specific needs of neurotoxicology, and these have remained within this domain or gained wider use.

This volume includes a collection of detailed protocols currently used in neurotoxicology. It encompasses some well-established methods that constitute gold-standards to approach key questions frequently addressed in neurotoxicological research. Other chapters describe protocols for new approaches that appear as important advancements in the field called to have an impact on its development. The number of chapters that could be included in this volume is enormous; therefore, a strategic selection has been made. This selection provides an overall vision of the field, illustrating the variety of available approaches, from the molecular level (Chapts .3, 12, 13, 14, and 20) to the organism level (Chaps. 5–11, 15–18), by way of methods at the cellular (Chaps. 1–4, 13, 19–22) and tissue/organ level (chapters 2, 4–7, 10, 18, and 19). For the sake of completeness of the view, both in vitro and in vivo methods have been included. Nevertheless, the selection has been biased favoring in vivo methods because in vitro methods in neurotoxicology have been addressed in other recent books. Therefore, in vitro methods are included to provide the intended broad coverage, but they are represented less than their present weight in the field. Among organisms, we have covered methods based on a broad range of species, including the most common laboratory animals, rodents, and alternative organism models, but also human cells, thus circumventing the need of interspecies extrapolation. Globally, we aim to provide a tool useful to many laboratories, either new or established in the field, which may need to implement new protocols to address neurotoxicological questions.

The 22 chapters included in the volume have been written by researchers that are recognized internationally for their expertise in the field and who are describing methods of current use in their laboratories. They have been peer-reviewed and revised according to the reviewer's criticisms. Therefore, each chapter provides an authoritative view and in-depth knowledge on the specific method being covered. Most chapters include detailed technical information for thorough reproducibility of the methods, as well as ample discussion of their applicability domain, and their advantages and limitations, with the aim of settling knowledge of the field and make it progress. This will make them useful to both young and experienced researchers, as an authoritative guidance for implementing any of these methods to the reader's laboratory or for a critical understanding of data generated using them.

Barcelona, Catalunya (Spain)

Jordi Llorens
Marta Barenys

Contents

PART I   CYTOLOGICAL AND HISTOLOGICAL EVALUATION IN RODENTS

PART II   PHYSIOLOGICAL EVALUATION IN RODENTS

PART III   BEHAVIORAL EVALUATION IN RODENTS

Contributors

ILARY ALLODI • *Department of Neuroscience, University of Copenhagen, Copenhagen, Denmark*

ORITOKE M. ALUKO • *The Neuro-Lab, School of Health and Health Technology, Federal University of Technology, Akure, Nigeria; Department of Physiology, School of Health and Health Technology, Federal University of Technology, Akure, Nigeria*

LEANDRO DE O. AMARA • *Postgraduate Program in Pure and Applied Chemistry, Federal University of Western of Bahia, Bahia, Brazil*

MICHAEL ASCHNER • *Departments of Molecular Pharmacology and Neurosciences, Albert Einstein College of Medicine, The Bronx, NY, USA*

ALEJANDRO BARRALLO-GIMENO • *Departament de Ciències Fisiològiques, Institut de Neurociències, Universitat de Barcelona, Hospitalet de Llobregat, Catalunya, Spain; Institut d'Investigació Biomèdica de Bellvitge, IDIBELL, Hospitalet de Llobregat, Catalunya, Spain*

KRISTINA BARTMANN • *IUF-Leibniz Research Institute for Environmental Medicine, Duesseldorf, Germany*

PERE BOADAS-VAELLO • *Research Group of Clinical Anatomy, Embryology and Neuroscience (NEOMA), Departament de Ciències Mèdiques, Facultat de Medicina, Universitat de Girona, Girona, Catalunya, Spain*

SARA BOLIVAR • *Department of Cell Biology, Physiology and Immunology, Institute of Neurosciences, Centro de Investigación Biomédica en Red sobre Enfermedades Neurodegenerativas, Universitat Autònoma de Barcelona, Bellaterra, Spain*

WILLIAM K. BOYES • *Public Health and Integrated Toxicology Division, Center for Public Health and Environmental Assessment, Office of Research and Development, U.S. Environmental Protection Agency, Research Triangle Park, NC, USA*

DONALD A. BRUUN • *Department of Molecular Biosciences, School of Veterinary Medicine, University of California, Davis, Davis, CA, USA*

VALENTINA A. CAROZZI • *Experimental Neurology Unit, Department of Medicine and Surgery, University of Milan-Bicocca, Milan, Italy; Young Against Pain Group, Milan, Italy*

GUIDO CAVALETTI • *Experimental Neurology Unit, Department of Medicine and Surgery, University of Milan-Bicocca, Milan, Italy; Experimental Neurology Unit, School of Medicine and Surgery, and NeuroMI (Milan Center for Neuroscience), University of Milano-Bicocca, Monza (MB), Italy*

MEGAN CHESNUT • *Center for Alternatives to Animal Testing (CAAT), Johns Hopkins Bloomberg School of Public Health, Baltimore, MD, USA*

GABRIELLE CHILDERS • *Neurotoxicology Group, National Toxicology Program Laboratory, National Institute of Environmental Health Sciences, Research Triangle Park, NC, USA; University of Alabama at Birmingham, Birmingham, AL, USA*

KAYLA D. COLDSNOW • *Department of Biological Sciences, Rensselaer Polytechnic Institute, Troy, NY, USA*

DEBORAH A. CORY-SLECHTA • *Department of Environmental Medicine, University of Rochester, Medical School, Rochester, NY, USA*

MERITXELL DEULOFEU • *Research Group of Clinical Anatomy, Embryology and Neuroscience (NEOMA), Departament de Ciències Mèdiques, Facultat de Medicina, Universitat de Girona, Girona, Catalunya, Spain*

DAVID C. DORMAN • *College of Veterinary Medicine, North Carolina State University, Raleigh, NC, USA*

SUSAN G. DORSEY • *Department of Pain and Translational Symptom Science, School of Nursing, University of Maryland, Baltimore, MD, USA*

MELANIE L. FOSTER • *Integrated Laboratory Systems LLC., Raleigh, NC, USA*

ELLEN FRITSCHE • *IUF-Leibniz Research Institute for Environmental Medicine, Duesseldorf, Germany; Heinrich-Heine-University, Duesseldorf, Germany*

GIULIA FUMAGALLI • *Experimental Neurology Unit, School of Medicine and Surgery, and NeuroMI (Milan Center for Neuroscience), University of Milano-Bicocca, Monza (MB), Italy*

DAVID V. GAUVIN • *Neurobehavioral Sciences Department, Charles River Laboratories, Inc., Mattawan, MI, USA*

NOELIA GRANADO • *Instituto Cajal (CSIC), Consejo Superior de Investigaciones Científicas, Madrid, Spain; CIBERNED, Instituto de Salud Carlos III, Madrid, Spain*

ERIN A. GREGUSKE • *Departament de Ciències Fisiològiques, Institut de Neurociències, Universitat de Barcelona, Hospitalet de Llobregat, Catalunya, Spain; Institut d'Investigació Biomèdica de Bellvitge, IDIBELL, Hospitalet de Llobregat, Catalunya, Spain*

ANA CRISTINA GRODZKI • *Department of Molecular Biosciences, School of Veterinary Medicine, University of California, Davis, Davis, CA, USA*

PRISCILA GUBERT • *Department of Biochemistry, Laboratório de Imunopatologia Keizo Asami, LIKA, Federal University of Pernambuco, Recife, Brazil; Postgraduate Program in Pure and Applied Chemistry, Federal University of Western of Bahia, Bahia, Brazil*

G. JEAN HARRY • *Neurotoxicology Group, National Toxicology Program Laboratory, National Institute of Environmental Health Sciences, Research Triangle Park, NC, USA*

JULIA HARTMANN • *IUF-Leibniz Research Institute for Environmental Medicine, Duesseldorf, Germany*

THOMAS HARTUNG • *Center for Alternatives to Animal Testing (CAAT), Johns Hopkins Bloomberg School of Public Health, Baltimore, MD, USA; Center for Alternatives to Animal Testing-Europe, University of Konstanz, Konstanz, Germany; AxoSim, Inc., New Orleans, LA, USA*

KATHERINE HARVEY • *Department of Environmental Medicine, University of Rochester, Medical School, Rochester, NY, USA*

ANDREW B. HAWKEY • *Department of Psychiatry and Behavioral Sciences, Duke University Medical Center, Durham, NC, USA*

JOAN M. HEDGE • *Office of Research and Development, Center for Computational Toxicology and Exposure, U.S. Environmental Protection Agency, Research Triangle Park, NC, USA*

MIREIA HERRANDO-GRABULOSA • *Department of Cell Biology, Physiology and Immunology, Institute of Neurosciences, Centro de Investigación Biomédica en Red sobre Enfermedades Neurodegenerativas, Universitat Autònoma de Barcelona, Bellaterra, Spain*

BRIDGETT N. HILL • *Oak Ridge Institute for Science and Education, Oak Ridge, TN, USA*

HELENA T. HOGBERG • *Center for Alternatives to Animal Testing (CAAT), Johns Hopkins Bloomberg School of Public Health, Baltimore, MD, USA*

ZADE HOLLOWAY • *Department of Psychiatry and Behavioral Sciences, Duke University Medical Center, Durham, NC, USA*

DEBORAH L. HUNTER • *Office of Research and Development, Center for Computational Toxicology and Exposure, U.S. Environmental Protection Agency, Research Triangle Park, NC, USA*

OMAMUYOVWI M. IJOMONE • *The Neuro-Lab, School of Health and Health Technology, Federal University of Technology, Akure, Nigeria; Department of Human Anatomy, School of Health and Health Technology, Federal University of Technology, Akure, Nigeria*

KIMBERLY A. JAREMA • *Office of Research and Development, Center for Public Health and Environmental Assessment, U.S. Environmental Protection Agency, Research Triangle Park, NC, USA*

JULIA KAPR • *IUF-Leibniz Research Institute for Environmental Medicine, Duesseldorf, Germany*

NILS KLÜVER • *Department of Bioanalytical Ecotoxicology, Helmholtz Centre for Environmental Research—UFZ, Leipzig, Germany*

DAVID KOREST • *Oak Ridge Institute for Science and Education, Oak Ridge, TN, USA*

EBERHARD KÜSTER • *Department of Bioanalytical Ecotoxicology, Helmholtz Centre for Environmental Research—UFZ, Leipzig, Germany*

JUSTIN G. LEES • *School of Medical Sciences, The University of New South Wales, Sydney, Australia*

PAMELA J. LEIN • *Department of Molecular Biosciences, School of Veterinary Medicine, University of California, Davis, Davis, CA, USA*

EDWARD D. LEVIN • *Department of Psychiatry and Behavioral Sciences, Duke University Medical Center, Durham, NC, USA*

JORDI LLORENS • *Departament de Ciències Fisiològiques, Institut de Neurociènces, Universitat de Barcelona, Hospitalet de Llobregat, Catalunya, Spain; Institut d'Investigació Biomèdica de Bellvitge, IDIBELL, Hospitalet de Llobregat, Catalunya, Spain*

RAÚL LÓPEZ-ARNAU • *Faculty of Pharmacy and Food Sciences, Department of Pharmacology, Toxicology and Therapeutic Chemistry, Pharmacology Section and Institute of Biomedicine (IBUB), Universitat de Barcelona, Barcelona, Spain*

ALBERTO F. MAROTO • *Departament de Ciències Fisiològiques, Institut de Neurociènces, Universitat de Barcelona, Hospitalet de Llobregat, Catalunya, Spain; Institut d'Investigació Biomèdica de Bellvitge, IDIBELL, Hospitalet de Llobregat, Catalunya, Spain*

LAUREN E. MATELSKI • *Department of Molecular Biosciences, School of Veterinary Medicine, University of California, Davis, Davis, CA, USA*

CHRISTOPHER A. MCPHERSON • *Neurotoxicology Group, National Toxicology Program Laboratory, National Institute of Environmental Health Sciences, Research Triangle Park, NC, USA*

CRISTINA MEREGALLI • *Experimental Neurology Unit, School of Medicine and Surgery, and NeuroMI (Milan Center for Neuroscience), University of Milano-Bicocca, Monza (MB), Italy*

ROSARIO MORATALLA • *Instituto Cajal (CSIC), Consejo Superior de Investigaciones Científicas, Madrid, Spain; CIBERNED, Instituto de Salud Carlos III, Madrid, Spain*

RHIANNA K. MORGAN • *Department of Molecular Biosciences, School of Veterinary Medicine, University of California, Davis, Davis, CA, USA*

AFOLARIN O. OGUNGBEMI • *Department of Bioanalytical Ecotoxicology, Helmholtz Centre for Environmental Research—UFZ, Leipzig, Germany*

COMFORT O. A. OKOH • *The Neuro-Lab, School of Health and Health Technology, Federal University of Technology, Akure, Nigeria*

STEPHANIE PADILLA • *Office of Research and Development, Center for Computational Toxicology and Exposure, U.S. Environmental Protection Agency, Research Triangle Park, NC, USA*

DAVID PAMIES • *Center for Alternatives to Animal Testing (CAAT), Johns Hopkins Bloomberg School of Public Health, Baltimore, MD, USA; Department of Biomedical Science, University of Lausanne, Lausanne, Switzerland; Swiss Centre for Applied Human Toxicology (SCAHT), Lausanne, Switzerland*

SUSANNA PARK • *Brain and Mind Centre, Faculty of Medicine and Health, University of Sydney, Sydney, Australia*

HÉLÈNE PASCHOUD • *Department of Biomedical Science, University of Lausanne, Lausanne, Switzerland*

BENOÎT POUYATOS • *National Research and Safety Institute for the Prevention of Occupational Accidents and Diseases (INRS), Vandoeuvre-lès-Nancy, France*

DAVID PUBILL • *Faculty of Pharmacy and Food Sciences, Department of Pharmacology, Toxicology and Therapeutic Chemistry, Pharmacology Section and Institute of Biomedicine (IBUB), Universitat de Barcelona, Barcelona, Spain*

CYNTHIA L. RENN • *Department of Pain and Translational Symptom Science, School of Nursing, University of Maryland, Baltimore, MD, USA*

ADRIÁN SANZ-MAGRO • *Instituto Cajal (CSIC), Consejo Superior de Investigaciones Científicas, Madrid, Spain; CIBERNED, Instituto de Salud Carlos III, Madrid, Spain*

MARTIN SCHMUCK • *Department of Molecular Biosciences, School of Veterinary Medicine, University of California, Davis, Davis, CA, USA*

STEFAN SCHOLZ • *Department of Bioanalytical Ecotoxicology, Helmholtz Centre for Environmental Research—UFZ, Leipzig, Germany*

MARISSA SOBOLEWSKI • *Department of Environmental Medicine, University of Rochester, Medical School, Rochester, NY, USA*

ELISABET TEIXIDÓ • *Department of Bioanalytical Ecotoxicology, Helmholtz Centre for Environmental Research—UFZ, Leipzig, Germany; GRET-Toxicology Unit, Faculty of Pharmacy and Food Sciences, Department of Pharmacology, Toxicology and Therapeutic Chemistry, University of Barcelona, Barcelona, Spain*

AURÉLIE THOMAS • *National Research and Safety Institute for the Prevention of Occupational Accidents and Diseases (INRS), Vandoeuvre-lès-Nancy, France*

ESTHER UDINA • *Department of Cell Biology, Physiology and Immunology, Institute of Neurosciences, Centro de Investigación Biomédica en Red sobre Enfermedades Neurodegenerativas, Universitat Autònoma de Barcelona, Bellaterra, Spain*

ALEXANDRE M. VARÃO • *Postgraduate Program in Pure and Applied Chemistry, Federal University of Western of Bahia, Bahia, Brazil*

THOMAS VENET • *National Research and Safety Institute for the Prevention of Occupational Accidents and Diseases (INRS), Vandoeuvre-lès-Nancy, France*

LUDIVINE WATHIER • *National Research and Safety Institute for the Prevention of Occupational Accidents and Diseases (INRS), Vandoeuvre-lès-Nancy, France*

HENRIK ZETTERBERG • *Department of Psychiatry and Neurochemistry, The Sahlgrenska Academy at the University of Gothenburg, Mölndal, Sweden; Clinical Neurochemistry Laboratory, Sahlgrenska University Hospital, Mölndal, Sweden; Department of Neurodegenerative Disease, UCL Institute of Neurology, Queen Square, London, UK; UK Dementia Research Institute at UCL, London, UK*

MARIE-GABRIELLE ZURICH • *Department of Biomedical Science, University of Lausanne, Lausanne, Switzerland; Swiss Centre for Applied Human Toxicology (SCAHT), Lausanne, Switzerland*

Part I

Cytological and Histological Evaluation in Rodents

Chapter 1

Amino-Cupric-Silver (A-Cu-Ag) Staining to Detect Neuronal Degeneration in the Mouse Brain: The de Olmos Technique

Rosario Moratalla, Adrián Sanz-Magro, and Noelia Granado

Abstract

Silver staining procedures have classically been used to study the structure of the nervous system. However, reduced silver staining methods, using substances that reduce silver ions against the natural reducing properties of the tissue, can also successfully reveal degenerative changes in the nervous system. It is not known how silver binds these degenerating elements. During degeneration, silver ions may form complexes with exposed amino acid chains in denatured proteins that are then seen as black-stained elements over an unstained background (of non-degenerating elements).

The reduced Amino-Cupric-Silver method developed by de Olmos, where cupric ions are added as an external reducer, provides the greatest contrast between the degenerating and non-degenerating neuronal elements when compared with other reduced silver staining protocols. This method is unique to study degenerative morphological changes and, when combined with other staining procedures, to identify which specific neuronal population is degenerating.

After a trauma, neurons undergo several physical changes that can be visualized with different markers, such as FluoroJade, caspases, or Hematoxylin and Eosin stains. Some neurons can overcome this damage and are restored, while others undergo irreversible changes and die. The Amino-Cupric-Silver method can reveal early irreversible neuronal damage before cell death. After these irreversible changes the neurons cannot regenerate, so detecting these degenerative changes will improve the understanding of pathological changes following certain injuries of the nervous system.

Key words Reduced silver stain procedures, A-Cu-Ag method, Neurotoxicity, Neurodegeneration, Cell death

1 Introduction

Cognitive function depends on the correct activity and integrity of neurons and their pathways. The loss of a neuron is the last step of a degenerative process that follows certain types of damage [1]. As a consequence of the plasticity of neural systems and their compensatory capacity, neurological symptoms frequently appear only when neuron death is advance, and the system has lost a significant

Adrián Sanz-Magro and Noelia Granado contributed equally to this work.

Jordi Llorens and Marta Barenys (eds.), *Experimental Neurotoxicology Methods*, Neuromethods, vol. 172, https://doi.org/10.1007/978-1-0716-1637-6_1, © Springer Science+Business Media, LLC, part of Springer Nature 2021

number of cells [2]. Also, considering that neurons cannot regenerate, it is essential to identify early irreversible neuronal damage [1]. Detecting degenerative changes before neuron death is a challenge in the field of neurotoxicology because it requires predicting where the neuronal loss will take place as a consequence of a certain process [1].

Based on the affinity of nervous system cells for silver (argyrophilia), silver staining procedures were largely used in neuroanatomy during the last century to study the structure of the nervous system. These methods were successfully employed to stain a large variety of neuronal elements and trace axonal pathways [1, 3, 4]. One of the first silver stain method was designed by Bielchowsky in 1904 [4]. This method stained all type of neural cells so, it was difficult to differentiate between normal and pathological neuronal elements [1]. Later on, Nauta and Gygax, improved this method introducing a pretreatment with phosphomolybdic acid-potassium permanganate solution, achieving a better contrast between normal and damaged elements [1, 5].

Although this modification can reveal neuronal degeneration [1, 3], the development of reduced silver methods further improved it, reaching its maximal strength and providing the optimal tool to approach neurodegeneration studies [1, 3]. With the reduced silver staining methods (in which silver ions are reduced to metallic silver against the natural silver-reducing properties of the nervous tissue), the degenerating neuronal elements are more intensely stained while non-degenerating elements (background) remain unstained [1, 3, 6, 7]. Despite the usefulness of these methods to study nervous system, the rise of new stains (e.g., horseradish peroxidase, fluorescent tracer methods) has caused them to be less commonly used [1, 3].

The main mechanism explaining how staining results from the interaction between the silver and the degenerating elements is still unclear [1, 8]. The most widely accepted hypothesis is that silver ions react with exposed single amino acids or short amino acid sequences [9, 10] and form molecular associations of a central metal cation (silver) surrounded by a certain number of amino acid ions [11, 12]. In a healthy cell, proteins have a globular conformation and amino acid groups are not exposed to react with silver ions. In a damaged cell, proteolytic mechanisms can alter protein conformations and expose these residues, which can then easily react with silver ions [1, 3, 8, 13].

The first known reduced silver staining is the "Golgi Stain," developed by Camillo Golgi in 1873, using potassium dichromate as a reducer. Santiago Ramón y Cajal used the Golgi Stain with small modifications to study the structure of the nervous system and could validate his "neuron doctrine" [1, 14]. Later on, de Olmos developed the reduced Amino-Cupric-Silver (A-Cu-Ag) method [6, 7, 15], where cupric ions decrease the argyrophilia of non-degenerating elements, but not in degenerating ones, providing the strongest contrast until present between degenerating and non-degenerating elements [1, 3]. The A-Cu-Ag method is based on the following steps: (1) pretreatment with a cupric solution that provides specificity to the staining of degenerating elements [3, 6, 7]; (2) silver impregnation of the samples with a silver nitrate ($AgNO_3$) solution; (3) reduction of silver ions to metallic silver around exposed amino acid groups in the tissue, forming microscopically visible nuclei; (4) bleaching to clear the background and increase the contrast between normal and degenerating elements; and (5) stabilization of the staining [1, 7].

Because silver can stain neuronal elements at different degenerative stages [7], the A-Cu-Ag method can be used not only to identify degeneration, but also to study the progression of degeneration, through the accumulation of silver deposits over time and the loss of the corresponding degenerating neurons or fibers associated with specific pathological changes, as shown previously [1, 7, 10, 16], see also Fig. 1. During a degenerative process, dendrites and axonal processes start to fragment in rows with slight silver staining, followed by the appearance of punctuate structures, and then their total disappearance [7]. Generally, in degeneration, cell somas undergo changes in size and shape, at first without the presence of silver deposits. As the degenerating soma get smaller the silver starts to accumulate. The neuron can then became dysfunctional and start losing the capacity to synthesize their phenotypic and domestic markers. Finally, small silver-stained debris remains until the soma completely disappears [1, 7, 13], see also Fig. 2. With methamphetamine, all these pathological changes can be observed as soon as 24 h after the administration (Fig. 2) although these changes do not necessarily represent sequential steps of the degenerating process, due to the severity of the insult. Nevertheless, we do observe all these degenerating changes [13].

Furthermore, as illustrated in Figs. 2 and 3, and mentioned above, because of the unstained background, the A-Cu-Ag method can be successfully combined with other stains [13, 17]. This approach allows the identification of specific types of degenerating neurons [13, 18–20]. Previous studies by our group, combining the A-Cu-Ag method with immunohistochemistry against the enzyme thyroxin hydroxylase (TH), a marker of dopaminergic neurons, or with green fluorescent protein or red-tomato protein, demonstrated that methamphetamine induces degeneration in dopaminergic cells of the substantia *nigra pars compacta* (SNpc) and in its projections to the striatum [13, 18]. Indeed, we showed that the peak of silver deposits after methamphetamine overlaps with that of striatal TH-immunoreactivity (TH-ir) fiber loss (Figs. 1 and 3) and furthermore, that silver deposits also occurred in the soma of dopaminergic neurons in the SNpc (Fig. 2).

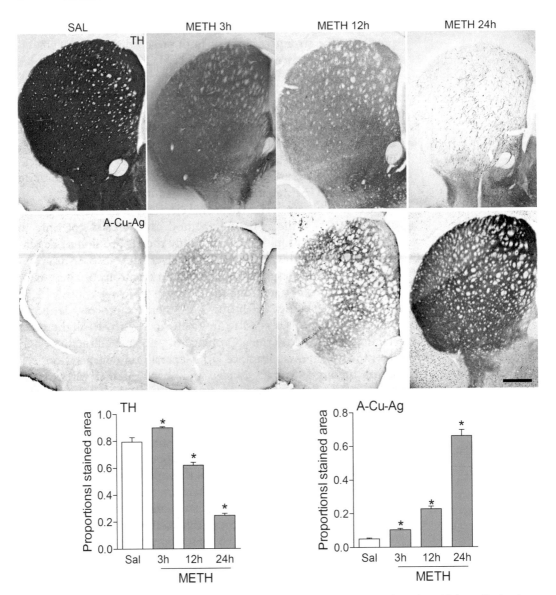

Fig. 1 Methamphetamine induces a progressive TH-ir fiber degeneration in the mice striatum. Photomicrographs of striatal coronal sections of mice treated with saline or methamphetamine (METH). Sections are stained against TH and adjacent section with the A-Cu-Ag method. Mice were sacrificed 3, 12 or 24 h after METH. Note the progression of dopaminergic terminal loss demonstrated by the decrease in TH-ir and its correlation with the progressive increase of silver deposits shown in adjacent sections. Histograms illustrate the quantification analysis, showing the proportional stained area with TH-ir or with the A-Cu-Ag method in the striatum. Data represent the mean ± SEM, $n = 4$–6 per group. Scale bar indicates 500 μm

An important factor to consider in the A-Cu-Ag method is the time between injury and when the stain is performed [1, 7]. Degenerative changes appear and remain at different times depending on the neuronal compartment where they are taking place or on the type of insult or toxin used [1, 7]. In our experiments with

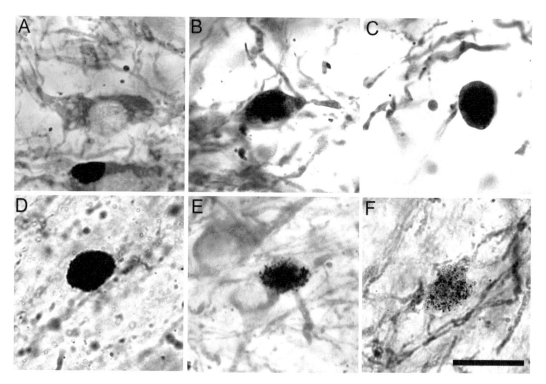

Fig. 2 Different stages of degeneration in dopaminergic neurons in the SNpc after methamphetamine. Photomicrographs of double TH/A-Cu-Ag-stained neurons of the SNpc of mice sacrificed 24 h after methamphetamine. Note that some degenerating neurons change their size and shape (**a**) and have silver deposits in their soma (**b** and **c**) but still express the dopaminergic marker, TH in this case. Also, neurons with only silver deposit indicating the irreversible degeneration process (**d**) and neurons with silver-stained debris that have already started to disappear (**e–f**). Scale bar indicates 20 μm

methamphetamine, silver deposits in dopaminergic terminals in the striatum appear few hours after treatment (around 3 h) and are still evident 3–7 days after the insult (Fig. 1) [13]. However, in the dopaminergic cell bodies of the SNpc, silver deposits appear later than in terminals (around 24 h) and can remain around 3–5 days [13]. Considering this, the best time to detect degenerative changes in the dopaminergic nigrostriatal pathway caused by methamphetamine administration is around 24 h after the insult because at this time the damage appears in both, somas and synaptic terminals, and it is also when it reaches the peak [13]. However, it must be taken into consideration that the appearance and duration of the degenerative changes could differ depending on the type of insult and the type of the neuronal target [1, 7].

Also, the A-Cu-Ag method is an ideal tool to anatomically localize neurodegeneration, allowing to determine its anatomical pattern [1, 7, 13]. In our experiments, we observed that dopaminergic degeneration in the striatum caused by methamphetamine is not homogeneous, and the same happens with the A-Cu-Ag

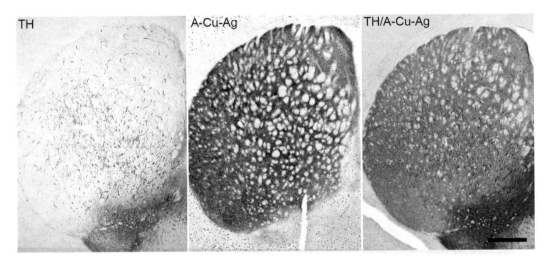

TH   A-Cu-Ag   TH/A-Cu-Ag

Fig. 3 Double A-Cu-Ag and TH staining. Photomicrographs of the coronal sections of single and double TH/A-Cu-Ag-stained sections of the striatum of mice sacrificed 24 h after methamphetamine. Because of the clear background of the A-Cu-Ag method, degenerating elements and residual TH-ir fibers can be observed in the same sections in the striatum. Scale bar indicates 500 μm

staining. Indeed, there is an overlap pattern of TH fibers loss and silver deposits (Figs. 1 and 3). Furthermore, we could even observe small striatal areas with greater TH-ir fiber loss that overlaps with greater silver deposits (Fig. 4). Curiously, these areas corresponded in shape and in numbers with striatal striosomes as we could demonstrate using μ-opioid receptor (MOR-1) as a marker for the striosomal compartment [21].

Despite the advantages of the A-Cu-Ag method [3, 6, 7, 15, 16], the development of new faster and more simple staining methods, that can indirectly show damage in the nervous system, has decreased its use. However, this A-Cu-Ag method can be complemented with other methods [13], such as the FluoroJade stain, that detects alterations in the cytoplasm or in the nucleus of degenerating neurons or hematoxylin and eosin (H&E) stain that detect eosinophilic necrotic neurons [1, 13, 22]. Nissl stain, that detects apoptotic cell bodies, can also be used in combination with the silver method as have been successfully shown in SNpc after methamphetamine [13, 23]. As inflammatory responses often accompany degeneration, specific immunohistochemistry assays such against astroglia and microglia markers can also detect degeneration when combined in adjacent section with the silver method as have shown before [18, 21]. DNA damage proteins or caspase activation can also provide information about the response to an injury [16, 23–27]. Nevertheless, all these methods fail to detect structural and pathological changes in small degenerating structures such as synaptic terminals, dendrites, and axons [1], but the A-Cu-Ag, due to the described characteristics, allows to successfully

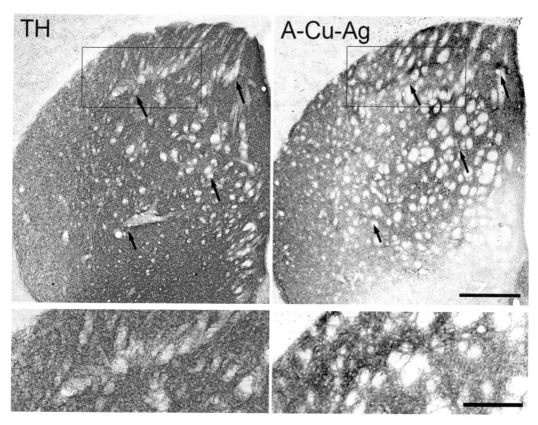

Fig. 4 Correlation between terminal degeneration and silver deposits: Striosomes are more vulnerable to methamphetamine and show higher silver deposits. Photomicrographs of adjacent striatal coronal sections stained against TH or with the A-Cu-Ag method. Mice were sacrificed 24 h after methamphetamine administration. Note that striatal striosomes undergo stronger degeneration after methamphetamine and show denser silver deposits. Arrows indicate striatal striosomes and boxes the magnified area. Scale bar indicates 500 μm at lower magnification and 250 μm at high magnification

studying these changes. Many studies have used the A-Cu-Ag method to specifically identify neuronal damage induced by neurotoxins [7, 17, 28, 29], drugs of abuse [13, 18–20, 30], physical damage [7, 31], or altered physiological processes [6, 7].

In summary, because of features such as (1) specificity, (2) power of contrast, and (3) the possibility to combine with other staining techniques, the A-Cu-Ag method may provide useful information on the degenerative changes after an injury to the nervous system.

The main goal of this chapter is to provide a comprehensive and simple protocol for the A-Cu-Ag method so it can be replicated in other laboratories. Besides the A-Cu-Ag method, we include a TH immunohistochemistry protocol that could be carry out in adjacent or even in the same sections that the A-Cu-Ag method, in order to gain anatomical depth and resolution and to identify the phenotype of the degenerating elements. We also provide protocols to quantify degeneration, studying fiber loss or silver deposits by measuring the proportional stained area or optical density and a stereological protocol to quantify the number of degenerating neurons (cell soma with silver deposits).

The protocol described below successfully detects dopaminergic degeneration after methamphetamine administration in the striatum and SNpc of mice of either sex, or different ages, even embryos or neonates and can be done in coronal, sagittal, or horizontal brain sections. Control brain sections from animals treated with saline should be processed at the same time with the experimental sections to serve as reference for the bleaching and other steps and for quantification studies.

The conditions described in this chapter are adapted to detect dopaminergic damage in the striatum and SNpc caused by methamphetamine administration. The experimenter is advised to test different conditions in order to obtain the optimal results with the specific insult under study.

2 Materials

2.1 Tissue Fixation and Brain Preparation

Subjects and drugs: This protocol uses 3–4 months old C57BL/6J mice. To induce dopaminergic degeneration, we use methamphetamine hydrochloride (SIGMA ref M-8750) dissolved in 0.9% sodium chloride (NaCl or saline; MERCK ref 1.06404.1000) given in three doses of 5 mg/kg at 3 h intervals. This dose regimen causes a clear degeneration in the brain, as previously demonstrated [23–25]. Animals are sacrificed at different time points after methamphetamine treatment (3, 12, and 24 h after methamphetamine administration), in order to assess different time points of dopamine degeneration in the striatum and SNpc. Before start with the protocol, carefully read **Notes 1–4**.

Perfusion: The following solutions are needed to perform the perfusion process:

- Anesthetic: Sodium pentobarbital (VETOQUINOL; 50 mg/kg).
- Rinse solution: 0.8% NaCl (MERCK ref 1.06404.1000), 0.4% glucose (SIGMA ref G7021), 0.8% sucrose (MERCK, ref 107687), and 0.5% sodium nitrite (NaNO$_2$; SIGMA ref S2252), as a vasodilator, in bi-distilled water (H$_2$Obd).
- Fixation solution (paraformaldehyde; PFA): 4% PFA (SIGMA ref 441244) with 50 mg sodium sulfite (Na$_2$SO$_3$; SIGMA ref. S0505) per L of H$_2$Obd; pH 7.6–7.8 (*see* **Note 5**).
- 0.2 M borate buffer pH 8.5: 50 mg Na$_2$SO$_3$ (SIGMA ref S2252) and 16.8 mg sodium tetraborate (Borax; SIGMA ref 71997) in 1 L H$_2$Obd (light sensitive).

Tissue processing: For rodent brains, cut 50 μm thick sections (*see* **Note 6**) with a Vibratome (Leica Microsystems GmbH, Wetzlar, Germany) on free-floating way using 0.2 M borate buffer, pH 8.5.

2.2 The A-Cu-Ag Method

Pre-impregnation solution

Prepare 100 mL of the pre-impregnation solution by combining the following reagents in this order:

1. 100 mg silver nitrate ($AgNO_3$; SIGMA ref 85228; light sensitive; never weigh with metallic instruments).
2. 100 mL H_2Obd.
3. 53 mg dl-α-amino-n-butyric acid (SIGMA ref 162663).
4. 46 mg dl-alanine (SIGMA ref A7502).
5. 2 mL 0.5% copper nitrate (Cu $(NO_3)_2$; SIGMA ref 229636; light sensitive).
6. 0.2 mL 0.5% cadmium nitrate ($Cd(NO_3)_2$; SIGMA ref 20911).
7. 1.5 mL 0.5% lanthanum nitrate ($La(NO_3)_3$; MERCK ref 2506).
8. 0.5 mL 0.5% Neutral Red (FLUKA ref 72210).
9. 1.0 mL pyridine (MERCK ref 1.097828.0500; toxic, always work in a fume hood and with the proper respiratory protection.
10. 1.0 mL triethanolamine (SIGMA ref T-58300).
11. 2.0 mL isopropanol (MERCK ref 1.09634.1011; work in a fume hood.

Impregnation solution

Combine the following compounds:

1. 5 mL H_2Obd.
2. 412 mg $AgNO_3$ (SIGMA ref 85228; completely dissolved before adding the 100% EtOH).
3. 4 mL 100% EtOH (MERCK ref 1.00983.2511).
4. 50 μL acetone (MERCK ref 1.00014.1011).
5. 3 mL 0.4% lithium hydroxide (LiOH; SIGMA ref I-4533; store at 4 °C).
6. 0.65 mL ammonium hydroxide (NH_4OH; FISHER ref 10552174; toxic, always work in a fume hood, use within 2–3 months after preparation).

Reduction solution:

This solution can be prepared several days before use. Store at room temperature and protect from light. To prepare 1 L of reduction solution, combine:

1. 800 mL distilled water (H_2Od).

2. 90 mL 100% EtOH (MERCK ref 1.00983.2511).

3. 11 mL 10% formalin (SIGMA ref 47608; store at 4 °C).

4. 6.5 mL 1% citric acid monohydrate (FISHER ref 16345410; freshly prepared).

Bleaching solution:
Two bleaching solutions are required.

- Solution 1: prepared in lactic acid (MERK ref 1.00366.2500)
  - 6% potassium ferricyanide ($K_3[Fe(CN)_6]$; MERCK ref 1.04973.0250).
  - 4% potassium chlorate ($KClO_3$; PANREAC ref 131493.1210).
- Solution 2: prepared in H_2Od
  - 0.06% potassium permanganate ($KMnO_4$; provided by Dr. R. Martínez, Cajal Institute, CSIC).
  - 5% sulfuric acid (FISHER ref S/9240/PB15).

Stabilization solution:
Two stabilization solutions are required:

- Stabilization solution: 2% Sodium thiosulfate ($Na_2S_2O_3$; SIGMA ref 217247) in H_2Od.
- Rapid Fixer solution: KODAK concentrated Rapid Fixer solution A + B (SIGMA ref P7542-1GA), diluted 1:6 in H_2Od.

2.3 TH Immuno-histochemistry

For immunohistochemistry, prepare the following buffers and solutions:

- 0.4 M phosphate buffer (PB) pH 7.4: 27.6 g sodium phosphate monobasic monohydrate ($NaH_2PO_4 \cdot H_2O$; SIGMA ref S-9638), and 106 g sodium phosphate dibasic ($Na_2HPO_4 \cdot 2H_2O$; MERCK ref 1.06580.1000) in 2 L H_2Od.
- 0.1 M PB: 500 mL PB4× in 1.5 L H_2Od.
- Phosphate-buffered saline (PBS) pH 7.4: 500 mL PB4× and 18 g NaCl in 1.5 L H_2Od.
- PBS-TX: 0.2% Triton X-100 (SIGMA ref T8787) in PBS buffer.
- 3% peroxide (H_2O_2; PANREAC ref 141076.1211) in PBS-TX.
- 10% Normal Goat Serum (NGS; MERCK ref S26-100ML) in PBS-TX.
- TH primary antibody: 1:1000 rabbit anti-TH (MERCK-MILLIPORE ref AB152) in PBS-TX, 1% NGS.
- Secondary antibody: 1:500 goat anti-rabbit biotinylated antibody (Vector Labs ref BA1000) in PBS-TX, 1% NGS.

- 1:5.000 Streptavidin (INVITROGEN ref 434323) in PBS-TX, 1% NGS.

- 3,3′-Diaminobenzidine (DAB; SIGMA ref D5637; light sensitive; toxic; 50 mg/100 mL of 0.1 M PB).

2.4 Mounting and Visualization

- *Gelatin solution for slides:* add 1.5 g gelatin to 50 mL hot H_2Od, wait until dissolved, then add 50–80 mL of ethanol.

- *Xylene* (VWR ref. 141769.2711).

- *DPX Mounting Medium* (MERCK ref 1.00579.0500).

2.5 Image Analysis Quantification and Stereology

- *Microscope:* to visualize the degenerating elements (both striatum and SNpc), we use an optic microscope equipped with $4\times$ lens and a Leica DFC 290 HD video camera to acquire the micro-pictures.

- *Image analysis system:* to study TH-ir and the silver impregnation, we use the Image J program (National Institutes of Health, Bethesda, Maryland, USA) that converts color intensities into a gray color scale. This allows us to quantify the stained area as the proportion of staining pixels in relation to the total pixels in the selected area [32].

  To count the number of degenerating and dopaminergic cell bodies, we use a *stereological software* (*see* Subheading 3.5) coupled to a Nikon Eclipse 80i microscope. The microscope is connected to (1) an interactive computer system comprising a high-precision motorized microscope stage, (2) a 0.5 μm resolution microadaptor (Heidenhain VZR401), (3) a solid-state Microbrightfield CX9000 videocamera, and (4) a high-resolution video monitor using the optical fractionator Stereo Investigator program (Microbrightfield Bioscience, Colchester, VT) [33].

3 Methods

The following A-Cu-Ag method protocol is taken from the one described by de Olmos et al. 1994 [7].

3.1 Tissue Fixation and Preparation

Good tissue fixation is a critical step in the A-Cu-Ag method (perfusion to remove all blood elements from the tissue and avoid a less clear background) [6, 16].

For the fixation solution (see Subheading 2.1), heat H_2Obd in a glass cup to 60 °C. Add Na_2SO_3 and a small amount of Borax (4 g for 1 L of solution). When the Borax is completely dissolved, add the PFA and stir at 50 °C until completely dissolved (10 mg boric acid per 1 L of solution). Allow the solution to cool for 45–60 min and adjust to a pH of 7.45 (*see* **Note 5**).

Anesthetize the animals (*see* Subheading 2.1) with sodium pentobarbital and perfuse transcardially with the fixation solution (PFA, *see* Subheading 2.1). Before and after the fixation, perfuse the animal with 10 mL (around one-third of the animal's weight) of the rinse solution (*see* Subheading 2.1). As a vasodilator, use 0.5% sodium nitrite.

Leave brains overnight in the skull at 4 °C to avoid any extra damage during dissection that will be detected by the silver stain, then remove the brain and store in 30% sucrose for 2 days. Place the brains in a solution of 3% agarose in 0.2 M borate buffer, pH 8.5, and cut into 50 μm thick coronal sections in a vibratome. Store in 4% PFA (*see* **Note 7**).

3.2 The A-Cu-Ag Method

Pre-impregnation: During pre-impregnation, the addition of cupric ions will selectively stain degenerating neuronal elements (*see* Subheading 2.2 and **Notes 8** and **9**) [3, 6, 7, 15, 17].

Before use, heat the pre-impregnation solution in the microwave to 48 °C. If the solution has precipitated, filter with glass fiber filters. Take the selected slices from the 4% PFA and wash with H_2Obd in a porcelain filter. Put the slices into the pre-impregnation solution and heat in the microwave to 47–48 °C. Allow to cool at room temperature for 2–3 h. Tissue must turn brown.

Impregnation: Take slices from the pre-impregnation solution and place in a porcelain filter. Wash for 1 min with H_2Obd and twice with acetone (45 s–1 min for each wash) to remove excess pyridine and pre-impregnation solution. Transfer the slices into the impregnation solution (*see* Subheading 2.2) for 45–50 min (*see* **Note 10**).

Reduction: Heat the reduction solution (*see* Subheading 2.2) to 30 °C. Transfer the slices from the impregnation solution to the reduction solution and incubate for 20 min (*see* **Note 11**). Tissue must get darker.

Wash the slices in H_2Od and place into 0.25% acetic acid for 1 min to stop the reduction reaction (do not move vigorously as the tissue can easily break). Perform two washes of 1 min with H_2Obd and keep the slices at 4 °C for at least 1 h before the bleaching treatment.

Bleaching is performed in two steps (*see* **Note 12**):

1. Put sections in the first bleaching solution (*see* Subheading 2.2) in a porcelain desiccator at room temperature until they become relatively transparent (approximately 60 s) to remove background silver deposits but not degenerating elements, then wash sections with H_2Od.

2. Transfer slices to the second bleaching solution (*see* Subheading 2.2) in a porcelain desiccator. Leave until the slices acquire a yellow color, around 15–20 s (if longer, the silver fixed in the

degenerating elements could disappear). Put slices into H_2Od and perform two washes of 5 min each before the stabilization step (*see* **Note 13**).

Stabilization: Transfer sections to the stabilization solution (*see* Subheading 2.2) for approximately 1 min (in slow agitation). Sections should lose their yellow color and become transparent. Wash with H_2Od for 5 min and put slices in the Rapid Fixer solution (*see* Subheading 2.2) for 1 min. Finally, put slices in H_2Od.

3.3   TH Immuno-histochemistry

Perform the immunohistochemistry on free-floating sections previously stained with the A-Cu-Ag method [13, 18]. Also, immunohistochemistry can be done alone in adjacent sections to the ones stained with the A-Cu-Ag method.

After stabilization, wash the slices in 0.2% PBS-TX (*see* Subheading 2.3). Quench endogenous peroxidase incubating slices in 3% H_2O_2 in 0.2% PBS-TX for 10 min. Wash again in PBS-TX and block nonspecific binding sites for 60–90 min with 10% normal goat serum in 0.2% PBS-TX. Incubate the primary antibody anti-TH (*see* Subheading 2.3) at room temperature overnight. Wash slices with 0.2% PBS-TX and incubate with the goat anti-rabbit biotinylated antibody (*see* Subheading 2.3) for 1–2 h at room temperature. Wash with 0.2% PBS-TX, incubate sections in streptavidin (*see* Subheading 2.3) for 1 h, and then stain with DAB.

3.4   Mounting and Visualization

After the DAB reaction, carefully mount the slices in gelatinized slides (*see* Subheading 2.4) [34]. When the tissue is completely dry, dehydrate the section through the following alcohol battery (10 min per solution): H_2Od, 70% ethanol, 96% ethanol, 100% ethanol, 100% ethanol, and xylene. Cover the slices with DPX as the mounting medium and dry them for around 24 h before visualization under a microscope.

3.5   Image Analysis Quantification and Stereology

Quantitative assessment of degenerating dopaminergic terminals or silver deposits is done by evaluating the proportional stained area or by optical density. First, take pictures of the region of interest (striatum, in this case) with a 4× lens in an optical microscope (*see* **Note 14**). Pictures should be taken using the same light conditions and filters in the microscope for all the samples. Analyze pictures using an ImageJ (*see* Subheading 2.5) as previous described [13, 18–20] (Fig. 1). There are many free programs available; one of the most robust and widely used programs is the Image J.

Stereological analysis estimate the number of degenerating cell somas (in our case, dopaminergic neurons), by counting the number of A-Cu-Ag-stained neurons in the SNpc using the optical fractionator Stereo Investigator program (Microbrightfield Bioscience, Colchester, VT), coupled to a Nikon Eclipse 80i microscope (*see* Subheading 2.4) (*see* **Note 15**). To assess methamphetamine toxicity in dopaminergic neurons in mice, around 4–6 animals per group are needed to obtain significant values. First, draw the outline of the SNpc (*see* **Note 16**) with a 2× lens in one out of four serial adjacent sections through the SNpc (usually around 9–12 sections per animal). Second, count the cells at higher magnification (60× or 100×). At this high magnification, cell somas can be identified as A-Cu-Ag-stained neurons (black color), TH-stained neurons (brown color), or TH/A-Cu-Ag double-stained neurons (silver-stained cell body surrounded by TH staining).

4 Notes

1. Always use glass materials; avoid the use of plastic because silver can strongly bind to plastic [7].

2. Use high quality water (H_2Obd or H_2Od) for all steps [7].

3. Never wash materials with soap or detergent because they can interact with silver and affect the staining quality. Wash glassware with a solution of 30% nitric acid in H_2Od [7].

4. Never use sodium hydroxide (NaOH) or hydrochloric acid (HCl) to adjust the pH of the solutions; sodium and chlorine ions can react with silver ions and interfere with the staining.

5. Use a fresh fixation solution for each experiment, prepared a few hours before the animals will be sacrificed [6].

6. Instead of cutting brains in coronal sections at 30–35 μm of thickness (as described in de Olmos et al. 1994), cut it in at 50 μm of thickness, to avoid that the slices break during tissue processing.

7. Pay attention to the post-fixation time of the slices in 4% PFA; less than 24 h can enhance the staining of non-degenerating elements, but longer than 24 h can reduce the silver signal in degenerating elements [7].

8. For an optimal silver precipitation, prepare the pre-impregnation solution 1 day before use and store it at 4 °C [7].

9. 100 mL of pre-impregnation solution should stain around 50 brain slices [7].

10. Incubation time in the impregnation solution is important; shorter times increase the staining of normal neuronal elements while longer times reduce the staining in degenerating areas [14]. Incubate in the impregnation solution in the dark, as the silver signal decreases with exposure to light.

11. During incubation in the reduction solution, add the impregnation solution (0.3 mL per 100 mL reduction solution) every 5 min to help the silver fix [7].

12. Perform the bleaching process 1 day after the reduction reaction [7].

13. Use oxalic acid to bleach the porcelain after the potassium permanganate has passed through it.

14. The magnification lens to use can vary depending on the size of the target structure. The striatum is a big nucleus that can be analyzed with the 4× objective. Smaller structures need bigger magnification.

15. This stereological estimation is an unbiased method not affected by the volume of the structure or the size of the elements [24]. Is important that the researcher is blind to experimental group conditions, in order to avoid subjective analyses.

16. Stereological counts in the SNpc are done in one hemisphere of the brain, and the data are multiplied by two. This approach can only be done in bilateral structures, if the region of interest is not bilateral, cells must be count in the entire nucleus.

Acknowledgments

This protocol was set up in the laboratory of Dr. Rosario Moratalla at Cajal Institute (CSIC) from the protocol described by de Olmos and colleagues in 1994 [14]. We thank Manuel Marquez-Rivera for his help with the revision of the manuscript. Preparation of this manuscript was supported by grants from the Spanish Ministries of Innovation, Science and Universities PID2019-111693RB-I00 and PCIN-2015-098 and Health, Social Services and Equality (PNSD 2016/033 and CIBERNED CB06/05/0055) and UE (H2020-SC1-BHC-2018-2020, grant agreement n° 848002).

References

1. Switzer RC (2000) Application of silver degeneration stains for neurotoxicity testing. Toxicol Pathol 28(1):70–83

2. Fearnley JM, Lees AJ (1991) Ageing and Parkinson's disease: substantianigra regional selectivity. Brain 114(Pt 5):2283–2301

3. Beltramino CA, de Olmos JS, Gallyas F et al (1993) Silver staining as a tool for neurotoxic assessment. NIDA Res Monogr 26:136–101; discussion 126–32

4. Bielschowsky M (1904) Silber impregnation der neurofibrillen. J Psychol Neurol 3:169–188

5. Nauta W, Gygax PA (1951) Silver impregnation of degenerating axon terminals in the central nervous system (1) technic (2) chemical notes. Stain Technol 26:5–11

6. de Olmos JS (1969) A cupric-silver method for impregnation of terminal axon degeneration and its further use in staining granular argyrophilic neurons. Brain Behav Evol 2:213–237

7. de Olmos JD, Beltramino CA, de Olmos-de Lorenzo S (1994) Use of an amino-cupric-silver technique for the detection of early and semiacute neuronal degeneration caused by neurotoxicants, hypoxia, and physical trauma. Neurotoxicol Teratol 16:545–561

8. Eiland MM, Ramanathan L, Gulyani S et al (2002) Increases in amino-cupric-silver

staining of the supraoptic nucleus after sleep deprivation. Brain Res 945(1):1–8

9. Breslow E (1973) Metal-protein complexes. In: Eichhorn GL (ed) Inorganic biochemistry. Elsevier, Amsterdam, The Netherlands, pp 227–249

10. Freeman HC (1973) Metal complexes of amino acid and peptides. In: Eichhorn GL (ed) Inorganic biochemistry. Elsevier, Amsterdam, The Netherlands, pp 121–166

11. Leigh GJ (1990) Nomenclature of inorganic chemistry (recommendations 1990)-the red book. Blackwell Science, London

12. Gallyas F (1982) Physico-chemical mechanism of the argyrophil III reaction. Histochemistry 74(3):409–421

13. Ares-Santos S, Granado N, Espadas I, Martinez-Murillo R, Moratalla R (2014) Methamphetamine causes degeneration of dopamine cell bodies and terminals of the nigrostriatal pathway evidenced by silver staining. Neuropsychopharmacology 39 (5):1066–1080

14. Ramón y Cajal S, de Castro F (1933) Elemento de técnica micrográfica del sistema nervioso. Tipografía Artística, Madrid

15. de Olmos JS, Ingram WR (1971) An improved cupric-silver method for impregnation of axonal and terminal degeneration. Brain Res 33:523–529

16. Tenkova TI, Goldberg MP (2007) A modified silver technique (de Olmos stain) for assessment of neuronal and axonal degeneration. Methods Mol Biol 399:31–39

17. de Olmos S, Bender C, de Olmos JS, Lorenzo A (2009) Neurodegeneration and prolonged immediate early gene expression throughout cortical areas of the rat brain following acute administration of dizocilpine. Neuroscience 164(3):1347–1359

18. Carmena A, Granado N, Ares-Santos S, Alberquilla S, Tizabi Y, Moratalla R (2015) Methamphetamine-induced toxicity in indusium griseum of mice is associated with astro- and microgliosis. Neurotox Res 27 (3):209–216

19. Mendieta L, Granado N, Aguilera J, Tizabi Y, Moratalla R (2016) Fragment C domain of tetanus toxin mitigates methamphetamine neurotoxicity and its motor consequences in mice. Int J Neuropsychopharmacol 19(8): pyw021

20. Granado N, Ares-Santos S, Tizabi Y, Moratalla R (2018) Striatal reinnervation process after acute methamphetamine-induced dopaminergic degeneration in mice. Neurotox Res 34 (3):627–639

21. Granado N, Ares-Santos S, O'Shea E, Vicario-Abejón C, Colado MI, Moratalla R (2010) Selective vulnerability in striosomes and in the nigrostriatal dopaminergic pathway after methamphetamine administration : early loss of TH in striosomes after methamphetamine. Neurotox Res 18(1):48–58

22. Fujikawa DG, Zhao S, Ke X, Shinmei SS, Allen SG (2010) Mild as well as severe insults produce necrotic, not apoptotic, cells: evidence from 60-min seizures. Neurosci Lett 469:333–337

23. Ares-Santos S, Granado N, Oliva I, O'Shea E, Martin ED, Colado MI, Moratalla R (2012) Dopamine D(1) receptor deletion strongly reduces neurotoxic effects of methamphetamine. Neurobiol Dis 45(2):810–820

24. Granado N, O'Shea E, Bove J, Vila M, Colado MI, Moratalla R (2008) Persistent MDMA-induced dopaminergic neurotoxicity in the striatum and substantia nigra of mice. J Neurochem 107(4):1102–1112

25. Granado N, Ares-Santos S, Oliva I, O'Shea E, Martin ED, Colado MI, Moratalla R (2011) Dopamine D2-receptor knockout mice are protected against dopaminergic neurotoxicity induced by methamphetamine or MDMA. Neurobiol Dis 42(3):391–403

26. Zamanian JL, Xu L, Foo LC et al (2012) Genomic analysis of reactive astrogliosis. J Neurosci 32:6391–6410

27. O'Callaghan JP, Kelly KA, VanGilder RL et al (2014) Early activation of STAT3 regulates reactive astrogliosis induced by diverse forms of neurotoxicity. PLoS One 9(7):e102003

28. Bender C, de Olmos S, Bueno A, de Olmos J, Lorenzo A (2010) Comparative analyses of the neurodegeneration induced by the non-competitive NMDA-receptor-antagonist drug MK801 in mice and rats. Neurotoxicol Teratol 32(5):542–550

29. Sigwald EL, Bignante EA, de Olmos S, Lorenzo A (2019) Fear-context association during memory retrieval requires input from granular to dysgranular retrosplenial cortex. Neurobiol Learn Mem 163:107036

30. Fernández MS, de Olmos S, Nizhnikov ME, Pautassi RM (2019) Restraint stress exacerbates cell degeneration induced by acute binge ethanol in the adolescent, but not in the adult or middle-aged, brain. Behav Brain Res 364:317–327

31. Rivarola ME, de Olmos S, Albrieu-Llinás G et al (2018) Neuronal degeneration in mice induced by an epidemic strain of Saint Louis encephalitis virus isolated in Argentina. Front Microbiol 7:9,1181

32. Darmopil S, Martín AB, De Diego IR, Ares S, Moratalla R (2009) Genetic inactivation of dopamine D1 but not D2 receptors inhibits L-DOPA-induced dyskinesia and histone activation. Biol Psychiatry 66(6):603–613

33. Espadas I, Darmopil S, Vergaño-Vera E, Ortiz O, Oliva I, Vicario-Abejón C, Martín ED, Moratalla R (2012) L-DOPA-induced increase in TH-immunoreactive striatal neurons in parkinsonian mice: insights into regulation and function. Neurobiol Dis 48 (3):271–281

34. Albrecht M (1954) Mounting frozen sections with gelatin. Stain Technol 29:89–90

Chapter 2

Assessment of Auditory Hair Cell Loss by Cytocochleograms

Aurélie Thomas, Thomas Venet, and Benoît Pouyatos

Abstract

The highly differentiated cells that compose the sensory epithelium of the cochlea, known as hair cells, are a wonder of refinement, but they are also highly vulnerable because of their location in the cochlea. These cells are indeed subjected both to mechanical stress, through the vibration of the basilar membrane at the base of sound transduction, and to chemical exposure, through the imperfectly protective blood–labyrinth barrier of the *stria vascularis*. Cochlear histology is therefore a critical step in the assessment of the consequences of ototoxic insults or noise overexposure in animal models. As opposed to other sensory systems, the cochlear neuroepithelium, the organ of Corti, can be harvested with simple tools and every single sensory cell can be observed on a microscope slide, thanks to their regular arrangement. Counting missing hair cells and locating them on the place-frequency map of the considered species provides a wealth of information, both quantitative and objective, on the type and severity of the exposure. This article describes a method to (1) harvest the cochlea, (2) stain and dissect the organ of Corti, (2) quantify the hair cell loss, and (3) construct a graphic representation of the distribution of hair cell loss along the length of the organ of Corti, the cytocochleogram.

Key words Cytocochleogram, Method, Dissection, Hearing loss, Cochlea, Organ of Corti, Hair cell count

1 Introduction

The hearing system of mammals is composed of a peripheral receptor, the cochlea, which transduces sound waves into neurochemical influxes, neuronal fibers that convey the information towards the brain, where dedicated structures treat the incoming electrical signal and extract meaningful information from it. Although impairment can occur at different levels within this functional chain, the cochlea is the structure most sensitive to environmental stressors because (1) it is directly exposed to mechanical stimuli and because (2) cochlear cells are limited in number and cannot regrow (at least in mammals). Therefore, to investigate the cause of hearing

Supplementary material: The online version of this chapter (https://doi.org/10.1007/978-1-0716-1637-6_2) contains supplementary material, which is available to authorized users.

Jordi Llorens and Marta Barenys (eds.), *Experimental Neurotoxicology Methods*, Neuromethods, vol. 172,
https://doi.org/10.1007/978-1-0716-1637-6_2, © Springer Science+Business Media, LLC, part of Springer Nature 2021

loss and/or the consequences of noise and/or chemical exposures, one must start by assessing cochlear health. This can be done by functional or histological methods, these two approaches being, of course, complementary. Functional assessment methods include Distortion Product Otoacoustic Emissions (DPOAE) [1], Auditory Brainstem Responses (ABR) [2] and reflex modification of the startle response [3], but only DPOAEs originate specifically from the cochlea, more precisely from outer hair cells (OHCs). The histology of the cochlea can be carried out by two main techniques: the cytocochleogram, which consists in counting the hair cells on a dissected organ of Corti placed between slide and coverslip [4], and the otic microscopy, which involves embedding the whole cochlea in a solid medium (resin or paraffin and cutting it in thin slices observable with a light or electron microscope. Another alternative is the block surface technique, which involves the slicing of plastic-(Araldite-) embedded whole-mount cochlea for observation with a phase-contrast microscope [5]. This method uses the perfusion of the perilymphatic spaces of the cochlea in vivo and avoids post-mortem autolytic changes. It is recommended for the assessment of subtle subcellular impairment. If your goal is to obtain an objective and quantitative assessment of hair cell loss (cell present or absent), then the cytocochleogram method described in this chapter is more adapted as it requires minimal equipment and is relatively simple, provided suitable dissecting skills. The method described below can be applied in most laboratory rodents, with varying degrees of difficulty, the cochleae of rats and guinea pigs being easier to process than that of mice for example. The cytocochleogram method allows the whole cochlea to be explored, as opposed to functional methods, which are commonly limited in frequency (the lowest and highest frequencies being, in most cases, inaccessible). Finally, the two ears constitute, in general, two identical replicates, which can save an experiment in case of the unsuccessful dissection of one of the cochleae.

There are also some inherent limitations to this method that should be taken into account before undertaking cytocochleograms: this technique is designed to count hair cells and does not allow the assessment of the state of stereocilia or subcellular structures. Also, one should wait a few weeks after exposure or treatment to be certain that all impaired cells have died and have been cleaned out from the organ of Corti before sacrificing the animals and undertaking cytocochleograms (*see* **Note 1**).

Although the dissection of the unstained organ of Corti is technically possible [6], it is much easier with the use of a cellular staining. The original cytocochleogram method involved the use of post-fixation with osmic acid [7], which turns the cellular lipids black by creating –C=C– double bonds in the unsaturated fatty acids. While osmic acid creates sufficient contrast to allow the dissection, and is compatible with intracardiac perfusion, it is also extremely toxic for the experimenter. Consequently, we favor an alternative method: the succinate dehydrogenase (SDH) staining [8]. The main advantage of this staining is that it the hair cells have a much stronger SDH activity than the surrounding cells (outer sulcus and *stria vascularis*), which facilitates grandly dissection and counting, given that they are the only "blue" cells in the organ of Corti. In addition, SDH staining does not render the organ of Corti brittle as the osmic acid staining does. The downsides are that the staining has to be immediately done right after cochlear harvest and that the 1-h incubation at 37 °C leaves time for the hair cell morphology to potentially evolve.

Once the organ of Corti has been harvested from the cochlea and the counting has begun, it is important to be able to place the missing cells in the cochlear place-frequency map (or tonotopic) map of the species considered. The exact knowledge of the cochlear place-frequency map is indeed a prerequisite for the interpretation of normal and abnormal structural and functional features of the inner ear. Such maps have been established in a number of species: rat [9], mouse [10], guinea pig [11], mole rat [12], cat [13], mustache bat [14], fat tailed gerbil [15], opossum [16], Mongolian gerbil [17].

In this chapter, we will focus on the rat, for which the mathematical function describing the place-frequency data is

$$x = 102.048 * e^{(-0,04357*f)} - 4.632$$

where x is the characteristic place expressed in percentage of the basilar membrane length from the base, and f is in the characteristic frequency in kHz [9].

Given the biological variation in the basilar membrane length within the same strain, it is important to standardize the length of the individual cytocochleogram (*see* Subheading 4.3). First, cochlear length has to be converted to percent distance from base or apex, and then averaged. Cytocochleograms should display both the distance and the characteristic frequency scales. The mathematical function describing the place-frequency data should be indicated in the text or the source given in the reference list.

The method described below works well in our laboratory, but might be adapted to the scientific context, (species, strain, age, type of exposure…), the available equipment and personnel in the reader's lab.

We provide a 20 min accompanying video showing the dissection of a rat cochlea. It can be found there: https://youtu.be/QaDFwHjYTbQ. Other videos showing the dissection of mouse cochleae are also available on the internet [6, 18].

2   Materials

- Equipment and tools:
  - laboratory balance (Mettler Toledo XP205).
  - pH-meter with 0.01 pH precision (Labo Moderne).
  - dentist's drill (NSK Volvere max NE120) equipped with diamond heads (Ø1.4 and 0.6 mm).
  - heating water bath.
  - light microscope equipped with a micrometric eyepiece (Olympus BX41).
  - calibration slide (2.5 µm/graduation).
  - friedman rongeur.
  - cutting edge.
  - bone scissors.
  - 23G needles.
  - fine forceps.
- Chemicals:
  - Nitro blue tetrazolium chloride (NBT) (Merck, Cat #1.24823).
  - Sodium succinate (Merck, Cat #8.20151).
  - Potassium phosphate monobasic (KH_2PO_4) (Merck, Cat #1.04873).
  - Dipotassium hydrogenphosphate (K_2HPO_4) (Sigma, Cat #60353).
  - Phosphate-buffered saline (PBS).
  - Paraformaldehyde 32% (EMS, Cat #15714-S).
  - Decalcifying solution (Cellpath, Cat #DC-LMR) (optional).

3   Sample Preparation

3.1   Preparing Solutions

Prepare the following stocks solutions and store at 4 °C:

- 0.1% Nitro blue tetrazolium chloride (NBT): 0.5 g for 500 mL of distilled water.
- 0.2 M sodium succinate: 27.015 g for 500 mL of distilled water.
- 0.2 M KH_2PO_4 solution: 13.6 g of KH_2PO_4 for 500 mL of distilled water.
- 0.2 M K_2HPO_4 solution: 34.84 g of K_2HPO_4 for 1000 mL of distilled water.

Mix 190 mL of KH_2PO_4 solution and 810 mL of K_2HPO_4 solution to obtain the final 0.2 M phosphate buffer, and adjust at pH 7.4.

The day of the sacrifice, prepare a solution of 4% paraformaldehyde in $1\times$ PBS and the SDH staining solution (0.1% NBT, 0.2 M sodium succinate, and 0.2 M phosphate buffer with a 2:1:1 ratio). Ensure the SDH solution has an osmolality of about 280 mmol/kg. Approximately 100 mL of staining solution is necessary for two rat cochleae.

3.2 Cochlea Harvest

Sacrifice the animal according to local ethical regulations for animal experiments (i.e., EU Directive 2010/63/EC). Cut the skull in the longitudinal axis with strong bone scissors and remove the brain (Fig. 1a). Locate the external auditory canal and cut the temporal bone around the cochlea with a large safety margin (Fig. 1b). Repeat for the other ear. Remove the masseter muscle by pulling the flesh with the rongeur (Fig. 1c) to reveal the otic capsule (Fig. 1d). Carefully break the capsule using the rongeur. Once the cochlea is visible in the open otic capsule, continue to clear the space around it by carefully removing the stapes, the basilar artery, and bone debris to access round and oval windows (Fig. 1f). Be sure you have a good hold on the temporal bone between your thumb and index finger to be comfortable for the dissection to come. To do so, remove as much flesh from the bone as you can.

3.3 SDH Staining

Immediately after cochlea collection, immerse the samples in the SDH staining solution. Access to both the round and oval windows must be clear. Gently perfuse both windows with the SDH staining

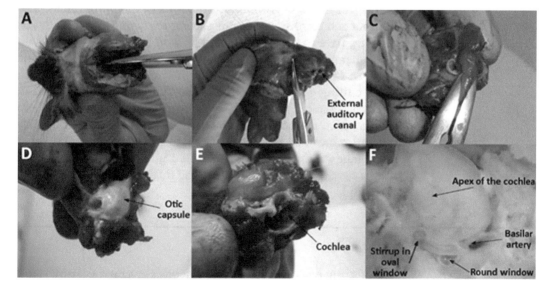

Fig. 1 Different steps of cochlea harvest

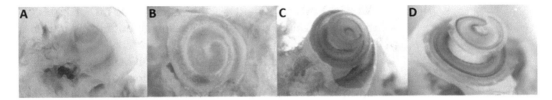

Fig. 2 Light microscope views of a cochlea at different steps of dissection. (**a**) Cochlea before drilling. (**b**) Cochlea drilled. (**c**) Cochlea without bone. (**d**) Cochlea without *stria vascularis*, Reissner, and tectorial membranes. Hair cells stained in dark blue are clearly visible

solution using a syringe and needle with a smoothed tip. Some blood usually escapes from both windows. This ensures that the cochlea is properly perfused. Then, immerse the cochleae in the same solution for 1 h at 37 °C. After staining, fix the cochleae by injecting 4% paraformaldehyde in both windows and keep 24 h at +4 °C. Finally, rinse with PBS and store at +4 °C until microdissection.

3.4   Cochlear Microdissection

Place the temporal bone in a large cup containing $1\times$ PBS under a light microscope (Fig. 2a). Before microdissection, carefully drill the cochlea using a dentist's drill equipped with a diamond head to leave only a thin layer of bone (*see* Fig. 2b, **Note 2** and video https://isicloud.inrs.fr/index.php/s/m57h4j4bC2QcdkJ).

Gently drill a small hole at the top (apex) of the cochlea. Carefully remove tiny chunks of bone using a cutting-edge starting from the apical hole and progressing slowly towards the base. Hair cells should be clearly visible with their dark blue color when the organ of Corti is exposed (Fig. 2c). With the tip of a needle (23G), cut the *stria vascularis* at the level of the outer sulcus. The Reissner and tectorial membranes, completely transparent with this staining, have to be removed entirely with fine forceps to facilitate subsequent counting (Fig. 2d). Using the 23G needle cut the organ of Corti and separate it from the osseous spiral lamina with a fine cutting edge (*see* **Note 3**). Then cut the organ of Corti in three parts (apical, medial, and basal turns) before immersing them in PBS. Trim the bone as close to the organ of Corti as you safely can and remove all remaining pieces to *stria vascularis* and other membranes. Now put the sample on a glass slide in glycerin and place a cover glass on top (Fig. 3). The three turns can be mounted on a single slide.

4   Data Collection and Analysis

The creation of a cytocochleogram involves several successive steps:

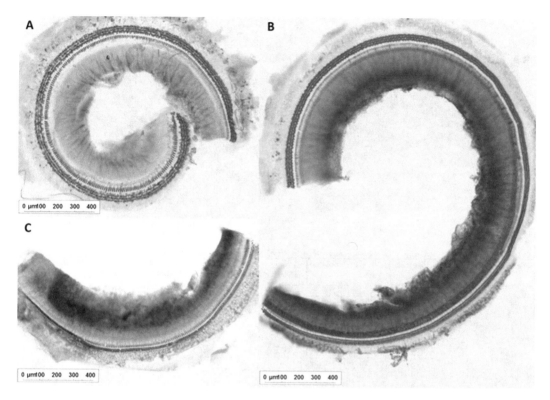

Fig. 3 The three turns of the rat cochlea after microdissection. (**a**) Apical turn, (**b**) Medial turn, and (**c**) Basal turn

1. Calibration.
2. Hair cell count (missing and/or present).
3. Normalization of the cochlear length (for averaging).
4. Calculation of percent loss of hair cells per unit of distance.
5. Plotting.

These steps require either a standardized Excel file or a dedicated computer interface (Fig. 4), with which you can enter the positions of the missing hair cells along the cochlea (*see* **Note 4**).

4.1 Microscope and Software Calibration

Place the calibration slide under the light microscope equipped with a micrometric eyepiece and a 40× objective. Record the distance covered by the 100 graduations. This step needs to be performed every time a new microscope or a new objective lens is used.

Now, place the surface preparation under the microscope. The first step is to determine how many outer and inner hair cells are visible within the 100 graduations of the eyepiece micrometric scale. Because of the difference in hair cell size between the apex

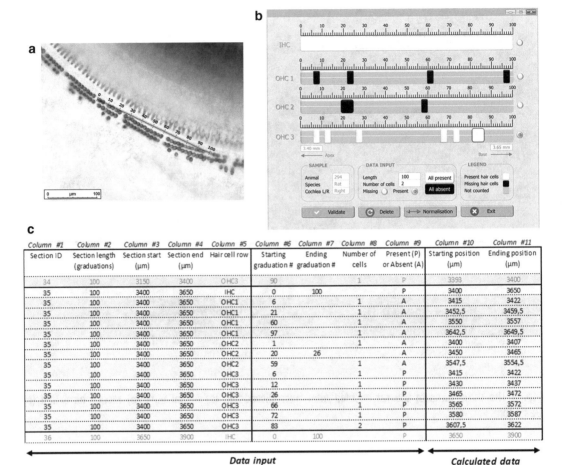

Fig. 4 (**a**) Microscopic view of dissected organ of Corti with the superimposed eyepiece micrometric scale places on top of the outer hair cell rows. For the two inner rows of outer hair cells (OHC1 and OHC2), missing hair cells appear in black on the software interface (**b**). For the third row of outer hair cells, extremely damaged, the remaining outer hair cells are counted and appear in white on the software (OHC3). No missing inner hair cell (IHC) is visible. In this case, the button "All present" is pressed. The data input can performed in simple Excel sheet (**c**). This example corresponds to the same cochlea partition visible on the photograph and analyzed in the software interface

Table 1
Number of hair cells for 100 graduations in the three turns of the cochleae of Brown-Norway rats (6–24 months old). Data are mean ($n = 82$) ± standard error

	Inner hair cells	Outer hair cells
Apical turn	28.6 ± 1.23	37.1 ± 1.63
Medium turn	31.0 ± 1.46	37.7 ± 1.39
Basal turn	31.1 ± 1.63	37.2 ± 1.65

and the base and between inner and outer hair cells, record six values, one for each cochlear turn and for each type of hair cell. For example, the average values obtained in adult Brown-Norway rats are depicted in Table 1.

At this point, you dispose, for each turn, of a correspondence table between the number of graduations, the number of hair cells, and the real distance on the organ of Corti.

Before starting to count the missing hair cells, you should record the correspondence table in the software interface/Excel Sheet, as well as the information concerning your sample (species/strain, left or right cochlea, reference number/code of the subject, treatment).

4.2 Hair Cell Count

Once the calibration is performed, counting can start. The aim of this step is to obtain the location of each missing cell, expressed as an absolute distance from the apical end of the organ of Corti. To do so, place the graduation at the extreme apex of the cochlea, zero on the first visible outer hair cell (*see* **Note 5**). This position will be considered as the "position zero," and the positions of all missing cells will be expressed as an absolute distance from this "position zero."

The apical turn of the organ of Corti being more curved than the other turns, you must count over a distance shorter than the full scale of the eyepiece, e.g., 50 graduations. In any case, the number of graduations should be registered (Fig. 4c; column #2) and the starting / ending graduations of the section should be converted as a distance from the "position zero" (Fig. 4c; columns #3 and #4) using your correspondence table (in our case: 100 graduations = 250 μm).

The locations of the missing hair cells should always be recorded in order (from apex to base). For each hair cell row (column #5), locate the starting graduation of each "hole" (Fig. 4c; column #6). It is then convenient to be able to record either the number of contiguous missing hair cells (for small "holes"; column #8) or the graduations of the end of the series of missing hair cells (for big "holes" for which the number of missing hair cell is difficult to know precisely; column #7). In either case, the software (or the Excel sheet) should convert the starting and ending graduations into a position relative to the "position zero" (column #10 and #11). In case of an intact hair cell row (Fig. 4a; inner hair cell) or an important number of missing cells (Fig. 4a; third row of outer hair cells), it is much easier and faster to count the remaining hair cells instead of the missing ones (Fig. 4b,

OHC3; Fig. 4c, column #9). When the counting is finished for the section considered, save the data and progress to the next section (*see* **Note 5**).

4.3 Cytoco-chleogram Construction and Display

Once the number of missing (or present) hair cells and their locations are registered in the computer interface of your choice, it is time to construct the cytocochleogram.

To do so, be sure to dispose of the following data:

- the total length of the cochlea.
- the reference length for an average cochlea (e.g., 9.3 mm for a rat and 18.5 mm for a guinea pig).
- the starting and ending positions of "holes" relative to "position zero".
- the mathematical function of the place-frequency map of the species considered.
- the number of cells per 100 graduations for each turn.

The cytocochleogram is generally represented as a bar chart or a line plot, with the percentage of hair cell loss in ordinate and the position along the length of the cochlea in abscissa. The abscissa should also include a second *x*-axis displaying the frequency calculated using the place-frequency map equation.

Therefore, the length of the cochlea has to be divided into "sections," which will be represented by each bar/point of the graph. The size of each section depends on the precision needed to depict the cochlear damage accurately. A default size of 100 μm is adequate in most cases.

To calculate the data to be plotted in the cytocochleogram, go through the following steps:

1. Read the positions of the "holes" in the different cell rows. In case the positions of "present cells" have been recorded, consider the intervals to get the positions of the "holes".

2. Normalize the positions of the "holes" against the reference length of the cochlea using a linear interpolation. Let us consider a series of missing cell between 7000 and 7050 μm on a cochlea measuring 9.6 mm. The normalized positions will be.

$$7000 * 9.3/9.6 = 6781.25 \text{ μm and } 7050 * 9.3/9.6 = 6829.69 \text{ μm.}$$

3. Affect each "hole" to the sections to be displayed in the cytocochleogram (Fig. 5).
   (a) *Case a:* If one or several "holes" are entirely contained within a section, the percentage of missing cells for this

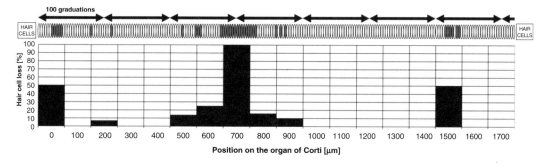

Fig. 5 Assignment of missing cells to the sections of a cytocochleogram. Red hair cells illustrate the missing cells. The cytocochleogram is plotted with 100 μm sections

section is directly calculated. Let us consider the section 1500–1600 μm (Fig. 5). The percentage of missing cells is [(1664–1630) + (1695–1679)]/100 = 50%.

(b) *Case b:* If the two extremities of the "hole" are not contained within a single section, distribute the distance in the sections overlapped by the "hole" and calculate the percentage of missing cells. Let us consider the "hole" overlapping sections 600–700, 700–800, and 800–900 μm in Fig. 5. The percentage of missing cells in section 600–700 μm is [(618–600) + (700–693)]/100 = 25%. One hundred percent of the cells are missing in section 700–800 μm. Then, the percentage of missing cells in section 800–900 μm is (916–900)/100 = 16%.

4. Use the place-frequency mathematical equation to convert the distance into frequencies and create a second *x*-axis (kHz).

5. Plot the different rows of hair cells in four different subplots (Fig. 6). Length standardization allows the calculation of average cytocochleograms and the application of statistical analyses. Cytocochleograms can be plotted with error bars.

5   Conclusions

There is no doubt that creating cytocochleograms takes a bit of effort. The drilling and the dissection of the cochlea require some training. The programming of the counting interface may also be intimidating. However, there is no unachievable step in this method, and it can be performed with the standard laboratory equipment and chemicals.

The cochlea is arguably the only sensory organ of the body, whose histological status can be assessed quantitatively, thanks to the quasi-linear organization of the organ of Corti. Although the cytocochleograms only provide information on the missing hair

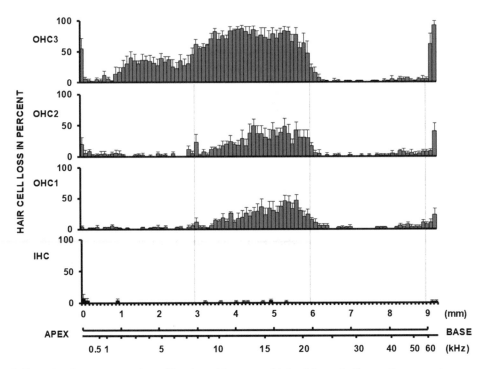

Fig. 6 Example of an average ($n = 5$) cytocochleogram obtained in male Brown-Norway rats exposed to 600-ppm styrene and an impulse noise with a L_{EX}, 8 h (equivalent continuous noise level averaged over 8 h) of 80 dB for 4 weeks, 5 days per week, 6 h per day by inhalation. Rats were sacrificed 4 weeks after the end of exposure. Error bars represent the standard error mean

cells and their location, and do not give any insight on the ganglion cells, it is possible (and wise) to use the contralateral ear to perform otic microscopy to observe there any additional neuronal damage.

Still, there are a few constraints and limitations inherent to the cytocochleogram. In case the goal is to assess cochlear damage caused by either a chemical treatment or noise, consider waiting at least 4 weeks after the end of the exposure to harvest the cochlea.

Logically, the cytocochleogram only allows the assessment of the permanent consequences of an exposure, and the exposure has to be severe enough to cause hair cell loss. If you are interested subtle temporary auditory effects, you should turn to functional methods (otoacoustic emissions, brainstem auditory evoked potential) which give you the possibility to perform repeated measures at different time points.

Ultimately, the mastery of cytocochleograms is a great benefit for a hearing research lab and only requires a bit of initial training and set up.

6 Notes

1. This process takes at least 4 weeks, but we recommend waiting a few weeks longer, if it is experimentally possible.

2. Take all the time needed for this step because the finer the bone layer is, the easier the dissection will be. It is better to make few holes with the drill in the bony shell, than leaving too much bone.

3. Please note that the "hook" of the basal turn of the organ of Corti is more difficult to dissect than the rest of the organ because the spiral lamina is thicker. You might want to soften the bone by immersing the cochlea for 5–10 min in the decalcifying solution before cutting the spiral lamina with the 23G needle. In case you are having difficulties with this dissection, it is also possible to cut the basal turn into reasonable segments with iris scissors to avoid tearing the organ of Corti although this method will cause more tissue loss than the dissection in one single piece.

4. If you plan to perform cytocochleograms on a regular basis, automatizing the different steps using a little programming (e.g., VBA Excel; Fig. 4b), is very helpful to save time and avoid mistakes. However, it also possible to use a simple Excel file, and manually apply the different calculations needed to obtain the data for plotting the cytocochleogram (Fig. 4c).

5. The hair cells of the extreme apical and basal ends of the cochlea appear often disorganized. In the rat, the apical outer hair cells are often widely spaced and unaligned. At the basal end, the third row of outer hair cells often stops before the second and first rows. Consider these features as normal. Note that these cellular disarrays worsen as the animals get older and can represent the first signs of presbycusis.

References

1. Venet T, Thomas A, Wathier L, Pouyatos B (2020) DPOAEs for the assessment of toxicant-induced cochlear damage in Neuromethods, Vol. 172, Jordi Llorens and Marta Barenys (Eds): Experimental Neurotoxicology Methods, ISBN 978-1-0716-1636-9 (Chapter 5)

2. Eggermont JJ (2019) Auditory brainstem response. Handb Clin Neurol 160:451–464. https://doi.org/10.1016/B978-0-444-64032-1.00030-8

3. Davis M (1984) The mammalian startle. In: Eaton RC (ed) Neural mechanisms of startle behavior. Springer, Boston, MA, p 287

4. Engström H, Ades HW, Andersson A (1966) Structural pattern of the organ of Corti: a systematic mapping of sensory cells and neural elements. Almqvist & Wiksell, Stockholm

5. Bohne BA (1986) The plastic-embedding technique for preparing the chinchilla cochlea for examination by phase-contrast microscopy. Washington University Laboratory Manual vol. 6; https://www.researchgate.net/profile/Barbara_Bohne/publication/237197190_THE_PLASTIC-EMBEDDING_TECHNIQUE_FOR_PREPARING_THE_CHINCHILLA_COCHLEA_FOR_EXAMINATION_BY_PHASE-CONTRAST_MICROSCOPY/

links/5755394308ae0405a5736685/
THE-PLASTIC-EMBEDDING-TECH
NIQUE-FOR-PREPARING-THE-CHIN
CHILLA-COCHLEA-FOR-EXAMINA
TION-BY-PHASE-CONTRAST-MICROS
COPY.pdf?origin=publication_detail

6. Liberman MC (2015) Cochlear dissection for whole mount immunostaining. Mass Eye and Ear. https://www.masseyeandear.org/research/otolaryngology/eaton-peabody-laboratories/histology-core

7. Anniko M, Lundquist PG (1977) The influence of different fixatives and osmolality on the ultrastructure of the cochlear neuroepithelium. Arch Otorhinolaryngol 218(1–2):67–78. https://doi.org/10.1007/bf00469735

8. Yang WP, Hu BH, Sun JH, Zhai SQ, Henderson D (2010) Death mode-dependent reduction in succinate dehydrogenase activity in hair cells of aging rat cochleae. Chin Med J 123(13):1633–1638

9. Muller M (1991) Frequency representation in the rat cochlea. Hear Res 51(2):247–254. https://doi.org/10.1016/0378-5955(91)90041-7

10. Muller M, von Hunerbein K, Hoidis S, Smolders JW (2005) A physiological place-frequency map of the cochlea in the CBA/J mouse. Hear Res 202(1–2):63–73. https://doi.org/10.1016/j.heares.2004.08.011

11. Tsuji J, Liberman MC (1997) Intracellular labeling of auditory nerve fibers in Guinea pig: central and peripheral projections. J Comp Neurol 381(2):188–202

12. Muller M, Laube B, Burda H, Bruns V (1992) Structure and function of the cochlea in the African mole rat (Cryptomys hottentotus): evidence for a low frequency acoustic fovea. J Comp Physiol A 171(4):469–476. https://doi.org/10.1007/bf00194579

13. Liberman MC (1982) The cochlear frequency map for the cat: labeling auditory-nerve fibers of known characteristic frequency. J Acoust Soc Am 72(5):1441–1449. https://doi.org/10.1121/1.388677

14. Kossl M, Vater M (1985) The cochlear frequency map of the mustache bat, Pteronotus parnellii. J Comp Physiol A 157(5):687–697. https://doi.org/10.1007/bf01351362

15. Muller M, Ott H, Bruns V (1991) Frequency representation and spiral ganglion cell density in the cochlea of the gerbil Pachyuromys duprasi. Hear Res 56(1–2):191–196. https://doi.org/10.1016/0378-5955(91)90169-a

16. Muller M, Wess FP, Bruns V (1993) Cochlear place-frequency map in the marsupial Monodelphis domestica. Hear Res 67(1–2):198–202. https://doi.org/10.1016/0378-5955(93)90247-x

17. Muller M (1996) The cochlear place-frequency map of the adult and developing Mongolian gerbil. Hear Res 94(1–2):148–156. https://doi.org/10.1016/0378-5955(95)00230-8

18. Fang Q-J, Wu F, Chai R, Sha S-H (2019) Cochlear surface preparation in the adult mouse. JoVE. https://doi.org/10.3791/60299. https://www.jove.com/video/60299

Chapter 3

Evaluation of Cellular and Molecular Pathology in the Rodent Vestibular Sensory Epithelia by Immunofluorescent Staining and Confocal Microscopy

Alberto F. Maroto, Erin A. Greguske, Alejandro Barrallo-Gimeno, and Jordi Llorens

Abstract

Hair cells in the vestibular and cochlear sensory epithelia are the main target of ototoxic drugs. Nevertheless, the synapses between the hair cells and the afferent terminals of the post-synaptic ganglion neurons have also been shown to be a target of ototoxic damage. In this chapter, we describe immunohistochemistry protocols adapted to the quantification of hair cells and synapses in the vestibular epithelia to assess ototoxic damage in rodents. Epithelia are immunolabeled intact and are used in whole-mount preparations for the quantification of hair cell numbers by confocal microscopy imaging. For synaptic assessment, the epithelia are first immunolabeled, embedded in a gelatin/albumin block, and then sectioned in a vibrating microtome before confocal microscopy imaging. The data thus obtained offer a thorough evaluation of the damage suffered by the vestibular sensory epithelia, including overt hair cell loss and subtle synaptic loss. Together, these pathological outcomes determine the loss vestibular input and the resulting behavioral alterations.

Key words Ototoxicity, Immunohistochemistry, Confocal microscopy, Vestibular epithelia, Hair cell, Synaptic uncoupling

1 Introduction

1.1 Hair Cell and Synapse Counts in the Vestibular Sensory Epithelia Following Ototoxic Damage

The vestibular system in the inner ear, also known as labyrinth, detects head accelerations resulting from body movements and gravity [1]. It provides both dynamic and static information for equilibrium and gaze control, as well as for the sense of orientation in space. Vestibular dysfunctions cause disequilibrium and loss of gaze control and are accompanied by vertigo, dizziness, impaired visual acuity, compromised motor competence, falls, and other disabilities. One possible cause of vestibular loss is exposure to toxic chemicals that target both the vestibular and the auditory systems, hence named ototoxic compounds [2]. Besides their shared location in the inner ear, the vestibular and the auditory

Jordi Llorens and Marta Barenys (eds.), *Experimental Neurotoxicology Methods*, Neuromethods, vol. 172,
https://doi.org/10.1007/978-1-0716-1637-6_3, © Springer Science+Business Media, LLC, part of Springer Nature 2021

systems rely for sensory transduction on subtypes of a common cellular type, the hair cell (HC). HCs are specialized epithelial cells characterized by bundles of apical specializations, known as stereocilia, that protrude into fluid-filled cavities and contain the molecular machinery for mechanoelectrical transduction [3]. Well known ototoxic compounds include clinically important drugs, such as aminoglycoside antibiotics and the chemotherapeutic drug cisplatin. Several different aminoglycosides have clinical use and all of them are ototoxic although some are more toxic to the auditory system and others, notably gentamicin and streptomycin, affect more the vestibular system. These drugs cause HC degeneration and often permanent loss of function because the ability of HCs to regenerate is null or very limited [4].

In animal models of ototoxicity, counts of sensory HC numbers are obtained for different purposes, such as to evaluate ototoxic potency of drugs, to assess the potential benefit of candidate protective agents that could have clinical utility by reducing the extent of damage, or to measure the efficacy of treatments aimed at activating HC regeneration [5]. In addition, there is growing interest in obtaining counts of synaptic contacts between the HCs and their post-synaptic afferents. Although the HCs are likely the primary target of ototoxic drugs, recent research has revealed that loss of synaptic contacts, referred to as synaptic uncoupling, may account for a significant part of the functional loss in particular ototoxicity models, depending on the ototoxic compound and mode and duration of exposure [6–9].

1.2 Functional and Structural Diversity in Vestibular Epithelia and Differential Vulnerability to Ototoxic Damage

Two types of HCs are identified in the vestibular sensory epithelia (Fig. 1). Type I HCs (HCI) have an amphora-like shape and are encased up to their neck by a remarkable structure, a calyx-shaped afferent terminal. In the lower two-third of the basolateral membrane of the HCI, the contact between the cell and the inner membrane of the calyx shows an adhesion complex, the calyceal junction, that is characterized at the transmission electron microscopy level as a symmetric densification of the plasma membrane, a greater than normal regularity in the distance between the membranes, and an electro-dense appearance of the extracellular space [10, 11]. Type II HCs (HCII) have a more columnar form and are contacted by bouton afferent terminals that are like regular synaptic boutons found elsewhere in the nervous system. Like HCs in the cochlea and photoreceptors and bipolar cells in the retina, both types of vestibular HCs are associated with a unique structure located in the pre-synaptic active zone, the synaptic ribbon. The ribbon likely contributes to the fast release and/or recycling of the synaptic vesicles thus enabling continuous, robust release of neurotransmitter [12]. The neurotransmitter in these synapses is glutamate, and post-synaptic sites contain ionotropic AMPA-type glutamate receptors clustered in post-synaptic densities [6, 8, 9,

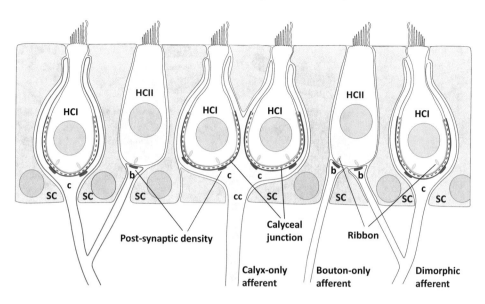

Fig. 1 Schematic of the various cell types and afferent endings in the vestibular epithelium. The epithelium consists of hair cells (HC), characterized by the stereocilia bundles on their apical surfaces, and supporting cells (SC). Type I HCs (HCl) have an amphora-like shape, are encased by calyceal afferent terminals (c), and show a prominent adhesion complex with this terminal, the calyceal junction. Type II HCs (HCII) are more cylindrical and are contacted by bouton terminals (b). Synaptic contacts are discrete structures, identified by pre-synaptic ribbons and post-synaptic densities. Three types of afferents are found: (1) calyx-only afferents, often forming complex calyces (cc) that encase 2 or 3 HCl; (2) bouton-only afferents, contacting only HCII; and (3) dimorphic afferents that form both calyceal and bouton type terminals

13]. Although there are two major types of HCs and two types of afferent terminals, three classes of afferents can be differentiated [14]. These are: (1) calyx-only afferents that form calyx terminals only and often form complex calyces that engulf two or more HCIs; (2) bouton-only afferents that form bouton terminals on HCII; and (3) dimorphic afferents that form both calyx terminals on HCI and bouton terminals on HCII.

Each human or rodent ear contains five vestibular sensory epithelia: three cristae, one in each of three semicircular canals, and two maculae in the utricle and saccule (Fig. 2). While the overall structures of the epithelia are similar in all vestibular end organs, there are regional differences within and among them, especially between the central and peripheral zones of each epithelium [14–17]. The stereocilia bundles are polarized structures, and their orientation reverses at approximately the central line of the utricular macula and the saccular macula, known as the striola. The peri-striolar (central) zone of these maculae is characterized by a higher density of HCI and denser innervation by calyx-only terminals. In the periphery of the maculae, HCII are more abundant and HCI are contacted by calyces that arise from dimorphic afferents. The crista receptors do not have a striola because all HCs in a crista

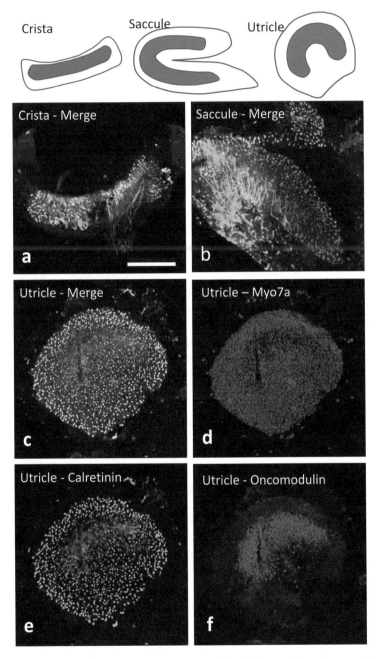

Fig. 2 Diversity of the vestibular sensory epithelia within specific end organs. The top drawings are schemas of the crista saccule and utricle, with the central part of the crista and peri-striolar part of the saccule and utricle shown in solid color. The bottom panels show confocal images of these epithelia (**a**: crista; **b**: saccule; **c–f**: utricle) labeled with antibodies against Myo7a (all HCs; red), calretinin (HCII and calyx-only afferents; green) and oncomodulin (HCs in central/peri-striolar zones; blue), as indicated within the figures. The scale bar in a is 100μm and applies to all panels

show the same orientation. They do have a central (apical) part with equivalent features to the peri-striolar zone of the maculae, and a periphery similar to the periphery of the maculae. One of the most striking differences between end organs is the length of the stereo-cilia, which are much longer in cristae than in maculae. These morphological differences associate with still poorly characterized biochemical and physiological differences.

One consistent observation across most experimental studies of vestibular toxicity is the differential sensitivity of the different end organs, zones, and cell types to different ototoxic compounds, including aminoglycosides and nitriles [18–21]. In dose–response or time-course studies, damage typically progresses in the following orders: crista > utricle > saccule, central > peripheral and HCI > HCII. Thus, HCI contacted by calyx-only afferents in the apical part of the crista show the highest sensitivity, degenerating at the minimally effective doses. Subsequently, at nearly maximal doses, only a few cells survive the ototoxic treatment, and these are typically HCII in the peripheral zones of the saccule.

1.3 Use of Immunofluorescent Labeling to Assess HC and Synaptic Loss

While scanning electron microscopy observation of the vestibular epithelia surfaces offers a direct overall view of vestibular HC loss [19, 22], immunofluorescent labeling, and subsequent assessment by confocal microscopy together offer the best approach to quantify loss of HCs by type and synaptic damage. This approach relies on knowledge of proteins selectively expressed in each type of HC or in discrete subcellular structures, and the availability of dependable antibodies against these proteins. The most common marker used to selectively label HCs for quantification is the unconventional myosin VIIa (Myo7a) [23]. HC nuclei have also been successfully labeled for HC quantification with antibodies against the transcription factor Gfi1b [5]. Antibodies against other proteins allow differentiation between HC types. HCIs have been shown to selectively express secreted phosphoprotein 1 (SPP1)/osteopontin [17, 24], while most HCII express calretinin [15]. Calyx-only afferents also express calretinin while bouton-only and dimorphic afferents do not [25]. Calyx terminals, and, indirectly HCI, can be visualized by labeling proteins characterizing the calyceal junction, either the adhesion molecule contactin-associated protein 1 (Caspr1), found in the afferent membrane [6, 9, 11, 16], or the extracellular matrix protein, tenascin-C [6, 9, 16].

Synaptic contacts can be recognized as close pairs of pre- and post-synaptic puncta. The core component of the pre-synaptic ribbon is ribeye, a protein encoded by the same gene encoding the transcription factor C-terminal binding protein 2 (CtBP2) [26]. The post-synaptic density is rich in the post-synaptic density protein 95 (PSD-95), a scaffold protein necessary for the clustering of the glutamate AMPA receptors [27]. Shank1A is another protein that characterizes the post-synaptic density [28, 29]. Although mRNA expression data [30] suggest that these may contain GluA2, GluA3, and GluA4, but not GluA1 subunits, so far only the GluA2 subunit has been clearly labeled in the vestibular synapses [6, 9].

1.4   Labeling of the Vestibular Sensory Epithelia

Immunofluorescent labeling is a routine procedure in many laboratories, and each laboratory optimizes the basic protocols for their needs, determined by the tissues to be examined and the properties of the antibodies to be used. This chapter will describe the protocols optimized in our laboratory to assess the number of HCs, including separate counts for HCI and HCII, and the number of synaptic contacts.

To study the vestibular epithelia, the traditional approach of labeling tissue sections on slides requires these tiny tissues to be embedded in an appropriate medium to form a block for sectioning. Embedding can be done with a medium for cryostat sectioning or with agar for sectioning at the vibrating microtome [31]. The vibratome sections can also be used for free floating labeling [32]. However, the vestibular sensory epithelia are thin enough to allow whole mount immunolabeling and imaging by confocal microscopy through their entire thickness. Therefore, this has become in recent years the most common approach to evaluate HCs and their synaptic densities [7, 8]. Nevertheless, imaging of epithelial cross-sections may be useful for a meticulous investigation of synapses and, for this purpose, we combine whole mount immunolabeling with sectioning of the epithelium labeling [6, 9]. To obtain the sections, we embed the labeled epithelia in gelatin/albumin blocks as often done with zebrafish and *Xenopus* embryos [33]. The immunolabeling protocol mostly follows that described by Lysakowski et al. in 2011 [16].

2   Materials

2.1   For Sample Processing and Immunohistochemistry

- Freshly depolymerized paraformaldehyde, 4% in 0.1 M phosphate-buffered saline (PBS), pH 7.2 (*see* **Note 1**).
- Cryoprotective solution: 34.5% glycerol, 30% ethylene glycol, 20% PBS, 15.5% distilled water.
- Triton-X-100 (Sigma-Aldrich).
- Gelatin from cold water fish skin, CAS #9000-70-8 (Sigma-Aldrich).
- Mounting medium suitable for fluorescence imaging, such as Mowiol.
- Microscope slides and coverslips.
- Primary antibodies (Table 1).
- Fluorochrome-conjugated secondary antibodies. These can be selected as appropriate for the primary antibodies from a large

Table 1
Antibodies

Target	Host and type	Reference and source	Working dilution
Calretinin	Guinea pig, polyclonal	214 104, Synaptic Systems	1/500
Calretinin	Rabbit, polyclonal	CR7697, Swant	1/1000
Caspr1	Mouse, monoclonal (IgG1)	Clone K65/35, Neuromab	1/400
GluA2	Mouse, monoclonal (IgG2a)	Clone 6C4, MAB397 Millipore	1/100
Myosin VIIa	Mouse, monoclonal (IgG1)	Clone 138-1-s, DSHB	1/100
Myosin VIIa	Rabbit, polyclonal	25-6790, Proteus Biosciences	1/400
Oncomodulin	Rabbit, polyclonal	OMG4, Swant	1/400
PSD-95	Mouse, monoclonal (IgG2a)	Clone K28/43, Neuromab	1/100
Ribeye	Mouse, monoclonal (IgG1)	Clone 16/CtBP2, BD Biosciences	1/200
SPP1	Goat, polyclonal	AF808, R&D Systems	1/200
Tenascin	Rabbit, polyclonal	AB19013, Millipore	1/200

variety of available choices. For the present protocols, we use the following secondary antibodies conjugated with Alexa Fluor fluorochromes: 488 goat anti-guinea-pig IgG H + L (catalog #A11073, Invitrogen/ThermoFisher), 555 donkey anti-rabbit IgGs H + L (#A-31572), 555 goat anti-mouse IgG2a (#A21137), 647 goat anti-mouse IgG1 (#A21240), and 654 donkey anti-mouse IgG H + L (catalog #A-21202). We also use the DyLight 405 donkey anti-rabbit IgG H + L (catalog #711-475-152, Jackson ImmunoResearch) (*see* **Note 2**).

– Appropriate nuclear fluorescent stain (i.e., 4′,6-diamidino-2-phenylindole/DAPI).

2.2 For Post-Labeling Sectioning

– Gelatin/albumin mixture for post-labeling inclusion and sectioning of the vestibular epithelia: 0.49 g alimentary gelatin, 30 g bovine serum albumin, 20 g sucrose, 100 mL of 0.1 M PBS (*see* **Note 3**).

– Glutaraldehyde 25%.

– Small molds. We use 1 cm long cylinders obtained by two transverse cuts of 2 mL Eppendorf tubes.

– Vibrating microtome.

2.3 For Confocal Imaging and Image Analysis

– Confocal microscope suitable for 4-color channel image acquisition. The images in this chapter were acquired in a Zeiss LSM880 spectral microscope.

– Microscopy image analysis software. We use both Image J (National Institute of Mental Health, Bethesda, Maryland, USA) and Imaris (Bitplane).

3 Methods

3.1 Immunohisto-chemistry Protocol

- Fix the vestibular epithelia in 4% PFA for 1 h.

- Rinse the samples in PBS (2 × 10 min) and use immediately or store at −20 °C in cryoprotective solution. For storage, the samples are placed in the solution at room temperature, then placed at 4 °C for 2 h for effective embedding before storage in the freezer. To use these samples, allow them to return to room temperature and rinse with PBS (2 × 10 min). We have successfully used samples stored in this way for up to 8 years.

- All the following steps are performed using slow agitation.

- Incubate samples for 1 h with 500μL of 4% Triton X-100 in PBS for permeabilization (see **Note 4**).

- Incubate samples for 1 h with 500μL of 0.5% Triton X-100 and 1% of fish gelatin in PBS for blocking (see **Note 5**).

- Incubate primary antibodies in 200μL of 0.1% Triton X-100 and 1% fish gelatin in PBS for 24 h at 4 °C. Details of the primary antibodies are shown in Table 1. We use two different combinations:

  – A: To obtain counts of HCI and HCII, we use the rabbit anti-Myo7a antibody to label all HCs, the guinea-pig anti-calretinin to label HCIIs and the mouse monoclonal anti-Caspr1 antibody to label calyceal junctions and thus identify HCI.

  – B: To quantify synaptic contacts, we use the rabbit anti-Myo7a antibody to label all HCs, the guinea-pig anti-calretinin to label HCII, the mouse IgG1 anti-ribeye antibody to label pre-synaptic ribbons and the mouse IgG2a anti-PSD95 to label post-synaptic densities.

- Rinse with 500μL of PBS (4 × 10 min).

- Incubate secondary antibodies in 200μL of 0.1% Triton X-100, 1% fish gelatin in PBS for 24 h at 4 °C in agitation. In this and following steps, the incubations are protected from light exposure by wrapping the culture plate with aluminum foil. We use the following combinations of Alexa Fluor conjugated secondary antibodies, selected for the corresponding combinations of primary antibodies:

  – A: 488 anti-mouse, 555 anti-rabbit and 647 anti-guinea pig.

  – B: 405 anti-rabbit, 488 anti-mouse IgG2a, 555 anti-mouse IgG1, and 647 anti-guinea pig.

- Rinse with PBS (2 × 10 min).

- For combination A only: Incubate with 1μL/mL DAPI in PBS (15 min).

- Rinse with PBS (4 × 5 min).

At this step, the vestibular epithelia can be mounted on slides as whole mounts or embedded for sectioning as explained in the next subsection. For mounting, we use Mowiol, but commercial mounting media suitable for immunofluorescence can also be used.

3.2 Post-Labeling Inclusion and Sectioning

- Embed epithelia in gelatin/albumin solution overnight at 4 °C.

- In an Eppendorf tube, start polymerization of a small volume (e.g., 250μL) of the gelatin/albumin solution by adding 9% (V/V) of 25% glutaraldehyde. Immediately transfer the mixture into a mold to form the bottom half of the block. Avoid the formation of bubbles. The solidification process can be slowed down if all the reagents are placed previously on ice. Cooling is recommended to facilitate the transfer of the polymerizing solution while avoiding the formation of bubbles.

- The sample should be placed appropriately oriented on the gelatin/albumin solution (Fig. 3a) once the top layer of the solution is firm but still sticky (usually within 5–10 min after adding the glutaraldehyde to the gelatin/albumin solution). Noting the orientation of the epithelia is important because it will determine the orientation for sectioning (*see* **Note 6**).

- Carefully absorb the excess liquid gelatin/albumin surrounding the sample with a paper towel. Pay special attention that movement of the gelatin/albumin solution does not shift the orientation of the epithelium and that the epithelium does not adhere to the paper towel.

- Draw a diagram with the location and orientation of the sample in the block to facilitate orientation for later sectioning.

- Cover the sample with a second polymerizing solution (Fig. 3b), prepared as described above, to form the upper half of the block. Pour the solution slowly down the walls of the mold to avoid undesired displacement of the sample.

- Let this second layer completely solidify for about 30 min.

- Take the solid block out of the mold and cut it into a rectangular pyramid that contains the sample oriented in the center and near the top (Fig. 3c) to allow transverse sections of the epithelia with the vibrating microtome.

- Store the blocks overnight at 4 °C in a 4% paraformaldehyde solution. Cooling and additional fixation increases their stiffness for better sectioning. If stored for 4 or more days, they risk becoming too rigid for proper sectioning.

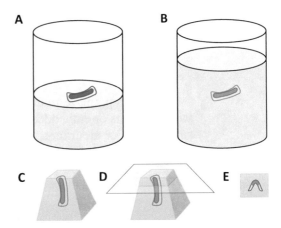

Fig. 3 Schematic of the procedure for post-labeling embedding and sectioning of the vestibular epithelia. (**a**) A crista placed on top of the bottom half of the gelatin/albumin block. (**b**) A second half of the block is formed on top of the first half, so the crista gets included into the block at the limit between the two halves. (**c**) After complete solidification, the block is cut into a pyramid with the sample in the appropriate orientation. (**d**) The block is sectioned horizontally using the vibratome. (**e**) Correctly oriented specimens yield transverse sections of the epithelia that contain both central and peripheral regions of the end organ, crista, or macula

- Section the specimens at 40μm in a vibrating microtome (Fig. 3d, e). Mount the sections with the appropriate medium (e.g., Mowiol), coverslip, and store protected from light at 4 °C until observation.

3.3   HC and Synapse Counts by Confocal Microscopy and Image Analysis

Sections are observed using a confocal microscope. For quantitative analysis, images are obtained from optimally oriented sections with the 63× (NA: 1.4) objective. For comparisons among groups of animals, the image acquisition settings must be maintained when observing the samples from the different groups of animals within the same batch. To avoid processing bias, the same number of samples from each experimental group must be processed in parallel. Generally, the entire epithelia can be scanned with a continuous Z-stack spanning around 25μm with an optical section thickness of around 0.3μm. Image processing software packages are used for the 3D visualization of stacks.

To obtain separate counts of HCI and HCII, we use combination "A" of primary antibodies, secondary antibodies, and DAPI (Fig. 4). Using this combination, all cell nuclei in the epithelia are labeled with DAPI and the cytoplasm of all HCs is labeled with Myo7a. In addition, the Caspr1 label of the calyceal junctions identifies HCI, and the cytoplasmic calretinin label (overlapping with Myo7a) identifies HCII. The calretinin label of calyx-only afferents in the central areas of the receptors is clearly distinguished

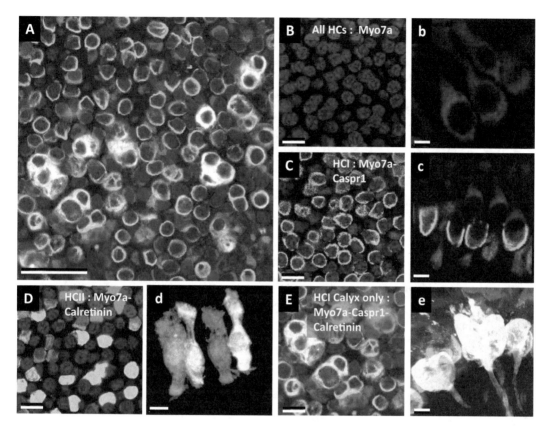

Fig. 4 Immunolabeling for HC counts in whole-mount preparations of the saccular macula. Images show labeling obtained with antibodies against Myo7a (red), Caspr1 (green), and calretinin (while), and the nuclear stain DAPI (blue). (**A**) General view of the striolar area. Other images are paired with letters to show a z-stack compression (capital letter) and a 3D reconstruction in higher magnification (lower case) of all HCs (**B, b**), HCI (**C, c**), HCII (**D, d**), and calyx-only complex afferents encasing two or three HCIs in the striolar zone (**E, e**). Scale bars = 30μm in **A**; 10μm in **B, C, D, E**; 5μm in **b, c, d, e**

from that of HCIIs. Since the whole vestibular epithelia is to be analyzed, thick stacks are obtained. To manage these, it is useful to use the Imaris blend option to reduce the opacity of channels and ease the task of counting. This blend option can be found in the 3D visualization mode of Imaris. The filtered images can be counted later with the cell counter plugin of ImageJ.

The "A" combination of primary antibodies is also useful to evaluate the integrity of the calyceal junction between HCIs and calyx afferents. The calyceal junction has been shown to be dismantled and rebuilt in rodent models of chronic ototoxicity and washout, consistent with the loss and recovery of function observed in the animals [6, 9]. Enduring damage of the calyces has also been recorded in models of partial ototoxic lesions [7].

For synaptic analysis, we use combination "B" of primary and secondary antibodies (Fig. 5). The Myo7a and calretinin labels allow identification of all HCs, HCII, and calyx-only afferent

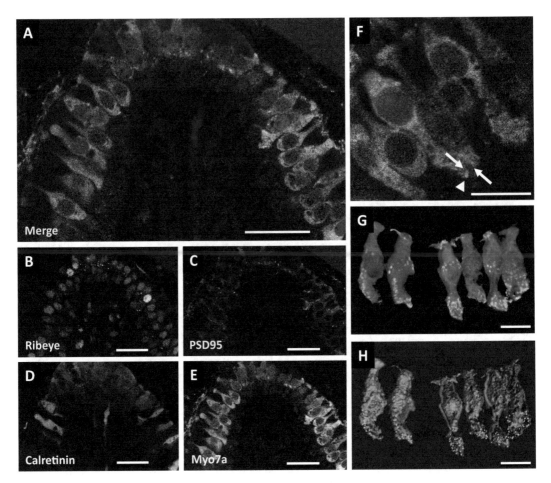

Fig. 5 Detail of a vestibular epithelium (crista) labeled for synapse counts. Confocal microscopy images are taken from 40μm thick transverse sections obtained using a vibrating microtome from a specimen embedded in a gelatin/albumin block. Four color images (**a**) are obtained with fluorescence channels (shown separate in **b–e**) adequate to the conjugate secondary antibodies recognizing the primary antibodies against ribeye (**b**; shown in red in **a**), PSD95 (**c**; shown in green in **a**), calretinin (**d**; shown in blue in **a**), and Myo7a (**e**; shown in white in **a**). Higher magnification of a raw 4-color image where pre-synaptic ribeye puncta (red, arrows) and post-synaptic PSD95 puncta (green, arrowhead) can be observed. (**g**) HCII with synaptic puncta as obtained by HCII segmentation using Imaris software. (**h**) Final 3D image obtained with Imaris of the same cells in G after cell and synaptic puncta recognition. Scale bars = 25μm in **a–e**; 10μm **f–h**

terminals. Pre-synaptic and post-synaptic puncta are labeled with ribeye and PSD-95, respectively. The anti-ribeye antibody also recognizes the transcription factor CtBP2, encoded by the same gene, and additionally labels the HC nuclei. Synaptic analysis is more complex than HC counting, and we use a semi-automated approach in Imaris to facilitate counting and to reduce potential biases. First, the diameters of pre- and post-synaptic components (puncta) are obtained with the measurement tool in the 2D slice visualization. These values are introduced as parameters into the program to aid detection of the puncta. Next, the spot function of the program is used in the 3D visualization mode. This function identifies the synaptic puncta based on the expected diameters measured previously for each fluorescence channel. Detection of puncta can be further optimized by adjusting the "quality threshold" setting of the spot function. Once selected, the acquisition parameters are maintained for the rest of the experiment to avoid bias during the analysis. To distinguish synapses on HCI and HCII, segmentation of the stack by each cell type is done before the spot analysis step. To this end, the cell functionality of Imaris is used. Adequate recognition of the cells is based on fluorochrome channels and cell diameters.

4 Notes

1. Fixative solutions, including paraformaldehyde and glutaraldehyde, are highly toxic and must be handled with protective equipment under a fume hood.

2. Colors shown in figures have been selected for best visualization and do not correspond to the emission colors of the fluorochromes.

3. Method to prepare the gelatin/albumin solution. In an Erlenmeyer flask, bring the PBS to a boil and dissolve the alimentary gelatin using a magnetic stir bar. Once dissolved, allow it to cool down to room temperature. Then, add the BSA in small fractions while stirring. Wait for each fraction to dissolve completely before adding the next fraction. Adding the BSA to a warm solution will cause it to denature, spoiling the mixture. After dissolving the albumin, add and dissolve the sucrose. The gelatin can be stored in aliquots at −20 °C. After thawing, it can be stored at 4 °C for up to 3 days.

4. The use of 4% Triton-X-100 for permeabilization of the vestibular epithelia exceeds largely the concentrations used for other tissues, but it has been demonstrated to be optimal for many antibodies in this tissue [16].

5. Fish gelatin has been found to be more effective in reducing background noise than other blocking agents, such as donkey serum, in these immunohistochemical protocols [16].

6. When making the gelatin/albumin blocks, the epithelia are placed flat and the pyramids made with the appropriate orientation to obtain transverse sections. Appropriate orientation of maculae will allow peri-striolar and peripheral regions (*see* Fig. 2) of the utricle or saccule to be obtained in the same sections. The orientation of the crista is critical, as only one orientation will permit representative transversal sections as shown in Fig. 5.

5 Discussion

The two protocols detailed here yield 4-color images suitable for HC and synapse quantification using image processing programs. The first combination of antibodies and protocol has been established as optimal in our hands to quantify HCs, providing a better assessment than initial approaches such as the use of anti-calbindin and anti-calmodulin antibodies [34]. Nevertheless, other alternative antibodies can be used for this aim. Thus, the rabbit anti-tenascin antibody listed in Table 1 offers a good alternative to the anti-Caspr1 antibody to identify calyx endings and hence, HCI. Also, the goat anti-SPP1/ostepontin (Table 1) offers a good label of HCI. However, using the indicated rabbit polyclonal anti-tenascin requires colabeling with a mouse monoclonal anti-Myo7a (Table 1). Labeling with the goat polyclonal anti-SPP1/ostepontin is not compatible with the use of secondaries raised in goat. Although alternate hosts for these primary and secondary antibodies are available, the combination we suggest has been found to yield the best results. For a distinct identification of the central zone of the receptors, labeling with an anti-oncomodulin antibody (Table 1, Fig. 2) is a good approach [35].

Labeling of the calyx afferents with anti-Caspr-1 or anti-tenascin antibodies is useful to identify HCI cells, but the integrity of the calyces is also an endpoint of interest in ototoxicity studies [6, 7, 9]. For further characterization of the calyces, several choices are available, including anti-beta III tubulin (Tuj-1 clone) [7, 13], anti-KCNQ5 [6, 16], and anti-Na+/K+ ATPase alpha-3 subunit [36] antibodies.

To quantify synaptic puncta, we previously used a rabbit anti-calretinin antibody (Table 1), anti-ribeye and either anti-PSD95 or anti-GluA2 (Table 1) [6, 9]. This triple labeling protocol missed a positive label for HCIs, which were suboptimally delineated using background staining. The use of the guinea-pig anti-calretinin allows the simultaneous use of the rabbit anti-Myo7a to label all HCs. Although theoretically this protocol should still allow the use of the anti-PSD95 and the anti-GluA2 antibodies, the quadruple labeling protocol yielded poor GluA2 staining. Labeling the GluA2 subunit of the AMPA receptors is a more direct measure of the capacity for glutamatergic synaptic signaling than labeling the scaffolding protein PSD-95 of the post-synaptic density. However, loss of PSD-95 immunopuncta has been demonstrated to be a result of ototoxic damage, and therefore this endpoint is adequate for ototoxicity assessment.

Acknowledgments

This study was supported by grants RTI2018-096452-B-I00 (Ministerio de Ciencia, Innovación y Universidades, Agencia Estatal de Investigación, Fondo Europeo de Desarrollo Regional, MCIU/AEI/FEDER, UE), and 2017 SGR 621 (Agència de Gestió d'Ajuts Universitaris i de Recerca, Generalitat de Catalunya). E.A.G. was supported by the Secretaria d'Universitats i Recerca del Departament d'Economia i Coneixement de la Generalitat de Catalunya (FI-DGR 2014 Program) and by the Ministerio de Educación, Cultura y Deporte de España (FPU 2014). A.B.G is a Serra-Húnter fellow. The confocal microscopy studies were performed at the Scientific and Technological Centers of the University of Barcelona (CCiT-UB). We thank Benjamín Torrejon-Escribano for advice on confocal imaging. We also thank Lara Sedó-Cabezón for her contributions to the establishment of these protocols within her PhD Thesis.

References

1. Bronstein AM (2013) Oxford textbook of vertigo and imbalance. Oxford University Press, Oxford

2. Boyes WK, Pouyatos B, Llorens J (2020) Sensory function. In: Pope CN, Liu J (eds) An introduction to interdisciplinary toxicology. Academic Press—Elsevier, London, pp 245–260

3. Michalski N, Petit C (2015) Genetics of auditory mechano-electrical transduction. Pflugers Archiv 467:49–72

4. Schacht J, Talaska AE, Rybak LP (2012) Cisplatin and aminoglycoside antibiotics: hearing loss and its prevention. Anat Rec 295:1837–1850

5. Wilkerson BA, Artoni F, Lea C, Ritchie K, Ray CA, Bermingham-McDonogh O (2018) Effects of 3,3'-iminodipropionitrile on hair cell numbers in cristae of CBA/CaJ and C57BL/6J mice. J Assoc Res Otolaryngol 19:483–491

6. Sedó-Cabezón L, Jedynak P, Boadas-Vaello P, Llorens J (2015) Transient alteration of the vestibular calyceal junction and synapse in response to chronic ototoxic insult in rats. Dis Model Mech 8:1323–1337

7. Sultemeier DR, Hoffman LF (2017) Partial aminoglycoside lesions in vestibular epithelia reveal broad sensory dysfunction associated with modest hair cell loss and afferent calyx retraction. Front Cell Neurosci 11:331

8. Cassel R, Bordiga P, Carcaud J, Simon F, Beraneck M, Le Gall A, Benoit A, Bouet V, Philoxene B, Besnard S, Watabe I, Pericat D, Hautefort C, Assie A, Tonetto A, Dyhrfjeld-Johnsen J, Llorens J, Tighilet B, Chabbert C (2019) Morphological and functional correlates of vestibular synaptic deafferentation and repair in a mouse model of acute-onset vertigo. Dis Model Mech 12(7):pii: dmm039115

9. Greguske EA, Carreres-Pons M, Cutillas B, Boadas-Vaello P, Llorens J (2019) Calyx junction dismantlement and synaptic uncoupling precede hair cell extrusion in the vestibular sensory epithelium during sub-chronic 3,3'-iminodipropionitrile ototoxicity in the mouse. Arch Toxicol 93:417–434

10. Seoane A, Demêmes D, Llorens J (2001) Pathology of the rat vestibular sensory epithelia during subchronic 3,3'-iminodipropionitrile exposure: hair cells may not be the primary target of toxicity. Acta Neuropathol 102:339–348

11. Sousa AD, Andrade LR, Salles FT, Pillai AM, Buttermore ED, Bhat MA, Kachar B (2009) The septate junction protein caspr is required for structural support and retention of KCNQ4 at calyceal synapses of vestibular hair cells. J Neurosci 29:3103–3108

12. Moser T, Grabner CP, Schmitz F (2020) Sensory processing at ribbon synapses in the retina and the cochlea. Physiol Rev 100:103–144

13. Sadeghi SG, Pyott SJ, Yu Z, Glowatzki E (2014) Glutamatergic signaling at the vestibular hair cell calyx synapse. J Neurosci 34:14,536–14,550

14. Eatock RA, Songer JE (2011) Vestibular hair cells and afferents: two channels for head motion signals. Annu Rev Neurosci 34:501–534

15. Dechesne CJ, Winsky L, Kim HN, Goping G, Vu TD, Wenthold RJ, Jacobowitz DM (1991) Identification and ultrastructural localization of a calretinin-like calcium-binding protein (protein 10) in the Guinea pig and rat inner ear. Brain Res 560:139–148

16. Lysakowski A, Gaboyard-Niay S, Calin-Jageman I, Chatlani S, Price SD, Eatock RA (2011) Molecular microdomains in a sensory terminal, the vestibular calyx ending. J Neurosci 31:10,101–10,114

17. Burns JC, Kelly MC, Hoa M, Morell RJ, Kelley MW (2015) Single-cell RNA-Seq resolves cellular complexity in sensory organs from the neonatal inner ear. Nat Commun 6:8557

18. Aran JM, Erre JP, Guilhaume A, Aurousseau C (1982) The comparative ototoxicities of gentamicin, tobramycin and dibekacin in the Guinea pig. A functional and morphological cochlear and vestibular study. Acta Otolaryngol Suppl 390:1–30

19. Llorens J, Demêmes D, Sans A (1993) The behavioral syndrome caused by 3,3′-iminodipropionitrile and related nitriles in the rat is associated with degeneration of the vestibular sensory hair cells. Toxicol Appl Pharmacol 123:199–210

20. Lopez I, Honrubia V, Lee SC, Schoeman G, Beykirch K (1997) Quantification of the process of hair cell loss and recovery in the chinchilla crista ampullaris after gentamicin treatment. Int J Dev Neurosci 15:447–461

21. Maroto AF, Barrallo-Gimeno A, Llorens J. BioRxiv preprint. https://doi.org/10.1101/2020.12.21.423804

22. Saldaña-Ruíz S, Boadas-Vaello P, Sedó-Cabezón L, Llorens J (2013) Reduced systemic toxicity and preserved vestibular toxicity following co-treatment with nitriles and CYP2E1 inhibitors: a mouse model for hair cell loss. J Assoc Res Otolaryngol 14:661–671

23. Hasson T, Gillespie PG, Garcia JA, MacDonald RB, Zhao Y, Yee AG, Mooseker MS, Corey DP (1997) Unconventional myosins in inner-ear sensory epithelia. J Cell Biol 137:1287–1307

24. McInturff S, Burns JC, Kelley MW (2018) Characterization of spatial and temporal development of type I and type II hair cells in the mouse utricle using new cell-type-specific markers. Biol Open 7(11):pii: bio038083

25. Desmadryl G, Dechesne CJ (1992) Calretinin immunoreactivity in chinchilla and Guinea pig vestibular end organs characterizes the calyx

unit subpopulation. Exp Brain Res 89:105–108

26. Schmitz F, Königstorfer A, Südhof TC (2000) RIBEYE, a component of synaptic ribbons: a protein's journey through evolution provides insight into synaptic ribbon function. Neuron 28:857–872

27. Chen X, Levy JM, Hou A, Winters C, Azzam R, Sousa AA, Leapman RD, Nicoll RA, Reese TS (2015) PSD-95 family MAGUKs are essential for anchoring AMPA and NMDA receptor complexes at the postsynaptic density. Proc Natl Acad Sci U S A 112:E6983–E6992

28. Braude JP, Vijayakumar S, Baumgarner K, Laurine R, Jones TA, Jones SM, Pyott SJ (2015) Deletion of Shank1 has minimal effects on the molecular composition and function of glutamatergic afferent postsynapses in the mouse inner ear. Hear Res 321:52–64

29. Sultemeier DR, Choy KR, Schweizer FE, Hoffman LF (2017) Spaceflight-induced synaptic modifications within hair cells of the mammalian utricle. J Neurophysiol 117:2163–2178

30. Niedzielski AS, Wenthold RJ (1995) Expression of AMPA, kainate, and NMDA receptor subunits in cochlear and vestibular ganglia. J Neurosci 15:2338–2353

31. Seoane A, Demêmes D, Llorens J (2003) Distal effects in a model of proximal axonopathy: 3,3′-iminodipropionitrile causes specific loss of neurofilaments in rat vestibular afferent endings. Acta Neuropathol 106:458–470

32. Gaboyard-Niay S, Travo C, Saleur A, Broussy A, Brugeaud A, Chabbert C (2016) Correlation between afferent rearrangements and behavioral deficits after local excitotoxic insult in the mammalian vestibule: a rat model of vertigo symptoms. Dis Model Mech 9:1181–1192

33. Gove C, Walmsley M, Nijjar S, Bertwistle D, Guille M, Partington G, Bomford A, Patient R (1997) Over-expression of GATA-6 in Xenopus embryos blocks differentiation of heart precursors. EMBO J 16:355–368

34. Cunningham LL (2006) The adult mouse utricle as an in vitro preparation for studies of ototoxic-drug-induced sensory hair cell death. Brain Res 1091:277–281

35. Hoffman LF, Choy KR, Sultemeier DR, Simmons DD (2018) Oncomodulin expression reveals new insights into the cellular organization of the murine utricle striola. J Assoc Res Otolaryngol 19:33–51

36. Schuth O, McLean WJ, Eatock RA, Pyott SJ (2014) Distribution of Na,K-ATPase α subunits in rat vestibular sensory epithelia. J Assoc Res Otolaryngol 15:739–754

Chapter 4

Morphometric Analysis of Axons and Dendrites as a Tool for Assessing Neurotoxicity

Rhianna K. Morgan, Martin Schmuck, Ana Cristina Grodzki, Donald A. Bruun, Lauren E. Matelski, and Pamela J. Lein

Abstract

Chemical perturbation of the temporal or spatial aspects of axonal or dendritic growth is associated with neurobehavioral deficits in animal models, and structural changes in axons and dendrites are thought to contribute to clinical symptoms associated with diverse neurologic diseases. Consequently, axonal and dendritic morphology are often quantified as functionally relevant endpoints of neurotoxicity. Here, we discuss methods for visualizing and quantifying axonal and dendritic morphology of neurons from the peripheral or central nervous systems in in vitro and ex vivo preparations. These methods include visualization of neuronal cytoarchitecture by immunostaining axon- or dendrite-selective antigens, transfecting cells with cDNA encoding fluorescent proteins, or labeling cells using membrane permeable small molecules that distribute throughout the cytoplasm, Golgi staining or Diolistics, as well as quantifying axonal and dendritic morphology using semi-automated or fully automated image analysis.

Key words Automated image analysis, Diolistics, Golgi staining, High-content imaging, Immunocytochemistry, LUHMES cells, Neurite outgrowth, Neurotoxicity, Primary neurons, Sholl analysis

1 Introduction

Most neurons in the vertebrate central and peripheral nervous systems extend two types of processes—axons and dendrites—which differ functionally, biochemically, and structurally [1, 2]. The primary function of axons is to convey electrical signals from the neuronal cell body to downstream cells in the neural circuit. Axons typically extend considerable distances from the neuronal cell body to the target tissue, are of uniform caliber throughout most of their length, and are mostly unbranched until they reach the target tissue. In contrast, dendrites comprise the primary site of synaptic input to the neuron. Dendrites are broad at their proximal end and taper over their length to fine distal tips.

Rhianna K. Morgan and Martin Schmuck contributed equally to this work.

Jordi Llorens and Marta Barenys (eds.), *Experimental Neurotoxicology Methods*, Neuromethods, vol. 172,
https://doi.org/10.1007/978-1-0716-1637-6_4, © Springer Science+Business Media, LLC, part of Springer Nature 2021

Dendrites tend to end proximal to the neuronal cell body and are highly branched throughout their length. Axons and dendrites are collectively referred to as neurites although the term "neurite" is also used to refer to neuronal processes that are not fully differentiated into an axon or dendrite, as occurs during early stages of neuronal differentiation [2], and is often the case with neuronal cell lines [3].

Axonal and dendritic morphology are critical determinants of neuronal function [1]. The shape of these processes affects signal processing within the neuron, while their number, length, and branching patterns determine the pattern of synaptic connections, which in turn regulates the distribution of information within the nervous system. Experimental evidence indicates that even subtle perturbations of temporal or spatial aspects of axonal and dendritic growth can cause persistent changes in synaptic patterning in the developing brain [4–7]. Clinical data confirm that altered axonal and dendritic structure is strongly associated with not only neurodevelopmental disorders but also neurodegenerative diseases [8–13]. Based on such observations, chemical-induced changes in axonal and dendritic morphology are considered functionally relevant endpoints of neurotoxicity [14–16]. Thus, there is considerable interest in using morphometric analysis of axons and dendrites to screen chemicals for neurotoxic potential and to elucidate mechanisms of neurotoxicity. Multiple approaches are used to measure axonal and dendritic morphology, but across all approaches, there are two key steps: (1) visualizing the axonal plexus and/or dendritic arbor of neurons and (2) quantifying axonal and dendritic morphology. Choosing which method to use depends upon the experimental model, e.g., in vitro or ex vivo, and on whether distinguishing between toxic effects on axons vs. dendrites is a desired outcome [17–19]. In this chapter, we present several methods for visualizing and analyzing axonal and dendritic growth of peripheral and central neurons in vitro and ex vivo.

2   Materials

2.1 Immunocytochemical Approaches for Visualizing Axons and Dendrites In Vitro

To identify axons in primary peripheral neurons, antibodies (Ab) specific for protein gene product 9.5 (PGP9.5, Thermo Fisher Scientific 38–1000/Invitrogen PA110011, RRID:AB_1088162) or the phosphorylated forms of heavy (NF-H) and medium (NF-M) neurofilament subunits (SMI-31; Sternberger Immunocytochemicals/MilliporeSigma NE1022, RRID:AB_2043448) are widely used. For primary central neurons, Ab specific for tau-1 (MilliporeSigma MAB3420, RRID:AB_94855) is used to label axons. To label dendrites in primary neurons from either the peripheral or central nervous system, Ab specific for microtubule-

associated-protein-2 (MAP2) (SMI-52; Sternberger Immunocyto-
chemicals/Synaptic Systems 188 004, RRID:AB_2138181/Invi-
trogen PA1-10005, RRID:AB_1076848) or non-phosphorylated
forms of NF-H and NF-M neurofilament subunits (SMI-32; Sternberger Immunocytochemicals/SigmaAldrich 559844, RRID:
AB_2877718) are effective. Additional materials needed include a
fixative, usually 4% paraformaldehyde (MilliporeSigma) in 0.1 M
phosphate buffer, and a permeabilization buffer [0.1% Triton-X-
100 (MilliporeSigma) in phosphate-buffered saline (PBS)] for antibodies to gain access to intracellular antigens. Also needed are a
blocking buffer (PBS supplemented with 1–10% bovine serum
albumin and/or 1–20% serum of the same species as the host
species of the secondary antibody) to decrease binding of antibodies to nonspecific binding sites, fluorophore-tagged secondary
antibodies that cross-react with the primary antibodies, and mounting medium [ProLong Gold Antifade Mountant (Invitrogen) for
slides].

2.2 Transfection of Cultured Neurons with Fluorescent Probes to Visualize Axonal or Dendritic Growth

Primary peripheral and central neurons can be transfected with low
efficiency (which is desirable to facilitate visualization of the axonal
plexus or dendritic arbor of individual neurons, particularly in high
density cultures) using Lipofectamine 2000 (Invitrogen) following
the manufacturer's protocol. However, due to cytotoxic effects of
lipid transfection reagents on primary neurons, we reduce the
incubation time for transfection to only 1–2 h with total replacement of the transfection solution with cell culture media at the end
of the incubation period. Dendrites are labeled by transfecting cultures with plasmids encoding microtubule-associated-protein-2B
fused to either enhanced green fluorescent protein (MAP2B-
eGFP) or red fusion protein (MAP2B-FusRed), which is under
the control of the neuron-specific CAG promoter [20]. Expression
of these MAP2B fusion proteins is restricted to the somatodendritic
compartment in cultured hippocampal neurons and does not alter
their intrinsic dendritic growth patterns [20]. To label all processes
(dendrites and axons), we transfect neurons with pCAG-tomato
fluorescent protein (TFP) constructs. TFP distributes throughout
the cytosol and does not alter neuronal morphology. We obtained
these plasmids, which are not commercially available, from
Dr. Gary Wayman (Washington State University, Pullman,
WA, USA).

2.3 Live Cell Staining to Visualize All Neurites in the Culture

Calcein-AM (C3100MP, Molecular Probes), a cell-permeant dye
used to determine cell viability in eukaryotic cells, will effectively fill
all neurites in live cells. In live cells, the non-fluorescent calcein-AM
is converted to fluorescent calcein-AM following acetoxymethyl
ester hydrolysis by intracellular esterases. The fluorescent calcein-
AM can be measured at excitation/emission (ex/em) wavelengths

of 494 and 517 nm, respectively. It is often used in conjunction with Hoechst 33342 (Thermo Fisher Scientific) and propidium iodide (PI; MilliporeSigma). Hoechst 33342 is a cell-permeable DNA stain that is excited by ultraviolet light and emits blue fluorescence at 460–490 nm. It preferentially binds adenine-thymine (A-T) regions of DNA and effectively labels nuclear chromatin. PI labels the nuclear membrane of compromised or dead cells and is used to identify non-viable cells in culture.

2.4 Diolistics to Label Processes of Neurons in Peripheral Ganglia or CNS Slice Cultures Ex Vivo

We use the Bio-Rad Helios Gene Gun Low-Pressure system for Diolistics labeling. This all-inclusive system contains a Tubing Prep Station, tubing cutter, cartridge storage vials and extractor tool, tungsten M-25 microcarriers, and Tefzel tubing. Additional supplies needed include polyvinyl pyrrolidone (PVP; MilliporeSigma), methylene chloride (Thermo Fisher Scientific), 1,1'-dioctadecyl-3,3,3',3'-tetramethylindocarbocyanine perchlorate (DiI) or other similar fluorescent dyes (Molecular Probes), hanging drop slides (Thermo Fisher Scientific), Sylgard 184 (Electron Microscopy Sciences), nitrogen, helium, three-way stopcock, ethanol, razor blades, and 35-mm plastic tissue culture dishes.

2.5 Golgi Staining to Visualize Dendrites in Intact Brain Tissue Ex Vivo

We use the FD Rapid GolgiStain kit (FD Neurotechnologies Inc.) for Golgi staining. Materials needed in addition to reagents supplied with this kit include plastic forceps to transfer tissues, plastic scintillation vials for incubation of tissue in Golgi solutions, 60 cm Whatman #1 paper to filter Golgi solutions, as well as standard histological supplies such as microscope coverslips and slides, staining dishes, slide holders, ethanol, chromium potassium sulfate, gelatin, sucrose (MilliporeSigma), and PBS.

3 Methods

3.1 Visualizing Neurites

Approaches for visualizing neurites include: (1) immunostaining for antigens selectively expressed in axons versus dendrites; (2) expressing cDNA encoding axon or dendrite-selective proteins linked to fluorescent tags or cDNA encoding intracellular fluorescent proteins that distribute throughout the entire cell; and (3) labeling neurons with cytoplasmic dyes or lipophilic membrane dyes (Table 1).

Immunostaining for antigens selectively expressed in axons versus dendrites is most often used to visualize neurites in primary cell culture and offers the advantage of distinguishing axons from dendrites. This method labels all axonal or dendritic processes in the culture. Therefore, in low cell density cultures or at very early times after plating in higher density cultures, this approach can be used to identify processes extended by individual cells.

Table 1
Biomarkers for visualizing neurite outgrowth in vitro and ex vivo within the peripheral and central nervous systems

	Axons	Dendrites
Peripheral nervous system, e.g., postganglionic sympathetic neurons	*In vitro:* ICC analyses of phosphorylated neurofilament (phospho-NF-H and phospho-NF-M) subunits or tau as early as DIV 1 (3–8 cells/mm^2) [31]	*In vitro:* ICC analyses of MAP2 or dephosphorylated forms of M and H neurofilament subunits in cultures ≥48 h post-induction of dendritic growth (25,000 cells/cm^2) [35, 36]
	Ex vivo: IHC analyses of phosphorylated neurofilament subunits or tyrosine hydroxylase (TH) in target tissues [72]	*Ex vivo:* Ballistic delivery system of fluorescent dyes (Diolistics) using 1,1′-dioctadecyl-3,3,3′,3′-tetramethyl-indocarbo-cyanine perchlorate (DiI)-coated tungsten beads [46, 50]
Central nervous system, e.g., hippocampal neurons	*In vitro:* ICC analysis of Tau-1 (33,000 cells/cm^2) performed at DIV 2 [19]	*In vitro:* MAP2B ICC or pCAG-MAP2B-TFP transfection (83,000 cells/cm^2) [19, 20]
	Ex vivo: "Brainbow" transgenic mice or injection of adenoviral or lentiviral vectors expressing cDNA encoding fluorescent proteins [21–23]	*Ex vivo:* Golgi staining [44]

ICC Immunocytochemistry, *IHC* Immunohistochemistry

Expressing cDNA encoding axon or dendrite-selective proteins linked to fluorescent tags or cDNA encoding intracellular fluorescent proteins that distribute throughout the entire cell are most useful when working with cultures of high cell density or mature cultures with extensive neurite outgrowth. In either case, it is often not possible to distinguish the dendritic arbor or axonal plexus of an individual cell from that of adjacent cells in the culture. The latter challenge can be overcome by using low efficiency transfection to label a small subpopulation of cells in the culture. Similarly, adenoviral or lentiviral infection with cDNA encoding fluorescent proteins can be used to label a subpopulation of neurons in the intact brain although this approach often involves infecting the living animal prior to harvesting of tissue for analysis [21–23].

Labeling neurons with cytoplasmic dyes or lipophilic membrane dyes can be used in vitro or ex vivo. Sometimes, there is incomplete labeling with this approach such that distal ends of processes are not labeled. Most membrane permeant cytoplasmic dyes and lipophilic membrane dyes label all neurites, so distinguishing axons from dendrites is only possible using structural criteria [2], which can be subjective. As with immunocytochemistry, in high density or mature low density cultures, it is challenging to identify the neurites associated with individual neurons. This issue can be overcome by measuring all neurites in a sample and then dividing by the number of nuclei (usually identified using a nuclear stain, such as Hoechst or DAPI). However, this approach is only feasible when cultures do not contain non-neuronal cells.

Specific examples of these different approaches to visualize neurites are provided below. This is not an inclusive list, but rather a description of methods that have worked well in our hands. While we describe their application using specific models, these methods can be readily adapted to other models.

3.1.1 Visualizing Neurites in the LUHMES Neuronal Cell Line

Neuronal cell lines are significantly easier to obtain and maintain than primary neuronal cell cultures and are particularly useful when large numbers of cultures are needed (e.g., for screening chemical libraries) [24]. The laboratory of Marcel Leist (University of Konstanz, Germany) has championed the use of the human LUHMES (*Lu*nd *H*uman *Mes*encephalic) cell line for neurotoxicity assays [25], and we have successfully used the LUHMES cell line to evaluate the effects of soluble immune mediators on neurite growth and retraction [26]. LUHMES cells were originally derived from 8-week-old female ventral mesencephalon. We obtained this cell line as a generous gift from the Leist laboratory, but LUHMES cells can also be purchased from the American Type Culture Collection (ATCC®; CRL-2927, RRID:CVCL_B056). This cell line is a sub-clone of the tetracycline-controlled, v-myc-overexpressing human mesencephalic-derived cell line MESC2.10 [27]. Like MESC2.10 cells, LUHMES cells can be differentiated into non-mitotic cells that recapitulate the morphological and biochemical characteristics of mature dopaminergic-like neurons by exposing undifferentiated, dividing LUHMES cells to tetracycline, glial cell line-derived neurotrophic factor (GDNF), and dibutyryl 3,5-'-cyclic adenosine monophosphate (db-cAMP).

LUHMES culture and differentiation are described in detail by Scholz and collaborators [3], and the quantification of neurite outgrowth using the Array-Scan II HCS Reader from Cellomics (Cellomics, PA, USA) has been described by Stiegler et al. [28]. Here, we describe measuring neurite outgrowth in LUHMES culture using a cell filling technique. While MAP2B immunocytochemistry has been used to image and quantify neurite growth in LUHMES cells [29], other studies have shown that the neurites extended by LUHMES cells express both dendritic and axonal properties [3]. Therefore, we generally use a cell filling technique to visualize neurites because it is quicker and less expensive.

To assay neurite growth and retraction in LUHMES cells:

- Detach LUHMES cells from the substrate with trypsinization 48 h after differentiation begins (day 2).

- Seed cells in 96-well plates at a density of 30,000 cells/well. Include both negative [cycloheximide (Sigma) at 3µM] and positive [p160ROCK inhibitor Y-27632 (Sigma) at 10µM] controls for neurite outgrowth in each 96-well plate [30]. These controls are most effective during the first 24 h after plating; when used later, they yield results that are more variable (Fig. 1).

- Chemical exposures can begin one or more hours after seeding as processes start to generate within hours after re-plating, which is optimal for assessing the effects of chemicals on initial neurite outgrowth.

- For neurite outgrowth exposures, prepare chemicals at twice the desired final concentration in tissue culture medium and add to each well at a volume equal to the volume of medium in the well (50µL) to minimize cell disruption.

- Remove the medium from each well 24 h later and replace it with 1× Dulbecco's PBS (with calcium and magnesium, ThermoFisher Scientific) containing 1µM calcein-AM, 1µg/mL Hoechst 33342 and 1.25µM PI.

- Incubate plates for 30 min at 37 °C while protected from light.

- Image the cells immediately after incubation. In live cells, calcein-AM is converted to a green fluorescent cytoplasmic product that fills the soma and neurites and can be observed using filters with ex/em spectra of 494/517 nm. Hoechst 33342, a nuclear dye, is imaged using ex/em 350/461 nm filters. PI labels dead cells and is imaged using ex/em 535/617 nm filters.

This protocol can be modified to measure neurite retraction at later times during differentiation by exposing cells to chemicals on day 5 (3 days after plating) via a complete exchange of media and imaging on day 6. Brefeldin-A (Millipore Sigma), an inhibitor of membrane trafficking, causes significant neurite retraction and is used as a positive control (at 10µM) for neurite retraction assays.

Chemical effects on neurite outgrowth can be influenced by plating surface, exposure time, and the method used for morphological analyses (Fig. 1). For example, 72, but not 48, h of exposure to PCB 95 (1 pM) increases neurite outgrowth in differentiated LUHMES cells plated on glass coverslips (*see* **Note 1**). Yet, neither a 72 nor a 48 h exposure to PCB 95 alters neurite outgrowth of LUHMES cells plated on tissue culture plastic. Moreover, the neurite growth promoting effects of PCB 95 were observed when cultures were immunostained for either MAP2B, a dendrite-selective cytoskeletal protein, or for phosphorylated neurofilaments, an axon-selective antigen, but not when using calcein-AM and Hoechst 33342. Since MAP2B is expressed only in more mature LUHMES cells, while neurites are extended by both mature and immature LUHMES cells [3], these data suggest that the

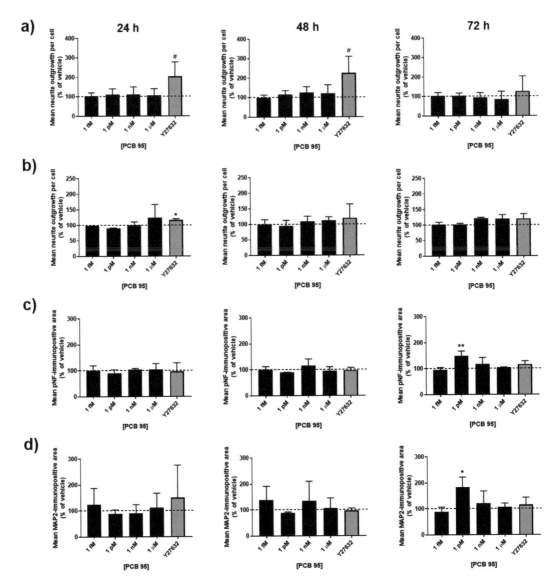

Fig. 1 *Substrate, exposure time, and method of morphometric analysis influence outcome in cultured LUHMES cells, a human neuronal cell line.* (**a**, **b**) LUHMES cells at similar passage number were plated at the same cell density onto (**a**) tissue culture plastic or (**b**) glass coverslips and exposed to vehicle (0.1% DMSO) or varying concentrations of PCB 95 for 24, 48, or 72 h. Y-27632, a p160ROCK inhibitor that increases neurite outgrowth during the first 24 h after plating, was used as a positive control. (Note: as indicated in this Figure, when used at later times after plating, Y-27632 does not increase neurite outrgrowth). Neurite outgrowth was imaged by labeling cells with calcein-AM. (**c**, **d**) LUHMES cells grown on glass coverslips were exposed to vehicle or PCB 95 for 24–72 h, then immunostained for axons or dendrites using antibodies specific for (**c**) phosphorylated neurofilaments or (**d**) MAP2B, respectively. Data are presented as the mean $\pm$ SD ($n = 3$ independent experiments with eight wells in each experiment). *Significantly different from DMSO vehicle control at $p < 0.05$; **$p < 0.01$ as determined by one-way ANOVA with Dunnett's post hoc test; #significantly different from DMSO vehicle control as determined by Student's t-test

morphogenic effects of PCB 95 are manifest only at latter stages of neuronal differentiation. The observation that immunostaining for either a dendrite or axon-selective antigen yields the same outcome is consistent with the observation that the neurites extended by LUHMES cells exhibit both dendritic and axonal properties [3].

3.1.2 Labeling Axons Versus Dendrites in Dissociated Primary Sympathetic Neurons

Primary cultures of sympathetic neurons dissociated from the superior cervical ganglia (SCG) of laboratory rodents are a robust model system for assessing chemical effects on axonal or dendritic growth [17, 31]. SCG are easily accessible at varying life stages of rodents and yield a homogeneous neuronal cell population consisting of principal sympathetic neurons that can be maintained for weeks to months in serum-free defined culture medium in the absence or presence of ganglionic glial cells [31]. Primary sympathetic neurons extend a single process that is axonal within 24 h of plating, and this unipolar morphology can be sustained for up to 3 months when cultures are maintained in serum-free medium in the absence of glial cells [32, 33]. Robust dendritic growth can be induced in primary sympathetic neurons cultured in serum-free media by co-culturing them with ganglionic glial cells [34]. In sympathetic neurons grown in the absence of glial cells, dendritic growth can be triggered by the addition of recombinant BMP-2, 4, 6 or 7 (30–100 ng/mL; R&D Systems) or Matrigel (50–75µg/mL, Corning Life Sciences) to the culture medium [35]. The processes induced by BMPs and Matrigel appear about 48 h post-exposure, and exhibit functional, biochemical, and morphological characteristics of dendrites [33, 36].

The protocol for setting up primary sympathetic neurons from the SCG of perinatal rat or mouse pups on glass coverslips (ideal for morphometric analyses that can be done as early as 2 days in vitro (DIV)) has been published [35]. Protocols for immunostaining these cultures have also been published [37, 38] and are briefly described below:

- Fix cultures with 4% paraformaldehyde in 0.2 M phosphate buffer for 10 min at room temperature.

- Permeabilize cultures with 0.1% Triton-X-100 in PBS for 10 min at room temperature.

- Incubate cells in blocking buffer (PBS with 10% bovine serum albumin) for at least 1 h at room temperature.

- To identify axons, incubate cultures for 1 h at room temperature or overnight at 4 °C with Ab specific for phosphorylated NF-H and NF-M neurofilament subunits or Ab that recognizes synaptophysin. To visualize dendrites, incubate cultures with Ab specific for MAP2B or with Ab that recognizes dephosphorylated NF-H and NF-M. Dilute antibodies in blocking buffer (*see* **Note 2**).

- Remove excess primary antibody by rinsing the cultures three times with PBS over 5 min.

- Incubate with a fluorescent-tagged secondary antibody that recognizes the primary antibody for 1 h at room temperature or overnight at 4 °C in the dark.

- Remove excess secondary antibody by gently rinsing cultures three times with PBS over 5 min.

- If grown on glass coverslips, cultures are mounted cell side down in mounting medium on glass slides and the edges sealed with nail polish.

- If grown in multi-well plastic tissue culture plates, PBS containing DAPI is added to the well for 15 min, the wells rinsed twice with PBS, and the culture dish sealed with Parafilm™ to prevent evaporation until cultures are imaged.

3.1.3 Labeling Axons and Dendrites in Dissociated Cultures of Primary CNS Neurons

Primary hippocampal and cortical neurons dissociated from perinatal rat or mouse pups are widely used in studies of neuronal morphogenesis and the effects of chemicals on axons and dendrites. In culture, these neurons exhibit a characteristic morphogenic program [39–41]. Stage 1 begins immediately after plating, when lamellipodia propagate around the somata signaling initial neurite formation. Growth cones develop in Stage 2 at the tips of neurites, enabling neurites to extend or retract short distances. Stage 2 lasts 12–36 h until one neurite begins to extend rapidly and its length exceeds that of the other "minor" neurites by 2- or 3-fold [40]. This single long neurite differentiates into the axon, while the other minor neurites slowly develop dendritic characteristics. As the culture ages, dendrites elongate and branch, and by approximately 21 DIV, synapses begin to form and dendritic spines appear [42]. This predictable morphogenic program makes primary hippocampal and cortical neurons a robust in vitro model for assessing chemical effects on axonal and dendritic morphologies of central neurons at varying stages of neuronal maturation.

Primary cortical and hippocampal neurons can be dissociated from embryonic day 18 or postnatal day 0 or 1 mouse or rat neocortices or hippocampi, as previously described [19, 43]. The former will contain significantly fewer glial cells than the latter. The neurons can be maintained at very low density for long periods of time using the "Banker" method of inverting the coverslip on which the neurons are plated over a monolayer of astrocytes [43]. Alternatively, cortical and hippocampal neurons can be cultured at high density in the presence of endogenous glia (mostly astrocytes) on the same coverslip [19]. With either method, cortical and hippocampal neurons develop a morphology that resembles that of their in vivo counterparts [20, 43]; however, we have found that the morphogenic response to neurotoxic chemicals can vary depending on which culture method is used. In our experience, the chemical-induced morphogenic response of neurons cultured as high density neuron-glia co-cultures more closely resembles the response observed in vivo [14]. Additionally, the substrate on which the cells are plated can influence the morphogenic response to chemicals. Plating on glass coverslips is the best practice because lipophilic compounds, such as PCB 95, stick to plastic, which prevents or minimizes their effect on neuronal morphogenesis (Fig. 2).

To quantify axons in primary hippocampal or cortical cultures:

- Plate hippocampal or cortical neuron-glia co-cultures at 33,000 cells/cm^2.

- To quantify chemical effects on the complete axonal plexus of individual neurons, expose cells to vehicle or chemicals for 48 h beginning at 4 h post-plating [38].

- Fix cultures with 4% paraformaldehyde in 0.2 M phosphate buffer for 10 min at room temperature.

- Permeabilize cultures with 0.25% Triton-X-100 in PBS for 5 min at room temperature.

- Incubate cells in blocking buffer (PBS with 10% bovine serum albumin and/or 10% goat serum) for at least 1 h at room temperature or overnight at 4 °C.

- To identify axons, incubate cultures for 1 h at room temperature with Ab specific for tau-1. Dilute antibodies in blocking buffer (see Note 2).

- Remove excess primary antibody by rinsing the cultures three times with PBS over 5 min.

- Incubate with a fluorescent-tagged secondary antibody that recognizes the primary antibody for 1 h at room temperature or overnight at 4 °C in the dark.

- Remove excess secondary antibody by gently rinsing cultures two times with PBS over 5 min.

- If grown on glass coverslips, cultures are mounted cell side down in mounting medium on glass slides and the edges sealed with nail polish.

- If grown in multi-well plastic tissue culture plates, PBS containing Hoechst or DAPI is added to the well for 15 min, wells are rinsed twice with PBS, and the culture dish sealed with Parafilm™ to prevent evaporation until cultures are imaged.

To quantify dendritic arborization (total length, number of terminal tips, neurite mass, and number of branching points) in high density neuron-glia co-cultures, transfect a subpopulation of neurons (0.1–0.5%) with cDNA encoding a fluorescent protein:

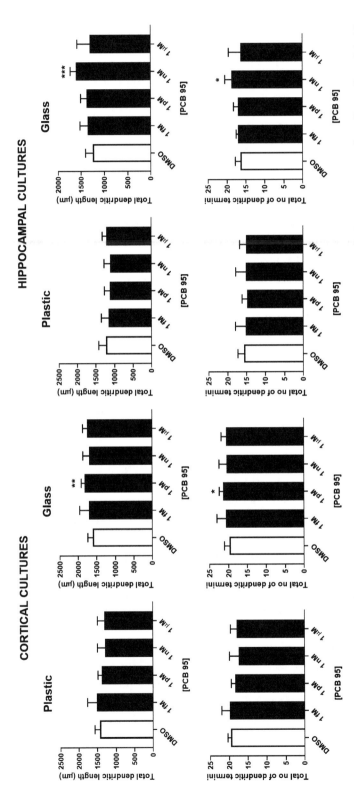

Fig. 2 *Substrate composition influences morphogenic response of primary rat central neurons to neurotoxic agents.* Primary neuron-glia cultures dissociated from perinatal rat hippocampi and cortices were established at high density and maintained as previously described [19] on either tissue culture plastic or glass coverslips. At day in vitro (DIV) 6, cultures were transfected with cDNA encoding a MAP2B-eGFP construct to label the dendritic arbors of approximately 10% of the neurons in the culture. At DIV 7, cultures were exposed to vehicle (0.1% DMSO) or PCB 95 at varying concentrations for 48 h. Dendritic arborization was quantified in eGFP-positive cells (10–30 cells per well) as the total dendritic length and the total number of dendritic tips per neuron. Data are presented as the mean ± SD (n = 5–9 wells from three independent dissections). *Significantly different from DMSO vehicle control at $p < 0.05$; ** $p < 0.01$; *** $p < 0.001$ as determined by one-way ANOVA with Dunnett's post hoc test

- Dissociate cells from the neocortices or hippocampi of postnatal mice or rats. These can be sex segregated [44].

- Plate cells on glass coverslips at 83,000 cells/cm^2.

- At 4 DIV, treat cells with the anti-mitotic agent, cytosine arabinoside (Ara-C; Sigma) to curb glial cell proliferation by replacing 50% of the conditioned media with medium supplemented with 5µM Ara-C to yield a final Ara-C concentration of 2.5µM.

- The optimal window to assess the effects of chemicals on dendritic growth in high density neuron-glia co-cultures is between 4–10 DIV because this is the period of peak dendritic growth [20] although chemicals can be added to more mature cultures to assess effects on dendritic retraction.

- Transfect cells with plasmid encoding MAP2B-eGFP, MAP2B-FusRed, or pCAG-TFP [20] using Lipofectamine 2000 on DIV 6. Exceptions to the manufacturer's protocol include leaving the transfection solution on the cells for only 1–2 h and immediately replacing it at the end of the transfection period with conditioned media.

- On 7 DIV, treat cultures with vehicle or varying concentrations of test chemicals for 48–72 h [45].

- Fix cultures with 4% paraformaldehyde on 9 DIV and mount to glass slides using ProLong Gold Antifade reagent with DAPI (Thermo Fisher Scientific).

3.1.4 Ex Vivo Labeling of Dendritic Arbors in Peripheral Autonomic Ganglia Using Diolistics

Diolistics uses pressure to deliver small beads coated with lipophilic carbocyanine dyes to stochastically label a subpopulation of cells within intact neural tissues [46]. The labeling of individual cells is rapid such that the entire dendritic arbor of neurons can be visualized within minutes after particles contact the cell membrane. This method is a variant of biolistics in which plasmids are delivered into hard-to-transfect cells, such as plant cells, by delivering small gold or tungsten particles coated with plasmids or other cDNA material using a gene gun. Diolistics was first described by Gan et al. [47] and uses fluorescent lipophilic dyes, such as DiI (1–1'-dioctade-cyl-3,3.3',3'-tetramethylindocarbocyanine perchlorate, Thermo Fisher Scientific). Because of their lipophilic nature, these dyes partition into the plasma membrane, thereby outlining neuronal processes and spines. Membrane staining is more efficient and allows for better visualization of small thin protrusions than cytoplasmic staining. DiI-labeled neurons can be observed by high-resolution imaging, such as confocal or two-photon microscopy, and can be digitally reconstructed in precise detail [48, 49].

Our laboratory has used Diolistics to successfully label individual neurons in intact rat and mouse superior cervical ganglia (SCG) using the Helios® Gene Gun system from Bio-Rad [50]; we and others have used this technique to label individual neurons in CNS organotypic slice cultures and cultured cells [20, 51]. The "bullets" used with the Helios gene gun are made of sections of Tefzel tubing filled with tungsten beads pre-coated with DiI.

To coat the Tefzel tubing with polyvinylpyrrolidone (PVP) (Sigma) to improve adherence of the tungsten beads to the interior walls of the tubing:

- Prepare a stock PVP solution at 20 mg/mL in 100% ethanol. Each day that bullets are prepared, the PVP solution is further diluted to 0.1 mg/mL in 100% ethanol.

- Cut approximately 30 inches of Tefzel tubing (Bio-Rad), rinse with 100% ethanol, and feed it into the Bio-Rad Prep station stopping just before the tubing enters the prep station gas outlet.

- Use a 10 mL syringe with a three-way stopcock mounted vertically on a ring stand next to the tubing prep station to avoid inserting air bubbles into the tubing and a short piece of silicone tubing to connect the stopcock to the Tefzel tubing.

- Pour the PVP solution into the syringe and open the stopcock to allow the PVP solution to fill the Tefzel tubing.

- Once the tubing is filled, close the stopcock to keep the PVP solution in the tubing for 5 min. Do not overfill the tubing and blot any excess with a Kimwipe™.

- Withdraw the PVP solution using the syringe.

- Gently push the Tefzel tubing into the gas outlet and connect it to the barbed hose connector on the prep station connected to a dry nitrogen gas source. Dry the tubing using a nitrogen gas rate of 0.4 L/min for approximately 5 min.

To prepare the tungsten particles:

- Suspend 15 mg of 1.1μm tungsten M-17 particles (Bio-Rad) in 150μL methylene chloride (99.9% pure, Fisher Scientific) in a 1.5 mL microfuge tube. Mix well and sonicate for 3 min to break up any bead clumps.

- Pipette the suspended particles onto a glass hanging drop slide and air dry for approximately 10–15 s to allow for even distribution of the particles.

- Place 3 mg of DiI in a microfuge tube, add 100μL of methylene chloride, and vortex.

- Pipette solution on top of the dried tungsten particles. Dried particles (<1 min) will appear as a dull gray powder. The edges of the hanging drop well where there is no tungsten will appear dark pink.

- Use a single edge razor blade to gently scrape the gray colored particles onto a piece of weigh paper, while avoiding the dark pink regions so as to not cause more clumping.

- Dice the particles into a fine powder and transfer to a 15 mL conical tube to which 5 ml of sterile MilliQ water is added.

- Vortex the tube containing the DiI-labeled beads in water for 30 s to allow larger particles to sink to the bottom of the tube. Do not sonicate because the dye will come off the particles.

- Place the vortexed particles in water in a new 10 mL syringe on the ring stand (a new syringe, stopcock, and silicone tubing should be used every time bullets are made).

- Open the stopcock and slowly push the dye-coated tungsten particle solution into the Tefzel tubing. Allow the particles to settle for 5 min. Leave the syringe connected to the tubing and do not rotate the tubing.

- Over a period of approximately 60 s, slowly remove the excess particles/water from the tubing with a syringe.

- Turn on the prep station motor to rotate the tubing and simultaneously dry it with nitrogen at a flow rate of 0.4 L/min for approximately 10–15 min for an even distribution of particles. As the water is removed, small droplets containing most of the particles will be left behind in the PVP-coated tube.

- Remove the tubing from the prep station and cut into 13-mm lengths using a tubing cutter.

- DiI bullets can be stored in a small scintillation type vial with a Dricap dehydrator cartridge (Sigma) in the dark for several weeks.

  To label dendritic arbors:

- Harvest superior cervical ganglia from euthanized mice or rats (ranging in age from the first few weeks postnatal to adult animals), dissect the ganglia free from the capsule, and post-fix in 4% paraformaldehyde (Sigma) in 0.1 M phosphate buffer for 24 h at 4 °C.

- Transfer ganglia to PBS and store at 4 °C until use (best results are obtained with ganglia stored no more than 2–3 weeks).

- Load bullets into the bullet cartridge and assemble the gene gun per the manufacturer's instructions. The diameter of the initial barrel of the gun is approximately 1 cm and the extension barrel is 1.5 cm (the greatest density of particles will be within the 1-cm diameter when the gun is fired).

- Attach the gene gun to a helium source as the propellant. Helium is used because it is inert and disperses quickly after firing. Adjust the helium pressure depending on the species: 80–90 PSI for rat ganglia and 60 PSI for mouse ganglia.

- Place ganglia to be labeled in the center of a 60 mm petri dish previously prepared with a layer of Sylgard (Sylgard 184 silicone elastomer kit, Electron Microscopy Sciences, Hatfield PA, USA) and labeled with a hand-drawn 1 cm circle in the center of the bottom of the dish and a 1.5 cm circle around that. Sylgard acts as a shock absorber to help minimize tissue damage when the gun is fired.

- Blot excess PBS from the ganglia with a Kimwipe™.

- Put a 70μm cell strainer (Falcon, Corning NY, USA) over the end of the gun barrel to reduce clumping and disperse the particles.

- Hold the barrel approximately 1 cm over the ganglia and fire.

- Immediately rinse the ganglia in PBS, place them on a microscope slide in a drop of PBS, and image under fluorescent microscopy using an excitation filter of 568 nm to check labeling. If a dye-coated particle comes into contact with a neuron, it will begin to fill the cell with the dye. If there are no labeled neurons present, the same ganglia may be shot again up to 4–5 times until cells are labeled.

Usually, 2–3 neurons per ganglia will be labeled and can be imaged by confocal microscopy. Images should be acquired within 30 min of labeling to minimize dye diffusion into adjacent cells (*see* **Note 3**). Labeling that is too sparse or too dense to identify individual cellular components can be adjusted by changing the concentration of the particles in the solution that is loaded into the Tefzel tubing. If the dye appears clumped, fresh bullets may need to be made with special attention to ensure that the dye is evenly dispersed before loading into the tubing.

3.1.5 Golgi Staining

In 1873, Camillo Golgi discovered a technique that would impregnate nervous tissue with potassium dichromate and silver nitrate and randomly label only a limited number of neurons in a tissue, revealing the dendrites and axons extended by a single neuron. The mechanism by which this happens is still largely unknown, but the technique has become the gold standard for visualizing the complex dendritic arbors of central neurons in intact brain tissue.

In our laboratory, we use the Rapid Golgi staining method developed by Glaser and Van der Loos [52] to label pyramidal neurons in the rat and mouse hippocampus and cortex (*see* **Note 4**):

- Using the FD Rapid GolgiStain Kit (FD NeuroTechnologies, Inc.), prepare the Golgi solution 24 h prior to harvesting brains (approximately 14 mL total solution per rat brain). Mix solutions A and B at a 1:1 ratio in a 50 mL plastic conical tube protected from light at room temperature to allow settling of metal precipitate.

- 24 h later, add 7 mL of Golgi solution to individual 15 mL plastic tubes or scintillation vials to accommodate one brain hemisphere each. If using scintillation vials, remove the metal insert on the inside of the lid prior to use.

- After euthanasia, remove the fresh brain and gently drop into a petri dish filled with chilled PBS to rinse away excess blood.

- Bisect the brain using rectangular glass coverslips. Metal tools cannot be used with the Golgi solution.

- Using plastic forceps, gently transfer one whole hemisphere to a plastic tube/ scintillation vial containing the Golgi A + B solution (5 mL/cm^3 of tissue) and incubate overnight at room temperature in the dark (brain tissue should be protected from light at all times.)

- Using a plastic transfer pipette, remove the Golgi A + B solution after 24 h of incubation (discard as hazardous waste) and replace with 7 mL of fresh Golgi A + B solution.

- Incubate brains at room temperature for a total of 14 days, including the initial 24 h. Incubation for longer than 14 days will result in higher penetrance of the Golgi stain, increasing the likelihood of labeling neurons with overlapping dendritic arbors.

- After 14 days, remove the Golgi solution and add 7 mL of solution C. Incubate overnight at room temperature.

- The next day, remove solution C and replace with 7 mL of fresh solution C. Store the brain at 4 °C for 6 more days.

- Following this incubation period, transfer the brains to a new tube containing 10% sucrose in 0.1 M phosphate buffer, pH 7.2, at 4 °C.

- 24 h later, replace the 10% sucrose solution with 30% sucrose in the same buffer. Brains can be held in 30% sucrose for months in the dark at 4 °C, but use a similar duration of incubation for all brains in a given project to minimize the confounding effects of shrinkage that occurs over time.

To sub slides with gelatin and chromium potassium sulfate prior to sectioning brains:

- Heat 10 mg/mL gelatin solution in deionized water (400 mL) while stirring until the gelatin is dissolved.

- Add chromium potassium sulfate (1 mg/mL or 0.4 g for 400 mL) and stir to dissolve.

- Cool the solution to room temperature. Note: The solution can be filtered through a 60 cm Whatman #1 paper, but because this can take a long time and the solution often breaks through the paper cone, we often omit this step.

- Pour the solution into clean staining dishes containing clean slides in slide holders for 10 min. To avoid fingerprints, wear gloves and handle slides by their edges.

- Cover slides to prevent dust accumulation and dry overnight at room temperature.

- The remaining subbing solution can be refrigerated and used the next day to sub additional slides.

  To section Golgi-stained brains with a Vibratome:

- Transfer Golgi-stained brains to 70% ethanol in a new plastic vial and keep at 4 °C for at least 4 h.

- Approximately 5 min before sectioning, cool the Vibratome basin to −20 °C.

- Remove the cerebellum and block the brain approximately 1 cm posterior to the olfactory bulb. Ensure the diameter of the brain cross-section is large enough to maintain tissue integrity during sectioning.

- Place the brain on a Kimwipe™ to remove excess ethanol on the face that was just blocked and glue (using Super Glue™) to the chuck with the cortex facing up. Keep the hippocampus above the glue to make the brain more stable. Allow the glue to harden for at least 5 min.

- Meanwhile, fill the Vibratome basin with ice-cold 70% ethanol with additional ice in the trough to keep the basin cool.

- Mount the chuck in the basin with the dorsal side of the brain facing the blade to reduce the force of the blade on the brain, which can break the glue and result in the brain cracking.

- Cut coronal sections of 100μm thickness and separate them from the whole brain using a paintbrush.

- Transfer sections to a petri dish containing cold solution C and allow each section to equilibrate until it is fully submerged.

- Transfer sections to subbed slides using a sterile transfer pipette with its tip cut off (to the wide diameter) so that suction can be applied to the whole brain slice. Note: Be sure to hold the cut tip pipette perpendicular to the petri dish and slide; otherwise, the entire volume of solution C may be ejected from the cut tip pipette onto the slide, resulting in shifting or loss of sections. For best results, move the section as close as possible to the pipette opening before transfer and keep parallel to the slide. Tap the cut tip pipette opening to the slide while gently ejecting a minimal volume. Ideally, there should be only a small volume of liquid covering the section. Alternatively, drops of solution C may be placed on the subbed slide, and once a slice has equilibrated in the petri dish of solution C, it can be moved with a paint brush tip to the subbed slide containing a light coating layer of solution C. Once the sections are on the slide, aspirate the excess solution and add one to two drops of fresh solution C to each section.

- After 10 min, remove excess solution C with a Kimwipe™ laid flat on top of the section and very gently pull it off. This may need to be repeated once or twice.

- Loosely cover sections with foil to protect from dust and light and air dry at room temperature until no visible moisture remains. This may take weeks. Dab sections with a Kimwipe™ daily to accelerate the process.

- Brain slices will look opaque with no evidence of liquid once completely dried at which point they can be coverslipped in Permount™ (Sigma) mounting medium.

3.2 Imaging

3.2.1 Optimization of Fluorescent Image Quality

The reproducibility and rigor of quantitatively analyzing neurite outgrowth is highly dependent on image quality. Starting with high-quality samples is key to obtaining a high-quality image, and there are a number of steps common to every imaging system that if followed will help to greatly increase the sharpness (correct focus) and contrast (high signal to noise ratio) of thin structures like dendrites and axons. We use the ImageXpress high-content imaging system (Molecular Devices), which is interfaced to MetaXpress software (Molecular Devices), to acquire images from cell cultures or tissue sections that have been immunostained using fluorophore-tagged secondary antibodies. Most of the steps are part of the Plate Acquisition Setup, but the basic steps described below are conserved across almost all imaging software.

In the left panel of the Plate Acquisition Setup tab, there are subtabs that help guide imaging acquisition. Some of these subtabs are for entering and keeping a record of details of a project (Names and Description) and some are for directing the software to specific journals to be performed during (Journals) or after (Post-Acquisition) acquisition of the images for analysis. The subtabs that require attention and are common to any imaging software are:

- **Microscopic lens and camera:** Camera binning can be chosen here. Binning refers to the combination of the information of adjacent detectors in a camera sensor to create one single pixel in the recorded image. For example, a binning of two gathers the charges from a square of four detectors (two horizontal and two vertical) to record them in just one of the image pixels. Thus, the intensity per pixel increases by a factor of about four, increasing the signal-to-noise ratio, but reducing the resolution. The microscopic lens will also determine the resolution based on its magnification. Higher magnification reduces the field of view.

- **Plate, wells, and sites to visit:** The settings for the specific plate or slides is set here. The user can choose to avoid the edge wells if an "edge effect" is expected and choose the number and location of sites to image.

- **Acquisition loop:** Users can choose the number of wavelengths used to acquire the images and perform shading correction. The interaction between objects being imaged, the illumination and the camera can produce shading across the field-of-view. In some cases, there is a bright center and decreased brightness on the edges or it is darker on one side and lighter on the other. Precalibration with selected filters and objectives can minimize this shading effect.

- **Autofocus:** MetaXpress has different options of autofocus that are dependent on the type of plate used (usually 96-well plates work well with added z-focus around a fixed offset). More important than which autofocus is used, is applying it to multiple sites to ensure accuracy of the defined autofocus offset throughout the samples.

- **Wavelengths:** The time of exposure can be manually set or calculated in each wavelength by the software using the auto expose feature. To manually define the exposure time, different samples are used as positive controls (bright samples, usually cells of interest) and negative controls (dark samples, usually samples reacted with secondary Ab only) to keep the average image histogram in the lower 70% of exposure (target maximum intensity at about 45,000), and prevent image saturation that would lead to indistinguishable levels of labeling or dye concentration.

- **Save tiff images:** Images should be saved at the maximum bit depth of the camera used in the tiff format to maintain the highest resolution. Other formats compress the image and discard the image gradients that are important for analysis.

3.2.2 Imaging of Golgi-Stained Sections and Ganglia Labeled Using Diolistics

The following describes how Golgi-stained sections and autonomic ganglia labeled with fluorescent dyes using Diolistics are selected and imaged on a confocal microscope in order to perform z-stacking:

- Visualize Golgi-stained sections under 10× magnification in brightfield to confirm staining is acceptable and identify suitable neurons.

- The brain region of interest must be clearly defined in the section (e.g., if analyzing pyramidal neurons in hippocampal CA1, sections should contain a clearly defined CA1 regions with highly aligned cell bodies (stratum pyramidale), distinguishable regions containing apical dendrites (stratum radiatum) and basilar dendrites (stratum origens)).

- Select individual neurons that: (1) are well-impregnated with no evidence of incomplete or artificial staining; (2) are not obscured by blood vessels, glia, or non-descript precipitate; (3) have a cell body located in the middle third of the thickness of the section; and (4) have a dendritic arbor clearly distinguishable from that of any adjacent labeled neurons. If a cell meets these criteria, the number of dendritic branches with cut ends should be ignored to prevent bias toward neurons with smaller dendritic arbors.

- When autonomic ganglia are labeled using Diolistics, select neurons for morphometric analysis if a tungsten particle is located within the cell body and the most distal aspects of the dendritic arbor are labeled as evidenced by tapering of fluorescence to very fine tips of dendritic processes.

Once neurons are identified that meet the inclusion criteria, sections are imaged at $20\times$ magnification with a large enough z-stack to ensure complete capture of the dendritic arbors with the slice/ganglia:

- Track dendritic arbors out to the ends to ensure that arbors are visible in their entirety.

- Dendritic arbors may have breaks in them. If both ends of a broken primary dendrite are in close proximity in all planes, consider them connected. Do not consider dendrites connected if breaks are more than a couple microns or the ends are clearly not in the same z-plane. Broken secondary, tertiary, and quaternary dendrites are generally not considered connected unless there is overwhelming (proximity-based) evidence to suggest otherwise.

- Capture image stacks in brightfield for Golgi-stained sections and fluorescence for Diolistics using a 0.2μm z-step. Generally, 250–350 steps are necessary to capture an entire dendritic arbor of a hippocampal CA1 pyramidal neuron.

- Begin imaging at the lowest point where any dendrite is in focus and extend up to the highest point where any dendrite goes out of focus to ensure that all dendritic arbors in a field are visible and overlapping arbors can be differentiated.

It is recommended to capture images of six to eight neurons per brain from a minimum of six individual animal brains per experimental condition. Neurons can be traced using Neurolucida version 11 (MBF Bioscience, Willston, VT, USA), and Sholl analysis can be performed using Neurolucida Explorer. Help files are available in both platforms and online. Also available is NeuroMorpho. Org, which is "a centrally curated inventory of digitally reconstructed neurons associated with peer-reviewed publications that contains contributions from laboratories worldwide and is continuously updated." Morphometric analysis of Golgi-stained images begins by manually outlining the cell body in Neurolucida and then manually tracing and marking dendrites. Manual tracing with Neurolucida is used to generate 3D Sholl plots, which can be exported into Excel files. For autonomic neurons labeled using Diolistics, three-dimensional images are reconstructed using Imaris software (Oxford Instruments) that are then compressed into two-dimensional images for morphometric analyses of dendritic lengths and somal diameter using NIH ImageJ.

3.3 Image Analysis

3.3.1 High-Content Image Analysis and High-Throughput Screening

High-content image analysis is a technique in which fluorescent or transmission images of entire cells or cell organelles are simultaneously analyzed for multiple parameters [53]. In high-content screening, the extracted parameters are used to describe phenotypic changes elicited during screening of chemical libraries or sets [54]. High-content image analysis and screening are usually performed on automated high-content platforms, such as the ImageXpress (Molecular Devices), which allow users to image entire multi-well and slide formats to increase throughput while ensuring unbiased image acquisition. Image analysis can be performed directly during acquisition if the image analysis speed matches the acquisition speed or can be performed post-acquisition.

High-content screening of chemical effects on neurite outgrowth has been performed using the ToxCast library of chemicals [55] on many different cellular systems [24, 56–58]. While there are different software approaches available, both commercial and open source, they share similar image pre- and post-processing steps, the number of which is highly dependent on the in vitro or ex vivo system used in the study. The more complex the system, the more complex the image analysis, e.g., analyzing neurite outgrowth is less complex in pure neuronal cell cultures than in 3-D cell aggregates or tissue sections since the latter requires that neuronal cells first be isolated from other cell populations in the culture [59]. Cell density also influences the complexity of analysis as morphometric analysis on a single cell level is only feasible if the processes of each cell can be assigned to a given cell somata. This is why cell densities are often adjusted to lower values, which is not necessarily ideal for the overall health and survival of cells in culture, but represents a compromise between accuracy of morphometric analysis and cell viability [60]. Table 2 summarizes endpoints often measured in high-content screens of neurite outgrowth using various in vitro and in vivo systems.

While many endpoints, such as total process (neurite, axonal, or dendritic) length, number of terminal tips, number of branching points, neurite and process mass, soma area, and synaptic density can be analyzed using fully automated systems [28, 56, 57, 59–62], the same endpoints can be assessed semi-automatically in 3D-utilizing software like Neurolucida, but with some limitations.

Table 2
High Content Screening (HCS) parameters that can be extracted from different in vitro and in vivo neuronal systems

Primary neuronal cell cultures	SY5Y, LUHMES, iPSCs	Neurospheres, organoid cultures, iPSCs-derived organoid cultures	Tissue section, cleared tissues imaged using CLARITY methodology
• Total neurite, axonal, and dendritic length • Number of terminal tips • Number of branching points • Synaptic density • Neurite, axonal, and dendritic mass • Soma area • Sholl analysis of dendritic or neuritic complexity	• Total neurite length • Number of terminal tips • Number of branching points • Synaptic density • Neurite, axonal, and dendritic mass • Soma area	• Total neurite, axonal, and dendritic length • Number of terminal tips • Number of branching points • Synaptic density • Neurite, axonal, and dendritic mass • Soma area	• Total neurite, axonal, dendritic length in 3D • Number of terminal tips in 3D • Number of branching points in 3D • Synaptic density in 3D • Neurite, axonal, dendritic mass in 3D • Soma area in 3D • Sholl analysis of dendritic or neuritic complexity in 3D

Sholl analysis is another example of a common analysis usually performed in a semi-automated approach [44] (e.g., the Sholl Analysis plugin in ImageJ [63]).

The flow chart in Fig. 3 depicts the principal steps of assessing neurite outgrowth in LUHMES cell cultures with sample images depicted from the MetaXpress software (Molecular Devices). Figure 4 depicts a semi-automated Sholl analysis utilizing the Sholl analysis plugin in ImageJ [63], and Fig. 5 shows preliminary results of 3D fiber tracings of CGRP-stained nerve fibers in mouse bladder using Neurolucida 360 (MBF Bioscience).

3.3.2 Image Preprocessing and Analysis

Acquired raw images of the nucleus and the biomarker used to identify processes (neurite, axon, or dendrite) are saved as 16 bit tiff images for further analysis. Most software for analyzing neurite outgrowth relies on dual-staining for nuclei and processes (neurites, axons, or dendrites). Cell nuclei are used to help identify cell somata and to separate cell soma that touch each other [59, 60, 64]. An exception is the widely used NeuronJ plugin [65] that allows the user to manually trace neurite structures by placing seeds on the image that are combined into lines by the software following the highest intensity gradient. Because this method is so tedious and time consuming, it is not suitable for screening assays.

Fluorescent images are almost exclusively converted into binary images through thresholding for consequent image segmentation (see Note 5). Thresholding aims to separate the raw image into two classes of pixels (foreground and background) and then computes the ideal threshold/intensity value to separate the two classes. If image intensities are comparable within a data set derived from a plate scanner with fixed exposure times and stained with the same antibody batch, a user-defined threshold can be set. The user defines foreground and background signal for the respective channels (Fig. 3b), and image information below this threshold is set to zero and only the foreground information (nuclei structures) is maintained. Since the signal-to-noise ratio for cell nuclei is usually very good, the fixed threshold method is robust among different experiments. In order to separate touching objects, a watershed segmentation [66] is performed (Fig. 3b), which separates touching cell nuclei. Henceforth, the image is treated as a surface with the height information encoded in the pixel intensity. The cell nuclei form catchment basins that are separated by the so-called watershed lines and can be used to separate the nuclei on the original image [67].

Somata identification is performed in a similar fashion as for nuclei and is required to separate processes from the neuronal cell body and assign processes to a single cell (*see* **Note 6**). Since the cell somata is usually much brighter than the cell processes [64], an initial high-fixed threshold value can be used to only maintain the

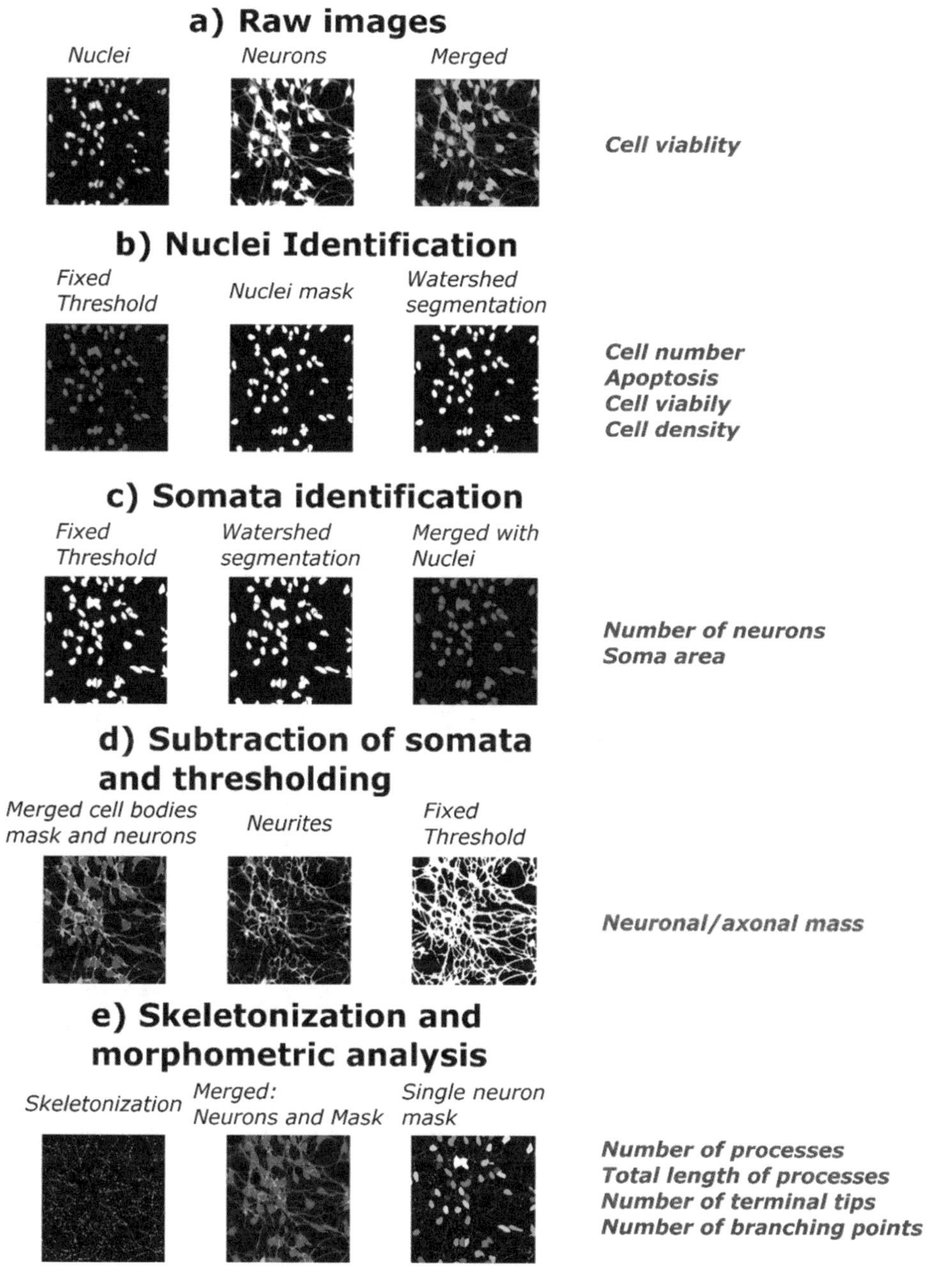

a) Raw images

Nuclei   Neurons   Merged

Cell viablity

b) Nuclei Identification

Fixed
Threshold   Nuclei mask   Watershed
                                 segmentation

Cell number
Apoptosis
Cell viabily
Cell density

c) Somata identification

Fixed   Watershed   Merged with
Threshold   segmentation   Nuclei

Number of neurons
Soma area

d) Subtraction of somata
and thresholding

Merged cell bodies   Neurites   Fixed
mask and neurons   Threshold

Neuronal/axonal mass

e) Skeletonization and
morphometric analysis

Skeletonization   Merged: Single neuron
                  Neurons and Mask mask

Number of processes
Total length of processes
Number of terminal tips
Number of branching points

Fig. 3 *High content analysis (HCA) workflow for morphometric analysis of LUHMES cells.* (*a*) Raw images of cell nuclei and neurons were acquired using an ImageXpress (Molecular Devices) high content imaging system and saved as 16 bit tiff files. Cell viability can be assessed by eye in the original images. (*b*) Cell nuclei are identified by applying a fixed threshold, followed by a watershed segmentation. The resulting binary mask of nuclei is used to determine cell number, cell density, and viability. Additionally, apoptosis can be assessed by counting bright cells with condensed chromatin. (*c*) Somata are identified by applying a fixed threshold on the original neuronal image, followed by a watershed segmentation. To further verify that the binary mask only contains neuronal somata, only areas containing a cell nucleus are considered to be neuronal somata. (*d*) In the next step, the neuronal somata are subtracted from the original image, and the remaining cellular

a) Manual Thresholding   b) Sholl mask   c) Sholl plot

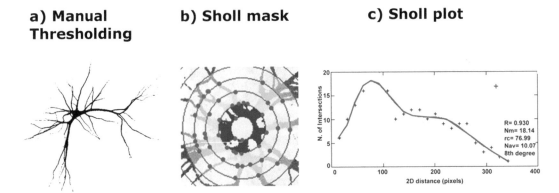

Fig. 4 *Sholl analysis of dendritic arborization in primary rat cortical neuron-glia co-cultures.* (**a**) Dendrites of individual rat cortical neurons transfected with MAP2B-eGFP are manually thresholded, and a centroid is placed at the cell somata. (**b**) The Sholl analysis ImageJ plugin [63] automatically creates a Sholl mask by superimposing the binary image with concentric rings at user-defined distances from each other centered on the neuronal somata and counts the number of dendritic intersections on each ring (purple dots on the rings). (**c**) The number of dendritic intersections are graphed in a Sholl plot and saved as png and xlsx files

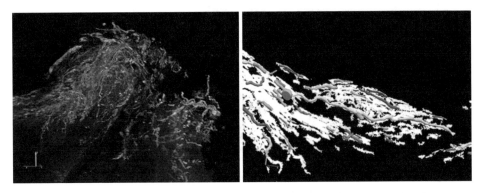

Fig. 5 *Nerve fiber tracings in a rat bladder.* The left panel illustrates immunohistochemical staining for calcitonin gene-related peptide (CGRP) in an optically cleared section of the bladder wall of a young adult rat. Images were captured by multiphoton confocal microscopy using a CLARITY objective, and Neurolucida 360 software (MBP Biosciences) was used to automatically identify CGRP-immunopositive processes (color-coded surfaces). The right panel is a magnification of the same image

←——

Fig. 3 (continued) processes are thresholded with a value calculated from the remaining image, which is lower than the initial threshold to maintain dimly stained neurites. (**e**) Finally, the resulting binary mask of processes is skeletonized. A maximum width is set in order to only skeletonize neurites, which can now be attributed to single cells (colored binary mask). The number of neurites, total neurite length, number of terminal tips, and branching points can be assessed on a per cell basis. For high density cultures, often the total length and/or total number of neurites of the skeletonized image of the entire field of view is normalized against the total number of cells because an accurate assignment of neurites to individual neuronal cells is not feasible

brightest part of the neuron (Fig. 3c). Touching objects can be separated using watershed segmentation and resulting areas can be verified by checking if an area overlaps with the cell nuclei. There is higher user bias with fixed thresholds because different users will choose different thresholding values resulting in different morphometric results. Therefore, in high-content image analysis, automated methods like the Otsu method [68], which is based on separating image pixels into classes (foreground and background) by minimizing inter-class variance, are often used [64]. The Iterative Self-Organizing Data Analysis Technique (isodata method) is another widely used automated method [59, 69]. Automated methods provide the advantage of allowing for adjustments to varying signal-to-noise ratios in between different staining batches and eliminate user bias introduced in manual thresholding. The resulting binary image can be analyzed for neuron number and area of cell somata (Fig. 3c).

Verifying cell somata by checking for underlying nuclei delivers robust results for pure neuronal cultures, but it is prone to identifying false-positive nuclei in mixed in vitro cultures since in many cases neuron processes lay on top of non-neuronal cell bodies. This is particularly problematic for high density neuron-glia co-cultures and organoid cultures. In these cases, additional algorithms are required to reduce the false discovery rate while maintaining a high detection power. Such an approach is described for the Omnisphero software, which is designed to analyze neurite outgrowth in the neurosphere culture system [59]. When differentiated, the 3-D organoid system forms a high density heterogeneous cell layer from which neuronal cells have to be isolated from glia cells. This is accomplished by not only checking for overlap between nuclear and neurite processes markers, but also by verifying nuclei by checking their connectivity to adjacent neuronal processes [59].

Neuron somata are excluded from original neuron images to maintain only process structures (Fig. 3d). Processes are then skeletonized (Fig. 3e), which is a procedure that reduces a binary shape to a skeletal remnant, preserving what is equidistant to its boundaries. The skeleton usually resembles topological characteristics of the shape, such as connectivity, topology, and width. Skeletonization is used to quantify neurite/axonal/dendritic length, number of processes, number of branching points, and number of terminal tips from the binary masks. While there are many different skeletonization algorithms available [70], a common problem is the skeleton's sensitivity to an object's boundary deformation [70], and thresholding often results in imperfect object boundaries, which then introduce artificial branches within the skeleton. One solution is to limit the number of allowed vertices of endpoints [70]. However, this relies on a good characterization of the in vitro/in vivo system because information on maximal number of end tips is required. In addition, while limiting the number of allowed endpoints in relatively immature neuronal cultures like LUHMES and neuronal progenitor cell-derived neurons [59], which are mainly bipolar with only a very low degree of branching, is feasible, this approach is not adequate for primary neuronal cell cultures. These cultures are comprised of different neuronal subtypes at different stages of maturation, which makes estimating maximal allowed endpoints challenging.

Another option is to post-process the obtained images in a pruning step that removes skeleton points that account for too high sensitivity. The two major pruning methods are based on significant measures assigned to each skeleton point and boundary smoothing algorithms performed on the original shape. The first method is often used to eliminate small branches originating from thresholding artifacts or artificial branching points. One method, described in [59], removes all branches from the skeleton and then checks the length of remaining sub-skeletons. Remaining sub-skeletons are reconnected and used for further analysis. This method removes artificial branching points as well as small branches originated from imperfect thresholding. Another method to describe denddritic structure is Sholl analysis [44]. This works well for mature primary hippocampal or cortical neurons. With this method, the user chooses individual neurons, defines a center point, and thresholds manually (Fig. 4a). The Sholl analysis plugin of ImageJ [63] superimposes concentric rings around the center point, determines the number of dendritic intersections at each ring, and generates a distribution plot (Fig. 4b, c). Because Sholl analysis is heavily dependent on the quality of thresholding, it is prone to thresholding biases from different researchers. Additionally, since it is only semi-automatic, it is not feasible for substance screening approaches. We recently published fully automated software for Sholl analysis that overcomes these limitations [62] by automatically determining thresholds, identifying cell somata, and correcting skeletons. This automated approach not only increases throughput, but also significantly reduces inter-researcher bias.

Analysis of dendritic and axonal morphology in the peripheral nervous system is mainly performed in a semi-automated fashion since nerve fibers have to be traced in 3D. Figure 5 shows automated fiber tracing results of Neurolucida 360 (MBF Bioscience). While some bright fibers are identified, the high auto-fluorescence background of the tissue prevents some fibers from being identified. However, auto-fluorescence can be subtracted by generating auto-fluorescence images derived from excitation wavelengths that are outside the used secondary antibody dyes. Furthermore, antibody penetration can be improved. This is a perfect example in which optimization of sample preparation and image acquisition will improve the results. Once the final skeleton is obtained, total neurite length, number of processes, number of endpoints, number of terminal tips, and number of branching points can be assessed on a cellular level.

3.3.3 Data Validation Following Image Analysis

After images have been analyzed, the data are evaluated for validity. This is usually done via manual inspection of the obtained cell mask (Fig. 3e) in cultures exposed to reagents that inhibit neurite outgrowth [60]. Another option is to compare different methods (different image analysis software) to one another to confirm results are qualitatively similar or to compare the results to a gold standard, which is often considered to be manual tracing of neurites [71]. One of the most common features to be compared between the gold standard and the automated method is neuronal identification and quantification since this is a binary classification. A neuron is either correctly identified (true positive, TP); not identified (false negative, FN); correctly not identified, such as non-neuronal cells (true negatives, TN); or a non-neuronal cell is wrongly identified as a neuron (false positive, FP). For process length, the same holds true. However, length values will differ between automated and manual evaluations in all instances because while the computer computes process length from binary images, manual length measurements are performed on the original image. When comparing measurements made by automated versus manual methods, the relative change in positive and negative control values is often used to inform results.

The following equations can be used to determine the accuracy and precision of neuron identification:

$$\text{Accuracy} = \frac{\text{TP} + \text{TN}}{(\text{TP} + \text{TN} + \text{FP} + \text{FN})}$$

$$\text{Precision} = \frac{\text{TP}}{(\text{TP} + \text{FP})}$$

While both accuracy and precision can be used to determine the validity of an automated method, it is important to remember that low accuracy or precision values can result from low neuron identification or high FP identification. Therefore, the false discovery rate (FDR) and the detection power (DP) should be analyzed:

$$\text{FDR} = \frac{\text{FP}}{(\text{FP} + \text{TP})}$$

$$\text{DP} = \frac{\text{TP}}{(\text{TP} + \text{TN})}$$

These quantities are directly related to either identification quality or the false identification rate. However, the next question is: What are acceptable FDR and DP values? This question can be answered by having two independent researchers analyze the same image set, with one researcher chosen as the "gold standard," and comparing the other researcher's results to this standard. The resulting FDR and DP may be considered as benchmark values [59]. See Tables 3 and 4 for open source and commercially available platforms (*see* **Note 7**).

4   Notes

1. *Morphometric analysis of cultured neurons:* Neural cells should be plated on glass coverslips for analyses of chemical effects on neurite outgrowth. We have found that German glass coverslips yield the most reproducible effects. In addition, we strongly advise that researchers consider whether the chemical properties of test compounds favor adsorption to tissue culture plastic or to glass.

2. *Visualizing axons* vs. *dendrites by immunocytochemistry:* It is critically important that axon and dendrite-selective antibodies be carefully titered prior to experimentation to ensure specificity of staining. If used at too high of a concentration, many of these antibodies will non-selectively label processes (e.g., at too high of a concentration, antibodies specific for MAP2B will label axons in addition to dendrites).

3. *Diolistics:* The lipophilic dyes coated on the tungsten beads spread very quickly into the cell membrane, labeling cells within minutes. With time, this dye will spread to adjacent cells, so it is critical that you have access to imaging equipment while you are labeling cells. If you wait more than several hours to image after labeling, it will become difficult to distinguish the dendritic arbor of an individual neuron from that of adjacent cells.

4. *Golgi staining:* In our hands, we have not had much success using the FD Rapid GolgiStain Kit to label central neurons in the brains of embryonic or early postnatal mice and rats. Postnatal day 21 is the earliest that we have been able to successfully Golgi stain brains using this particular kit. We have also not had much success in using the FD Rapid GolgiStain kit to label neurons in autonomic ganglia. Please note that the metals used in the Golgi stain stick to many surfaces. In particular, if you use your vibratome for applications in addition to Golgi staining, it is advisable to have a separate blade and basin reserved exclusively for processing Golgi-stained brains.

5. *Imaging: Thresholding.* Thresholding results are highly dependent on the signal-to-noise ratio and are thus dependent on image quality. In many cases, optimization of image quality

Table 3

Examples of open source software packages and their applicability for analyzing various morphological features of differentiated neurons

Tool	Software platform	Total neurite length	Total axonal length	Total dendritic length	Number of terminal tips	Number of branching points	Synapse density	Neurite Mass	Axonal Mass	Dendritic Mass	Cell body area	Sholl analysis
NeuronJ, AnalyzeSkeleton SYNAPCOUNTJ Sholl Analysis [63]	ImageJ	+	+	+	+	+	+	+	+	+	+	+
GAIN [64]	MATLAB	++	++	++	-	-	-	-	-	-	-	-
NeuronCyto II [73]	MATLAB	++	++	++	-	++	-	-	-	-	++	-
Omnisphero [59, 62]	MATLAB	++	++	++	++	++	++	++	++	++	++	++
NeuriteQuant [74]	ImageJ	++	++	++	++	++	-	-	-	-	++	

– Not available, + Manual or semi-automatic, ++ Fully automated

Table 4
Examples of commercial software packages and their usefulness for analyzing various morphological features of differentiated neurons

Tool	Software Platform	Total neurite length	Total axonal length	Total dendritic length	Number of terminal tips	Number of branching points	Synapse density	Neurite Mass	Axonal Mass	Dendritic Mass	Cell body area	Sholl analysis
Neurite Outgrowth Assay [60] Synaptogenesis Assay [75]	HCS Studio 2.0 (Thermo Fisher)	++	++	++	++	++	++	++	++	++	++	–
MetaXpress 5	Molecular Devices	++	++	++	++	++	++	++	++	++	++	–
Neurolucida Neurolucida 360 [76]	Neurolucida	+	+	+	+	+	+	+	+	+	+	+

– Not available, + Manual or semi-automatic, ++ Fully automated

is the most important step. Image quality is highly dependent on the optical system. The higher the resolution and the lower the noise of the detector, the easier the analysis of images. This, however, will sometimes come at the cost of longer acquisition time, thus slowing down screening approaches. When fixed optical systems are used, the algorithms can be made more robust by using deep learning approaches for image semantic segmentation. In order to do so, masks of correctly segmented neurons would be used to train convolution neuronal networks. During training, image augmentation can be introduced to generalize brightness and contrast fluctuations. Image augmentation will artificially alter image brightness and contrast so that the algorithms become more robust. Of note, deep learning approaches require training images in the range of 1000–2000 depending on the difficulty of the task (variation within the training data, such as brightness variations and artifacts like bubbles), which will require a high manual workload on checking masks. To reduce the time to generate training masks, it is common practice to refine masks obtained by classic thresholding methods. Using this approach, only some pixels need to be deleted or added rather than annotating the entire cell. Thresholding methods can vary for different systems, and available options should be examined prior to establishing a new analysis pipeline. It is important to test the thresholding method on non-stained samples and controls reacted with only secondary antibody to ensure nonspecific signal is not detected. A complete new thresholding algorithm might be needed for classical image analysis. In contrast, neuronal networks can be retrained with the new images, by refining masks obtained from the initial segmentation of the network by a researcher (filling or deleting parts of the mask).

6. *Imaging:* Cellular morphology requires the assignment of processes to a given cell somata. While this works well for low density cell cultures, assignment in high density cultures is not feasible due to the high intersection of processes from different cells. In this case, the more accurate approach is to normalize the total skeleton length to the number of identified cell somata. Another option is to use transfection agents with low transfection efficiency so that the density of labeled cells is low.

7. *Imaging:* Tables 3 and 4 give an overview of some open source and commercially available software platforms and their endpoint spectra.

Acknowledgments

The authors thank Dr. Suzette Smiley-Jewell (University of California, Davis) for reviewing and editing the manuscript. This work was funded by the National Institute of Aging (R01 AG056710), the National Institute of Child Health and Human Development (P50 HD103526), and the National Institute of Environmental Health Sciences (R01 ES014901).

References

1. Lasek R (1988) Studying the intrinsic determinants of neuronal form and function. In: Lasek R, Black MB (eds) Intrinsic determinants of neuronal form and function. A.R. Liss, New York, pp 3–58

2. Higgins D, Burack M, Lein P, Banker G (1997) Mechanisms of neuronal polarity. Curr Opin Neurobiol 7:599–604

3. Scholz D, Poltl D, Genewsky A, Weng M, Waldmann T et al (2011) Rapid, complete and large-scale generation of post-mitotic neurons from the human LUHMES cell line. J Neurochem 119:957–971

4. Berger-Sweeney J, Hohmann CF (1997) Behavioral consequences of abnormal cortical development: insights into developmental disabilities. Behav Brain Res 86:121–142

5. Cremer H, Chazal G, Goridis C, Represa A (1997) NCAM is essential for axonal growth and fasciculation in the hippocampus. Mol Cell Neurosci 8:323–335

6. Maier DL, Mani S, Donovan SL, Soppet D, Tessarollo L et al (1999) Disrupted cortical map and absence of cortical barrels in growth-associated protein (GAP)-43 knockout mice. Proc Natl Acad Sci U S A 96:9397–9402

7. Rice D, Barone S Jr (2000) Critical periods of vulnerability for the developing nervous system: evidence from humans and animal models. Environ Health Perspect 108(Suppl 3):511–533

8. Copf T (2016) Impairments in dendrite morphogenesis as etiology for neurodevelopmental disorders and implications for therapeutic treatments. Neurosci Biobehav Rev 68:946–978

9. Engle EC (2010) Human genetic disorders of axon guidance. Cold Spring Harb Perspect Biol 2:a001784

10. Garey L (2010) When cortical development goes wrong: schizophrenia as a neurodevelopmental disease of microcircuits. J Anat 217:324–333

11. Penzes P, Cahill ME, Jones KA, VanLeeuwen JE, Woolfrey KM (2011) Dendritic spine pathology in neuropsychiatric disorders. Nat Neurosci 14:285–293

12. Robichaux MA, Cowan CW (2014) Signaling mechanisms of axon guidance and early synaptogenesis. Curr Top Behav Neurosci 16:19–48

13. Supekar K, Uddin LQ, Khouzam A, Phillips J, Gaillard WD et al (2013) Brain hyperconnectivity in children with autism and its links to social deficits. Cell Rep 5:738–747

14. van Thriel C, Westerink RH, Beste C, Bale AS, Lein PJ et al (2012) Translating neurobehavioural endpoints of developmental neurotoxicity tests into in vitro assays and readouts. Neurotoxicology 33:911–924

15. Stamou M, Streifel KM, Goines PE, Lein PJ (2013) Neuronal connectivity as a convergent target of gene x environment interactions that confer risk for autism spectrum disorders. Neurotoxicol Teratol 36:3–16

16. Costa LG (2017) Overview of neurotoxicology. Curr Protoc Toxicol 74:11.1.1–11.1.11

17. Howard AS, Bucelli R, Jett DA, Bruun D, Yang D et al (2005) Chlorpyrifos exerts opposing effects on axonal and dendritic growth in primary neuronal cultures. Toxicol Appl Pharmacol 207:112–124

18. Yang D, Kania-Korwel I, Ghogha A, Chen H, Stamou M et al (2014) PCB 136 atropselectively alters morphometric and functional parameters of neuronal connectivity in cultured rat hippocampal neurons via ryanodine receptor-dependent mechanisms. Toxicol Sci 138:379–392

19. Sethi S, Keil KP, Chen H, Hayakawa K, Li X et al (2017) Detection of 3,3'-dichlorobiphenyl in human maternal plasma and its effects on axonal and dendritic growth in primary rat neurons. Toxicol Sci 158:401–411

20. Wayman GA, Impey S, Marks D, Saneyoshi T, Grant WF et al (2006) Activity-dependent

dendritic arborization mediated by CaM-kinase I activation and enhanced CREB-dependent transcription of Wnt-2. Neuron 50:897–909

21. Cai D, Cohen KB, Luo T, Lichtman JW, Sanes JR (2013) Improved tools for the Brainbow toolbox. Nat Methods 10:540–547

22. Chung K, Wallace J, Kim SY, Kalyanasundaram S, Andalman AS et al (2013) Structural and molecular interrogation of intact biological systems. Nature 497:332–337

23. Feng G, Mellor RH, Bernstein M, Keller-Peck C, Nguyen QT et al (2000) Imaging neuronal subsets in transgenic mice expressing multiple spectral variants of GFP. Neuron 28:41–51

24. Radio NM, Mundy WR (2008) Developmental neurotoxicity testing in vitro: models for assessing chemical effects on neurite outgrowth. Neurotoxicology 29:361–376

25. Delp J, Funke M, Rudolf F, Cediel A, Bennekou SH et al (2019) Development of a neurotoxicity assay that is tuned to detect mitochondrial toxicants. Arch Toxicol 93:1585–1608

26. Matelski L, Morgan RK, Grodzki AC, Van de Water J, Lein PJ (2021) Effects of cytokines on nuclear factor-kappa B, cell viability, and synaptic connectivity in a human neuronal cell line. Mol Psychiatry 26:875–887

27. Lotharius J, Barg S, Wiekop P, Lundberg C, Raymon HK et al (2002) Effect of mutant alpha-synuclein on dopamine homeostasis in a new human mesencephalic cell line. J Biol Chem 277:38,884–38,894

28. Stiegler NV, Krug AK, Matt F, Leist M (2011) Assessment of chemical-induced impairment of human neurite outgrowth by multiparametric live cell imaging in high-density cultures. Toxicol Sci 121:73–87

29. Ilieva M, Della Vedova P, Hansen O, Dufva M (2013) Tracking neuronal marker expression inside living differentiating cells using molecular beacons. Front Cell Neurosci 7:266

30. Roloff F, Scheiblich H, Dewitz C, Dempewolf S, Stern M et al (2015) Enhanced neurite outgrowth of human model (NT2) neurons by small-molecule inhibitors of rho/ROCK signaling. PLoS One 10: e0118536

31. Lein PJ, Fryer AD, HIggins D (2009) Cell culture: autonomic and enteric neurons. In: Squire LR (ed) Encyclopedia of neuroscience. Academic Press, Oxford, pp 625–632

32. Bruckenstein DA, HIggins D (1988) Morphological differentiation of embryonic rat sympathetic neurons in tissue culture I. Conditions under which neurons form axons but not dendrites. Dev Biol 136:324–336

33. Lein PJ, Higgins D (1989) Laminin and a basement membrane extract have different effects on axonal and dendritic outgrowth from embryonic rat sympathetic neurons in vitro. Dev Biol 136:330–345

34. Tropea M, Johnson MI, Higgins D (1988) Glial cells promote dendritic development in rat sympathetic neurons in vitro. Glia 1:380–392

35. Ghogha A, Bruun DA, Lein PJ (2012) Inducing dendritic growth in cultured sympathetic neurons. J Vis Exp (61):3546. https://doi.org/10.3791/3546

36. Lein P, Johnson M, Guo X, Rueger D, Higgins D (1995) Osteogenic protein-1 induces dendritic growth in rat sympathetic neurons. Neuron 15:597–605

37. Lein PJ (1991) The NC1 domain of type IV collagen promotes axonal growth in sympathetic neurons through interaction with the alpha 1 beta 1 integrin. J Cell Biol 113:417–428

38. Lein PJ, Guo X, Shi GX, Moholt-Siebert M, Bruun D et al (2007) The novel GTPase Rit differentially regulates axonal and dendritic growth. J Neurosci 27:4725–4736

39. Dotti CG, Sullivan CA, Banker GA (1988) The establishment of polarity by hippocampal neurons in culture. J Neurosci 8:1454–1468

40. Goslin K, Banker G (1989) Experimental observations on the development of polarity by hippocampal neurons in culture. J Cell Biol 108:1507–1516

41. Goslin K, Banker G (1990) Rapid changes in the distribution of GAP-43 correlate with the expression of neuronal polarity during normal development and under experimental conditions. J Cell Biol 110:1319–1331

42. Fletcher TL, Banker G (1989) The establishment of polarity by hippocampal neurons: the relationship between the stage of a cell's development in situ and its subsequent development in culture. Dev Biol 136:446–454

43. Kaech S, Banker G (2006) Culturing hippocampal neurons. Nat Protoc 1:2406–2415

44. Keil KP, Sethi S, Wilson MD, Chen H, Lein PJ (2017) In vivo and in vitro sex differences in the dendritic morphology of developing murine hippocampal and cortical neurons. Sci Rep 7:8486

45. Wayman GA, Bose DD, Yang D, Lesiak A, Bruun D et al (2012) PCB-95 modulates the calcium-dependent signaling pathway

responsible for activity-dependent dendritic growth. Environ Health Perspect 120:1003–1009

46. Grutzendler J, Tsai J, Gan WB (2003) Rapid labeling of neuronal populations by ballistic delivery of fluorescent dyes. Methods 30:79–85

47. Gan WB, Grutzendler J, Wong WT, Wong RO, Lichtman JW (2000) Multicolor "DiOlistic" labeling of the nervous system using lipophilic dye combinations. Neuron 27:219–225

48. O'Brien JA, Holt M, Whiteside G, Lummis SC, Hastings MH (2001) Modifications to the hand-held gene gun: improvements for in vitro biolistic transfection of organotypic neuronal tissue. J Neurosci Methods 112:57–64

49. Noterdaeme M, Mildenberger K, Minow F, Amorosa H (2002) Evaluation of neuromotor deficits in children with autism and children with a specific speech and language disorder. Eur Child Adolesc Psychiatry 11:219–225

50. Kim WY, Gonsiorek EA, Barnhart C, Davare MA, Engebose AJ et al (2009) Statins decrease dendritic arborization in rat sympathetic neurons by blocking RhoA activation. J Neurochem 108:1057–1071

51. Seabold GK, Daunais JB, Rau A, Grant KA, Alvarez VA (2010) DiOLISTIC labeling of neurons from rodent and non-human primate brain slices. J Vis Exp (41):2081. https://doi.org/10.3791/2081

52. Glaser EM, Van der Loos H (1981) Analysis of thick brain sections by obverse-reverse computer microscopy: application of a new, high clarity Golgi-Nissl stain. J Neurosci Methods 4:117–125

53. Haney S (2008) High content screening: science, techniques and applications. Wiley Online Library

54. Gasparri F (2009) An overview of cell phenotypes in HCS: limitations and advantages. Expert Opin Drug Discov 4:643–657

55. Radio NM, Breier JM, Reif DM, Judson RS, Martin M et al (2015) Use of neural models of proliferation and neurite outgrowth to screen environmental chemicals in the ToxCast phase I library. In: McKim JM (ed) Applied in vitro toxicology. Mary Ann Liebert, Inc.

56. Anderl JL, Redpath S, Ball AJ (2009) A neuronal and astrocyte co-culture assay for high content analysis of neurotoxicity. J Vis Exp (27):1173. https://doi.org/10.3791/1173

57. Harrill JA, Freudenrich TM, Robinette BL, Mundy WR (2011) Comparative sensitivity of human and rat neural cultures to chemical-induced inhibition of neurite outgrowth. Toxicol Appl Pharmacol 256:268–280

58. Radio NM, Freudenrich TM, Robinette BL, Crofton KM, Mundy WR (2010) Comparison of PC12 and cerebellar granule cell cultures for evaluating neurite outgrowth using high content analysis. Neurotoxicol Teratol 32:25–35

59. Schmuck MR, Temme T, Dach K, de Boer D, Barenys M et al (2017) Omnisphero: a high-content image analysis (HCA) approach for phenotypic developmental neurotoxicity (DNT) screenings of organoid neurosphere cultures in vitro. Arch Toxicol 91:2017–2028

60. Harrill JA, Freudenrich TM, Machacek DW, Stice SL, Mundy WR (2010) Quantitative assessment of neurite outgrowth in human embryonic stem cell-derived hN2 cells using automated high-content image analysis. Neurotoxicology 31:277–290

61. Harrill JA, Robinette BL, Freudenrich T, Mundy WR (2013) Use of high content image analyses to detect chemical-mediated effects on neurite sub-populations in primary rat cortical neurons. Neurotoxicology 34:61–73

62. Schmuck MR, Keil KP, Sethi S, Morgan RK, Lein PJ (2020) Automated high content image analysis of dendritic arborization in primary mouse hippocampal and rat cortical neurons in culture. J Neurosci Methods 341:108793

63. Ferreira TA, Blackman AV, Oyrer J, Jayabal S, Chung AJ et al (2014) Neuronal morphometry directly from bitmap images. Nat Methods 11:982–984

64. Long BL, Li H, Mahadevan A, Tang T, Balotin K et al (2017) GAIN: a graphical method to automatically analyze individual neurite outgrowth. J Neurosci Methods 283:62–71

65. Meijering E, Jacob M, Sarria JC, Steiner P, Hirling H et al (2004) Design and validation of a tool for neurite tracing and analysis in fluorescence microscopy images. Cytometry A 58:167–176

66. Roerdink JBTM, Meijster A (2000) The watershed transform: definitions, algorithms and parallelization strategies. Fundamenta Informaticae 41:187–228

67. Gonzalez RC, Woods RE (2008) Digital image processing. Prentice Hall Inc., Upper Saddle River, NJ

68. Otsu N (1979) A threshold selection method from gray-level histograms. EEE Trans Syst Man Cybern 9:62–66

69. Ridler TW (1978) Picture thresholding using an iterative selection method. IEEE Trans Syst Man Cybern 8:630–632

70. Bai X, Latecki LJ, Liu WY (2007) Skeleton pruning by contour partitioning with discrete curve evolution. IEEE Trans Pattern Anal Mach Intell 29:449–462

71. Versi E (1992) "Gold standard" is an appropriate term. BMJ 305:187

72. Blacklock AD, Cauveren JA, Smith PG (2004) Estrogen selectively increases sensory nociceptor innervation of arterioles in the female rat. Brain Res 1018:55–65

73. Ong KH, De J, Cheng L, Ahmed S, Yu W (2016) NeuronCyto II: an automatic and quantitative solution for crossover neural cells in high throughput screening. Cytometry A 89:747–754

74. Dehmelt L, Poplawski G, Hwang E, Halpain S (2011) NeuriteQuant: an open source toolkit for high content screens of neuronal morphogenesis. BMC Neurosci 12:100

75. Harrill JA, Robinette BL, Mundy WR (2011) Use of high content image analysis to detect chemical-induced changes in synaptogenesis in vitro. Toxicol In Vitro 25:368–387

76. Abate G, Colazingari S, Accoto A, Conversi D, Bevilacqua A (2018) Dendritic spine density and EphrinB2 levels of hippocampal and anterior cingulate cortex neurons increase sequentially during formation of recent and remote fear memory in the mouse. Behav Brain Res 344:120–131

Part II

Physiological Evaluation in Rodents

Chapter 5

DPOAEs for the Assessment of Noise- or Toxicant-Induced Cochlear Damage in the Rat

Thomas Venet, Aurélie Thomas, Ludivine Wathier, and Benoît Pouyatos

Abstract

Distortion Product Otoacoustic Emissions (DPOAEs) are acoustic responses generated by the inner ear, more specifically the outer hair cells (OHCs), when stimulated simultaneously by two pure tone frequencies. DPOAEs are useful because they are reduced or absent when the OHCs are impaired by noise or ototoxic agents, and they offer an objective and noninvasive assessment of the OHC function, which is not mitigated by central plasticity. Here, we provide a detailed method to measure DPOAEs in rats. We emphasize especially on the calibration of the sound stimulation and the analysis of the otoacoustic emissions, which ensure the reliability of the method.

Key words Otoacoustic emissions, Distortion products, Cochlea, Inner ear, Outer hair cells, Fast Fourier transform

1 Introduction

Cochlear outer hair cells (OHCs), which are distributed along the organ of Corti, have the unique property of being motile. This *electro*motility, achieved by the motor protein called prestin [1], drives cochlear amplification. It is a major contributing factor to the high sensitivity and the extremely sharp frequency tuning of mammalian audition [2]. The cochlea is a nonlinear amplified system, which means that the output is not proportional to the change in the input. This nonlinearity is due to the activity of the above-mentioned "cochlear amplifier" and other cochlear processes, such as the nonlinearities in stereocilia hair bundle deflection [3, 4], the asymmetries in stereocilia stiffness [5], and the opening/closing of mechanically activated ion channels [6].

The mechanical activity of OHCs generates reverse traveling waves that makes the middle ear ossicles and tympanic membrane vibrate, thereby producing sound waves in the ear canal: the acoustic otoemissions (OAEs) [7]. OAEs can be evoked by a sound stimulation (e.g., clicks, single tone, or tone pairs) or occur

Jordi Llorens and Marta Barenys (eds.), *Experimental Neurotoxicology Methods*, Neuromethods, vol. 172, https://doi.org/10.1007/978-1-0716-1637-6_5, © Springer Science+Business Media, LLC, part of Springer Nature 2021

spontaneously. Evoked OAEs are efficient, noninvasive, and objective indicators of healthy cochlear function, especially that of OHCs. Indeed, numerous studies have shown that OHC damage reduces or cancels evoked OAEs [8–11].

When two pure tones ($f1$ and $f2$, with $f2 > f1$), continuous and closely spaced in frequency, are presented to the ear canal, the acoustic response of the cochlea is called a Distortion-Product Otoacoustic Emission (DPOAE). When the two traveling waves induced by these two frequencies interact, they produce a family of intermodulation distortions at different mathematically related frequencies, e.g., the sum and difference of $f1$ and $f2$, as well as the sums and differences of multiples of those frequencies. In addition to the distortion phenomenon, a coherent-reflection mechanism also occurs. These complex mechanisms are not the topic of the current chapter and detailed descriptions of the physics of DPOAE generation can be found elsewhere [12].

The most prominent DPOAE that can be measured in both humans and laboratory rodents is often the cubic distortion product $2*f1 - f2$, which explains why it has been the most used in clinical settings and laboratory experiments.

The chapter focuses on the description of a method to record cubic DPOAEs in rats. However, it is worth mentioning that an excellent methodological description of DPOAE recordings in mice is also available [13]. Actually, DPOAEs have been measured in almost all common and exotic laboratory species, including amphibians and birds.

All DPOAE measurement systems are composed of four parts:

1. a frequency synthesizer, coupled to an amplifier, which produces the sinuses for the two pure tones.

2. two speakers, which transduce those frequencies.

3. one microphone, also coupled to an amplifier, which converts the cochlear response into an electrical signal.

4. a spectral analyzer, which allows the averaging and the Fourier transform of electrical signal.

The critical step in recording DPOAEs is the separation of the meaningful DPOAE response from the acoustic noise coming from the environment and the animal. This separation requires the averaging of numerous acquisitions, a pertinent choice of $f1$ and $f2$ settings, as well as a proper setup of spectral analysis parameters. Since the reliability of DPOAE measurement depends mostly on these parameters, they are detailed in this chapter.

Although DPOAEs can be, in theory, measured in non-anesthetized animals, using habituation and contention (e.g., guinea pigs) or a surgically implanted cranial anchor (e.g., rats), the procedure is much easier and the DPOAEs more stable in ketamine/xylazine anesthetized animals. In addition, the ketamine/

xylazine mixture does not significantly alter DPOAE levels, as opposed to isoflurane [14].

It is important to bear in mind that the OHC function of the low-frequency region of the cochlea (apex) cannot be investigated using DPOAEs. Indeed, the gain of the cochlear amplifier is only a few-fold at the apex of the cochlea against a 1000-fold at the base [15]. Consequently, DPOAEs can only be considered as reliable with $f2 > 3000$ Hz in rats. The rat being able to hear sounds up to 60 kHz, we do not advise to use human commercial DPOAE systems dedicated to the clinic, which are generally limited to 8 kHz. However, some off-the-shelf systems specifically designed for the measurement of DPOAE in rats are commercially available (*see* **Note 1**).

2 Materials

- Otoscope.
- Ketamine.
- Xylazine.
- Data acquisition unit (e.g., Brüel & Kjær Type 3560D).
- Two high-sampling, low-distortion signal generators (e.g., Brüel & Kjær Type 3110).
- High-frequency emitters (e.g., Brüel & Kjær Type 4192 microphone capsules) to produce pure tones up to 60 kHz.
- Low-noise microphone (e.g., Knowles FG-23329-CO5).
- Low-distortion amplifier(s) (e.g., Femto DLPVA-100-B Series).
- PC equipped with a software to control the data acquisition system (e.g., Brüel & Kjær PULSE interface).
- Multifunction acoustic calibrator (e.g., Brüel & Kjær Type 4226).
- Sound-attenuating chamber (*see* **Note 2**).
- Heating pad (e.g., Harvard Apparatus).
- 1/8 inch reference microphone (e.g., Brüel & Kjær Type 4138).
- Home-made cylindrical calibration cavity (0.045 cm^3).
- Small curved forceps and elastic bands.
- Laboratory stand.
- Articulated holder.

3   Methods

3.1   Equipment Setup (Fig. 1)

3.1.1   Generators and Analyzer

DPOAE measurements are performed using a Brüel & Kjær Type 3560D data acquisition unit equipped with two 3110 modules, which offer two signal generators with 204.8 kHz bandwidths and 24-bit D/A convertors, as well as two measurement channels (*see* **Note 3**). This configuration is therefore perfectly suited for measuring rat's hearing whose audible range extends up to 60 kHz. PULSE software can be controlled by a Visual Basic OLE interface to design automatized measurement sequences. The use of this interface speeds up the measurement and increases the reliability of the data recording. Calibration and periodic checks carried out with a multifunction acoustic calibrator (Brüel & Kjær Type 4226) guarantee a maximum error of ± 0.15 dB up to 4000 Hz and ± 0.20 dB beyond.

3.1.2   DPOAE Measurement Probe

Most DPOAE measurement probes available on the market are dedicated to humans. The upper frequency limit is generally less than 8000 Hz, which is insufficient for the rat. Therefore, we manufactured a probe in the laboratory, which is detailed hereafter (Fig. 2).

This probe uses two and a half inch microphone capsules (Brüel & Kjær Type 4192) as a sound transmitter for $f1$ and $f2$. These capsules require a 200 V polarization, which is superimposed on the sinusoidal signal in a conditioner near the probe. Two cascaded amplifiers (voltage gains: 20 + 5 dB) amplify the sinusoidal signal, allowing an alternative effective voltage of 45 Vrms to be reached. The usable bandwidth ranges from 500 to 30,000 Hz. The microphone used to measure the sound signal is a Knowles FG 23329-C05 miniature microphone. The signal delivered by this microphone is amplified by 40 dB by a conditioner placed near the probe (homemade voltage amplifier whose characteristics are close to those of commercial amplifiers; *see* Subheading 2 above) and operating on battery (± 12 V). The bandwidth of this microphone ranges from 100 to 10,000 Hz (± 3 dB). However, by judiciously choosing the frequencies, this custom probe fitted with this microphone allows DPOAE measurements up to 25,000 Hz after calibration. Beyond that, the sensitivity of the assembly (probe + microphone) is too low and this device is no more usable.

The different elements of this device (probe, conditioner, amplifiers) are periodically checked and calibrated. The quality of the electrical connections and even more of the mechanical

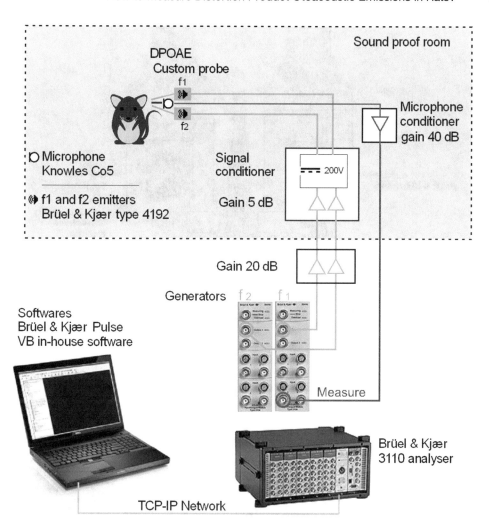

Fig. 1 Schematic description of the DPOAE measurement system for the rat

assembly have a great influence on the performance of the system, especially at high frequencies (*see* **Note 4**).

3.1.3 Probe Calibration and Periodic Verification

In order to calibrate the sound emissions for the rat's hearing system, a cavity with a volume equivalent to that of the ear canal is used (Fig. 3). For the adult (6-month-old) Brown Norway and Long-Evans type rat, the volume is approximately 0.045 cm³. The custom-made cylindrical calibration cavity (length = 6.5 mm; diameter = 2.9 mm) allows the insertion of the probe to be fitted with an audiometric plug with a diameter of 6 mm at one end. The other side of the cavity is fitted with a Teflon ring, which allows the reference microphone to be inserted (Brüel & Kjær type 4138, 1/8 inch).

The probe is calibrated between 500 and 30,000 Hz. The probe calibration allows you to define the levels emitted in dB

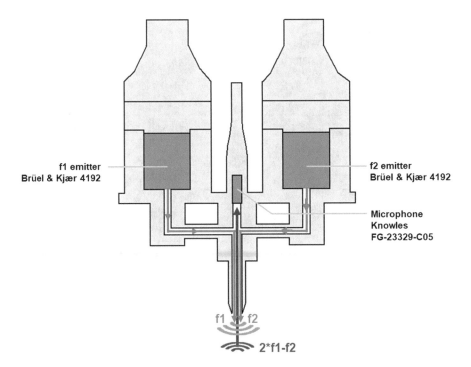

Fig. 2 Probe diagram

Fig. 3 [Probe/cavity/reference microphone] assembly for calibration

SPL by the $f1$ and $f2$ transmitters (*see* **Note 5**). It allows choosing the frequencies that can be used according to the response curve measured at the reference microphone and at the microphone of the probe.

A simplified procedure has been defined for a rapid verification that can be carried out just before each series of recordings. This verification consists in launching a DPgram measurement (*see* Subheading 3.3.2), leaving the probe free (without rat, cavity, and audiometric plug). For each $f1$ and $f2$ frequencies of the DPgram, the levels $L1$ and $L2$ measured by the probe microphone are compared to the reference levels measured under the same conditions following calibration of the probe. The maximum permissible error is fixed at ± 2 dB. This cavity-free procedure provides good reproducibility even at high frequencies.

3.2 DPOAE Measurements

3.2.1 Animal Preparation and Experimental Setup

General anesthesia is induced by an intraperitoneal ($i.p.$) injection of a mixture of Ketamine (Clorketam®, 45 mg/kg) and Xylazine (Rompun®, 5 mg/kg). Otoscopic examination has to be performed before the measurements in order to ensure the absence of infection, obstruction of the external auditory canal, or anomaly of the tympanic membrane. It is very important to stabilize the rat's body temperature at 38 °C throughout the whole procedure using a heating blanket (*see* **Note 6**). Additionally, the maintenance of a steady atmospheric pressure around the animal is critical because anesthetized animals cannot regulate the middle ear pressure and the tympanic tension strongly influences the DPOAE level (*see* **Note 7**).

On the bench of the soundproof chamber, place a laboratory stand with a horizontal rod and an articulated holder on which you can secure the DPOAE probe. This holder needs to be easily adjustable for the convenient placement of the probe into the ear canal of the animal. Be sure to have two small clamps linked to two elastics bands at your disposal. Place a 6-mm diameter audiometric plug on the probe.

3.2.2 Probe Placement

Place the anesthetized rat on its side (with the test ear facing up) on the heating blanket. Approach the probe until it touches the entrance of the ear canal. Use the small clamps to gently pull up the pinna while carefully inserting the probe microphone (Fig. 4). Note that it is better to pull up the pinna than to push down the probe.

3.2.3 Stimulation Parameters

Table 1 summarizes the parameters used for the primary frequencies $f1$, $f2$, and their respective intensities $L1$ and $L2$.

The DPOAE application should allow two types of paradigms, DPgram or input-output (I/O). The DPgram is a plot of DPOAE amplitude as a function of stimulus frequency with the stimulus level held constant. By contrast, the I/O consists of DPOAE amplitude as a function of the level of the stimulus for a given frequency. It is also convenient to be able to perform more complex protocols using a mix of DPgrams and I/Os, or to monitor the time course of a DPOAE response using a single frequency/level combination.

The optimal stimulation parameters yielding maximum DPOAEs have been defined experimentally by adjusting the intensity difference between $L1$ and $L2$ for Long-Evans ($L1 = L2 + 14$ dB) and Brown Norway ($L1 = L2 + 11$ dB) rats. They might differ

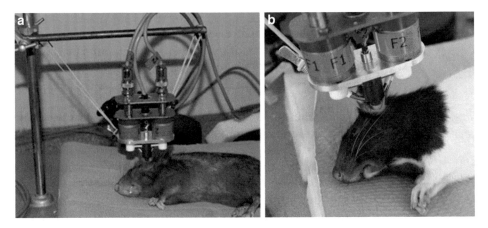

Fig. 4 Brown Norway (**a**) and Long-Evans (**b**) rats during DPOAE measurements

Table 1
Value of the main parameters used for DPOAE measurements in rats

Frequency ratio ($f2/f1$)	1.2
Level difference ($L1 - L2$)	11–14 dB depending on the strain
Frequency range ($f2$)	3600–25,000 Hz
Intensity range ($L1$)	25–70 dB

slightly in your specific conditions. Indeed, a value of 10 dB ($L1 = L2 + 10$ dB) is often mentioned in the literature. Alternatively, a symmetric protocol can be applied, in which the primary tones, $f1$ and $f2$, are presented at equal levels ($L1 = L2$). The best practice is to try out several intensity parameters and to select the one that yields the optimal DPOAE amplitudes.

Our standard measurement protocol includes five frequencies: $f2 = 3600$ Hz, $f2 = 6000$ Hz, $f2 = 9600$ Hz, $f2 = 17,520$ Hz, $f2 = 25,440$ Hz. The ratio $f1/f2$ is fixed at 1.2. Because the origin of DPOAEs on the organ of Corti is mainly located near the $f2$ frequency and because DPOAE values depend primarily on the $L1$ level, our DPOAE measurements are expressed and plotted as a function of the $f2$ frequency and $L1$ intensity.

3.2.4 DPOAE Signal Acquisition

The acoustic signal to be analyzed consists of three periodic signals:

- $f1$ and $f2$ are the sinusoidal signals emitted by the probe to generate DPOAEs in the inner ear.

- $f_{\text{DPOAE}} = 2 * f1 - f2$ is the sinusoidal signal corresponding to the cubic product of acoustic distortion emitted by the inner ear and transduced by the middle ear (*see* **Note 8**).

These three signals are superimposed over a background noise. This background noise (or random noise) is produced by the ambient acoustics of the room and the endogenous noises of the animal. It mainly consists of low and medium frequencies. At high frequencies, electronic background noise is preponderant. Background noise can also contain periodic signals such as electromagnetic interference (telephone, chopper circuits on power lines). The measured signal is analyzed by fast Fourier transformation algorithms (real-time FFT) on Brüel & Kjær analyzers. The acquisition parameters, useful frequency bandwidth, and the number of lines of the spectrum are linked by the following formula:

$$N = 2.56 \ N_{\text{line}}$$

$$F_{\text{e}} = 2.56 \ F_{\text{span}}$$

$$dF = F_{\text{span}} / N_{\text{line}}$$

$$T = NF_{\text{e}}$$

with

N_{line}	Number of FFT lines composing the frequency spectrum
F_{span}	Maximum frequency of the frequency analysis bandwidth (Hz)
N	Number of acquisition block samples
F_{e}	Sampling frequency (Hz)
dF	Frequency resolution (Hz)
T	Duration of an acquisition block (s)

The type of average (temporal or spectral, linear, or exponential), the number of time blocks averaged as well as the temporal and frequency weights are adapted according to the type of measurement carried out: DPgram, I/O or follow-up of the time course.

The acquisition parameters common to the different types of measurement are as follows:

- Average acquisition blocks

    The averages are calculated on the auto-spectrum. This corresponds to the average of the signal energy. This type of analysis provides a good representation of the total energy contained in the signal, including random signals composing the background noise, which is important for judging the signal/noise ratio of the measurement. A temporal average of the blocks would not allow a good evaluation of the background noise since only the signals in phase with the acquisition frequency of the blocks are correctly taken into account.

- Time weighting of acquisition blocks

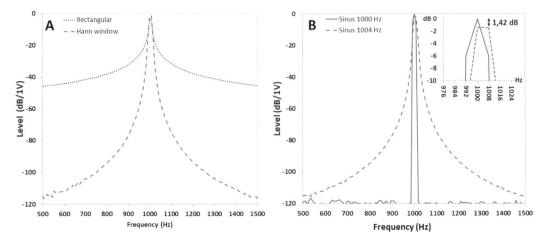

Fig. 5 FFT analysis with sampling frequency $F_e = 65{,}530$ Hz. Size and period of the time block: 8192 points/ 125 ms. Number of FFT lines: 3200. Frequency resolution: 8 Hz. (**a**) Effect of Hanning (dashed line) vs. rectangle (dotted line) windowing on a sinusoidal signal of 1 Vrms at 1004 Hz. (**b**) Picket fence effect on a sinusoidal signal of 1 Vrms at 1000 or 1004 Hz. Solid curve, signal at 1000 Hz or 125 periods per time block. Dashed curve, signal at 1004 Hz or 125.5 periods per time block

Hann windowing (Fig. 5a) is systematically used with an overlap of 66% between blocks. This analysis is well suited for continuous and pseudo-random signals. The Hann window eliminates the edge effects of the acquisition blocks, which gives good frequency resolution. The 66% overlap is optimal for the Hann window because it allows for uniform overall weighting without losing time data (Fig. 6). Without overlap, in case an event occurred at the junction of two blocks, its intensity would be attenuated and the contained information lost.

- Choice of frequencies

By picket fence effect, an FFT analysis can also cause an error related to digitization depending on the position of the true frequency of the phenomenon analyzed in relation to the frequency resolution of the FFT. With a Hann window, this error can be up to 1.4 dB (Fig. 5b). To avoid this problem, the DPOAE application adjusts the frequencies $f1$ and $f2$ so that the f_{DPOAE} is a multiple of the frequency resolution of the FFT.

3.2.5　Specific Signal Processing Parameters

Depending on the type of measurement performed, the bandwidth (sampling frequency), the number of lines (frequency resolution) and the number of averages are different. Table 2 summarizes the configurations suitable for obtaining an average value (DPgram or I/O) or monitoring the DPOAE responses over time.

The rejection of the FFT spectra disturbed by the background noise is carried out in real time. The DPOAE application rejects any acquisition according to the following criterion:

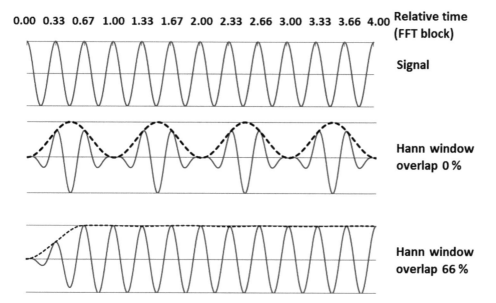

Fig. 6 Effect of overlap on the weight of time samples with Hann windowing

Table 2
Parameters of the FFT analysis for DPgrams, I/O measurements, and time course at f2 = 9600 Hz

Parameter	DPgram or I/O	Time Course (at f2 = 9600 Hz)
Sampling frequency	131,079 Hz	32,765 Hz
Bandwidth	0–51,200 Hz	0–12,800 Hz
Number of lines	6400	800
Frequency resolution	8 Hz	16 Hz
Duration of an elementary block	125 ms	62.5 ms
Number of means	4	22
Type of average	Linear	Linear
Duration for one recording	250 ms	500 ms

- If the difference between the DPOAE and the background noise is less than 3 dB, the emergence of the DPOAE signal compared to the noise is considered insufficient; as a result, the spectrum is not used for the calculation of the average DPOAE level. This difference is calculated by subtracting to the DPOAE value obtained on the FFT line "n" (corresponding to $2*f1 - f2$), the background noise calculated by averaging eight close FFT lines ($n - 6$ to $n - 3$ and $n + 3$ to $n + 6$). If 75% or more of the instantaneous spectra are rejected, the DPOAE application indicates that the entire acquisition is unusable for determining the average DPOAE value reliably.

Spectra averaged over 250 or 500 ms are stored in memory. They can be plotted in waterfall or time / frequency diagrams. One frequency can also be extracted and displayed in the form of a two-dimensional time/intensity graph.

During DPgram or I/O measurements, the average DPOAE level is determined by the linear average of 20 average spectra of 250 ms (i.e., 80 instantaneous spectra), which corresponds to a 5-s acquisition time (excluding rejection). This duration is long enough to obtain a reliable average value.

3.3 DPOAE Analysis and Interpretation

3.3.1 Input/Output (I/O) Measurement

The I/O measurements allow the response of the cochlea at both low and medium intensities to be explored. This type of measurement provides a large amount of information, but requires a significant acquisition time per frequency, which has for consequence a limitation of the number of frequencies that can be measured. The example below shows how the choice of the level of stimulation can modify the measured effect and affect the interpretation.

Figure 7 shows the average I/O DPOAEs obtained for 17 untreated Brown Norway rats. These animals were measured at 6 and 24 months. The variations recorded correspond to the effect of aging on hearing performances called presbycusis. Presbycusis includes several mechanisms, including a loss of hair cells in apical and basal extremities of the organ of Corti. These results clearly illustrate that the measured age effect depends on the intensity of the primaries $L1$ and $L2$. The age-related DPOAE decrease would be of 3.2 dB for a measurement at $L1 = 50$ dB, whereas the difference would be close to zero with $L1 = 70$ dB. In this experiment, $L1 = 40$ dB was the lowest level usable due to the proximity of the background noise, notably for the 24-month measurements (*see* **Note 9**).

From our experience, DPOAEs measurements are more sensitive to damage to OHCs for $L1$ intensities below 60 dB. Therefore, DPgram measurements (*see* below) should preferably be carried out between 50 and 60 dB in order to obtain a good compromise between sensitivity to cochlear damage, the intensity of the measured signal, and the emergence from the noise floor.

3.3.2 DPgram Measurement

Performing a DPgram consists in recording DPOAEs at different $f2$ frequencies with the same level of stimulation. This procedure is commonly called DPgram because, graphically, it resembles a clinical audiogram. The DPgram is very convenient because it allows the exploration of a wide frequency range, and therefore a large portion of the organ of Corti, in a minimum amount of time.

Figure 8 presents different ways to display DPgrams obtained in control and exposed Brown Norway rats [16]. The treatment

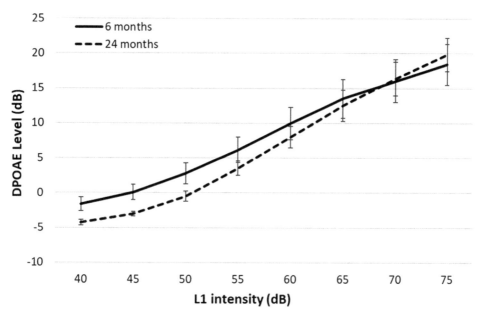

Fig. 7 DPOAE levels measured in a non-exposed group ($n = 17$) of Brown Norway rats at 6 and 24 months of age. $f2 = 25{,}440$ Hz with $f2/f1 = 1.2$ and $L1 = 40$–75 dB with $L1 = L2 + 11$ dB. The error bars represent the standard deviation

was a co-exposure to impulsive noise and to 600-ppm styrene (6 h/day, 5 days/week, for 4 weeks). The impulsive noise was an 8-kHz octave band burst lasting 10 ms at 112 dB, which was repeated every 15 s. This noise exposure was designed to ensure that each noise burst was absorbed by the cochlea without being reduced by the middle ear reflex. In terms of acoustic energy, the equivalent noise exposure was an 80 dB continuous noise during 8 h ($L_{\text{EX},8h} = 80$ dB).

The DPgram (Fig. 8a, b) was performed with $L1 = 55$ dB, which yields DPOAE levels between 15 and 25 dB SPL depending on the frequency considered. These levels are largely sufficient not to be disrupted by the noise floor, which is comprised between -6 and -1 dB SPL at high and medium frequencies, respectively. They are also sufficient to evaluate the temporary and/or permanent impact of exposure on the cochlea that will result in a decrease in DPOAE amplitudes. In the rat, the DPOAE magnitudes are relatively low at 3600 Hz (15 dB), compared to those obtained at higher frequencies. This occurs because the lower frequency limit measurable with otoacoustic emissions (about 3000 Hz) is close. It might also be convenient to plot the DPOAEs as differences from baseline data (Fig. 8c, d). The dashed line at the "0" on the ordinate indicates no change from baseline levels. One might notice that unexposed animals display a slight shift in DPOAE levels between 3600 and 9600 Hz ($f2$). This is likely due to aging (*see* **Note 10**). To take into account this drift in DPOAE amplitudes

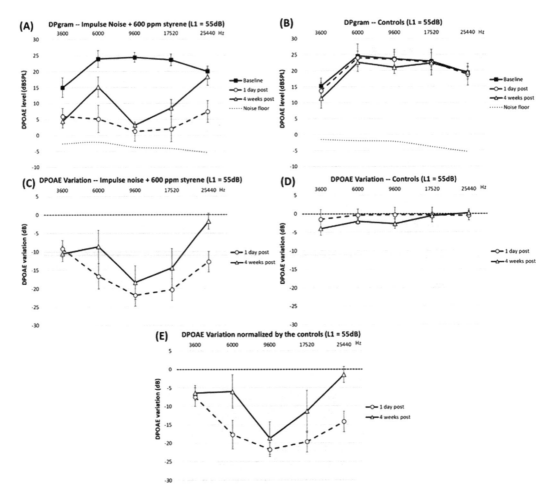

Fig. 8 DPgrams (**a**, **b**) and DPOAE variations (**c**, **d**) obtained in control rats (**b**, **d**) and rats exposed to impulse noise (octave band noise centered at 8 kHz, $L_{EX,8h}$ = 80 dB) and 600-ppm styrene for 4 weeks (6 h/day, 5 days/week) (**a**, **c**). Exposure was followed by a 4-week recovery period. DPOAE measurements were performed with $L1$ = 55 dB before exposure (baseline) and 1 day (1 day post) and 4 weeks post-exposure (4 weeks post). DPgrams (**a**, **b**) display the raw DPOAE values at different time-points for a given experimental group. DPOAE variations (**c**, **d**) are defined as [post-exposure DPOAE − baseline] for a given experimental group. DPOAE variations normalized by the controls (**e**) are calculated as follows: [(post-exposure DPOAE$_{exposed}$ − post-exposure DPOAE$_{controls}$) − (baseline DPOAE$_{exposed}$ − baseline DPOAE$_{controls}$)] and take into account the drift in DPOAE amplitudes observed in the control group

observed in the control group, one might consider normalizing the DPOAE value of the exposed animals by the values obtained in control animals (Fig. 8e) by doing the following calculation:

$$[(post - exposure \ DPOAE_{exposed} - post - exposure \ DPOAE_{controls}) - (baseline \ DPOAE_{exposed} - baseline \ DPOAE_{controls})].$$

The present analysis highlights the wide frequency range affected by the noise and styrene co-exposure although the impulsive noise was filtered band-pass filtered between 5600 and 11,200 Hz (octave band noise 8 kHz). A clear permanent impairment was obtained at all tested $f2$ frequencies but the highest, 25,440 Hz. Moreover, this DPgram analysis shows that the post-exposure recovery is highly dependent on the frequency considered (full at 25,440 Hz, limited at 9600 Hz).

4 Conclusion

DPOAE measurements are particularly attractive for assessing cochlear function in laboratory animals such as rats because they are objective, robust, easily measured and offer the possibility to test a wide range of frequencies within the species' hearing range. In addition, measuring DPOAEs in rats is fast, noninvasive, and can be performed repeatedly in the same subjects. The downside of DPOAE measurements in rats is the anesthesia, which might introduce in some cases an additional variable in the experiment. It is also of importance to keep in mind that, in rats, frequencies below 3000 Hz cannot be investigated using this method, and that it is necessary to maintain steady body temperature and atmospheric pressure. Despite these limitations, DPOAEs are extremely valuable because OHCs are very sensitive to chemical impairment, noise and oxidative stress, among other nuisances. For a thorough investigation of the health of the cochlea and central auditory pathways, additional auditory measurements can be performed, such as cochlear microphonic response and auditory brainstem-evoked responses (ABR). The cross analysis of these two measurements can help to localize the origin of the impairment. DPOAEs are specific to OHCs and can be measured even if IHCs are damaged [9, 17]. In contrast to DPOAEs, ABR thresholds are dependent on the function of the whole auditory system, i.e., OHCs, IHCs, and ascending auditory pathways and nuclei [18]. In addition, depending on your scientific question, you might consider complementing these functional techniques by cochlear histology, such as cytocochleograms (see Chap. 2), scanning electron microscopy of the stereocilia bundles, or mid-modiolar sections of resin-embedded cochleae.

5 Notes

1. At the time this text is being written, two companies (Tucker-Davis Technologies and Intelligent Hearing systems) provide integrated commercial systems allowing DPOAE measurements up to 32–40 kHz in rodents. Although these systems have not been tested by the authors, they have been previously

used in several published studies [14, 19] and should be considered as viable and simple alternatives to our custom setup.

2. Ideally, DPOAE measurements should be performed within a double-walled sound-attenuation chamber to reduce background noise. The chamber should include a small double-pane window or a video camera for observing the animal during testing and should be large enough to stand inside it. In case such an equipment is not an option, you might consider a single-walled chamber, or tabletop sound booth. However, always attempt to maintain the noise floor as low as possible by eliminating/concealing all surrounding noise sources.

3. Using two different generators, each connected to a transmitter, reduces the risk of acoustic distortion generated by the equipment setup itself. Indeed, electrical and acoustic signal distortions easily occur in a measurement system. Such signal interferences, mixed to the physiological signal, might render the recording unusable. It is very important to ascertain that the signal is purely of physiological origin.

4. All the elements composing our prototype probe are glued together with cyanoacrylate glue so that the manipulation of the probe does not create any air-leak, which could modify sound-wave impedance. Electrical ground loops must be eradicated and shield must be thorough.

5. As an indication, the uncertainty widened to a level of confidence of 95% (widening factor 2.26) is 0.5 dB at 4000 Hz and 1.0 dB at 16000 Hz for the levels $L1$ and $L2$ of the primaries $f1$ and $f2$. The uncertainty increased to a 95% confidence level (2.26 enlargement factor) of the sound pressure measured by the probe's microphone is 0.9 dB at 4000 Hz and 1.8 dB at 16000 Hz.

6. When anesthetized, the body temperature of the rats drops rapidly. It is very important to maintain it stable using a heating blanket with a feedback probe (38 ° C). If the rat gets cold, DPOAE responses are profoundly reduced and become unusable. This occurs because the eardrum gets "depressed." A "depressed" eardrum is easily visible by otoscopy as it gets very close to the ossicles and let them appear by transparency. In this case, warm up the rat for a few minutes and check that eardrum is slightly "inflated" towards you.

7. You should ascertain that there is no difference of pressure between the room where the animal has been anesthetized and the measurement booth. The anesthetized rat does not regulate the balance of pressures on either side of the eardrum. Performing a cochlear paracentesis [20] can solve the problem for a single measure of DPOAE, but is not recommended for

repeated measurements because of the risks of eardrum damage and infection.

8. When stimulated by a two-tone stimulus, the organ of Corti produces a family of distortion products having specific arithmetic relationships with the frequencies of the stimulus tones ($3*f1 - 2*f2$, $4*f1 - 3*f2$; $2*f2 - f1$; $3*f2 - 2*f1$, $4*f2 - 3f1$; $f2 - f1$, etc.). However, the most robust and easily recorded in both laboratory animals and human is the DPOAE at $2*f1 - f2$.

9. With $L1 < 40$ dB (data not shown), the difference in DPOAE levels between 6 and 24 months decreases because DPOAE levels hardly emerge from background noise. Only rats with the highest DPOAEs can be measured with such a low $L1$, especially at 24 months. Therefore, the data are no longer representative of the group but only of the rats having the best hearing performances.

10. During an experimental protocol of DPOAE measurements with rats, it is important to carry out the measurements of a control group simultaneously with the exposed group. The two groups must be measured under the same conditions in order to be able to analyze any variations and thus reinforce the accuracy of these measurements.

References

1. Mahendrasingam S, Beurg M, Fettiplace R, Hackney CM (2010) The ultrastructural distribution of prestin in outer hair cells: a post-embedding immunogold investigation of low-frequency and high-frequency regions of the rat cochlea. Eur J Neurosci 31 (9):1595–1605. https://doi.org/10.1111/j.1460-9568.2010.07182.x

2. Fettiplace R (2017) Hair cell transduction, tuning, and synaptic transmission in the mammalian cochlea. Compr Physiol 7 (4):1197–1227. https://doi.org/10.1002/cphy.c160049

3. Jaramillo F, Markin VS, Hudspeth AJ (1993) Auditory illusions and the single hair cell. Nature 364(6437):527–529. https://doi.org/10.1038/364527a0

4. Strelioff D, Flock A (1984) Stiffness of sensory-cell hair bundles in the isolated Guinea pig cochlea. Hear Res 15(1):19–28. https://doi.org/10.1016/0378-5955(84)90221-1

5. Khanna SM (2002) Non-linear response to amplitude-modulated waves in the apical turn of the Guinea pig cochlea. Hear Res 174 (1–2):107–123. https://doi.org/10.1016/s0378-5955(02)00645-7

6. Howard J, Hudspeth AJ (1988) Compliance of the hair bundle associated with gating of mechanoelectrical transduction channels in the bullfrog's saccular hair cell. Neuron 1 (3):189–199. https://doi.org/10.1016/0896-6273(88)90139-0

7. Kemp DT (2002) Otoacoustic emissions, their origin in cochlear function, and use. Br Med Bull 63:223–241. https://doi.org/10.1093/bmb/63.1.223

8. Subramaniam M, Salvi R, Spongr V, Henderson D, Powers N (1994) Changes in distortion product otoacoustic emissions and outer hair cells following interrupted noise exposures. Hear Res 74(1–2):204–216

9. Trautwein P, Hofstetter P, Wang J, Salvi R, Nostrant A (1996) Selective inner hair cell loss does not alter distortion product otoacoustic emissions. Hear Res 96(1–2):71–82

10. Hofstetter P, Ding D, Powers N, Salvi RJ (1997) Quantitative relationship of carboplatin dose to magnitude of inner and outer hair cell loss and the reduction in distortion product otoacoustic emission amplitude in chinchillas. Hear Res 112(1–2):199–215

11. Emmerich E, Richter F, Reinhold U, Linss V, Linss W (2000) Effects of industrial noise exposure on distortion product otoacoustic emissions (DPOAEs) and hair cell loss of the cochlea–long term experiments in awake Guinea pigs. Hear Res 148(1–2):9–17

12. Shera CA (2004) Mechanisms of mammalian otoacoustic emission and their implications for the clinical utility of otoacoustic emissions. Ear Hear 25(2):86–97. https://doi.org/10.1097/01.aud.0000121200.90211.83

13. Martin GK, Stagner BB, Lonsbury-Martin BL (2006) Assessment of cochlear function in mice: distortion-product otoacoustic emissions. Curr Protoc Neurosci. Chapter 8: Unit8.21C-Unit28.21C. https://doi.org/10.1002/0471142301.ns0821cs34

14. Sheppard AM, Zhao DL, Salvi R (2018) Isoflurane anesthesia suppresses distortion product otoacoustic emissions in rats. J Otol 13(2):59–64. https://doi.org/10.1016/j.joto.2018.03.002

15. Robles L, Ruggero MA (2001) Mechanics of the mammalian cochlea. Physiol Rev 81(3):1305–1352. https://doi.org/10.1152/physrev.2001.81.3.1305

16. Venet T, Campo P, Thomas A, Cour C, Rieger B, Cosnier F (2015) The tonotopicity of styrene-induced hearing loss depends on the associated noise spectrum. Neurotoxicol Teratol 48:56–63. https://doi.org/10.1016/j.ntt.2015.02.003

17. Le Calvez S, Avan P, Gilain L, Romand R (1998) CD1 hearing-impaired mice. I: distortion product otoacoustic emission levels, cochlear function and morphology. Hear Res 120(1–2):37–50. https://doi.org/10.1016/s0378-5955(98)00050-1

18. Kujawa SG, Liberman MC (2009) Adding insult to injury: cochlear nerve degeneration after "temporary" noise-induced hearing loss. J Neurosci 29(45):14077–14085. https://doi.org/10.1523/JNEUROSCI.2845-09.2009

19. Dong W, Stomackin G, Lin X, Martin GK, Jung TT (2019) Distortion product otoacoustic emissions: sensitive measures of tympanic-membrane perforation and healing processes in a gerbil model. Hear Res 378:3–12. https://doi.org/10.1016/j.heares.2019.01.015

20. Bergin M (2013) Systematic review of animal models of middle ear surgery. World J Otorhinolaryngol 3(3). https://doi.org/10.5319/wjo.v3.i3.71

Chapter 6

Use of Visual Evoked Potentials to Assess Deficits in Contrast Sensitivity in Rats Following Neurotoxicant Exposures

William K. Boyes

Abstract

This chapter describes a procedure for recording pattern-elicited visual evoked potentials from experimental animals, focused primarily on pigmented rats. When recorded over a range of visual pattern contrast values, the results can be used to derive estimates of visual contrast threshold, contrast sensitivity, and contrast gain. Visual contrast is defined as the difference between the bright and dark regions of a visual pattern, adjusted for the overall luminance. Contrast encoding is an important feature of the neurological processes underlying spatial vision and is dependent on integrated processing within defined neurological circuits. This chapter describes procedures to measure contrast-related parameters that have been developed over years of experience and trial and error approaches. They involve electrophysiological recordings from visual cortex while animals view modulating visual patterns. The resulting evoked potentials are signal averaged, subjected to spectral analysis and interpreted relative to the contrast of the eliciting visual patterns. The resulting parameters include measurements of response amplitude, contrast threshold, contrast sensitivity, and contrast gain. The data from experimental animals are highly analogous to those from human subjects and have shown similar responsivity to neurotoxicant exposures.

Key words Visual evoked potentials, Contrast threshold, Contrast sensitivity, Contrast gain, Developmental neurotoxicity

1 Introduction

Testing visual function for potential toxic effects is important as a primary adverse outcome and is also an approach to assess neurological function. The retina is part of the central nervous system and projects to the visual centers of the brain. The visual system is accessible for well-controlled stimulation that enables careful assessments of many aspects of neurological function. It is important to distinguish optical from neurological aspects of vision. A typical examination of visual acuity, in which a patient observes black letters on a white background and reports the letters of the smallest font that are discernable, is primarily an evaluation of the

Jordi Llorens and Marta Barenys (eds.), *Experimental Neurotoxicology Methods*, Neuromethods, vol. 172,
https://doi.org/10.1007/978-1-0716-1637-6_6, © Springer Science+Business Media, LLC, part of Springer Nature 2021

optical apparatus (cornea and lens) at the front of the eye. Visual acuity measurement depends on the ability to make a well-focused image on the retina and works well for determining refraction values for glasses or contact lenses. However, a measure of visual acuity does not evaluate much of the neural aspects of visual function. Visual perception is important over a range of visual pattern sizes, not just the smallest letters identifiable. Vision also requires encoding features of the visual world such as pattern size, shape, motion, color, and depth. One of the most important features of visual pattern perception is contrast, expressed as the luminance difference between the bright and dark parts of a visual pattern, adjusted for the overall luminance. Contrast thresholds are defined conceptually as the lowest contrast level which is readily visible. In psychophysical experiments, contrast threshold might be defined by the lowest contrast target that is detected statistically on half or more of the trials presented. Using visual evoked potentials (VEPs), (*see* **Note 1**), contrast thresholds can be thought of as the lowest contrast stimulus to elicit a response with an amplitude above zero, or alternatively, above the recording noise level. Visual contrast sensitivity, the inverse of contrast threshold, turns out to be highly sensitive to disruption by exposure to a range of neurotoxic compounds.

Prior to describing a procedure for testing visual contrast perception in experimental animals, a few basic concepts relating to the neurological processes encoding spatial and temporal vision will be considered. Figure 1 presents a simplified and stylized scheme of the retina. The photoreceptors (rods and cones; P) lie at the back of the retina with cell bodies in the outer nuclear layer. The photoreceptors receive photons of light and generate neurological signals that feed directly into retinal bipolar cells (B), which in turn project to retinal ganglion cells (G). The retinal ganglion cells send axons via the optic nerve to the visual areas of the brain. At the retinal outer plexiform layer, where the photoreceptors meet the bipolar cells, there are also synaptic contacts with horizontal cells (H). The horizontal cells form inhibitory connections across neighboring photoreceptor-bipolar cell channels, so that when one channel is activated by light, the neighboring channels are correspondingly deactivated, and vice versa. This has the effect of giving bipolar cells and retinal ganglion cells what are referred to as "center-surround" receptive fields (Fig. 2). The receptive field of a neuron is that portion of the receptor surface (in this case the retina) which, when stimulated, causes the neuron to increase or decrease its firing rate. Light falling on the center of the receptive field might activate a cell, while stimulation of the surrounding area would inhibit it, forming a center-surround topography. The retina contains cells with both on-center and off-center receptive fields. For off-center cells, stimulation of the center inhibits, and the surround activates, the cell. The receptive fields also vary in size, with small receptive

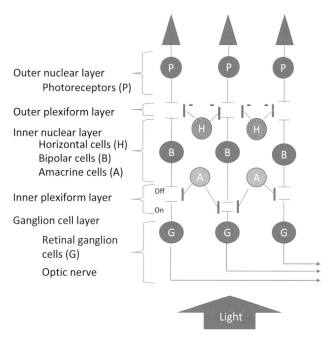

Fig. 1 A schematic representation of retinal connectivity. The three photoreceptors (P) are depicted with their cell bodies located in the outer nuclear layer. The photoreceptors make synaptic connections in the outer plexiform layer with retinal bipolar cells (B) and horizontal cells (H). Horizontal cells make inhibitory contact between the photoreceptor/bipolar cell interface so that when a photoreceptor and bipolar cell become activated the neighboring bipolar cells are inhibited. Conversely, when one of the photoreceptors depicted on the side is activated, the bipolar cell in the middle is inhibited. The bipolar cells which in turn contact retinal ganglion cells (G) and amacrine cells (A) in the inner plexiform layer. Synaptic connections in the outer portion of the inner plexiform layer primarily signal stimulus offsets, whereas synaptic contacts in the inner portion of the outer plexiform layer primarily signal stimulus onsets. The axons of retinal ganglion cells form the optic nerve, leave the eye, and project to visual areas of the brain. With this simplified diagrammatic arrangement, the primary spatial and temporal features of visual pattern perception are established

fields tuned for fine patterns and large receptive fields for large patterns. The size of visual patterns can be measured as spatial frequency in units of the cycles of a repetitive pattern per degree of visual angle (cpd). The luminance difference between the center and surround portions of the receptive field governs the neuronal firing rate, and hence cellular sensitivity to contrast. Therefore, in the combination of spatial frequency and contrast, the basic elements of perceiving visual patterns are encoded in the outer plexiform layer of the retina. Although sensitivity to contrast begins in the retina, it is tuned and refined at later stages of the visual system.

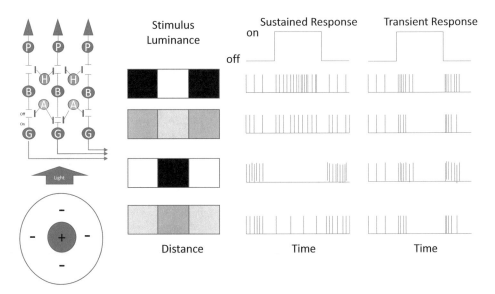

Fig. 2 Left Column. The retinal cellular connectivity from Fig. 1 is reproduced along with a depiction of an excitatory center-surround receptive field. Receptive fields reflect region of the innervated surface (in this case the retina) for which the neuronal firing rate changes (increase or decrease) with stimulation. The horizontal cell inhibitory interactions with neighboring photoreceptor/bipolar cell connections has the result of producing center receptive fields as shown below. Therefore, if the neural activity pattern of a retinal ganglion cell is measured it has a concentric "center-surround" spatial organization as depicted from the retinal surface perspective in the circles below. Stimulation in the central area of the receptive field causes the firing rate of this retinal ganglion cell to increase, while stimulation in the surround region causes the firing rate to decrease. The size of the receptive field reflects the extent of the reach of horizontal cell processes across the retinal surface. The amount of increase or decrease in firing rate is a function of the total luminance difference between the center and surround regions

Second column from left: the three-part panels represent the stimulus luminance impinging on the center or surround regions of the receptive field shown at bottom left. The four panels at right depict the neuronal firing rate recorded in response to the associated stimulation

Top: a high-contrast stimulus with the bright part over the excitatory region and the dark part over the inhibitory region causes the neuron to display a maximum firing rate

Second panel: the contrast between the bright and dark parts of the stimulus is reduced

Third panel: a high-contrast stimulus in reverse phase to the stimulus in the top row

Bottom: a lower contrast, reverse-phase stimulus

Third column from left shows hypothetical "sustained" firing rate patterns elicited by stimuli on the same row at the left. Top: The neuron shows a sustained increase in firing rate while the stimulus is presented. Second panel: firing rate proportionally less than maximal because of the lower contrast stimulus. Third panel: the high-contrast stimulus is reversed. The reverse-phase stimulus causes a reduction of firing rate that is proportional to the summed stimulation of the center and surround portions of the receptive field. Bottom. Reverse-phase lower contrast stimulus causes a lesser degree of inhibition of the normal firing rate

Right column. Hypothetical responses of cells firing in a "transient" manner to the stimuli shown in second column. Top: Sustained stimulation with high-contrast stimuli causes robust responding when turned on or off, which fades after the stimulus transition. Second panel: Transitions of lower contrast stimulus cause transient bursting responses of lesser intensity than the higher contrast stimulus. Third panel: Reverse-phase transitions cause transient responses similar to top row. Low--contrast stimulus transitions cause response firing patterns similar to the second row

The retina also encodes important temporal aspects of visual stimuli, which happens primarily at the inner plexiform layer. Temporal tuning involves synaptic activity of amacrine cells (A) as they influence transmission between bipolar cells and retinal ganglion cells. Off-signals are transmitted in the outer portion of the inner plexiform layer, while on-signals are processed in a distinct layer in the inner portion of the inner plexiform layer. Electrophysiological recordings from retinal ganglion cells reveal two distinct temporal profiles of neuron activity (Fig. 2). In response to light falling on their receptive fields, some neurons show a steady firing rate, either an increase for on-center cells or a decrease for off-center cells. These are sometimes referred to as "sustained" responses. Other cells show bursts of firing whenever the light is either turned on or turned off, but the change in firing rate rapidly adapts if the stimulation is prolonged. These are referred to as "transient" responses. Some think of sustained responders as "pattern detectors" since they respond whenever a pattern is present, and transient responders as "motion detectors" since they respond whenever a pattern is changing. Another terminology for sustained and transient response patterns is "linear," and "nonlinear," respectively, relating to the fidelity in which neuronal response rates follow temporal and spatial aspects of visual pattern modulation. There are variations in the proportions of cells with linear or nonlinear responses among species. Human, non-human primates, and many mammalian retinas have both sustained and transient responding neurons. Rats tend to have a preponderance of transient, "motion-detector" neurons, and under steady-state sinusoidal modulating patterns, the rat visual system responds primarily as a frequency-double ("nonlinear") rate, reflecting bursts of responses to both the on and off cycles of a temporally modulated visual pattern (Fig. 3).

This chapter describes procedures to record VEPs as a method to assess visual pattern perception and sensitivity to visual contrast. These procedures were originally adopted from human clinical and psychophysical practices and have proved to be sensitive measures of potential neurotoxicity. Literature on the basic theory and practice of electrophysiological recordings, including sensory evoked potentials and their use in neurotoxicology, are available elsewhere (1–5). Rats were selected as the primary subjects for study because, at the time these procedures were developed, the rat was a primary subject of experimental neurotoxicology research. Sensory evoked potential procedures also work in mice (6, 7) and other species such as cat, pig, or primate (8–10). It is important, however, that a pigmented strain of rat or mouse is used (see Note 2), because albino strains of rodents, lacking melanin in the retinal pigment epithelium and other ocular structures, have poor pattern vision and give inferior pattern VEPs (11). The procedures involve surgically implanting indwelling recording electrodes in the skull located

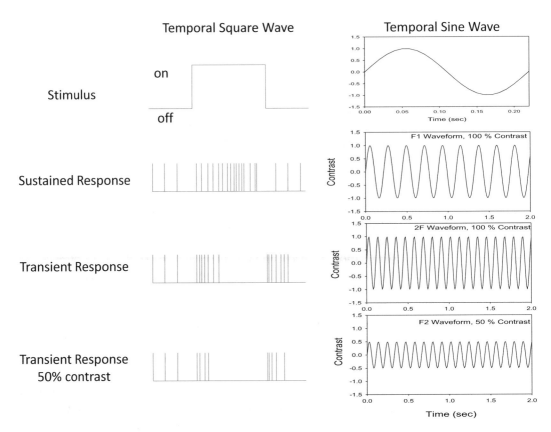

Fig. 3 Illustration of how hypothetical sustained and transient responses translate into steady-state VEPs at the stimulus rate (F1) and twice the stimulus rate (F2), respectively. Basically, the firing rate of sustained neurons follow the temporal modulation of the pattern (F1), while the transient neurons, responding to both on and off phases of the stimulus transitions, respond at twice the stimulation rate (F2). Stimulation of high-contrast patterns causes a strong F2 evoked potential (third row). Stimulation with a pattern 50% reduced in contrast causes a lower amplitude F2 response (bottom row)

over visual cortex (*see* **Note 3**) (Fig. 4), followed by approximately 1 week for surgical recovery prior to VEP recording. The recordings are performed using awake rats (Fig. 5) in order to avoid the effects of anesthesia on the cortical VEPs and potential interactions between the anesthetics and the actions of potential neurotoxic substances being studied. The awake animals are mildly restrained (*see* **Notes 11, 14, 16**) during testing to assure that they are watching the visual stimuli and are at a constant distance (15 cm) and orientation to the stimulus screen. Computer-generated stimulus patterns are presented on a video monitor for the subject to observe, modulated over time, and electrical activity from visual cortex is concurrently recorded. The electrophysiological activity is amplified, filtered, and artifacts from movement and other causes are detected and removed. The cleaned electrophysiological data are signal averaged in synchrony with the temporal modulation of the visual pattern in order to enhance neuronal activity related to

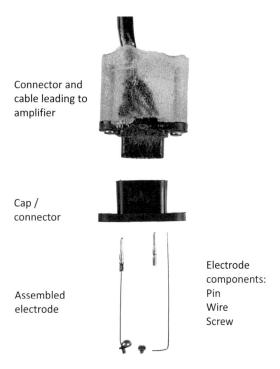

Connector and
cable leading to
amplifier

Cap /
connector

Assembled
electrode

Electrode
components:
Pin
Wire
Screw

Fig. 4 Photograph of components for construction of cranial screw electrodes and connection to amplifier input cables. Lower left: assembled electrode. Lower right: electrode components including: stainless-steel screw, insulated wire with insulation scrapped from both ends for soldering to screw, and gold pin to crimp on the top of electrode wire. The wire is soldered to the screw leaving the screw slot opened. The screw component is threaded to a matched size hole in the skull over visual cortex to provide the recording surface. The pin is placed into a designated hole in the cap/connector (an Amphenol 9 pin connector) for top of headset. The headset assembly (also including ground and reference electrodes not pictured) is cemented in place. After surgical recovery, the cap is temporarily attached to the matched connector and cable leading to the amplifier for a recording session

visual stimulation ("signal") and reduce non-synchronized activity which presumably is unrelated to visual signal processing ("noise"). The resulting VEP is submitted to post hoc analyses, in our case Fast Fourier Transformation (FFT) processing, that yields the signal amplitude at the stimulus rate and harmonic frequencies. Stimulus amplitude data are then expressed as a function of the visual stimulus contrast to determine contrast sensitivity and contrast gain parameters. Response amplitudes, contrast sensitivity, and contrast gain values can be compared across groups of subjects that were treated with vehicle (i.e., control) or different doses of the test material.

The experimenter has multiple options regarding temporal and spatial parameters of the visual stimuli, data recording, and analyses. We will discuss those parameters and the rationale for selections below.

Pattern Evoked Potential Recording Procedure

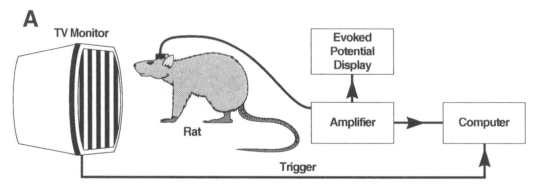

Fig. 5 Schematic illustration of procedure for recording pattern-elicited VEPs from awake rats (reproduced with permission from Boyes, 1994)

1.1 Stimulus Temporal Parameters

Several different modes of temporal modulation of the visual patterns have been used. The preferred option in our laboratory is steady-state pattern-reversal modulation. The steady-state term implies that the stimulus modulation and the resulting evoked potentials are continuous, in contrast to "transient" evoked potentials, in which the stimuli change abruptly and discrete responses to each stimulus presentation are observed. With transient stimulation, the evoked potential waveforms are typically complex, with multiple positive and negative peaks, and are usually scored as the latency and amplitude of each positive and negative going peak. In some cases, the underlying neurological generators of these peaks have been identified which can be an advantage for interpretation of results. In other cases, however, the peak generators are unknown or disputed, and neurotoxicological treatments may alter the waveform shape so that the identification of individual peaks becomes problematic. In addition, scoring the entire waveform with multiple peaks and valleys produces a large number of dependent variables (twice the number of peaks and troughs if the latency and amplitude of each is measured). Statistically, recording multiple dependent measures from individual subjects causes an increased probability of incorrectly rejecting the null hypothesis, which should be addressed through a correction of the alpha level using appropriate statistical adjustments.

In steady-state recordings, in contrast, the stimulus modulation is continuous, and the stimulus rate is adjusted so that the response from one transition blends into another and the evoked potential takes on a sinusoidal characteristic (11). In this case, it is possible to submit the waveform to a Fourier analysis and obtain amplitude of the primary Fourier component(s) of the waveform. This is an unambiguous measure of the strength of the response that is not dependent on the semi-subjective selection of peaks for scoring.

Temporal modulation (1F)

Hypothetical transient response (2F)

Hypothetical transient response, (2F); 50% contrast

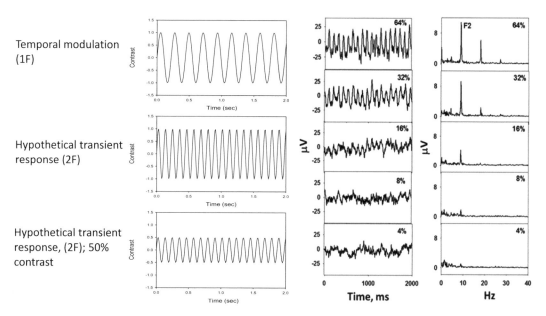

Fig. 6 Left. Sine waves at the stimulus rate (Top), a high-contrast and twice the stimulus rate (second row) and at 50% contrast, and twice the stimulus rate (Bottom row)
Center: group averaged VEP waveforms from rats in response to stimuli of varying contrast as noted. Right column. Spectral analysis of the waveforms to the left showing predominant F2 amplitude peak, which increases as a function of the stimulus contrast. Higher frequency harmonic responses are also evident in the spectral analysis of the higher contrast responses

The rat visual system responds primarily in a transient mode—rats have much more visual neurons that respond in a transient than a sustained manner, and the rat waveform shows a strong frequency—double (nonlinear) response component labeled F2. The amplitude of the F2 component therefore becomes an unambiguous, objectively scored, single response measure (Fig. 6).

In addition to transient and steady-state presentation, there is the choice of pattern-reversal ("counterphase") or pattern on-off (appearance–disappearance) modulation (Fig. 7). For pattern-reversal modulation, the dark and light portions of the stimulation alternate fully in a counterphase fashion so that every part of the screen changes over time from dark to light and back again. In on-off, the pattern modulates from a mean luminance screen to one phase of the spatial stimulus and back to a mean luminance screen. On-off has an advantage in that steady-state responses to on-off modulation contain both F1 ("linear") and F2 ("nonlinear") components (12). Pattern-reversal stimulation reveals only F2 ("nonlinear") responses in the VEP waveform. Counterphase has the advantage that the contrast modulation is effectively twice that of on-off modulation, yielding stronger, and more robust evoked potentials. The F1 (linear) responses of rats are very small in comparison to humans, which reflects the high proportion of the visual

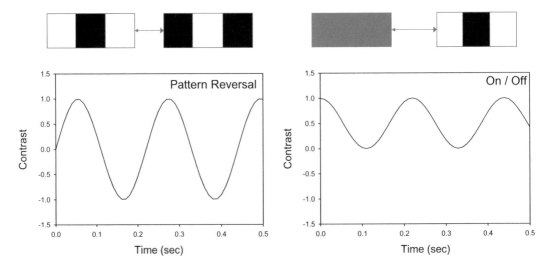

Fig. 7 An illustration of pattern-reversal ("counterphase) vs On/Off ("appearance–disappearance") modulation of the visual patterns. In pattern-reversal (left side), the stimulus pattern alternates so that each portion of the stimulus reverses from light to dark phase of the pattern and back again with each temporal cycle. In On/Off modulation (right side), the stimulus transitions between a mean luminance, zero contrast pattern, and the full pattern and back on each temporal cycle

neurons of the rat responding in a transient fashion. In practice, we have never seen a robust toxic effect in the rat F1 response. Therefore, we now prefer using pattern-reversal modulation for most testing applications due to the F2 responses being larger, less variable and having a higher signal-to-noise ratio.

1.2 Stimulus Spatial Parameters

Spatial Frequency. Spatial frequency refers to the size of a repetitive visual grating and is expressed as cycles of the pattern/degree visual angle (cpd). Human subjects are typically tested over a range of spatial frequencies. The range of spatial frequency sensitivity in rats, however, is more limited. We use stimuli of about 1.6 cpd, which is the approximate peak of the contrast sensitivity function of pigmented rats (13).

Contrast. Contrast (C) is defined as the difference between the maximum and minimum luminance (L_{max} and L_{min}, respectively) of a visual pattern, adjusted for the overall background luminance.

$$C = (L_{max} - L_{min})/(L_{max} + L_{min})$$

Contrast values range from 0 (no pattern) to 1 (all the light from the bright parts of a pattern). Frequently, contrast values are multiplied by 100 to convert them into percent contrast (ranging from 0 to 100%).

The F2 amplitude of the rat steady-state pattern-reversal VEP shows a predictable relationship to the contrast of the visual stimulus. The F2 amplitude increases in a linear fashion with the log of the stimulus contrast (14, 15). This relationship follows Weber-

Fechner law of sensory psychophysics, which contends that the perceived intensity of sensory stimuli is a linear function of the log stimulus intensity. The perceived intensity of sensory stimulation is encoded by the firing rate of sensory neurons. Therefore, the log-contrast F2 amplitude functions of pattern VEPs reflect original coding of differential stimulation of the center-surround receptive fields of retinal neurons, translated into the firing rates of visual neurons, and carried on ascending sensory pathways to the visual areas of the brain.

Another concept is contrast gain, which refers to the increased neuronal firing rate with increased contrast of visual patterns. Contrast gain adjustment refers to the adaptation of the firing rate of sensory neurons in the face of altering visual contrasts. Adaptation enables the sensitivity to be adjusted so that subtle stimuli can be perceived in a low-contrast environment, such as a dim-light overcast day, and also not overwhelmed on a bright sunny high-contrast day. The firing rates of visual neurons are adjusted to changing visual contrast to maintain an optimal firing range. Contrast gain adjustments of visual neurons have been observed to some extent in neurons of the retina and lateral geniculate nucleus of the thalamus, but the primary site of contrast gain adjustment is thought to be neurons of visual cortex (16). In VEPs, the slope of the amplitude log-contrast functions has been interpreted to reflect contrast gain (15). VEP log-contrast amplitude functions therefore reflect the sensitivity of the visual system to contrast as well as the adjustments of the system to changing contrast levels.

We recently reported that adult rats, who were deficient in thyroid hormone during their pre- and postnatal development due to perinatal exposure to propylthiouracil (PTU), had dose-related decreases in the slope of pattern-reversal VEP log-contrast F2 amplitude functions (Fig. 8) (17). Propylthiouracil causes a deficit in thyroid hormones by inhibiting iodine transport in the thyroid gland. Thyroid hormones are critical for neurodevelopment, including for neurons in the visual system. There was evidence that PTU-treated rats had both retinal deficits (decreases in green flicker electroretinograms), as well as deficits in the superficial layers of visual cortex (decreases in visual contrast gain). A reduction of thyroid hormones during pre- and postnatal neurodevelopment altered the visual system in a way that persisted into adulthood, long after the thyroid hormone levels of the grown pups had returned to normal. Many environmental contaminants may alter expression of thyroid hormones, but the level of thyroid hormone inhibition and the consequences for neurodevelopment are often unclear. It is possible that assessment of contrast sensitivity and VEP contrast amplitude functions could serve as one approach to evaluate residual neurological effects of developmental thyroid disruption.

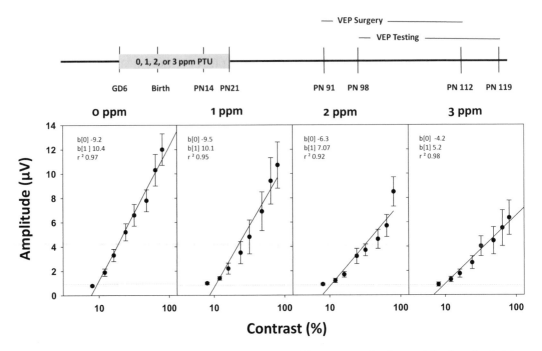

Fig. 8 The effects of perinatal dose-related inhibition of thyroid hormone synthesis on adult rat contrast vision. The time line of the experiments is depicted along the top of the figure. Pregnant rats were treated with 0, 1, 2, or 3 ppm propylthiouracil (SP) PTU in drinking water beginning on gestational day (GD) 6. The treatment continued through the duration of gestation and postnatally until postnatal day (PN) 21 when the pups were weaned and received control drinking water thereafter. PTU inhibits uptake of iodine into the thyroid gland and caused a graded amount of developmental insufficiency of thyroid hormone. The pups showed dose-related reductions of serum T4 during the treatment period, but T4 levels had returned to normal by the time of VEP testing. When the rats were adults, they were surgically implanted with recording electrodes and tested for VEPS about a week later. The four panel graphs depict the pattern VEP F2 amplitude measured from FFT analysis of the individual rat waveforms. The horizontal axes of each panel depict stimulus contrast on a repeated log scale. The vertical axis depicts F2 amplitude and is the same for each dose group. Linear functions were fit to the F2 amplitude log-contrast functions from each dose group. The slope of the functions (i.e., contrast gain) was significantly reduced following either 2 or 3 ppm PTU (Reproduced with permission from Boyes et al., 2018)

The neurons in the visual pathway operate by the same set of biochemical and neurophysiological processes as other neurons of the central nervous system. The primary excitatory and inhibitory neurotransmitters are present. They operate in sophisticated and integrated networks. Knowledge of the neural mechanisms for encoding visual pattern and contrast and the ability to precisely stimulate visual pathways with computer-generated visual patterns, make visual system testing a sophisticated and sensitive approach to evaluate potential neurotoxicity.

The following sections describe protocols to record pattern-reversal evoked potentials from awake rats, as was reported previously (17). These procedures were built on and evolved from those used before. Interested readers may wish to find additional applications of these and related procedures (18–22).

2 Materials

- Subjects: Adult pigmented rats. They should be about 60 days of age or older at the time of surgery so that growth of the skull has slowed, and the implanted electrode assembly will remain intact. Pigmented strains are preferred because albino strains of rats have aberrant retinas and give poor pattern evoked potentials (11). Toxicology studies typically require a vehicle treated control group and one or preferably more dose groups for dose–response studies. As a general rule of thumb, about 10 rats per group is a usually sufficient, although it is preferable to calculate sample sizes using statistical power calculations.
- Construction of cranial electrodes (Fig. 4)
  - Stainless-steel screws (00–90 × 1/16 inch).
  - Formvar-coated Nichrome wire (0.10″ diameter).
  - Soldering gun, flux, solder.
  - Amphinol gold pins.
- Implantation of cranial recording electrodes
  - A protocol approved by the institutional committee responsible for oversight of ethical and humane animal care and use.
  - Stereotaxic device.
  - Anesthesia: our current protocol involves administration of injectable 5 mg/kg Rimadyl at a 1 mL/kg volume (sc) before beginning surgery, followed by Isoflurane inhalation anesthesia.
  - Electric shaver.
  - Ophthalmic ointment.
  - Betadine surgical scrub.
  - Surgical tools (scalpel, retractors, wound staples, etc.).
  - Small drill matched to size of screws.
  - Screw driver.
  - Amphenol 9-hole connector with associated gold pins for electrode connectors (Fig. 4).
  - Dental acrylic (cranioplastic cement).

- – Impedance meter.
- – Analgesic.
- Creating Visual Stimuli
  - – Computer software capable of generating temporally modulated sine-wave gratings of different spatial frequency and contrast values. Our software package is custom written (22, 23). Other options are available (e.g., http://psychtoolbox.org/overview.html).
  - – Video monitor (*see* **Note 4**).
  - – Photometer.
- Electrophysiological recording
  - – Faraday cage (grounded).
  - – Shielded cables and connector (paired to headset).
  - – Animal restrainer (head and eyes free for unobstructed view).
  - – Electrophysiological amplifiers and bandpass filters (22).
  - – Computer for experimental control, data collection, signal averaging, and data analysis (23). Other options are available (e.g., https://www.mathworks.com/help/signal/ref/tsa.html).
  - – Ruler.
  - – Black cloth for covering.

3 Methods

- Construction of electrodes (Fig. 4).
  - – Cut nichrome wire to desired length and scrape off the insulation coating from ~2 mm of the both ends. Crimp a gold pin to one scrapped end.
  - – Bend the other scraped end of the wire at a 90° angle and solder this end to the top of a stainless-steel screw, leaving the screw slot opened.
  - – Check continuity from screw to pin with an ohmmeter.
- Surgical procedures (*see* **Notes 5–10**)
  - – Establish survival surgical procedures and obtain approval from the institutional committee responsible for oversight of ethical and humane animal care and use in compliance with applicable guidelines for humane animal experimentation. Approved animal care and use procedures have evolved substantially over time. It is recommended that investigators establish surgical procedures in collaboration with their appropriate institutional authorities (e.g., veterinarian and/or institutional animal care and use committee) to assure

compliance with current practices rather than rely on previously established procedures.

– Anesthetize rat in accordance with approved procedures.

– Apply ophthalmic ointment to both eyes to avoid drying of the cornea.

– Shave hair from top of head and clean the area with Betadine surgical scrub.

– Place head into a stereotaxic frame so that position is secured.

– Expose the skull, remove periosteal connective tissue, stop any bleeding using sterile gauze pads and slight pressure. Wipe the skull with a saline-soaked gauze pad and dry the skull with a clean gauze pad.

– Mark electrode locations.

visual cortex: 4 mm lateral to midline and 1 mm anterior to lambda.

reference and ground electrodes: 2 mm anterior to bregma and 2 mm right and left of midline, respectively.

– Using sterile drill, gently drill holes in skull for electrode placement. Additional holes may be included for anchor screws to better secure headset to skull. Be careful to gently drill the holes, especially at the base of the skull. The drill should not puncture the dura or damage the underlying brain tissue.

– Gently screw the electrodes and anchor screws into the drilled holes. Being careful not to extend below the base of the skull and compress brain tissue.

– Using hemostats, snap pin ends of electrodes into designated holes of the Amphenol connector (Fig. 4).

– Mix up dental cement and use to build a skull cap securing the electrode assembly. Smooth the exterior of the headset while the cement is still soft.

– Once the cement has hardened close the wound using sutures or wound clips.

– Remove the animal from the stereotaxic device and allow to recover from anesthesia in a warm and secure chamber. Apply post-surgical analgesia as designated in approved laboratory animal protocols.

– Check impedance of electrode assembly using an impedance meter.

– House animals singly after surgery to prevent cage-mates from damaging electrode assemblies or electrodes.

– Allow approximately 1 week after surgery for recovery and wound healing prior to sensory function testing.

3.1 Visual System Calibration (see Note 4)

The visual monitor used to present stimuli needs to be calibrated so that there is a linear relationship between input signal voltage and screen luminance. We have used just the green gun of a three-gun color video monitor for presenting visual stimuli. This is (1) in order to simplify calibration (only one of the three cathode ray guns needs to be calibrated) and (2) because the green spectral distribution closely matches the color frequency response spectrum of M-cones and rod photoreceptors, the most common photoreceptors in the rat retina.

The voltage-luminance response function (referred to as the gamma function) of most phosphors used in traditional video monitors is typically nonlinear. Therefore, a gamma correction needs to be computed based on the input voltage—screen luminance function established during system calibration. Calibration involves using the computer to generate a steady (unmodulated) spatial square-wave pattern on the screen. The spatial frequency of the stimulus should be low enough (bars wide enough) so that the measurement head of the photometer measures only a single light or dark bar of a square-wave grating stimulus pattern. Alternatively, a non-patterned uniform luminance screen could be used. The calibration involves measuring the luminance at several steps of input signal voltage. Our calibration system interfaces the stimulus computer with the photometer to automatically record the luminance value for each stimulus voltage level presented. Alternatively, the values can be recorded for separate entry. After collecting stimulus voltage-screen luminance values over the range of stimulus levels to be used, a function is fit to the luminance voltage function. Then a mathematical model is generated to linearize the input/output function (23). The resulting adjusted values should provide a linear relationship between input voltage and output screen luminance. The brightness and contrast adjustments of the monitor can be adjusted to achieve an optimal range of stimulation. It typically is preferred to set the contrast adjustment high and then adjust the brightness to the optimal value for mean screen luminance. Determining a linear range of the gamma-corrected input/output function defines the range of useable luminance and contrast values.

3.2 Visual Stimuli

3.2.1 Spatial Properties

- Spatial pattern. There are numerous possible choices. We use a vertical sine-wave grating, which is a pattern of vertical bars with a sinusoidal spatial luminance profile. A spatial sinusoid means that there is a gradual transition between the dark and light parts of the screen which would reflect a sine wave if the head of a point photometer measuring luminance were passed along the screen. This is opposed to a square wave (dark and light bars with sharp edges) which would show the sharp transitions from dark to light and back with a moving photometer head. Square waves contain multiple Fourier components and, due to the center-surround receptive fields of visual neurons and their consequent spatial frequency tuning, stimulate multiple spatial frequency channels. Spatial sine waves, therefore, are considered more selective and interpretable.

- Spatial frequency. Select a spatial frequency (bar width) appropriate for the test species or a range of spatial frequency values. The rat has a peak sensitivity to contrast of about 0.16 cycle/ degree visual angle (cpd), and therefore we have used this spatial frequency for most testing applications.

- Select contrast. In order to construct VEP log-contrast amplitude functions, it is important to test multiple levels of stimulus contrast. The range of contrast levels selected will be limited at the upper end by the highest values of the linear range of the input voltage/output luminance calibration curve. At the lower end, VEPs recorded to low-contrast values (near the perceptual threshold) have low amplitudes and low signal-to-noise ratios. Because of the eventual plotting of VEP amplitudes on linear (amplitude) vs log (contrast) axes, it is beneficial to have log spacing on the contrast values selected.

3.2.2 Temporal Properties

- Mode of temporal modulation. There are several possible modes of temporal modulation. We typically prefer sine-wave temporal modulation (the stimulus transitions are gradual) over square-wave temporal modulation (the stimulus transitions are abrupt) because of the same reason to use sine instead of square-wave spatial profiles. The square-wave stimuli contain higher temporal frequency Fourier components that may complicate interpretation of the results.

- The rate of temporal modulation should be high enough to elicit a semi-sinusoidal response profile. For the rat, this rate is above about 4–5 Hz. This rate yields a strong frequency-double (F2) response amplitude. The temporal modulation rate should not be a direct multiple/divisor of either the line frequency (60 Hz) or the video monitor refresh rate in order to avoid inadvertently signal averaging artifactual electrical noise into the VEP waveform.

3.2.3 Additional Stimulus Parameters

- Luminance level. The overall mean luminance of the pattern should not change during on-off or pattern-reversal modulation. This ensures that the responses reflect changing pattern and not changing luminance (although evoked potentials to changing luminance can also be recorded if that is the goal of the study). We selected a value of about 50 lux for overall mean luminance.

- Pattern adaptation. Visual neurons adapt rapidly to repeated or continuous visual stimuli. This leads to a rapid reduction in the amplitude of VEPs with repeated pattern stimulation. Electrophysiological data are averaged over multiple trials to improve signal-to-noise ratios, yielding an averaged evoked potential. It is recommended to have the subject view mean luminance non-patterned stimuli for adaptation intermixed with stimulus sessions. A general rule of thumb is to have equal amounts of time devoted to adaptation as stimulation. There is a tradeoff between limiting stimulation time to avoid adaptation and increasing stimulation time to provide more trials for signal averaging and improving signal-to-noise levels in the recordings (signal averaging is discussed below).

3.2.4
Electrophysiological
Procedures

Signal epoch length. Evoked potentials from steady-state visual modulation take on a sinusoidal form that is readily analyzed with FFT routines. The frequency resolution of an FFT analysis is inversely proportional to the length of the recording epoch. We find that recording 5 s epochs, which gives frequency resolution of 0.2 Hz, gives sharply defined and unambiguous F2 response components. We typically publish only a 1–2 s portion of the 5 s waveform epoch for clarity.

Signal averaging. The brain has a considerable amount of spontaneous and ongoing electrical activity—generally referred to as the electroencephalogram (EEG). One approach to separate sensory-related activity from the background ongoing EEG is signal averaging. This involves averaging the electrophysiological data in synchrony with the repeated temporal changes (sometimes accomplished via a "trigger pulse") so that random ongoing activity is averaged out and activity that regularly accompanies the stimulus is averaged in. The signal-to-noise ratio is increased as the square root of the number of trails averaged. Others have used Fourier analysis of steady-state responses without signal averaging (3). We, however, have had the best results combining signal averaging with FFT analysis of the averaged waveform. We have typically averaged 20 individual sweeps for a steady-state VEP waveform. The decision on how many sweeps to include is a tradeoff between increased signal-to-noise ratios with higher numbers of sweeps averaged, against increased recording time, increased pattern adaption over time, and limiting the number of different stimulus contrast levels (or other parameter of interest) that can be included within any recording session.

- Noise level. Evoked potentials recorded to zero percent contrast (i.e., a non-modulated, non-patterned, mean luminance screen) are thought to reflect the ongoing electrical activity of the brain independent of visual stimuli, and are considered as "noise." Noise levels can be established by averaging the same number of sweeps and analyzing the data with the same parameters and procedures as for the VEP recording. A noise level value is important in understanding the signal-to-noise level, especially of low-contrast responses.

VEP recording (*see* **Notes 11–16**)

- VEP recording parameters
  - Filter Bandpass (Hz): 0.1–300
  - Sample epoch: 5 s
  - Analog/digital conversion rate: 1 kHz
  - Trials averaged: 20
- Placement relative to screen. A restrained rat (*see* **Notes**) is placed inside the Faraday cage with its eyes 15 cm from the screen. The optics of the rat eye provide a great depth of field of focus such that everything from 15 cm to infinity can be in focus at the same time. The video monitor is placed outside the Faraday cage to minimize electrical interference in the recordings.
- Connection. The headset is connected to the shielded cable leading to the electrophysiological amplifiers. The oscilloscope display of the ongoing EEG is checked for connectivity and signal quality.
- The Faraday cage is covered with a black cloth so that the only light available to the rat comes from the video monitor, which is showing a non-patterned mean luminance adaptation display.
- Adaptation. The rat is allowed a few minutes in front of the screen in a quiet room to calm and adapt to the screen. Most rats quietly watch the screen.
- Recording. The stimulus session is started. Several levels of stimulus contrast are presented in a random order.
- Session monitoring. The operator quietly monitors the rat during the recording session in case eyes are closed or undue amounts of movement artifact are observed. The recording progress is paused, and corrective measures are taken such repositioning the animal or reconnecting the electrode cable as needed.

3.3   Data Analysis

Spectral Analysis Using FFT. Averaged VEP waveforms can be submitted to readily available FFT routines for spectral analysis. We have used published routines programmed into the signal collection and analysis system (23, 24). The spectral analysis provides frequency components across a range of frequencies, including relatively high frequencies which may not be biologically relevant in these recordings. We typically visualize the frequencies to slightly above 60 Hz in order to assure that the recording is not contaminated by 60 Hz (in the US) line frequency artifact. Spectral amplitude is expressed as the square root of spectral power. A prominent spectral peak should be apparent for high-contrast stimuli at twice the rate of temporal modulation (F2), and absent (or nearly so) for the noise level recordings. Higher harmonics (multiples) of F2 may also be apparent in the spectral analysis. Some investigators sum the amplitude at all the harmonic response frequencies. If this approach is selected, then the noise level calculation should also sum noise level recordings at those same frequencies. We have not found an advantage to analyzing the summed harmonics rather than simply using the F2 amplitude.

Contrast Amplitude Functions. The VEP F2 amplitude is plotted on a linear scale as a function of log stimulus contrast. Prior to fitting the VEP amplitude log-contrast linear function, it is important to eliminate VEP recordings that are indistinguishable from or below the recording noise level. We eliminate VEP data that are less than the recording noise level plus three times the noise level standard error prior to fitting F2 amplitude log-contrast functions (17). After elimination of noise/level recordings, a linear regression equation is fit to the amplitude log-contrast function.

Contrast Thresholds. Contrast threshold may be defined as the zero-amplitude intercept of the contrast amplitude function. Alternatively, threshold may be defined as contrast value where the log-contrast amplitude function crosses the VEP recording noise level. Low-contrast noise level amplitude values can distort slope of regression lines fit to log-contrast amplitude data, driving the intercept of the function to the left on the contrast axis. Because contrast is plotted on a log scale, slight deviations in the slope of the function can have a large influence on the derived intercept value and provide artifactual low measures of contrast thresholds.

Contrast Sensitivity. Contrast sensitivity (CS) is defined as the inverse of contrast threshold (CT).

$$CS = 1/CT.$$

Contrast sensitivity functions are typically plotted as the contrast sensitivity value derived for each spatial frequency, over a range of spatial frequencies. Spatial frequency is often plotted on a log scale. Contrast sensitivity functions expressed in this way are typically inverse U-shaped functions with the contrast sensitivity being highest to intermediate spatial frequencies, and lower contrast sensitivity to both lower and higher spatial frequency stimuli (12).

Contrast Gain. The slope of the linear function fit to the F2 amplitude/log-contrast function has been interpreted as the "contrast gain." We fit the log-contrast amplitude function using a linear regression:

$$Y = \beta_0 + \beta_1(X)$$

In which Y is the $F2$ amplitude, X is log contrast, β_0 is the intercept, and β_1 is the slope parameter.

4   Notes

General Notes:

1. The terms "evoked potential" and "evoked response" are synonymous.

2. Rat strain. Albino strains of rodents are not advised for this procedure. They have poor spatial vision and yield low-quality pattern-elicited VEPs (11).

3. We are describing here the electrode materials and configurations used in our laboratory for over two decades. However, recent descriptions are available for alternative electrodes (25). We have not had experience with these devices, but they may provide better performance than the procedures described here.

4. The present description is based on the tube monitor screens used in our laboratory. Monitors based on other technologies may have different operating characteristics.

Problems from surgery:

5. Lost headsets: The headsets may become dislodged for a variety of reasons.
   (a) Poor adhesion of the acrylic due to not completely cleaning and drying the skull prior to application.
   (b) Infection: insufficiently sterile procedures.
   (c) Torque: too much lateral pressure, especially when plugging or unplugging cables for recording.
   (d) Insufficient anchoring (with screws).
   (e) Time: the more time after surgery, the more likely headset loss becomes.

6. Broken electrodes, lost connectivity
   (a) Loose crimping of pin to wire.
   (b) Poor soldering of wire to screw.
   (c) Too much pressure on electrode during surgery, especially when inserting pin to headset connector.

7. Electrodes too deep—focal lesions
   (a) Electrode should not extend below the bottom of the skull as this can damage the cortex.

8. Infection
   (a) Review surgical procedures for cleanliness.
   (b) Keep area surrounding surgical procedures clean and draped.
   (c) Sterilize instruments between procedures.

9. Damage from group housing
   (a) Rats that are group housed will chew on cage-mate's headsets and destroy the electrodes. It is imperative to single-house rats after cranial implantation.

10. Loss of body weight/ill health
    (a) Rats should be followed post-surgery for general health and body weight. It is normal to lose a few grams body weight after surgery. Excessive weight loss is a sign of infection or other poor consequences of surgery. Animals that do not thrive after surgery should be eliminated from the study.

Problems during recordings:

11. Eyes closed. Sometimes rats close their eyes during recordings. This is especially true if they are stressed. It is possible to wait for them to calm down. It may be necessary to adjust their restraints to improve comfort. It is helpful to be calm and confident when handling the rats to lower their initial stress response to the restraint.

12. Unplugged/lost headset. Monitor the EEG on an oscilloscope during all recording sessions. The signal from intact electrodes and a connected animal will be a familiar EEG pattern. Noise in the recording is immediately apparent as large 60 Hz patterns due to induction and amplification of the building electrical activity.

13. Escape restraint. Rats are motivated escape restraint. We have found that plastic de-capitation cones work well as disposable single-use restrainers. The narrow end of the plastic cone is cut open to expose the head, (including eyes and ears), prior to placing the rat into the cone. The rat's toes can be covered with small pieces of tape to prevent clawing their way out of the device. The rat is placed into the cone and the area around the neck is taped to be snug, but not to impair breathing. A rectal temperature probe is inserted and taped in place. The rear end of the cone is taped shut with the rat's rear legs facing backwards. The rat in the cone is placed on a specially designed holder made from a half-longitudinal section of PVC plumbing pipe, mounted on an adjustable stand.

14. Excessive movement/artifact rejections. Some rats display movements, such a struggling against the restraint or chewing during recording sessions which cause large electrical artifacts in the EEG recordings. The data acquisition program should include artifact rejection capability such that signals which exceed the digital-analog converter input window, or a designated percentage of it, are rejected prior to inclusion in the average evoked potential. Calming the rat, reconfiguring the

restraint, or waiting can reduce movements and their associated artifacts.

15. 60 cycle noise. As for any neurophysiological recording, care must be taken to eliminate 60 Hz noise. The methods to do this are well-described elsewhere, and involve use of a grounded Faraday cage, proper grounding of equipment, avoiding ground loops and shielding of electrode cables. The use of 60 Hz notch filters is discouraged as they can distort the signal and are a poor substitute for proper grounding and elimination of other sources of artifacts.

16. Chromodacryorrhea. Some rats exhibit a red porphyrin containing lacrimal discharge around their eyes referred to as chromodacryorrhea. This can indicate a stress response. Because it can interfere with vision if covering the eye, chromodacryorrhea should be gently removed with a cotton swab while recordings are paused. Measures to reduce stress are advisable.

Acknowledgments

The author thanks David Herr, Jordi Llorens, and an anonymous reviewer for helpful comments on an earlier version of the manuscript. The author thanks Mary Gilbert for initiating the experiments depicted in Fig. 8. The author also thanks Garyn Jung for assistance and Chuck Gaul for photography in creating Fig. 4. The surgical and electrophysiological procedures were based on those originally developed by Robert S. Dyer with technical innovations by Mark Bercegeay.

This document has been subjected to review by the National Health and Environmental Effects Research Laboratory and approved for publication. Approval does not signify that the contents reflect the views of the Agency, nor does mention of trade names or commercial products constitute endorsement or recommendation for use.

References

1. Herr DW, Boyes WK (1995) Chapter 9—Electrophysiological analysis of complex brain systems: sensory-evoked potentials and their generators. In: Chang LW, Slikker W (eds) Neurotoxicology. Academic Press, San Diego, pp 205–221. https://doi.org/10.1016/B978-012168055-8/50013-3

2. Norcia AM, Appelbaum LG, Ales JM, Cottereau BR, Rossion B (2015) The steady-state visual evoked potential in vision research: a review. J Vis 15:4

3. Regan DEDE (1989) Human brain electrophysiology : evoked potentials and evoked magnetic fields in science and medicine. Elsevier, New York

4. Kothari R, Bokariya P, Singh S, Singh R, Comprehensive A (2016) Review on methodologies employed for visual evoked potentials. Scientifica (Cairo) 2016:9852194

5. Creel DJ (2019) Visually evoked potentials. Handb Clin Neurol 160:501–522

6. Tseng HC et al (2015) Visual impairment in an optineurin mouse model of primary open-angle glaucoma. Neurobiol Aging 36:2201–2212

7. Demyanenko GP et al (2011) NrCAM deletion causes topographic mistargeting of thalamo-cortical axons to the visual cortex and disrupts visual acuity. J Neurosci 31:1545–1558

8. Strain GM, Tedford BL, Gill MS (2006) Brainstem auditory evoked potentials and flash visual evoked potentials in Vietnamese miniature pot-bellied pigs. Res Vet Sci 80:91–95

9. Mitzdorf U (1987) Properties of the evoked potential generators: current source-density analysis of visually evoked potentials in the cat cortex. Int J Neurosci 33:33–59

10. Schroeder CE, Tenke CE, Givre SJ, Arezzo JC, Vaughan HG Jr (1991) Striate cortical contribution to the surface-recorded pattern-reversal VEP in the alert monkey. Vis Res 31:1143–1157

11. Boyes WK, Dyer RS (1983) Pattern reversal visual evoked potentials in awake rats. Brain Res Bull 10:817–823

12. Boyes WK (1994) Rat and human sensory evoked potentials and the predictability of human neurotoxicity from rat data. Neurotoxicology 15:569–578

13. Birch D, Jacobs GH (1979) Spatial contrast sensitivity in albino and pigmented rats. Vis Res 19:933–937

14. Silveira LC, Heywood CA, Cowey A (1987) Contrast sensitivity and visual acuity of the pigmented rat determined electrophysiologically. Vis Res 27:1719–1731

15. Bobak P, Bodis-Wollner I, Marx MS (1988) Cortical contrast gain control in human spatial vision. J Physiol 405:421–437

16. Bonds AB (1991) Temporal dynamics of contrast gain in single cells of the cat striate cortex. Vis Neurosci 6:239–255

17. Boyes WK, Degn L, George BJ, Gilbert ME (2018) Moderate perinatal thyroid hormone insufficiency alters visual system function in adult rats. Neurotoxicology 67:73–83

18. Boyes WK et al (2003) Dose-based duration adjustments for the effects of inhaled trichloroethylene on rat visual function. Toxicol Sci 76:121–130

19. Boyes WK et al (2014) Neurophysiological assessment of auditory, peripheral nerve, somatosensory, and visual system functions after developmental exposure to ethanol vapors. Neurotoxicol Teratol 43:1–10

20. Boyes WK et al (2005) Momentary brain concentration of trichloroethylene predicts the effects on rat visual function. Toxicol Sci 87:187–196

21. Boyes WK et al (2016) Toluene inhalation exposure for 13 weeks causes persistent changes in electroretinograms of long-Evans rats. Neurotoxicology 53:257–270

22. Herr DW et al (2016) Neurophysiological assessment of auditory, peripheral nerve, somatosensory, and visual system function after developmental exposure to gasoline, E15, and E85 vapors. Neurotoxicol Teratol 54:78–88

23. Hamm CW, Ali JS, Herr DW (2000) A system for simultaneous multiple subject, multiple stimulus modality, and multiple channel collection and analysis of sensory evoked potentials. J Neurosci Methods 102:95–108

24. Bergland GD, Dolan MT (1979) Fast Fourier transform algorithms. In: Weinstein CJ (ed) Programs for digital signal processing. John Wiley & Sons, Inc, New York, pp 1.2-1–1.2-18

25. Tian L et al (2019) Large-area MRI-compatible epidermal electronic interfaces for prosthetic control and cognitive monitoring. Nat Biomed Eng 3:194–205

Chapter 7

Electrophysiological Assessments in Peripheral Nerves and Spinal Cord in Rodent Models of Chemotherapy-Induced Painful Peripheral Neuropathy

Susanna Park, Cynthia L. Renn, Justin G. Lees, Susan G. Dorsey, Guido Cavaletti, and Valentina A. Carozzi

Abstract

Chemotherapy-induced peripheral neuropathy (CIPN) is a severe side effect related to anticancer treatment, typically characterized by sensory symptoms including numbness, tingling in the distal extremities and neurophysiological impairments. CIPN is often painful, which is identified by adding a second "P" to the acronym. The incidence of CIPN is variable depending on the drug, pre-existing neuropathy, and clinical history, but generally increases with the cumulative dose and can also persist after treatment discontinuation. In the last 30 years, many rodent models of CIPN have been developed reproducing the clinical features of the pathology, useful to study the mechanisms of pathogenesis and test neuroprotective strategies.

In this chapter, we will focus our attention on sensitive and reproducible methods to study the pathophysiology of chemotherapy-induced painful peripheral neuropathy (CIPPN), in animal models. In particular, we describe the techniques to record nerve conduction velocity and nerve excitability parameters in peripheral nerves and the electrical activity of wide dynamic range neurons of the dorsal horn of the spinal cord in mice, as parameters of evaluation of nerve function and painful neuronal sensitization, respectively. Our intent is to provide the reader with guidelines on how to prepare and manage the animals according to the 3Rs (Reduction, Refinement, and Replacement) principles, how to record neuronal activity and analyze resulting data and describe common technical problems and appropriate solutions. These protocols can also be useful to study peripheral nerve damage and pain of other origins, such as traumatic injury, inherited, or acquired neuropathies.

Key words Chemotherapy, Neuropathic pain, Peripheral neuropathy, Electrophysiology, Peripheral nerves, Wide dynamic range neurons, Spinal cord

---

Susanna Park and Cynthia L. Renn equally contributed to this work.

Jordi Llorens and Marta Barenys (eds.), *Experimental Neurotoxicology Methods*, Neuromethods, vol. 172,
https://doi.org/10.1007/978-1-0716-1637-6_7, © Springer Science+Business Media, LLC, part of Springer Nature 2021

1   Introduction

1.1   Clinical Features of Chemotherapy-Induced Painful Peripheral Neuropathy (CIPPN)

Chemotherapy-induced peripheral neuropathy (CIPN) is a prominent side effect of cancer treatment, affecting patients treated with commonly used chemotherapies. Chemotherapy classes including taxanes, platinum-based agents, vinca alkaloids, thalidomide, and bortezomib analogs are associated with the development of CIPN. CIPN typically produces sensory symptoms, most prominently including numbness and tingling in the distal extremities of the hands and feet and can include pain (CIPPN) [1]. In severe cases, these symptoms produce functional disability, leading to difficulty with walking, balance, fine motor skills, and ultimately activities of daily life [2]. Severe CIPN often leads to dose reduction or premature cessation of treatment, which may affect long-term outcomes. Further, CIPN can produce long-lasting symptoms, leading to reduced quality of life in cancer survivors [3]. There remains no neuroprotective therapy to reverse peripheral nerve damage due to CIPN.

The presentation of CIPN can vary between chemotherapies and can include motor or autonomic nerve involvement in addition to sensory involvement [1]. Motor nerve involvement can produce weakness while autonomic nerve involvement can include gut dysfunction or orthostatic hypotension. The most prominent electrophysiological finding on nerve conduction studies (NCS) is the reduction or loss of sensory compound action potentials [4], highlighting an axonal, sensory predominant neuropathy with most chemotherapy types. While the most common reported sensory symptom of CIPN is tingling and numbness, depending on the agent, neuropathic pain can also occur in 25–40% of patients [5, 6]. Neuropathic pain in CIPPN, characterized by burning sensations, may have an additional adverse effect on quality of life. Typically, the incidence and severity of CIPPN increases with increasing cumulative dose of neurotoxic chemotherapy [2]. However, some chemotherapies are also associated with acute neurotoxic syndromes—most notably oxaliplatin, which produces acute cold-triggered tingling, dysesthesia, and cramps immediately following infusion [7]. However, there is significant variability in clinical expression and severity between individuals, suggesting that patient-specific or pharmacogenetic risk factors are also important [8].

1.2 Animal Models and Methods of Chemotherapy-Induced Painful Peripheral Neuropathy (CIPPN) Investigation

1.2.1 Animal Models of CIPPN

Rodent models have been the most commonly utilized experimental models for the study of CIPPN to date. Many preclinical rodent (both rat and mouse) models [9–18] have been established during the last 30 years faithfully mimicking the clinical features of CIPPN.

Because the majority of neurotoxic chemotherapy drugs do not cross the blood–brain barrier, most preclinical studies have been focused on the study of the function and structure of peripheral nerves and dorsal root ganglia (DRG). Direct toxic effects of chemotherapy drugs have been observed in peripheral axons, in primary afferent sensory neurons, and in peripheral support cells such as satellite cells in the DRG and the Schwann cells in peripheral nerves. Preclinical animal studies showed that chemotherapy-induced cellular abnormalities in both the nerve and glial cells lead to structural damage, to alteration in neuronal-glial cross talk, and finally to the loss of peripheral nerve function. The specific mechanisms producing nerve degeneration vary in relation to the different mechanism of action of each chemotherapy drug. Moreover, animal studies have shown that chemotherapy-induced neurotoxic damage in the peripheral sensory fibers (A-Beta, A-Delta, and C fibers) is not peripherally confined but is able to induce indirect alterations of transmission through the somatosensory system and spinal cord [9, 10, 12, 13]. This generates an excess of spinal neuron excitability, which coupled with a decrease of inhibition, results in spinal hyper-excitability, increased mechanical and/or thermal sensitization typical of neuropathic pain conditions.

As a general consideration, acute models of CIPPN (those in which the chemotherapy drugs are injected only once or few consecutive times) faithfully reproduce the early and transient painful symptoms after chemotherapy treatment (e.g., the cold hyperalgesia reported by patients few hours after oxaliplatin injection). Chronic models of CIPPN (those in which the chemotherapy drugs are repeatedly injected for several weeks) are characterized by established peripheral nerve lesions that underlie the typical features of painful peripheral neuropathy (neurophysiological abnormalities, decrease in intra-epidermal nerve fiber density, neuropathological alterations into peripheral nerves, DRG, spinal roots and mechanical allodynia, hyperalgesia or hypoesthesia, thermal hyper- or hypoalgesia). These models are characterized by ectopic discharges and spontaneous action-potential firing in primary sensory afferents, which accordingly produce mechanical and/or thermal sensitization in the central nervous system. In recent decades, many mouse and rat models of acute and chronic CIPPN have been developed and phenotypically characterized [9–28].

1.2.2 *Methods of Experimental CIPPN Investigation*

In addition to the establishment of reliable rodent models of CIPPN, the setup of sensitive and reproducible methods of investigation of neurotoxicity and pain are mandatory for the success of an experimental paradigm.

Morphological Methods

Light and electron microscopy analysis are useful tools to investigate the presence and the subcellular origin of the structural damage induced by chemotherapy on its peripheral target sites (DRG, peripheral nerve fibers). Briefly, most chemotherapy drugs produce axonal degeneration with occasional additional damage to myelin, as well as alterations in the structure of DRG sensory neurons with organelle vacuolization and cytoplasmic dark inclusions. In addition, satellite cells can be damaged and increased in number. Sensory neurons can also undergo atrophy or hypertrophy depending on the chemotherapy drug used. Similarly, the density of peripheral nerve fibers as well as the intra-epidermal nerve fibers density can be reduced [10, 14, 15]. Apoptotic and/or degenerative processes induced by chemotherapy can be identified via staining (i.e., TUNEL, FluoroJade) as well as immune-histochemical labeling against specific markers (together with molecular biology and biochemical assays) to help in the identification of the cellular pathways leading to neurotoxicity [12, 15].

Electrophysiological Methods: Nerve Electrophysiology

Nerve electrophysiological methods are minimally invasive techniques which enable assessment of peripheral nerve function in situ. Nerve conduction studies (NCS) are widely utilized in the clinical neurology setting as the gold standard technique to assess peripheral neuropathy [29]. NCS assess the properties of the fastest and largest conducting axons but are insensitive to small, unmyelinated axonal dysfunction. The major components of NCS include assessment of the size of the compound action potential (amplitude) and assessment of the speed of conduction (nerve conduction velocity NCV). These parameters can be collected from different peripheral nerves, stimulating directly through the skin or using needle electrodes. Compound muscle action potentials (CMAPs) can be recorded from innervated muscles following electrical stimulation of a motor nerve. Similarly, compound sensory action potentials (CSAPs) can be recorded following electrical stimulation of sensory fibers from a site within the distribution of the nerve [30]. NCS can reveal evidence of axonal damage or demyelination [31]. Broadly, reduction in compound action potential amplitudes can demonstrate the development of axonal neuropathy and changes in conduction velocity can indicate demyelination. Conduction slowing can be demonstrated via slowed conduction velocity, prolonged latency, or increased temporal dispersion. Axonal damage can be characterized by decreased distal amplitudes without prominent conduction slowing although mild conduction velocity slowing can occur in moderate to severe polyneuropathy [32]. Importantly, very similar or analogous techniques can also be utilized in animal models of CIPPN, which provides an important way to compare disease models to the clinical setting.

While NCS examine the amplitude and conduction velocity of the largest conducting fibers, nerve excitability studies utilize patterns of stimulation to examine modulation of excitability in response to impulse conduction. These patterns have been linked to ion channel function and membrane potential [33, 34]. Nerve excitability studies have also been established in animal models, examining both sensory and motor nerve excitability in rats and mice [35–37] and accordingly can be compared between animal models and the clinical setting.

Neurophysiological studies can be undertaken in sensory, motor, or mixed nerves in rodent models. Motor nerve conduction studies are often undertaken in the sciatic nerve, with the stimulating electrodes at the sciatic notch and the recording from the foot or leg [38]. To generate conduction velocity, a second, more proximal stimulation site is selected. Sensory nerve conduction studies are often undertaken in the tail in rodent models, stimulating at the base of the tail and recording more distally along the caudal nerve [39]. However, there are many nerves and recording sites which have been utilized for neurophysiological recordings across different animal models of peripheral neuropathy. In the animal model setting, it is important to note that there is some overlap between these phenomena and the time required for the development of axonal loss means that this is not always demonstrated in models. It is often more difficult to reliably measure action potential amplitude compared to NCV, so it can be difficult to dissociate between a primary effect on myelin compared to axonal damage [40].

Electrophysiological Methods: Spinal Cord Electrophysiology

As mentioned previously, an increased activity in the peripheral sensory neurons leads to an increased activity in the spinal dorsal horn neurons, which then translates to an increased neuronal activity in the higher centers of the CNS [41]. There are two types of neurons in the spinal dorsal horn that transmit nociceptive information. One type is represented by the nociceptive-specific neurons that respond only to noxious stimuli in the tissue-damaging range of intensity and will remain silent in response to innocuous stimuli [42]. The second type is represented by the wide dynamic range (WDR) neurons that are located in the deep layers of the spinal dorsal horn, predominantly lamina V [43, 44]. The WDR neurons were first identified by Mendell in 1966 [45] and later were well characterized in the 1980s [46–48]. Unlike the nociceptive-specific neurons that have a very high threshold of activation, the WDR neurons are responsive to innocuous low-intensity stimuli from A-Beta fibers as well as to noxious high-intensity stimuli from A-Delta fibers [43]. WDR neurons are also activated by stimuli from intrinsic interneurons that transmit signals from C-fiber terminals in laminae I and II [43]. In a persistent pain state, the ongoing nociceptive stimuli from primary afferent neurons lead to the development of wind-up, which induces hyperexcitability of the WDR neurons, neuroplasticity, and eventually central sensitization [44]. When WDR neurons become hyperexcitable, they produce an exaggerated response to innocuous stimuli from A-Beta fibers. These stimuli are then perceived as painful rather than innocuous, which is the basis for the phenomenon of allodynia that is associated with neuropathic pain [43, 44].

WDR neurons represent a good target for electrophysiological examination of whether central changes have occurred in response to persistent pain, as in CIPPN [49, 50]. In the absence of central neuronal plasticity, WDR neurons generate action potentials in a graded response to varying stimulus intensity [43]. Thus, if changes in the function of primary afferent fibers are detected, but the stimulus-response of WDR neurons remains within the normal range, then it is not likely that central neuronal plasticity or central sensitization have developed [43]. However, if the WDR evoked response and after discharge to innocuous stimuli is significantly increased, then that is evidence that neuronal plasticity and possible central sensitization have developed [51]. Electrophysiological recording of WDR neurons is an excellent tool for studying the neurophysiological development of CIPPN since changes in the response properties of WDR neurons persist after a painful stimulus similar to the persistence of the psychophysical ratings of pain intensity and unpleasantness [42]. Also, given that the response properties of WDR neurons are similar across a variety of mammalian species [42], recording the response of WDR neurons in the development of CIPPN in rodents can be an indication for what may occur in humans.

2   Protocols Described

There are many experimental approaches to investigate CIPPN. This chapter describes the protocols for the recording of NCV and nerve excitability parameters in peripheral nerves and the electrical activity of WDR neurons of the spinal cord of mice. Other experimental approaches are beyond the scope of this chapter but are addressed in detail by Bruna and collaborators [52] and Hoke and collaborators [38].

All the procedures described in this chapter comply with local governmental regulation concerning the care of laboratory animals, in vivo research protocols and the 3Rs (Reduction, Refinement, and Replacement) principles.

Although different instrumentation is described in the two protocols, some of the equipment can be employed for both

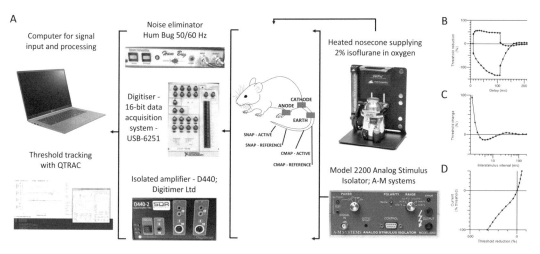

Fig. 1 (**a**) Equipment and materials for nerve excitability experiments. Panels (**b–d**) depict example traces of nerve excitability recordings from the mouse caudal nerve, illustrating (**b**) threshold electrotonus, (**c**) recovery cycle and (**d**) current-threshold relationship waveforms. Excitability traces provided by PG Makker

protocols (including the anesthesia system, amplifier, noise eliminator). Similarly, supplies including alligator clips, connection cables, aluminum foil, syringes, hemostatic sponges, cotton swabs, and animal shavers can be useful for both procedures.

2.1 Materials and Methods for Electrophysiology in Peripheral Nerves

In this section, we describe the procedures to record NCV and nerve excitability parameters from mice [25, 37]. The instrumentation required is illustrated in Fig. 1.

2.1.1 Materials

Preparation of the animal

- Isoflurane 2% anesthetic system with vaporizer and oxygen supply equipped with an anesthetizing box and nose-cone mask fixed to heating block.
- Homeothermic blanket system adjusted by rectal probe to maintain core body temperature at 37 °C.

Electrophysiological recordings

- QtracS software (Copyright Institute of Neurology, UCL, available from Digitimer Ltd.) for data acquisition.
- Isolated amplifier—D440-2; Digitimer Ltd. (Welwyn Garden City, UK).
- Digitizer—16-bit data acquisition system—USB-6251; National instruments (Austin, Texas).
- Isolated stimulator Model 2200 Analog Stimulus Isolator; A-M systems (Carlsborg, Washington).

- Noise eliminator 50/60 Hz, eliminate 50/60 Hz noise and harmonics without filtering (KF Technologies).
- Disposable non-polarizable Ag/AgCl ring electrodes (The Electrode Store; Buckley, Washington).
- Platinum needle electrodes (Natus; San Carlos, California).

2.1.2 *Methods*

Preparing the mouse

1. Mouse (*see* **Note 1**) is removed from the holding cage and placed into anesthetizing box with intake of 2% isoflurane in oxygen. Wait until mouse is completely unconscious (~4 min).

2. Remove mouse from anesthetizing box and transfer to nose-cone apparatus supplied with intake of 2% isoflurane in oxygen.

3. Used toe-pinch test to ensure that mouse is completely unconscious.

4. Ensure mouse is in a supine position on top of a heating mat covered with paper toweling. Heating mat is temperature adjusted by rectal probe which is inserted into mouse rectum for duration of procedure maintaining core body temperature at 37 °C.

5. Immobilize mouse by placing a folded paper towel over upper body torso and secure to underlying mat with tape.

Electrophysiological study setup

1. To stimulate the caudal nerve, Ag/AgCl ring electrodes (*see* **Note 2**) connected to the stimulator are attached. The anode electrode is placed on the left rear limb around the ankle and the cathode electrode is placed around the base of the tail.

2. To record from the caudal nerve, platinum needle recording electrodes connected to amplifier are fitted to record compound muscle action potentials (CMAPs) and sensory nerve action potentials (SNAPs):

   (a) For CMAP recording the active electrode is inserted into the tail muscle 20–30 mm distal to the cathode. The reference electrode is inserted into the tail 8 mm distal to the active electrode.

   (b) For antidromic SNAP recording the active electrode is inserted into the tail-skin 65–70 mm distal to the cathode. The reference electrode is inserted into the tail-skin 8 mm distal to the active electrode.

3. For both CMAP and SNAP recording a grounded ring electrode is placed around the tail distal to the cathode and proximal to the active recording electrode.

Recording of action potential amplitude and conduction velocity (NCV)

1. The optimal stimulation parameters should be determined to generate a maximal response, for example, 10 mA stimulus intensity, 0.75 s frequency of stimulation, and 0.04 ms stimulus duration from Renn et al. 2011 [13].

2. To record the maximal compound action potential amplitude, increase the stimulus intensity until the amplitude of the compound action potential does not further increase. In order to ensure that supramaximal stimulation has been reached, further increase the stimulus intensity by 20%, ensuring that there is no further increase in the compound action potential amplitude. Record the values as the maximal CMAP or SNAP as appropriate.

3. Caudal nerve NCV can be recorded and calculated as the ratio of the distance between the stimulating and recording electrodes and the latency between the stimulus artifact and the onset of the action potential (Fig. 2). Supramaximal stimulation, as above, should be utilized to determine NCV. The NCV is calculated in m/s.

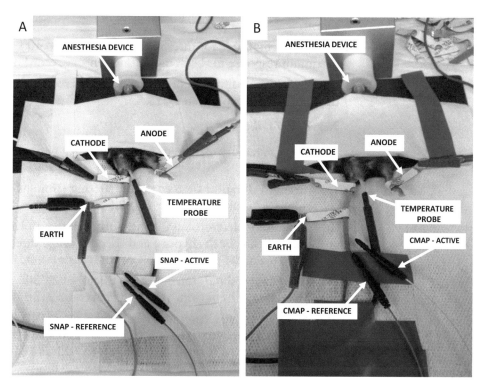

Fig. 2 Experimental setup for nerve excitability experiments for (**a**) sensory nerves and (**b**) motor nerves. Photographs taken by PG Makker

It should be noted that for the calculation of sensory NCV, only one stimulating site is required. However, for determination of motor NCV, two stimulating sites are required, one distal and one proximal. Distal motor latency is subtracted from the proximal motor latency to calculate the nerve conduction time between sites. In conjunction with the distance between the two sites, this can be used to determine NCV for motor nerves.

Initiating recording of excitability

1. Excitability studies are undertaken using a constant-current stimulator, with computer software QTRAC (Institute of Neurology) used to control the stimulation, recording and for processing data.

2. Multiple excitability measurements can be recorded with specialized semi-automated protocols, (e.g., the TROND protocol, developed to record multiple excitability measurements via threshold tracking; Kiernan et al. 2019 [33]).

3. Initially, the stimulus current (stimulus duration 1 ms motor, 0.5 ms sensory) is manually increased until a maximal response is obtained.

4. A stimulus-response curve is automatically generated, utilizing the pre-set maximum current level and incrementally decreasing the stimulus current by 2% steps until the response is absent.

5. The target amplitude for threshold tracking is set to 40% of the maximum response for both motor and sensory recordings.

6. For sensory recordings, additional averaging is required, using the small sensory option in QTracS.

Nerve excitability protocols

1. Multiple excitability parameters are semi-automatically collected, including:

   (a) Strength duration time constant: altering stimulus durations (0.2–1 ms for motor; 0.1–0.5 for sensory).

   (b) Threshold electrotonus: using subthreshold polarizing currents in both hyper- and depolarizing directions ($\pm20\%$, $\pm40\%$, -70%, -100% of control threshold).

   (c) Current-threshold relationship: Effect of a 200 msec polarizing current ($+50\%$ to -100% of control threshold).

   (d) Recovery cycle: Changing interstimulus intervals between an initial supramaximal conditioning stimulus and the target potential (from 1.3 to 200 msec; *see* **Note 3**).

2.1.3 Data Analysis	Nerve excitability study data can be analyzed and visualized with the QTracP program utilizing multiple excitability measure files (MEM).
2.1.4 Notes	1. Both male [53] and female [54] rodents have been shown to demonstrate changes in NCV due to chemotherapy treatment. However, few studies have included both sexes to directly compare the sexes using the same chemotherapy dosage regimen. However, the strain of mouse does appear to make a difference in the degree of electrophysiological changes with chemotherapy treatment—with DBA2J mice showing significant alterations following paclitaxel treatment but C57BL/6 mice remaining unaffected [55]. Further, most of these experiments are performed in young adult rodents. The obvious advantage of using more mature mice is that the tail is longer and therefore when recording from caudal nerve there is more distance to separate the stimulation location and the sensory nerve recording location giving better separation from the stimulation artifact [25].
	2. In the majority of studies, needle electrodes are used for both stimulation and recording of CMAP and SNAP [55]. An advancement for stimulating the caudal muscle and nerve is the incorporation of ring electrodes that create an electrical field around the entire section of the tail being stimulated and therefore provide and more even and stronger stimulation for distal recording [25].
	3. In the recovery cycle when the interstimulus intervals are close together (<15 ms sensory, <30 ms motor), the conditioning stimulus-response must be subtracted from the target response as it temporally contaminates the target response. When sensory and motor recordings are occurring simultaneously, the CMAP response may contaminate the SNAP response at short interstimulus intervals (<5 ms). In this scenario, the supramaximal conditioning stimulus is reduced from 170% to 150% [25].
2.2 Materials and Methods for Electrophysiology in the Dorsal Horn of Spinal Cord	In this protocol, we describe the basic procedures to record the electrical activity of WDR neurons in the dorsal horn of the spinal cord of deeply anesthetized mice. The aim of this analysis is to investigate the effects on the central nervous system of neuropathic pain of peripheral origin. Specifically, the baseline electrical activity of these neurons, as well as the activity in response to the mechanical and thermal stimulation of the hind paw is recorded in the dorsal horn. The procedure here described for mice [9, 10, 12, 13] is applicable also to rats with some specific precautions, properly highlighted in the following description (*see* **Note 1**).

Preparation of the animal and surgery

All the procedure of animal preparation and recordings have to be performed under a Faraday chamber, equipped with a stable workstation protected from vibration and electric interference.

- Trinocular stereomicroscope equipped with $8\times$–$50\times$ magnification (Vision Engineering).

- Spinal Robot stereotaxic system, Dual including Retractors, model 986C, with Vertebrae Clamps to support the animal during the surgery and recordings and to support the recording microelectrodes (Kopf Instruments).

- Isoflurane 1–3% anesthetic system (Ugo Basile Instruments) similar to that previously described. In addition, pentobarbital is intraperitoneally injected at a concentration of 80 mg/Kg.

- Homeothermic blanket system, similar to that previously described.

- Agar 5% solved in saline solution.

- Surgical instruments (scalpel, dissection scissor, iris scissor, needle holder, silk suture (3/0, 5/0), silk needle (0–2), 0.3 mm angled micro-forceps jeweller type (two are needed), 0.4 mm angled micro-forceps jeweller type, 1×2 teeth (one is enough), self-retaining retractor, curved bone rongeur with fine tips (Aesculap Surgical Instruments). Supplementary instruments can be required for the surgery in the rat (*see* **Note 1**).

Electrophysiological recordings

For a better understanding of the electrophysiological recording system, the instrumentation required is illustrated in Fig. 3 as well as described here:

- SciWorks Experimenter Premier Acquisition software with Datawave 16 channel, 1 MHz 16 Bit USB Acquisition instrument for data capture and data analysis (SciWorks Data Acquisition Software), (Datawave Technologies), equipped with a computer.

- Audiomonitor (KF Technologies).

- Oscilloscope 400 MHz (Tektronics).

- Noise eliminator (KF Technologies), similar to that previously described.

- AC differential amplifier, similar to that previously described, with Active Probe (World Precision Instruments).

- Three-axis micromanipulator equipped with an interactive control of the motorized stereotaxic apparatus (Ugo Basile Instruments).

- MouseOx® Plus Small Animal Pulse Oximeter for a noninvasive physiological monitor of the animal (Starr Life Sciences Corp.).

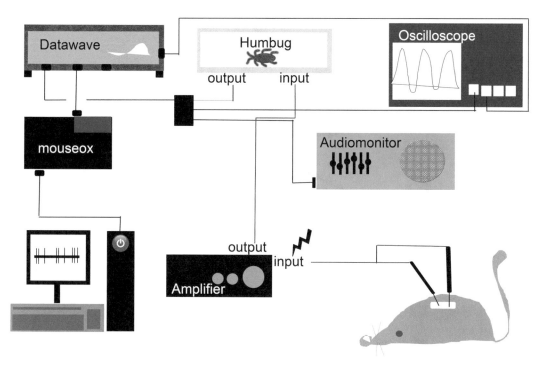

Fig. 3 Instruments and connections for spinal electrophysiological recordings

- Tungsten microelectrodes with a fine tip (<1.0 μm), (FHC Neural Microtargeting Worldwide).

  Equipment for hind paw stimulation

- Hind paw bone wax support to minimize interfering movements of the paw.

- Von Fray hairs (0.4, 1, and 4 for mice; 4, 15, 26 for rats, *see* **Note 1**) for mechanical stimulation.

- Sable brush for mechanical stimulation.

- Wooden probe (10 mm diameter) for the "punch" mechanical stimulation.

- Surgical forceps for the "pinch" mechanical stimulation.

- "Hamilton" syringe filled with acetone 100% for thermal stimulation.

2.2.2 Methods

Preparing the mouse

1. The mouse is prepared as previously described (see methods 1–4 of electrophysiology in the nerve). Additionally, fix the collar of the "mouseOx" instrument to monitor the blood pressure and the heartbeat of the mouse.

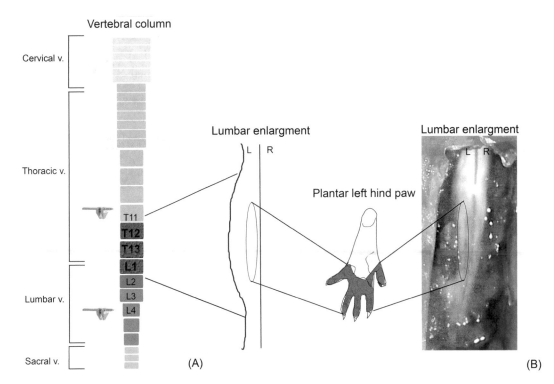

Fig. 4 (a) Schematic representation of the vertebral column of the mouse: vertebral clamps are fixed at vertebrae T11 and L3. Vertebrae T13, L1 and L2 are removed to expose the lumbar enlargement. **(b)**. Illustration and live picture of the optimal region of the spinal cord for positioning the microelectrode B and the corresponding receptive field on the plantar surface of the ipsilateral hind paw

Laminectomy

1. Make a rostro-caudal incision with the scalpel through the skin of the animal back, from the base of the neck to a position rostral to the hips. Then make parallel incisions through the connective tissue and muscles on either side of the vertebral column from T13 to L6 to allow the vertebral clamps to be fixed to the vertebrae.

2. Position two vertebral clamps at vertebrae T11 and L3 (Fig. 4). Use a cotton swab with saline solution to keep the tissues wet. Take care not to crush the vertebrae. *Trick:* pull back the clamps on the frame to keep the animal slightly raised to help him to breathe normally and avoid interfering signal of breathing during the recordings. Be careful to keep the animal low enough to be warmed on the heat pad.

3. Under the stereomicroscope: cut and scrape the connective tissues and muscles overlying the vertebrae with dissection forceps and angled micro-forceps until the vertebrae and their joints are clearly visible. *Trick:* leave the lateral part of the spinal

column intact to avoid mechanical damage of DRG and spinal nerves.

4. Perform a laminectomy by removing the lamina bone tissues between vertebral T13 and L2 segments to have access to the lumbar portion of the spinal cord (Fig. 4). *Trick:* to avoid damage to the spinal cord, insert the tips of the bone rongeur at the base of each vertebra (up to the transverse processes) and make a light and constant pressure to weaken the spinous processes on its median axis. Then, by using two angled micro-forceps, remove the dorsal part of the vertebrae. The excessive bleeding can be contained by using small pieces of cotton swab or hemostatic sponge. Be careful not to let the exposed portion of spinal cord dry out by applying some drops of saline solution.

5. In order to make the system more stable, put four ties of silk suture in the skin at four lateral point of the animal body and fix them to the vertebral clamps at both sides. Place a self-retaining retractor and fill the space with liquid warm agar (38–40 °C) creating a "pool" and let it solidify (Fig. 5).

6. By using a blade, cut a rectangular portion out of the solid agar to expose the spinal segment (Fig. 5).

7. If still intact, remove the Dura Mater from the dorsal surface of the spinal cord by using two dedicated sharp micro-forceps. Drop saline solution in the well above the exposed portion of the spinal cord (Fig. 5).

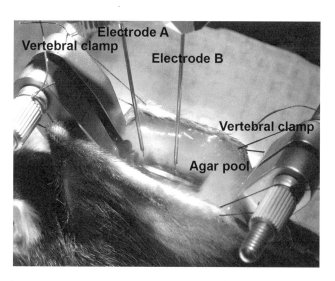

Fig. 5 The mouse is ready for the recording session: vertebral clamps are fixed and the spinal segment is exposed. A pool of solid agar keeps the system more stable. A rectangular piece of agar is displaced to have access to the lumbar enlargement and the space is filled with saline solution. Dura mater is gently removed, and tungsten microelectrodes positioned

8. Reduce the level of isoflurane 3–1% or, alternatively, to the minimum level required to keep the animal unconscious. Check the physiologic condition of the animal at the end of the surgery step and pull out the mouseOx collar to avoid interference during recordings.

9. Place the left hind paw in a natural position with the dorsal surface facing down and the paw plantar surface upward exposing the receptive field to the stimulation. *Trick:* fixing the paw to a bone wax can limit the movements of the paw that originate during the stimulation of the receptive field under the hind paw. The activation of proprioceptive cells can be disturbing in identifying the activity of WDR neurons in the dorsal horn.

Setup of the recording system and microelectrodes positioning

1. Place a silver-grounded wire into the saline pool in a position away from the dorsal horn.

2. As reproduced in Fig. 3, the acquisition system receives a differential electrical signal between the "reference" microelectrode (microelectrode A) and the "recording" microelectrode (microelectrode B) allowing a decrease in the background interference. Then, the signal is amplified by an AC amplifier, filtered, displayed on an oscilloscope and made audible by an audiomonitor. The quantification of the signal is made by a data acquisition system and a data analysis software.

3. Position the microelectrodes and start to search for dorsal horn neurons: under the stereomicroscope, carefully move both the tungsten microelectrodes until the surface of the dorsal horn is reached. Place the microelectrode B on the surface of the spinal region of the lumbar enlargement close to the central blood vessel, as reported in Fig. 5; then vertically advance it by using the micromanipulator until a depth of 350 (or 700 for rats, *see* **Note 1**). Manually insert the tip of the microelectrode A on the opposite side respect to microelectrode B. *Trick:* ensure the point of insertion of microelectrode A is devoid of spontaneous electrical activity and its corresponding receptive field on the hind paw is far away from the receptive field of microelectrode B. Use aluminum foil to reduce external interference (*see* **Note 2**).

Isolating wide dynamic range (WDR) neurons

1. While electronically moving the microelectrode B between 350 and 700 μm in depth (or 700–1000 for rats, *see* **Notes 1** and **3**), begin the source of dorsal horn neurons by producing a continuous stroking stimulation of the plantar hind paw ipsilateral to microelectrode B with a sable-hair brush (*see* **Note 4**).

2. Once identified, the spikes belonging to evoked neuronal activity are distinguishable from the background and their amplitude is uniform. Different amplitudes suggest the activity of more than one neuron. *Trick:* slowly move microelectrode B ventrally or dorsally (5–10 μm) to achieve a uniform signal belonging to a single-cell unit.

3. When a single unit is identified, optimize the amplitude of the signal by making small dorsal–ventral adjustments of the position of the microelectrode B and by refining the precise position of the stimulation on the receptive field on the plantar hind paw. Then ascertain this dorsal horn neuron is a WDR by testing its clear response to increasing mechanical stimulations with sable-hair brush, three increasing Von Fray hairs (0.1, 1, 4 g for mouse and 4, 15, 26 g for rats, *see* **Note 1**), "punching" of a wooden probe and "pinching" of a surgery forceps. WDR neurons are responsive to both high- and low-intensity stimulation (*see* **Note 5**). *Trick:* it may be helpful to ground the forceps because it can be charged. If no neurons are encountered, up to 700 μm-in depth (or 1000 for rats, *see* **Note 1**), reposition the microelectrode B and repeat the procedure.

Recording

1. Once a WDR neuron has been identified, use SciWorks software of acquisition and start the recording session following this scheme:
   (a) 20 s of background activity.
   (b) Three trials of 10 s of sable-hair brush separated by 20 s of elapsed time.
   (c) Three trials of 2 s of 0.1 g Von Fray hair (4 g for rats) separated by 20 s of elapsed time.
   (d) Three trials of 2 s of 1 g Von Fray hair (15 g for rat) separated by 20 s of elapsed time.
   (e) Three trials of 2 s of 4 g Von Fray hair (26 g for rat) separated by 20 s of elapsed time.
   (f) Three trials of 2 s of punching with a wooden probe separated by 20 s of elapsed time.
   (g) Three trials of 2 s of pinching with a surgery forceps separated by 20 s of elapsed time.
   (h) Three trials of 1 s leaving an acetone drop on the hind paw, separated by 20 s of elapsed time.

2. Perform two or three recordings along the rostral-caudal axis by repeating the procedure in different sites.

3. Terminate the experiment by administering a lethal dose of pentobarbital (*see* **Note 6**).

2.2.3 *Data Analysis*

The WDR data are acquired, digitized, recorded, and analyzed off-line after the recording session is complete using specialized software programs such as that produced by DataWave Technologies (Longmont, CA). The number of animals per group varies from 6–10 on average depending on the testing paradigm being used and the number of neurons isolated per animal. The data analysis is based on single unit activity ($n = 10$–20 per condition) and expressed as the mean ($+/-$ S.E.M.) number of spikes per second. Some experimental paradigms allow only a single recording per animal. For example, when studying the effect on neuronal activity of a drug that is applied in the saline bath overlying the dorsal horn. In this case, a neuron is isolated, and its response is recorded in the absence of the drug. Then, the same neuron is recorded after application of the drug in the bath. Because the bath application of the drug would affect all of the neurons in the dorsal horn, it is not possible to obtain a subsequent baseline recording of another neuron in the same animal. In most other paradigms, multiple recordings can be made in each animal. When the recoding electrode is inserted into the spinal cord, a search stimulus is done to isolate a neuron. However, the activity of other neurons near the isolated neurons will also be acquired by the electrode and recorded. The first step in the analysis process is to perform a principle components analysis (PCA) based on the positive and negative amplitudes of various spikes to determine how many neurons have been recorded. After the neurons are sorted by PCA, the next step is to identify those that are WDR neurons, which is based on the neurons responding to all of the innocuous and noxious stimuli. The data from each neuron are sorted into 1-s bins and peristimulus histograms are generated for each of the test stimuli administered. Only those neurons that respond to stimuli throughout the entire range of intensities from innocuous to noxious are included in the data analysis. The data are analyzed using standard parametric statistics and post hoc testing. Repeated measures analysis is also used when appropriate. The experimental design and research questions dictate which statistical tests will be used. Figure 6 reported an example of data analyzed for mice treated with oxaliplatin.

2.2.4 *Notes*

1. The protocol of spinal electrophysiological recording of WDR neurons is here described for mice but can also be applicable to rats, with particular precautions. As an example, the laminectomy in the rat requires additional surgical instruments, able to break up more resistant tendons and bones. Moreover, the neuronal searching requires different coordinates in rats: WDR neurons are located at 700–1000 μm-in depth. Finally, the Von Fray hairs employed for the plantar hind paw stimulation are different in mice and rats (0.1, 1, 4 g for mice; 4, 15, 26 g for rats).

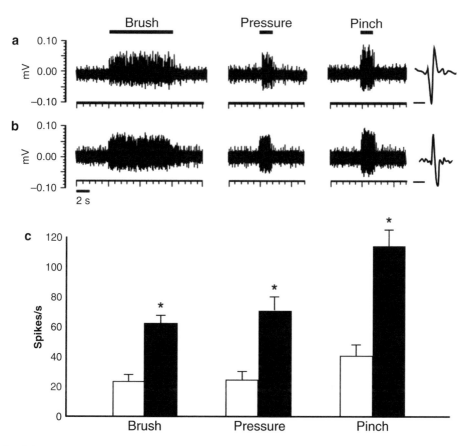

Fig. 6 Oxaliplatin increases activity of WDR neurons in the spinal dorsal horn. (**a** and **b**) Example raw data used to construct a histogram of the stimulus-response to brush, press and pinch of an individual WDR neuron from (**a**) a naïve mouse and (**b**) a mouse 48 h after the final dose of oxaliplatin (3.5 mg/kg/iv twice weekly × 4 weeks). The waveform to the right of each trace shows a representative spike that was analyzed to generate each histogram (bar = 1 ms). (**c**) The number of spikes per second was higher in the oxaliplatin-treated mice (black bars; $n = 15$ neurons) compared with naive (white bars; $n = 12$ neurons; $*p < 0.001$ vs. naive)

2. Interference and/or noise can be a real concern during electrophysiology. They can perturb the neuronal electrical baseline making any recording impossible. External interference (often 50 Hz) can originate from other switched-on instruments located in laboratories close to the electrophysiology room (on the same floor or in the same building) and for such reason, the faraday chamber is not enough to protect the system. External interference can also come from the researcher him (her) self: *The technical suggestion* is not to wear bracelets, rings, watch on the hand used to stimulate the plantar hind paw. Moreover, use aluminum foil and insulator tape to isolate the microelectrodes and be sure to have grounded everything with insulator cables, alligator clips and copper bars. The recordings can be also complicated by

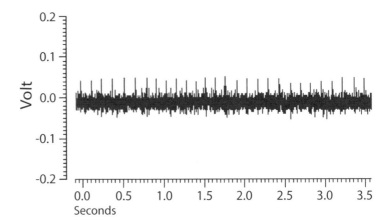

Fig. 7 Example recording with the electrodes detecting activity from the heart, which is shown by the regularly spaced spikes that have a higher amplitude than the background signal

electrical noise from the animal's breaths and heartbeat, which can interfere with the identification of the electrical activity of WDR neurons. In this case, try to change the position of the reference microelectrode (*see* Fig. 7).

3. The study of the electrical activity of WDR neurons is of particular interest in investigating neuropathic pain. These neurons receive the input from both low-threshold A-fibers and from nociceptive C fibers elaborating a response to both innocuous (brush, light Von Frey hairs) and painful stimuli (punching and pinching). These neurons are located predominantly in deeper laminae of the dorsal horn (V and VI). *Technical suggestion*: in order to limit tissue damage during the search of WDR neurons, a quick and short advance of the microelectrode B with the micromanipulator (1 μm-step) is better than a slow but considerable (20–50 μm-step) advance.

4. Laminectomy has to be performed in order to expose that portion of the dorsal horn where the neurons responding to the receptive field of stimulation on the plantar surface of the hind paw are located. Anatomical considerations suggest that these cells are located where the dorsal roots of the lumbar dorsal root ganglia enter into the spinal cord. Technically, this area can be found few millimeters rostral to the equivalent vertebral bones. Realistically, the standardization can be quite complicated due to anatomical variabilities: small variations in the position of the peripheral nerves of the paws correspond to a different localization of their receptive neurons in the relative lumbar spinal segments. *Technical suggestion*: during every experiment, you can annotate the position of "nice responding" WDR neurons and their corresponding receptive field of stimulation of the hind paw (*see* Fig. 4).

5. Once a neuronal signal has been identified and its receptive field mapped on the hind paw by using the brush, the identification of the type of neuron is of biological importance. Low-threshold and nociceptive-specific neurons have to be excluded in favor of WDR neurons. Low-threshold neurons only receive A-fibers input and respond only to low-intensity stimuli, such as brush stimulation. On the other hand, nociceptive-specific neurons respond only to high-intensity stimulation. Nociceptive-specific neurons are quite rarely encountered because of their position (predominantly in lamina I). However, be careful to exclude all those cells responding only to punch and/or pinch stimulation. As a general suggestion, check for a vigorous electrical response to both low- and high-intensity stimulation to ensure you are dealing with WDR neurons.

6. The time taken for completing the experiment depends on: (a) the ability of the researcher to perform the surgery, (b) the time required for identifying WDR neurons and for the stabilization of their electrical activity, (c) the time for eliminating or reducing eventual electrical interference and/or noise. As a general technical indication, the surgery could take about 60 min, followed by 30 min of animal acclimation before positioning the microelectrodes in the dorsal horn. The time required for each neuron search can ranges between 20 min and 1 h. Authors discourage prolonged neuron search to avoid any risk of plantar overstimulation and induction of spontaneous continuous discharge. A complete experiment in which three or four recordings are performed, can take on average about 3–4 h.

3   Final Considerations

Electrophysiological tools such as nerve conduction studies and nerve excitability studies are useful in the examination of CIPPN in animal models. The direct clinical translatability of these techniques is an important factor, with identical noninvasive electrophysiological tools being employed in the clinical setting. Because these techniques are typically noninvasive, they are also well suited to serial longitudinal studies to enable assessment of neuropathy progression over time [56]. This is an important consideration as the development of CIPPN occurs progressively with increasing cumulative dose.

As in the clinical setting, it is important that multiple assessment tools be utilized to assess neuropathy [57]. Electrophysiological testing provides quantitative information about the extent of nerve injury and recovery profile. Other types of neurological assessment such as behavioral or functional testing methods, provide complementary information. However, these methods to assess properties such as pain, withdrawal thresholds and sensation, may also be affected by factors unrelated to nerve function and therefore there is additional benefit from the inclusion of objective electrophysiological assessment tools.

However, there are also limitations to electrophysiological techniques. Conventional electrophysiological studies only assess large nerve fiber properties, and do not provide insight into small fiber function. Further, while nerve conduction studies can identify the presence of nerve dysfunction, they do not provide specific information about the underlying pathophysiological mechanisms. Nerve excitability studies are an emerging technique in animal models and have been less commonly implemented than nerve conduction studies. Nerve excitability studies can provide information about ion channel function and membrane potential, and may provide better insights into pathophysiological mechanisms [33]. However, further studies comparing nerve excitability findings in animal models to clinical findings are required.

An important consideration is the strain-specific susceptibility to CIPPN and in particular the lack of electrophysiological changes in some strains. Other more general considerations also apply, such as the need for standardisation of dosing and administration protocols used to induce CIPPN, and careful selection of time points to enable comparison between studies utilising electrophysiological variables. In the clinical setting, there is considerable variation between individuals related to the development and severity of CIPPN, which is dependent on genetic factors as well as dosing factors. Accordingly, it is important to consider the potential effects of strain and dose on the development of CIPPN in animal models.

In authors' expertise, spinal electrophysiology is helpful to investigate the neuropathic pain component of CIPPN. Since it is time-consuming, expensive and quite invasive for the animals, the employment of well-established animal models is mandatory. In vivo behavioral validation of pain can be performed at regular time points during the chemotherapy experiment as well as temporally close to the time of electrophysiological analysis. As an example, chronic oxaliplatin-induced painful peripheral neuropathy in the mouse is established by treating the animals twice a week for 4 weeks with the drug. Every week animals can be behaviorally tested for their mechanical and thermal threshold, identifying the timing of the development of neuropathic pain. On the basis of these validations, the researcher can identify the optimal time points in which spinal electrophysiology can be performed.

The organization of the experiment when spinal electrophysiology is planned has particular implications. For example, this analysis requires dedicated animals that do not undergo other stressful tests during the experiment. Moreover, since it is a terminal analysis, the animals after electrophysiology are euthanized. As happens also for behavioral assessment of pain, spinal recordings have to be organized at the same time points in each animal. Considering that only two animals can undergo spinal electrophysiology in 1 day, the pharmacological treatments have to be arranged accordingly. Moreover, it is recommended that, within the same experiment, the same researcher, blind to the animal condition and history, perform the recordings.

The spinal electrophysiology of WDR neurons is a useful technique to investigate the processes involved in the development of neuropathic pain providing information related to functional indirect changes in central neurons and in their environment produced by a direct damage of peripheral nerves. However, this analysis do not provide mechanistic evidences or progresses in the knowledge at a molecular level.

4 Expected Results

4.1 Electro-physiology in Peripheral Nerves

Nerve conduction studies have been undertaken in many different rodent models of painful (or not) CIPN (Table 1). Broadly, studies demonstrate dose-dependent reductions of NCV with taxanes, platinum-based drugs, vincristine and bortezomib (see Table 1). These effects have been demonstrated across both rat and mice models [56]. However, neurophysiological effects of chemotherapy seem to be different across species—with greater effects of paclitaxel in DBA2J mice than in C57 mice [55]. Similarly, the neurophysiological effects of oxaliplatin were much greater in Balb-c and FVC mice, with less evidence of neurotoxicity in C57Bl6 mice [10].

In line with the predominant sensory clinical expression of CIPN, several studies have demonstrated a prominent effect on sensory nerves in the absence of effect on motor nerves for bortezomib [56], paclitaxel [53, 57] and oxaliplatin [65]. Interestingly, vincristine treatment in rodent models has been demonstrated to affect both motor and sensory caudal nerves, similar to the clinical picture [15].

While NCV is the predominant NCS measure examined, some studies have also demonstrated reduced sensory amplitudes in animal models of CIPN [54, 55]. For example, both sensory amplitudes and NCV were reduced in the tail in paclitaxel-treated mice [57]. However, in oxaliplatin-treated rat models there has been some discrepancy—with some models demonstrating effects on both sensory NCV and amplitude [80] and others solely on SNCV [65]. However, while prominent declines in NCV are typically observed, examination of myelin structure directly revealed no changes with cisplatin, paclitaxel or bortezomib treatment, suggesting that NCV changes in animal models may be more reflective

Table 1
Examples of nerve conduction study findings in rodent models of CIPN or CIPPN

Chemotherapy	Findings	Studies
Paclitaxel	Reduced SNCV	Cavaletti et al. [58], Persohn et al. [16], Leandri et al. [55], Wozniak et al. [54], Gilardini et al. [40], Carozzi et al. [20, 21], Lauria et al. [59], Boehmerle et al. [60], Pisano et al. [61], Resham and Sharma [57], Beh et al. [62]
Paclitaxel	Reduced SNAP	Cliffer et al. [63], Leandri et al. [55],Wozniak et al. [54], Boehmerle et al. [60], Park et al. [53], Huehnchen et al. [64]
Epothilone-B	Reduced SNCV	Carozzi et al. [20]
Oxaliplatin	Reduced SNCV	Xiao et al. [65], Kroigard et al. [66], Marmiroli et al. [10], Areti et al. [67], Wozniak et al. [68], Renn et al. [13], Cerles et al. [69], Chiorazzi et al. [23]
Oxaliplatin	Reduced SNAP	Cerles et al. [70], Wozniak et al. [68]
Cisplatin	Reduced SNCV	Canta et al. [71], Lauria et al. [59], Pisano et al. [61], Gilardini et al. [40], Carozzi et al. [20, 21], Boehmerle et al. [60], Khasabova et al. [72], Cavaletti et al. [73]
Cisplatin	Reduced SNAP	Boehmerle et al. [60]
Bortezomib	Reduced SNCV	Ghelardini et al. [74], Meregalli et al. [75], Gilardini et al. [40], Carozzi et al. [20, 21], Boehmerle et al. [60], Chiorazzi et al. [23]
Bortezomib	Reduced SNAP	Ale et al. [56], Boehmerle et al. [60]
Vincristine	Reduced SNAP and SNCV	Chine et al. [76], Gong et al. [77]
Vincristine	Reduced SNAP, NCV, CMAP	Meregalli et al. [14, 15]
Vincristine	Reduced MNCV	Khalilzadeh et al. [78]
Vincristine	Reduced MNCV and CMAP	Greeshma et al. [79]
Vincristine	Reduced SNCV	Boehmerle et al. [60]

of axonal than myelin pathology [40]. Accordingly, reductions in tail sensory NCV were associated with reduced intra-epidermal nerve fiber density on skin biopsy following cisplatin or paclitaxel treatment [59].

Some models have demonstrated length-dependent effects, with caudal NCV reduced to a greater extent than digital NCV with bortezomib, paclitaxel, cisplatin and epothilone-B treatment in mice [20]. Similarly, vincristine affected caudal nerve but not digital nerves [15]).

Fewer studies have been undertaken utilizing nerve excitability analysis in models of CIPN. Sensory and motor nerve excitability

Table 2
Examples of WDR electrophysiological study findings in rodent models of CIPPN

Chemotherapy	Findings	Studies
Oxaliplatin	Increased WDR activity	Renn et al. [13]
Bortezomib	Increased WDR activity	Carozzi et al. [12]
Oxaliplatin	Increased WDR activity	Marmiroli et al. [10]
Cisplatin	Increased WDR activity	Lessans et al. [9]

studies have also been undertaken in oxaliplatin-treated mice in caudal and sciatic nerves, with a suite of changes attributed to hyper-excitability [69, 70]. Motor nerve excitability studies demonstrate similar acute changes following oxaliplatin treatment in humans and rats [80].

4.2 Electrophysiology in Dorsal Horn of Spinal Cord

Extracellular electrophysiological recordings of WDR neurons in the spinal dorsal horn have been done in several different mouse models of CIPPN (Table 2). These studies show an increase in WDR neuron activity following treatment with the platinum-based drugs oxaliplatin and cisplatin as well as with bortezomib (see Table 2). The increase in WDR neuron activity after chemotherapy treatment with oxaliplatin has been demonstrated in a variety of mouse strains [10]. However, while all of the strains developed increased sensitivity to nocifensive testing, the magnitude of change in WDR neuron activity varied across the strains. Of the strains that were tested, the Balb-c and A/J strains were the most sensitive to the nocifensive testing and had the greatest increase in WDR neuron activity. In contrast, the FVB mice were moderately sensitive in response to nocifensive testing and did not demonstrate increased WDR neuron activity. The increase in WDR neuron activity in the other strains fells between the Balb-c/A/J strains and the FVB strain of mice [10]. These data suggest that changes in WDR neuron activity are not merely the result of increased activity from damaged primary afferent neurons. Other factors that are strain specific such as genetic differences between the strains may also play a role in the function of WDR neurons after chemotherapy treatment.

5 Applications

The applications of these techniques in experimental models of CIPPN are immense. Since nerve electrophysiology is a "quick" analysis (45 min for each animal assessing both sensory and motor nerves), the information that can be obtained can be diagnostic and can be useful to screen animals for the development of peripheral neuropathy: it is a gold standard analysis useful to identify the onset of peripheral neurotoxicity and follow the development of functional damage in peripheral nerves. By contrast, since spinal electrophysiology is a time-consuming analysis (4 h total for each animal), it cannot be considered a "screening technique". However, it provides an objective evaluation of central sensitization due to neuropathic pain, complementary to the behavioral methods, affected by the operator subjectivity. The data that return from the electrophysiological analysis can also be useful as preliminary guidelines for molecular investigations of specific pathogenic pathways as well as to test neuroprotective strategies against neurotoxicity and neuropathic pain.

Finally, it is important to underline that these electrophysiological techniques can be employed to study peripheral nerve damage and pain of other origins then chemotherapy treatment, such as traumatic nerve or spinal injury and all kinds of inherited or acquired neuropathies.

References

1. Hausheer FH et al (2006) Diagnosis, management, and evaluation of chemotherapy-induced peripheral neuropathy. Semin Oncol 33:15–49

2. Park SB et al (2013) Chemotherapy-induced peripheral neurotoxicity: a critical analysis. CA Cancer J Clin 63:419–437

3. Cavaletti G et al (2019) Chemotherapy-induced peripheral neurotoxicity: a multifaceted, still unsolved issue. J Peripher Nerv Syst 24 (Suppl 2):S6–S12

4. Kandula T et al (2017) Neurophysiological and clinical outcomes in chemotherapy-induced neuropathy in cancer. Clin Neurophysiol 128:1166–1175

5. Wolf SL et al (2012) The relationship between numbness, tingling, and shooting/burning pain in patients with chemotherapy-induced peripheral neuropathy as measured by the EORTC QLQ-CIPN20 instrument, N06CA. Support Care Cancer 20:625–632

6. Perez C et al (2015) Prevalence of pain and relative diagnostic performance of screening tools for neuropathic pain in cancer patients: a cross-sectional study. Eur J Pain 19:752–761

7. Wilson RH et al (2002) Acute oxaliplatin-induced peripheral nerve hyperexcitability. J Clin Oncol 20:1767–1774

8. Chan A et al (2019) Biological predictors of chemotherapy-induced peripheral neuropathy (CIPN): MASCC neurological complications working group overview. Support Care Cancer 27:3729–3737

9. Lessans S et al (2019) Global transcriptomic profile of dorsal root ganglion and physiological correlates of cisplatin-induced peripheral neuropathy. Nurs Res 68(2):145–155

10. Marmiroli P et al (2017) Susceptibility of different mouse strains to oxaliplatin peripheral neurotoxicity: phenotypic and genotypic insights. PLoS One 12(10):e0186250

11. Quartu M et al (2014) Bortezomib treatment produces nocifensive behavior and changes in the expression of TRPV1, CGRP, and substance P in the rat DRG, spinal cord, and sciatic nerve. Biomed Res Int 2014:180428

12. Carozzi VA et al (2013) Bortezomib-induced painful peripheral neuropathy: an electrophysiological, behavioral, morphological and mechanistic study in the mouse. PLoS One 8(9): e72995. https://doi.org/10.1371/journal. pone.0072995. eCollection 2013

13. Renn CL et al (2011) Multimodal assessment of painful peripheral neuropathy induced by chronic oxaliplatin-based chemotherapy in mice. Mol Pain 7:29

14. Meregalli C et al (2018a) High-dose intravenous immunoglobulins reduce nerve macrophage infiltration and the severity of bortezomib-induced peripheral neurotoxicity in rats. J Neuroinflammation 15(1):232

15. Meregalli C et al (2018b) Neurofilament light chain as disease biomarker in a rodent model of

chemotherapy induced peripheral neuropathy. Exp Neurol 307:129–132

16. Persohn E et al (2005) Morphological and morphometric analysis of paclitaxel and docetaxel-induced peripheral neuropathy in rats. Eur J Cancer 41(10):1460–1466

17. Tredici G et al (1999) Effect of recombinant human nerve growth factor on cisplatin neurotoxicity in rats. Exp Neurol 159(2):551–558

18. Cavaletti G et al (1992) Morphometric study of the sensory neuron and peripheral nerve changes induced by chronic cisplatin (DDP) administration in rats. Acta Neuropathol 84 (4):364–371

19. Potenzieri A et al (2019) Oxaliplatin-induced neuropathy occurs through impairment of haemoglobin proton buffering and is reversed by carbonic anhydrase inhibitors. Pain. https:// doi.org/10.1097/j.pain.0000000000001722

20. Carozzi VA et al (2010) Neurophysiological and neuropathological characterization of new murine models of chemotherapy-induced chronic peripheral neuropathies. Exp Neurol 226(2):301–309

21. Carozzi VA et al (2015) Chemotherapy-induced peripheral neurotoxicity in immune-deficient mice: new useful ready-to-use animal models. Exp Neurol 264:92–102

22. Cavaletti G et al (2002) Circulating nerve growth factor level changes during oxaliplatin treatment-induced neurotoxicity in the rat. Anticancer Res 22(6C):4199–4204

23. Chiorazzi A et al (2018) Ghrelin agonist HM01 attenuates chemotherapy-induced neurotoxicity in rodent models. Eur J Pharmacol 840:89–103

24. Carozzi VA et al (2009) Effect of the chronic combined administration of cisplatin and paclitaxel in a rat model of peripheral neurotoxicity. Eur J Cancer 45(4):656–665

25. Makker PGS et al (2018) A unified model of the excitability of mouse sensory and motor axons. J Peripher Nerv Syst 23:159–173

26. Tonkin RS et al (2018) Attenuation of mechanical pain hypersensitivity by treatment with Peptide5, a connexin-43 mimetic peptide, involves inhibition of NLRP3 inflammasome in nerve-injured mice. Exp Neurol 300:1–12

27. Feather CE et al (2018) Oxaliplatin induces muscle loss and muscle-specific molecular changes in mice. Muscle Nerve 57(4):650–658

28. Livni L et al (2019) Dorsal root ganglion explants derived from chemotherapy-treated mice have reduced neurite outgrowth in culture. Neurosci Lett 16(694):14–19

29. Huynh W, Kiernan MC (2011) Nerve conduction studies. Aust Fam Physician 40:693–697

30. Mallik A, Weir AI (2005) Nerve conduction studies: essentials and pitfalls in practice. J Neurol Neurosurg Psychiatry 76(Suppl 2): ii23–ii31

31. Kimura J (2017) Nerve conduction studies. In: Mills K (ed) Oxford textbook of clinical neurophysiology. Oxford University Press, Oxford, pp 49–66

32. Franssen H (2017) Generalized peripheral neuropathies. In: Mills K (ed) Oxford textbook of clinical neurophysiology. Oxford University Press, Oxford, pp 223–234

33. Kiernan MC et al (2019) Measurement of axonal excitability: consensus guidelines. Clin Neurophysiol 131:308–323

34. Bostock H, Cikurel K, Burke D (1998) Threshold tracking techniques in the study of human peripheral nerve. Muscle Nerve 21:137–158

35. George A, Bostock H (2007) Multiple measures of axonal excitability in peripheral sensory nerves: an in vivo rat model. Muscle Nerve 36:628–636

36. Maurer K, Bostock H, Koltzenburg M (2007) A rat in vitro model for the measurement of multiple excitability properties of cutaneous axons. Clin Neurophysiol 118:2404–2412

37. Boerio D, Greensmith L, Bostock H (2009) Excitability properties of motor axons in the maturing mouse. J Peripher Nerv Syst 14:45–53

38. Hoke A, Ray M (2014) Rodent models of chemotherapy-induced peripheral neuropathy. ILAR J 54:273–281

39. Schaumburg HH et al (2010) The rat caudal nerves: a model for experimental neuropathies. J Peripher Nerv Syst 15:128–139

40. Gilardini A et al (2012) Myelin structure is unaltered in chemotherapy-induced peripheral neuropathy. Neurotoxicology 33:1–7

41. Boadas-Vaello P et al (2016) Neuroplasticity of ascending and descending pathways after somatosensory system injury: reviewing knowledge to identify neuropathic pain therapeutic targets. Spinal Cord 54:330–340

42. Coghill RC, Mayer DJ, Price DD (1993) Wide dynamic range but not nociceptive-specific neurons encode multidimensional features of prolonged repetitive heat pain. J Neurophysiol 69:703–716

43. D'Mello R, Dickenson AH (2008) Spinal cord mechanisms of pain. Br J Anaesth. https://doi.org/10.1093/bja/aen088

44. Feizerfan A, Sheh G (2015) Transition from acute to chronic pain. Contin Educ Anaesthesia Crit Care Pain 15(2):98–102

45. Mendell LM (1966) Physiological properties of unmyelinated fiber projection to the spinal cord. Exp Neurol. https://doi.org/10.1016/0014-4886(66)90068-9

46. Willis WD (1965) The pain system. Karger, New York

47. Maixner W et al (1986) Wide dynamic range dorsal horn neurons participate in the encoding process by which monkeys perceive the intensity of noxious heat stimuli. Brain Res 374:385–388

48. Price DD (1988) Psycho logical and neural mechanisms of pain. Raven, New York

49. Guan Y et al (2010) Spinal cord stimulation-induced analgesia: electrical stimulation of dorsal column and dorsal roots attenuates dorsal horn neuronal excitability in neuropathic rats. Anesthesiology 113:1392–1405. https://doi.org/10.1097/ALN.0b013e3181fcd95c

50. Zain M, Bonin RP (2019) Alterations in evoked and spontaneous activity of dorsal horn wide dynamic range neurons in pathological pain: a systematic review and analysis. Pain 160(10):2199–2209

51. Dickenson AH, Sullivan AF (1987) Evidence for a role of the NMDA receptor in the frequency dependent potentiation of deep rat dorsal horn nociceptive neurones following c fibre stimulation. Neuropharmacology 26 (8):1235–1238

52. Bruna J et al (2019) Methods for in vivo studies in rodents of chemotherapy induced peripheral neuropathy. Exp Neurol 325:113154

53. Park JS, Kim S, Hoke A (2015) An exercise regimen prevents development paclitaxel induced peripheral neuropathy in a mouse model. J Peripher Nerv Syst 20:7–14

54. Wozniak KM et al (2011) Comparison of neuropathy-inducing effects of eribulin mesylate, paclitaxel, and ixabepilone in mice. Cancer Res 71:3952–3962

55. Leandri M et al (2012) Electrophysiological features of the mouse tail nerves and their changes in chemotherapy induced peripheral neuropathy (CIPN). J Neurosci Methods 209:403–409

56. Ale A et al (2014a) Treatment with anti-TNF alpha protects against the neuropathy induced by the proteasome inhibitor bortezomib in a mouse model. Exp Neurol 253:165–173

57. Resham K, Sharma SS (2019a) Pharmacological interventions targeting Wnt/beta-catenin signaling pathway attenuate paclitaxel-induced peripheral neuropathy. Eur J Pharmacol 864:172714

58. Cavaletti G et al (1995) Experimental peripheral neuropathy induced in adult rats by repeated intraperitoneal administration of taxol. Exp Neurol 133:64–72

59. Lauria G et al (2005) Intraepidermal nerve fiber density in rat foot pad: neuropathologic-neurophysiologic correlation. J Peripher Nerv Syst 10:202–208

60. Boehmerle W et al (2014) Electrophysiological, behavioral and histological characterization of paclitaxel, cisplatin, vincristine and bortezomib-induced neuropathy in C57Bl/6 mice. Sci Rep 4:6370

61. Pisano C et al (2003) Paclitaxel and Cisplatin-induced neurotoxicity: a protective role of acetyl-L-carnitine. Clin Cancer Res 9:5756–5767

62. Beh ST et al (2019) Preventive hypothermia as a neuroprotective strategy for paclitaxel-induced peripheral neuropathy. Pain 160:1505–1521

63. Cliffer KD et al (1998) Physiological characterization of Taxol-induced large-fiber sensory neuropathy in the rat. Ann Neurol 43:46–55

64. Huehnchen P, Boehmerle W, Endres M (2013) Assessment of paclitaxel induced sensory polyneuropathy with "Catwalk" automated gait analysis in mice. PLoS One 8:e76772

65. Xiao WH, Zheng H, Bennett GJ (2012) Characterization of oxaliplatin-induced chronic painful peripheral neuropathy in the rat and comparison with the neuropathy induced by paclitaxel. Neuroscience 203:194–206

66. Kroigard T et al (2019) Protective effect of ibuprofen in a rat model of chronic oxaliplatin-induced peripheral neuropathy. Exp Brain Res 237:2645–2651

67. Areti A, Komirishetty P, Kumar A (2017) Carvedilol prevents functional deficits in peripheral nerve mitochondria of rats with oxaliplatin-evoked painful peripheral neuropathy. Toxicol Appl Pharmacol 322:97–103

68. Wozniak KM et al (2012) The orally active glutamate carboxypeptidase II inhibitor E2072 exhibits sustained nerve exposure and attenuates peripheral neuropathy. J Pharmacol Exp Ther 343:746–754

69. Cerles O et al (2017) Niclosamide inhibits oxaliplatin neurotoxicity while improving colorectal cancer therapeutic response. Mol Cancer Ther 16:300–311

70. Cerles O et al (2019) Preventive action of benztropine on platinum-induced peripheral neuropathies and tumor growth. Acta Neuropathol Commun 7:9

71. Canta A et al (2011) In vivo comparative study of the cytotoxicity of a liposomal formulation

of cisplatin (lipoplatin). Cancer Chemother Pharmacol 68:1001–1008

72. Khasabova IA et al (2012) Cannabinoid type-1 receptor reduces pain and neurotoxicity produced by chemotherapy. J Neurosci 32:7091–7101

73. Cavaletti G et al (1994) Protective effects of glutathione on cisplatin neurotoxicity in rats. Int J Radiat Oncol Biol Phys 29:771–776

74. Ghelardini C et al (2014) Spinal administration of mGluR5 antagonist prevents the onset of bortezomib induced neuropathic pain in rat. Neuropharmacol 86:294–300

75. Meregalli C et al (2012) CR4056, a new analgesic I2 ligand, is highly effective against bortezomib-induced painful neuropathy in rats. J Pain Res 5:151–167

76. Chine VB et al (2019) Targeting axon integrity to prevent chemotherapy-induced peripheral neuropathy. Mol Neurobiol 56:3244–3259

77. Gong SS et al (2016) Neuroprotective effect of matrine in mouse model of vincristine-induced neuropathic pain. Neurochem Res 41:3147–3159

78. Khalilzadeh M et al (2018) The protective effects of sumatriptan on vincristine-induced peripheral neuropathy in a rat model. Neurotoxicol 67:279–286

79. Greeshma N, Prasanth KG, Balaji B (2015) Tetrahydrocurcumin exerts protective effect on vincristine induced neuropathy: behavioral, biochemical, neurophysiological and histological evidence. Chem Biol Interact 238:118–128

80. Alberti P et al (2019) Topiramate prevents oxaliplatin-related axonal hyperexcitability and oxaliplatin induced peripheral neurotoxicity. Neuropharmacology 164:107905

Part III

Behavioral Evaluation in Rodents

Chapter 8

The Functional Observation Battery: Utility in Safety Assessment of New Molecular Entities

David V. Gauvin

Abstract

The rat functional observational battery is the regulatory standard assay for the assessment of the central nervous system prior to the first dose administration in man. This chapter describes the utility and scientific foundation for the general adoption of this technique by both European and the United States drug regulatory agencies.

Key words Functional observational battery, Irwin Screen, Safety pharmacology, ICH 57A, Neurotoxicity

1 Introduction

The purpose of this chapter is to highlight the importance, impact, and implications of quantifying behavioral change as a critical biomarker predictive of neurotoxicity. In the early development of safety and risk assessment testing procedures histopathological changes were considered the essential piece of the puzzle that defined the field of neurotoxicology [1]. But with the growth of the scientific discipline of behavioral toxicology regulatory agencies worldwide have come to accept that behavior itself represent the integration and complexity of the central nervous system. The diversity of behavioral change that is observed following the intentional exposure of animals to a new chemical entity is considered the most sensitive biomarker and perhaps the ultimate assay of neural function [2–4].

In 2020, the functional observational battery (FOB) has now become the "gold standard" for testing CNS functions required for regulatory risk-to-benefit analysis required of all drugs prior to licensure approval [5]. This chapter highlights the methodology used to assess qualitative and quantitative changes in neurological function induced by new drugs in purpose-bred research laboratory

Jordi Llorens and Marta Barenys (eds.), *Experimental Neurotoxicology Methods*, Neuromethods, vol. 172,
https://doi.org/10.1007/978-1-0716-1637-6_8, © Springer Science+Business Media, LLC, part of Springer Nature 2021

rats. US Federal regulatory agencies and the National Academy of Sciences (NAS) have strongly endorsed the use of rats as the "preferred species" to use in safety, toxicity, teratogenicity, and carcinogenicity studies [6, 7]. However, if there is overriding predictive advantages to other species (i.e., nonhuman primates in biologics screening), other species can be used with justification. In particular, it is the rat FOB that is used to ensure the integrity and safety of all new drugs and environmental chemicals intended for the human consumer in today's world drug development programs.

The FOB is a type of neurological examination of the integration of sensory and motor systems in an intact animal. In the long time-line from proof of efficacy to licensure of a new drug, the use of FOBs are usually initiated almost a decade before a mountain of other scientific evidence is summarized and submitted for review to the drug approval agency (US FDA, UK Home Office, Health Canada, EmEA, etc.) and subsequently licensed for marketing to the general public. By the adoption of the FOB by the International Commission on Harmonization's "Guidance on Nonclinical Safety Studies for the Conduct of Human Clinical Trials and Marketing Authorization for Pharmaceuticals" [4] drug companies are required to include a neurobehavioral examination into the standard set of studies needed to best estimate the relative safety of both short-term and long-term exposure to the drug. FOBs are used to predict the relative safety of new drugs on the CNS of animal subjects prior to the first dose administration in man [8] and to assess the acute, chronic, and residual (hangover) effects of both drugs and chemical exposures prior to licensure for manufacturing and sales by the federal health care agencies. The specific logistical details on how to conduct the rat FOB have been detailed elsewhere and the reader is directed to those sources for further reading [9–14]. Moser [1] has highlighted a key point relative to FOBs:

> there is no one single FOB protocol, but rather the guidelines describe more general aspects and experimental tests that should be included. Over the years several protocols for behavioral assessments have been published (see Moser for references) and each test laboratory generally uses its own version.

The key value of the FOB is not only the selected observations that are conducted but the fact that these parameters are conducted in purpose-bred laboratory animal subjects of known hereditary lineage by animal breeding facilities controlled, maintained, and regulated by federal (USDA, US FDA) agencies, and reviewed by ancillary civil organizations focused on animal welfare, such as the voluntary participation of the research laboratory with the American Association of Laboratory Animal Care Certification Programs (AALAC-accreditation). Additionally, through protocol development the FOBs are conducted in animals of equivalent age, identical experimental- and drug-histories, environmental exposures, nutrient/dietary histories, and in the same laboratory at the same time-of-day, and by the same technicians proficiently trained under Good Laboratory Practice Guidelines to conduct the FOBs. The tight control on these critically important experimental factors provide strong evidence that any differences documented between- and within-treatment groups in a study can be confidently attributed to the dose of drug administered to the animals. These types of nonclinical laboratory studies are critical for determining safe levels of investigational drugs that may be used in human subjects participating in clinical studies [15].

2 The General Battery of Tests

A clinical neurological examination is usually described as a complicated process that is wholly separate from a standard veterinary physical examination. For most veterinarians, much of the neurologic examination is done as a routine part of the physical examination and is a basic checklist of nerves checked for system integrity and integration [16–18]. That is not what an FOB in intended to be.

There is a continuing academic discussion as to what constitutes the "proper" test for neurotoxicity under all of the administrative requirements of health care agencies such as the FDA, EPA, OECD, and ICH. As reviewed by the EPA in 1998 [19], many tests that can measure some aspect of neurotoxicity have been used in the field of neurobiology in the past 70 years (*see* Table 1). The FDA and EPA have published animal testing guidelines that highlighted the test endpoints that serve to be the focus of both agencies used as a framework for interpreting data collected in tests frequently used by neuro-toxicologists. Both FDA and EPA have made it clear that a standard neurological examination is neither necessary a nor sufficient test to assess safety of new molecular entities—more is needed:

These regulatory guidelines highlight the importance of behavior as a critical endpoint in neurotoxicity risk assessment plans. The behavioral output in animal subjects is a reflection of changes in neural communication and integration of sensory and motor pathways (efferent and afferent pathways) as well as morphological or architectural modification of tissue measured by the pathologist postmortem. Behavioral pharmacology/toxicology represents the scientific discipline(s) that assume(s) that observed behavior is the product of the integration and integrity of the nervous system and that behavior change, itself, represents the most sensitive indicator, and perhaps the ultimate assay, of neuronal functional change [1–3, 20]. The FOB is not a standard neurological examination described in veterinary textbooks; it is a unique test designed to go beyond an assessment of the presence and absence of basic stimulus-response reflexes or tests to assess if a nerve remains intact after an insult. The

Table 1
Examples of possible indicators of a neurotoxic effect (https://www.epa.gov/sites/production/files/2014-11/documents/neuro_tox.pdf)

Behavioral and neurological endpoints
• Increases or decreases in motor activity. • Changes in touch, sight, sound, taste, or smell sensations. • Changes in motor coordination, weakness, paralysis, abnormal movement or posture, tremor, ongoing performance. • Absence or decreased occurrence, magnitude, or latency of sensorimotor reflex. • Altered magnitude of neurological measurement, including grip strength, hindlimb splay. • Seizures. • Changes in rate or temporal patterning of schedule-controlled behavior. • Changes in learning, memory, and attention.

FOB is a functional behavioral test to assess the integrity of neuronal pathways of the central and peripheral nervous system.

In Section 3.1.2.4 of the neurotoxicity guidelines for industrial and agricultural chemicals [19], the EPA highlights the adoption of the agencies' policy that behavior reflects the integration of the various functional components of the nervous system and changes in behavior can arise from a direct effect of a toxicant on the nervous system, or indirectly from its effects on other physiological systems. In risk assessment plans, it is vital then to understand the interrelationship between systemic toxicity and behavioral changes (e.g., the relationship between liver damage and motor activity). The presence of systemic toxicity may complicate, but does not preclude, interpretation of behavioral changes as evidence of neurotoxicity. In addition, several behaviors (e.g., grip strength or schedule-controlled behavior) may require a motivational component for successful completion of the task. With respect to chemical safety, the screening for neurotoxicity potential cannot be based solely on a simple veterinary neurological examination. Experimental paradigms designed to assess the motivation of an animal during behavior are also necessary to interpret the meaning of some chemical- or drug-induced changes in behavior. Examples of measures obtained in a typical risk assessment plan using an FOB are presented in Table 2 [21, 22]:

For safety risk assessments of new drug therapeutics under the International Committee on Harmonization [8, 23], it must be acknowledged that other behaviors besides those listed in the table above could be affected by test article exposures. For example, alterations in food and water intake, reproduction, sleep, temperature regulation, and circadian rhythmicity are all controlled by specific regions of the brain, and test article-induced alterations in these behaviors could be indicators of neurotoxicity that *are not included* in a standard veterinary neurological examination.

Table 2
Examples of measures in a representative functional observational battery. Section 3.1.2.4. EPA Guidelines

Home cage and open field	Manipulative	Physiological
Arousal	Approach response	Body weight
Autonomic signs	Click response	Body temperature
Convulsion, tremor	Foot display	
Gait	Grip strength (response)	
Mobility	Righting reflex	
Posture	Tail pinch response	
Rearing		
Stereotypy		
Touch response		
Cognitive effects: Habituation Open field: Thigmotaxis Manipulative: Startle, click, touch, etc.		

3 CNS Safety Assessments

Clinical trial failures are a financial nightmare for drug development companies. Clinical trial failures occur when a company terminates a study because the new product is not effective or worse and causes unexpected harm or off-target symptoms not predicted in nonclinical (animal) testing [24]. The earliest dose administration of a new drug to human patients is planned in the Phase I Clinical Trial. Phase I trials recruit a relatively small number of healthy volunteers to determine the full spectrum of "intended" and "unintended" effects of the new drug. Both intended and unintended effects of a drug are a direct result of dose administrations. The unintended drug effects are usually referred to as "side effects," they remain a direct effect of the drug—just not the target of interest for the drug developer. The patient population in the earliest clinical trial may identify adverse effects of the drug that cause a rapid patient dropout rate or initiates medical interventions to ameliorate unintended effects of the drug (side effects). Table 3 lists some of the most common adverse events characterizing clinical trials:

Illness-inducing effects of dramatic subjective effect changes following dosing in Phase I trial will jeopardize continued patient compliance or threaten the likelihood of fulfilling census requirements in future clinical trial programs. It is impossible to maintain a subject pool in clinical trials if the subjective or physiological response to drug administration is so aversive that "normal,

Table 3
Common adverse events and symptoms occurring in patients who received at least one dose of study drug (Consolidated Standards of Reporting Trials [CONSORT] Harms Extension; [25])

Vertigo/Dizziness	Confusion
Headache	Hallucinations
Insomnia/fatigue	Diarrhea
Anxiety	Nausea/dyspepsia/vomiting
Depression/sense of doom or regret	Heart palpitations

healthy, altruistic patient volunteers" are not willing to remain in the programs. Additionally, the Institutional Review Board (IRB) may refuse to allow a clinical trial to continue or progress to the next phase of development if it perceives the study as not safe enough or not providing any relative therapeutic benefit to the patient [24]. There is a general consensus among many drug developers that not all preclinical trials (animal studies) are effective in predicting safety and efficacy of a drug and more relevant to the topic of this chapter, animal data do not reliably predict "side effect" profiles in humans. For example, Bettge et al. [26] published the results of a systematic analysis of published clinical trial data studying glucagon-like peptide-1 receptor agonists (e.g., liraglutide, Victoza™) for the treatment of Type-2 diabetes. In that report gastrointestinal adverse events such as nausea and vomiting were significant clinical trial adverse events produced by all long-acting GLP-1 receptor agonists. But which of the CNS safety assessment parameters used in standard rat FOBs should have predicted this toxicity? None of them because rats do not show signs of nausea or vomiting.

This lack of "rigor" between nonclinical and clinical findings was a major impetus in shifting the regulatory mandate from strict harmonized study protocols of the 1950s through 1990s to the more flexible guidelines and tier-testing approaches of today. By encouraging diversity as a major factor included in the design of all animal and human studies and encouraging "adaptive trial designs" the FDA has established a more meaningful government-industry partnership in the drug development process.

The FDA uses the rat FOB data to estimate an initial safe starting dose and dose range of a new drug candidate for the human trials (safety pharmacology) and to identify parameters for clinical monitoring for potential adverse effects (general toxicology). The nonclinical safety studies, although usually limited at the beginning of clinical development, need to adequately characterize the potential adverse effects that might occur under the conditions of the clinical trial to be supported. Some thought leaders in the industry do not believe the FOB is the "best strategy." Early in the investigation pathway some take moves to diminish or underplay the scope of the study design or compromise the statistical power of the assay to substantiate their own claims that the FOB is an unnecessary economic investment. However, the current *zeitgeist* at most drug approval (regulatory) agencies retains the FOB examination as an "optimized strategy." The difference between "best" and "optimized" strategies is that the latter is simple, effective, efficient, and the most economical with respect to both financial investment and functional utility of the assay to the drug approval agencies in both Europe and North America. The critical nature of the optimization principal is always linked to statistical power of the test. In today's drug development industry, there is a speed and cost tradeoff with reliability and validity in early stages of the drug discovery.

The extrapolation of animal data to humans is tenuous, at best [27, 28]. However, for obvious reasons human experimentation with new drugs is not socially or morally acceptable. Animal testing has been established by national (US) and international statutes (law) but, unfortunately, there is no single animal model in which drug effects correlate perfectly with toxicity in humans. It is up to the drug development industry itself to select an "industry standard" animal model that is suitable for regulatory submissions. The selection of the specific assays to include in the respective FOBs cannot be based on economics alone. While there are economic factors to wrestle with housing, maintenance conditions, and husbandry costs it is the Institutional Animal Care and Use Committee (IACUC) that must accept the selection of behavioral tests to include in the FOBs. The use of a single animal species is not allowed under current governmental regulatory administrative policies; however, the specific strain of rat used throughout the preclinical testing phase of the drug development program will facilitate comparisons among measures and will help determine the reliability and sensitivity of the behavioral assay targeting the mode and mechanisms of action of the new drug [28].

To place the importance of simple behavior on nervous system safety assessment, MacPhail [29] reminds us of the critical nature of the nervous system on the well-being of all animals. And it has been over 30 years since Evans [30, 31] highlighted the fact that it was simple alterations in locomotor behavior in home cages of laboratory rats that best indicated toxicity for half of a broad group of chemicals tested in early proof-of-concept validation studies conducted in five different research laboratories in the United States and Europe. The old adage of "keep it simple" exemplifies the fact that it was the changes in the total number of "rearings" in a rat's home cage that provided one of the best predictors of methyl mercury toxicity because of its impact on sciatic nerve functions. It is the nervous system that transmits information about the environment, integrates that information, and communicates the appropriate sequences of behavior in order to respond and operate on that environment. The nervous system exhibits executive control over all behavioral and physiological responses to environmental change—both internal and external environment. Whether it is sensory or motor function, autonomic function or adaptive functions (emotional or cognitive responses), it is the executive control of behavior in response to drug or chemical exposures that provide the most essential framework for risk-to-benefit analysis of the NME. Motor function is totally unified; the body moves as a whole in a highly coordinated manner.

Like the simple neurological examination of the human patient, Table 4 shows the basic sections of a standard neurologic examination of the rat. One big difference in the FOB examination is that the rat cannot verbally communicate issues of concern, no mental status exam is conducted, no subjective reports of what is occurring are communicated by the subject, and no medical history can be reported. The second difference, and probably the most important feature between animal and human neurological evaluation, is that prey species have an innate genetic pressure to mask their sickness [32] or any weakness [33–36], or drug-induced response that serves to complicate even a standard veterinary diagnosis [37]. For example, it is generally accepted that the vagus nerve acts as a key interface between the brain and the peripheral internal organs. Kobrzycka et al. [38] have provided convincing evidence that suggests animals without a vagus input to the CNS, communicate information about the internal milieu (illness, pain, etc.) to the CNS through a pathway independent of that vagus nerve. The genetic pressure to survive initiates alternate pathways of communication to permit restoration of CNS activity associated with peripheral inflammation control and to mask any signal to cohorts of weakness or vulnerability that may alter dominance/submissive behaviors or susceptibility to other species. The FOB is a systematic and ordered series of observations and measurements of behavioral, physiological, and sensory changes that occur in response to drug administrations.

The FOB screening exam is an amalgam of the function and systems approaches of animal observation that focuses on speed and efficiency by trained technicians in research laboratories that conduct studies under federal regulatory agency Good Laboratory Practice guidelines (GLPs; 21 USC, Ch 13, §58.1). The conceptual framework for using observable behavior to screen for toxicity has been attributed to Irwin [39]. The screening examination of motor and sensory functions and the coordination of upper and lower extremities can be conducted easily by a trained laboratory technician with even a brief history of handling and observing purpose-bred laboratory rats. The examination of freely moving laboratory animals requires the observation of animals performing at a high

Table 4
Major sections of the standard examination (some redundancy exists in biomarker for deficits)

Target	Standard FOB parameter
Mental status examination	Demeanor of animal, response to cage removal, cage cohorts, human handling
Cranial nerves	Pupillary reflex, ptosis, bilateral symmetry of facial movements, grimacing, vibrissae movements, tongue movements (home cage lixit or food), chewing, swallowing, pinna reflex, response to auditory stimulus
Motor function	Open field, rearing counts, ambulation counts, automated locomotor activity monitor, grip strength, air-righting reflex, pulmonary function (diaphragm and abdominal muscles)
Sensory	Presentation of auditory stimulus (click), observation in home cage feeding and drinking, orienting response to stimuli, touch, tail pinch, pinna reflex, handling by observer, limb placement during ambulation and air-righting reflex test, body temperature
Reflexes	Imbedded in above targets, plus hot-plate test, tail flick response or hindlimb splay tests, pinna reflex, air-righting reflex
Cerebellar function, coordination	Open field, automated locomotor activity chamber counts (LMA), air-righting reflex, hindlimb splay, ataxia, laterality of motor function and coordination, prosody of movements in open field.
Gait and station	Open field, grip strength, ataxia, left front paw, right rear paw positioning

level. A simple task that relies heavily on sensitive observable signs of behavioral change can identify the flawless execution of complex functions (e.g., air-righting reflex and hindlimb splay measurements). If a laboratory rat suspended in the air can orient itself, turn around in midair and land on its feet all within 18 inches of "air space" you can rest assured that the integration of vestibular function, motor cortex, and basal ganglia and cerebellar functions are intact.

4   Neurobehavioral Screening in Rodents

Neurobehavioral screening for nervous system safety in purpose-bred laboratory animals prior to delivery of the first drug dose to humans is required by law. It is an important component for the neurotoxicity potential for all new drugs, environmental chemicals, and food additives. A protocol that includes a framework for the systematic recording of observations and controlled assessments of standardized performance measures that have been validated within each research laboratory using prototypic drugs (positive control comparators) from the same therapeutic or pharmacological class as the new drug of interest will serve as the basis for determining if enough information is available following the study to answer the question, "do we know enough?" relative to human safety. Tier I study completion will determine if further testing is required (Tier II testing) in order to identify or characterize the risk, hazard, or threat to humans in clinical trials suggested by the initial study results. A nervous system safety protocol is composed of observations of (1) home cage behaviors (social order, dominance, submissive postures, general health status), (2) the FOB (manipulative and non-manipulative test items), and (3) an assessment of locomotor activity.

A variety of endpoints (parameters) should be chosen to assess an array of neurological functions, including autonomic, neuromuscular, sensory, and physiological. Most, if not all, of these protocol endpoints should be used in the context of a broad, full spectrum, neurobehavioral test battery where judicious selection of parameters is based on the known or expected pharmacology of the new drug.

Moser [40] has previously defined a set of endpoints typical of what is expected in a standardized regulatory-based functional test for safety assessments:

5 Regulatory Perspectives on FOBs

Safety evaluation has been defined as the scientific process of assessing the potential for drug administration to adversely affect human health [42]. Similar studies are conducted in purpose-bred laboratory animals with chemicals regardless of their commercial target such as pharmaceuticals, pesticides, consumer products, and/or industrial or agricultural products [43]. Current risk assessment plans have remained largely unchanged for decades [44]. In 2007, the U.S. National Research Council of the National Academy of Sciences [45] published a blueprint for migration of all regulatory agencies away from a "high-dose" animal testing approach (e.g., lethal dose [LD50] determinations) to a more tailored "pathway" approach from in vitro testing to predicting the extent of exposure to consumers that are safe, effective, and provide a minimal risk for harm. The NRC recommended a universal layered or tiered testing approach that focused the attention of all in vitro tests and all preclinical (animal) testing to one common goal—to establish human exposure guidelines on risk avoidance not harm identification (death or organ failure). The new testing method shifted primary focus from "mechanism of action" to "mode-of-action" information. It was not that the mechanism of action had limited importance, the mode-of-action information was deemed critical for identifying the dose–response component of the risk assessment paradigm. Risk assessment superseded the hazard paradigm that predominated regulatory programs in the 1950s and 1970s [45]. The "tiered testing" approach to regulatory drug testing shifted away from a focus of demonstrating adverse health effects in experimental animals toward a more detailed understanding of biological disruptions in key toxicity pathways that may lead to adverse health outcomes; for safety pharmacology, these pathways are pulmonary, cardiovascular, neurological, and if identified by experiential data other pathways became of interest (i.e., renal and gastrointestinal pathways). The tiered testing of toxicity testing utilizes a step approach to generate pertinent data for more efficient assessments of potential health risks posed by an new drug taking into consideration available knowledge of the drug, similar drugs in the same pharmacological class, its mode or mechanism of action, its intended use, and estimated exposures [45, 46]. The systematic development of tiered decision-type selections of more limited "suites" of animal tests would conceivably allow for reduced animal use and allow flexibility in testing based on risk-management information needs at every step of product development. Under the ICH-guidelines for safety pharmacology testing [23, 47], the risk assessment questions are simply focused on the relative risk imposed by a single dose of drugs prior to its administration in the initial trials in humans. After completion of the safety assessments, the Sponsor asks a single simple question, "do we know enough to continue?" Tables 5 and 6 lists, in no implied order of importance, some of the safety concerns we would like to know before we administer the dose to healthy, clinically screened volunteers with respect to immediate health crisis (Table 6A) but also critical symptom-inducing effects of the drug in well-screened healthy normal volunteers that may jeopardize the initial patient enrollment, continued participation once a single dose is administered, or future clinical trial enrollments (Table 6B) compared to future repeat-dose studies (Table 7).

The term "weight of evidence" (WoE) is used to convey the current thinking of the FDA and other drug regulatory agencies to accept a diverse family of approaches that can provide multiple lines of evidence in support of a drugs' safety or adversity [43, 48, 49], No single approach is favored and no approach is automatically rejected in the current drug approval environment. The WoE approach is a process of assembling evidence, giving a regulatory value (scientific valence or "weight") to the data and, in the final review process giving "weight" to the body of evidence in its entirety. The process of weighting evidence is not always simple since the specific focus of interests of each FOB may be different. That is, the targets of interest for the assessment of relative adversity or safety assessments of a new drug prior to the first dose delivery in medically screened, healthy patients volunteering to participate in Phase I clinical trials may be different from the targets of interest for safety assessments needed to allow repeated dosing of drugs in a clinically relevant patient population of Phase III clinical trials. The parameters that make up the FOBs in early safety pharmacology

Table 5
Endpoints for inclusion in a nervous system safety screening battery (derived from Moser [40]; and Baird, Catalano, Ryan, and Evans [41]

Observational assessments	Manipulative tests
Activity/Arousal levels	**Neurological Reflex**
Home Cage Observations	Pupillary response
Open-Field Observations	Palpebral reflex (blink response)
Rearing Counts (Assisted, Unassisted)	Pina reflex
Thigmotaxis (wall walking)	Extensor thrust reflex
Time Spent in Center of Field	
Number of Quadrants Crossed	**Neuromuscular tests/ Postural Reactions**
Reactivity/Excitability	
Reactivity to Environmental Change	Grip Strength (forelimb, hindlimb)
Reactivity to Handling	Landing Foot Splay
Arousal (State or Trait Changes)	Hopping
	Air Righting Reflex
Gait, Stance, and Posture	Platform Righting Reflex
Descriptive or Rank Scoring	Stride Length / Stride Angle
Laterality of Function	
	Sensory Integration
Involuntary/Abnormal Motor Movements	Visual Test: approach response
Tremors	Visual Test: visual placing
Fasciculations	Somatosensory test: touch response
Clonus	Auditory Test: Click response and Pinna orientation reflex
Tonus Stereotypy (repetitive behaviors with no functional significance) Bizarre behaviors Convulsions	Nociceptive Test: Tail/Toe pinch Flexor Reflex Tail Flick (thermal stimulus) Hot Plate (paw lick latency) Tail Immersion (hot water) Hargreaves Test (thermal) Whisking (Vibrissae movements)
Clinical Signs	Auditory Evoked Seizure Susceptibility
Lacrimation	
Salivation	
Hair coat condition	

(continued)

Table 5
(continued)

Observational assessments	Manipulative tests
Palpebral Closure	
Ocular Abnormalities	
Muscle Tone	
Body Mass/Body Weight	
Grooming Status (complete, partial)	
Pica (eating of non-nutritive substances)	
Gaping (oral, facial)	
Facial expressions	

Table 6
Partial list of identifiers of unexpected findings following first-dose in man

A. Immediate Health Hazards (S7A)	
Cerebrovascular accident (induction of Torsades de Pointes)	Distorted reality, hallucinations,
Seizure/tremor/convulsion	Syncope
Respiratory arrest or depression	Thermal homeostasis (hyperthermia/hypothermia)
Hemodynamic instability	
B. Health Risks that Predict "Drop Outs" or Hinder Continued Participation	
Unexpected subjective experiences:	Unexpected symptom inductions:
Anxiety	Headaches/migraine
Panic	Jitteriness, unsteadiness in gait

Agitation	Visual disturbances
	    double vision: diplopia
	    dry eyes: keratoconjunctivitis sicca
	    nystagmus

Dizziness	Vertigo
Nausea	Vomiting
	Diarrhea
	Acute compensatory response as plasma concentrations approach zero (hangover)

Table 7
Partial list of findings that develop over the course of repeated dose administrations (not single dose)

Tolerance to target therapeutic effects
  Need to increase dose with treatment length
Sensitization to off-target adverse signs
  Unmasking of deleterious effects with exposures (seizure, tremor, fasciculations)
	Development of Metabolic Injury
	Renal damage
	Liver damage
	Hypoglycemia/Pancreatic dysfunction

Oxidative stress injury, apoptosis, necrosis	Axonopathies, synaptic injury, CNS pathology
Free radical-induced damage (nitrones, hydroxyl radicals, etc.)	

protocols may be different from the parameters included in FOBs in standard toxicology IND-enabling study protocols. The weighting evaluation of each parameter included in the FOB requires both knowledge (information), beliefs (weights), and a systematic, logical, and transparent scientific method to integrate them [43, 49–52]. According to Kraft et al. [53], the lack of data examining silent neurotoxicity increases the possibility that drug approval agencies are being provided an incomplete description of potentially sensitive animal subjects who were exposed to a complex combination of drug and environmental factors that are capable of unmasking neurotoxicity. Overall, this continues to represent a critical and controversial topic for neurotoxicity risk assessment in the drug development industry.

6   Flexibility Provided by FOBs

Using the tiered testing approach, there may be selected parameters that confer an advantage to Tier I FOBs that may have less utility in Tier II FOBs and vice versa. Specific measures of neurotoxicity may add value and flexibility to target-specific types of toxicities, populations, or risk assessments expected in a single dose study design which then may be more readily interpreted for "first dose in man" decisions [42]. FOB parameters could be pre-sorted into levels according to sensitivity or selectivity with respect to the "weight of evidence" required for each tier of testing [43].

Important to note here is that the ICH safety pharmacology (S7A) guidelines [4] recommends a "Modified Irwin or FOB" screen for CNS safety assessments, but neither the FDA or the ICH identifies any particular set of parameters of interest that would adequately address the concept of "safety prior to the first dose administration in man" in Phase I Clinical Trials. Most published studies of nervous system safety testing report a list of parameters from the US Environmental Protection Agency

Neurotoxicity Guidelines OPPTS 870.6200 Neurotoxicity screening battery [54]. The stated purpose of that FOB is:

...The functional observational battery consists of noninvasive procedures designed to detect gross functional deficits in animals and to better quantify behavioral or neurological effects detected in other studies...*This battery is designed to be used in conjunction with general toxicity studies* and changes should be evaluated in the context of both the concordance between functional neurological and neuro-pathological effects, and with respect to any other toxicological effects seen. This test battery is not intended to provide a complete evaluation of neurotoxicity, and *additional functional and morphological evaluation may be necessary to assess completely the neurotoxic potential of a chemical.*

It seems intuitive then that over the years there has been some concern about the utility, predictive validity, and usefulness of the CNS Safety FOBs within the pharmaceutical drug development industry in papers appearing in peer-reviewed scientific journals (for example, refer to [55–60]). Based on reviews of safe and effective drugs approved for licensure by the FDA, few ICH S7A FOB reports identified clinical signs that were subsequently documented in Phase 1 through III clinical trials for the compound or failed to predict reported adverse events that were identified following post-marketing approvals. Three decades ago, Sette [61] and Moser [62] were two of many public and vocal defenders for the use of "behavioral change" to assess drug adversity in live purpose-bred animals. As described by Sette, the determination of safety requires:

a comprehensive description of all the available data, consideration of the severity and the duration of the effect[s], the homology between the measured effect and a toxic effect, and the social values of those making the judgment ([61]; p. 415).

Evans [63] has defined an "apical test" for neurotoxicity as one that reflects a change at any of the many integrated centers within the nervous system. Two examples presented by Evans are body weight and spontaneous locomotor activity. Irwin [39] first promoted the use of several simple endpoints (a test battery), rather than one or two very complex parameters to use as an apical test for Tier 1 testing since the combined observational platform casts a wide net over a set of common activities of daily living. As Moser [10] has suggested, the behavioral effects of drugs may be a result of disruptions in nerve cell communication and integration as well as a morphological change in the CNS that represents the integration and integrity of the neural pathways. Behavioral change is generally considered to be a sensitive indicator (a biomarker) and perhaps the most sensitive assay of neuronal function following drug dose administration [2–4, 11, 20].

At the "first go-around" of testing, it is not imperative to decide all alternative mechanisms which might explain an observed behavioral change. Identification of specific mechanisms of the nervous system modified by drug administration is the focus for higher tiers of screening. If that same rat is placed into the center of an open field and shows a rapid scurrying to walls and within a 3-min observation period returns to the center of the box, you can rest assure that a simple form of cognitive processing (habituation) is intact. If a Tier I assay can demonstrate that the CNS can perform a complex task perfectly, it is very unlikely that there is significant pathology present, and going through a more complex or expensive series of Tier II functional tests is not likely to prove productive.

The FOB is an observation plan that represents a "shot-gun" approach to identify safety-related (single dose) and toxicology-related (repeated doses) assessment and signs of "behavioral disruptions" that are sensitive biomarkers of pathology that is both time- and cost-efficient. The parameters of the FOB "cast a wide net" [63] in order to collect evidence of nervous system change at a global level of analysis. The screening examination of motor pathways, sensory inputs, and the integration of these functions to produce coordination of purposeful, directed, and unrestricted movements can be done in a single multifaceted observation period by a trained animal technician under the GLPs.

There *should not* be a unitary or single harmonized FOB protocol that is used during the full development pathway from drug discovery (efficacy) to safety assessment, developmental and reproductive toxicology (DART), through repeat-dose toxicology assessments, and carcinogenicity risk assessments [64]. Each phase of drug development should be designed to answer specific questions regarding "risk" or "hazard" predictions. There are no expectation of severe nervous system changes following a single acute dose of a new drug during ICH S7A safety assessments [8] required prior to dose administration in humans when compared to the possible panorama of nervous system-related changes induced by repeated high-dose administrations required for licensure by the drug regulatory agencies. For example, a single dose of drug in a safety assessment study most likely will not induce CNS axonopathies that might occur over 3 or 9 months of daily dosing in a required toxicology assessment study. Under the current government, regulatory environment published guidelines that delineate the parameters of the FOB have the regulatory weight of a guidance that is not legally binding on the agency or the drug company.

Drug development programs often select an animal model without any concerns for the predictive utility to the human condition [27]. More critical to the drug approval process, this lack of relating animal models to human responses leaves drug regulatory agency reviewers in a state of confusion in trying to establish permissible exposure levels for humans when only animal data are available [27]. For decades, antivivisectionists have rallied around the position that "mouse is not a rat is not a man" [65]. For example, one common adverse event reported in human patients voluntarily participating in Phase I clinical trials is nausea and vomiting, but the ICH S7A CNS safety study [8] is recommended to be conducted in rats—a species that does not vomit, regurgitate, or express "vulnerability" in the well-established dominance/submissive social order of socially housed rodents. While observable bouts of vomiting or the induction of malaise suggestive of nausea are not possible in the purpose-bred rat, there are other observable homologous behaviors expressed by rats that can be scored in the CNS safety FOB that would predict vomiting in humans. Predictive validity of the behavior is the scientific foundation for risk and safety assessments, not face validity.

The changes in the expression of normal ongoing behaviors from a documented baseline set of behaviors are used to identify and quantify the magnitude of a drug's impact on the neurological integration of the nervous system. A trained neurologist can conduct a four-minute neurologic screening examination of a human patient and accurately identify the functional integrity of the nervous system and identify further finer-grained analyses that may be required to make an accurate diagnosis of the patients' complaint [66]. Due to the temporal relationship between dose administration and neurological evaluations in preclinical safety assessment assays used for agency submissions the primary diagnosis of concern is "drug-induced toxicity." The rat FOB takes a well-trained animal technician approximately the same time (4–5 min) to conduct as the human neurological examination and provides credible, reliable, and valid changes of the functional integrity of the animal subject's nervous system when critical sample size requirements are maintained.

A fundamental tenet of pharmacology is that all drugs have multiple effects. Behavior is adaptive. It is behavior that is the major mechanism by which an animal adapts to changes in the environment, including experimenter-administered drug doses. The behavioral changes induced by drug administration will vary with experience. It is the change in behavior that objectively verifies that the effects of drugs are impinging on the nervous system of the animal. So while the core elements of the FOB may remain the same across the course of the drug development program (safety pharmacology, acute single dose studies, repeat-dose studies, etc.), the significance of any behavioral change in one of the many single parameters quantified in the FOB may have a differential valence of concern.

Our laboratory has previously defended the position that the "Irwin Screen" [39] and the FOB are the same assay [67]. The rat FOB is mandated as a standard test for:

1. The relative safety of a new drug entity prior to the first dose administration in man (safety pharmacology).

2. Demonstrative evidence of adverse effects of the drug to normal physiological functions when administered acutely.

3. When repeatedly administered at greater than expected human exposure levels (general toxicology).

4. To actively search for any long term or enduring changes in homeostasis following both acute and long-term exposures to the drug of interest.

The reliability of the nervous system like its flexibility, speed, and predictability is used in judging the relative safety of new drugs prior to being approved for general consumption. The CNS itself is not always "foolproof." Sensory inputs are quick, reflexes almost instantaneous, and behavioral responses are as fast as they need to be for a given situation. Nerve centers seldom make mistakes, and pharmacological actions of a new drug on a substantial portion of a population of neurons in the brain may not lead to a disruption of function because a residual set of "like-minded" circuits provide redundancy and reliability to maintain homeostasis. However, the CNS of any living organism can ill afford to misdirect or inadequately address emergency messages pertaining to the maintenance of the well-being of the body (respond to pain, react to threat, etc.). The CNS of all living animals represents more than information processor, integrator, and regulator of the behavioral output to survive in its environment. The CNS takes an active role in learning and planning; it assesses and weighs the physical, psychological, and behavioral demands based on need and within the context of past, present, and even the likelihood of future stimulus-response contingencies. The FOB is intended to assess if and how a new drug alters these primary functions for survival. The FOB is a critical part of the preclinical armamentarium that is used to make decisions in the standard risk-to-benefit ratio that is sensitive in identifying the missed or inaccurate signal that initiates or maintains the rate or strength of ongoing behavior. The consequence of a single, acute drug administration may be a minor "glitch" in the integration, such as a bit of clumsiness or ataxia in the open field, or a slowness in responding to a tail pinch, but the animal maintains its ability in the activities of daily living—eating, drinking, interacting with cage cohorts, defecation, urination, and sleep.

For example, in the New Drug Application for Victoza™ (NDA#22–341; liraglutide) described the "Neurological and behavioral effects" from single acute dose administrations in the Safety Pharmacology assessments (Section 2.6.2.3, page 43 of 513):

> Neurobehavioral effects of subcutaneously administered 0 (vehicle), 0 (saline) **0.02, 0.2, or 2 mg/kg** NNC 90-1170 (liraglutide, sic) in **male NMRI mice** (6/dose) were assessed in a functional observational battery performed 0.5, 1, 2, 4, and 24 h after dosing and observations in the home cage, and during handling...**There were no liraglutide-related findings**.

Increased incidence of exploratory activity at 0.02 mg/kg (low dose, sic) occurred 0.5–4 h after dosing was considered equivocal because it didn't occur at higher doses. . . .

And in Section 2.6.6.2 of the same pharmacology review, the effects of a single acute *intravenous or subcutaneous dose of 10 mg/kg* liraglutide in *CD-1 mice* were reported to have no clinical signs of toxicity, no unscheduled deaths, and no necropsy findings. In contrast, a single acute *10 mg/kg of subcutaneously* administered liraglutide in Sprague-Dawley (SD) rats induced "hunched postures on Day 2, and hair loss occurred on the limbs, ventral, dorsal, and sacral regions.". What is the most scientifically prudent or most predictive conclusion of risk assessment based on these Tier I single dose summary statements? From a safety pharmacology perspective does Day 2 post-dose findings of "hunched posture in rats, hair loss (alopecia areata) following a single test article dose meaningful? Are these findings suggestive of "risk" when combined with reports of "increased incidence of activity" in mice? In determining the "ethical weight" of evidence to clinical signs observed during toxicity testing, Ringblom et al. [68] rated "hunched posture" a higher threat score when compared to signs of tremor, piloerection, porphyria, and convulsion with respect to severity of threat. Observations of "hunched postures" in rats have been considered predictive of: (1) a premonitory sign in drug-induced seizures in laboratory animal species [69] and (2) drug-induced nociception (pain), distress, stress, and /or toxicity [70]. Observations of alopecia areata following a single dose administration in rats has been reported with cancer chemotherapy drugs and is supportive of a risk response of the immune system [71], characterized by antigen presentation and co-stimulation of lymphocytes in the lymph nodes and skin, a deficiency of CD4+/CD25+ regulatory cells, and an action of activated lymphocytes on hair follicles via Fas/FasL signaling and cytokines—all suggestive of drug-induced stress/distress. A rapid onset of hair loss following a single dose of drug may also suggest a direct toxic effect on functional immune system regulatory cells such as Th1 cells [72]. Type 1 T helper (Th1) cells produce interferon-gamma, interleukin (IL)-2, and tumor necrosis factor (TNF)-beta, which activate macrophages and are responsible for cell-mediated immunity and phagocyte-dependent protective responses [73]. Th1 cells are involved in the pathogenesis of organ-specific autoimmune disorders. While technicians are observing behavior, it is behavioral change that is predictive of significant organ toxicity well before pathological changes are found histologically. The FDA-approved Victoza™ in 2010 as a safe and effective drug for the treatment of Type 2 diabetes. The drug has enjoyed 10 years of relative safety and has achieved subsequent approval for pediatric patients. This example of interpreting nonclinical research findings highlights the critical value of rodent FOBs in the initial nervous system safety assessments conducted under the ICH.

Some high-profile safety-related issues have emerged for FDA-approved and widely prescribed drugs, and there is an estimated 30% of developmental drug failures related to safety/toxicity issues [74]. FDA and the drug industry itself have provided scientists an avenue of innovative new strategies and tools through which they can make better early-stage predictions regarding a drug's ultimate safety.

7   Adversity: Seen and Unseen

Weiss and Reuhl [75] warned of the concept of "silent damage." Kraft et al. [52] have defined "silent neurotoxicity," also referred to as silent toxicity or silent damage to the nervous system, as a persistent biochemical change or morphological injury to the nervous system that does not induce overt evidence of toxicity (i.e., remains clinically unapparent) unless unmasked by experimental or natural processes [76, 77]. Reuhl [76] has suggested that exposure to neurotoxic compounds may be followed years or decades later by clinically evident neurological disease. An example of silent nervous damage is the Syndrome of Irreversible Lithium-Effectuated Neurotoxicity (SILENT) which presents with permanent neurologic deficits that persist long after the acute toxic reaction. These sequelae are more often cerebellar symptoms but dementia, parkinsonian syndromes, choreoathetosis, brain stem syndromes and peripheral neuropathies defined as *irreversible* if they persist more than 2 months after the interruption of lithium treatment [78, 79]. Kraft et al. [52] suggests that silent neurotoxicity overlaps with other commonly used terms, including, perhaps most closely, the "2-hit" or "multiple hit" models from developmental and reproductive toxicology (DART) journals. Each of these terms describe the occurrence of initial insults that can alter cellular function and prime a system to either make it vulnerable to a subsequent insult(s) or progressively lead to loss of normal function with additional insult(s), with the effects of one insult in isolation being insufficient to induce disease. Paradoxically, nonclinical animal studies conducted with FOBs according to regulatory agency guideline protocols for use in risk assessments are not designed to detect these types of silent neurotoxic effects.

Evidence has emerged linking environmental factors to several neurodegenerative diseases, including amyotrophic lateral sclerosis and the Parkinsonism-dementia complex, and while these diseases have not been definitively demonstrated to arise from exposure to a specific toxin, the possibility exists that xenobiotic exposure could lead to neurological conditions possessing a period of clinical "silence" prior to observable signs of a compromised CNS. For example, a single, high, acute dose of alcohol can induce perturbations of the CNS appearing approximately 12–14 h later (ethanol hangover) [80] and is similar to signs of CNS damage related to a classic ethanol withdrawal syndrome seen after years of daily consumption [81]. Silent or delayed neurotoxicity is a progressive, cumulative progression of damage that eventually "blooms" into a compromised state. Silent CNS damage has been demonstrated by an FDA-approved antibiotic. Cefepime is a widely used antibiotic with neurotoxicity attributed to its ability to cross the blood–brain barrier and exhibit concentration-dependent gamma-aminobutyric acid (GABA) antagonism that was not identified at the time of license approval (US FDA; New Drug Application [NDA] #0505679, FDA@Drugs, page 59 of 1256). Neurotoxic symptoms include depressed consciousness, encephalopathy, aphasia, myoclonus, seizures, and coma. Payne et al. [82] have suggested that up to 15% of intensive care unit (ICU) patients treated with cefepime may experience these CNS neurotoxicity effects. Many other common FDA-approved chemotherapeutic agents may cause silent CNS toxicity manifested as encephalopathy of various severities. Additionally, involvement of the peripheral nervous system manifested as distal peripheral neuropathy results after other FDA-approved therapies (e.g., cisplatin, vincristine, taxanes, suramin, and thalidomide) [83]. The combined action of drug exposure and another "causative factor" such as normal aging or a pharmacological challenge may unmask the underlying damage, such as purported changes in seizure thresholds following long-term repetitive voluntary use of benzodiazepines, or the effects of lead or mercury in youth and expression of cognitive deficits later in life. It is the fact that after any drug-induced injury the CNS undergoes a series of adaptive and compensatory responses to attempt to return to "normal." The nature of the CNS change is critically dependent on the dose and duration of the drug effect [75].

On April 30, 2019 the FDA issued a safety announcement advising that rare but serious injuries have happened with certain common prescription insomnia medicines because of sleep behaviors, including sleepwalking, sleep driving, and engaging in other activities while not fully awake [79]. These complex somnambulistic behaviors have also resulted in deaths. These behaviors appear to be more common with three FDA-approved drug products: (1) eszopiclone (Lunesta), (2) zaleplon (Sonata), and (3) zolpidem (Ambien™, Ambien CR™, Edluar™, Intermezzo™, Zolpimist™) when compared to other prescription medicines used for sleep [84] and occur as plasma concentrations approached their nadir. Two of the publicly available NDA pharmacology review documents used for drug license approvals (NDA# 021476—Lunesta, and NDA#020859—Sonata) on the FDA@Drugs website do not identify a single nervous system predictive marker in any of the nonclinical standard FOBs that should even suggest the presence of this "silent adverse effect" of these drugs. It may be interesting to note that the standard regulatory-preferred species used for CNS safety assessments is the purpose-bred laboratory rat that characteristically have four basic sleeping postures [85, 86]: 2 daytime sleeping postures and 2 night-time sleeping postures. Standard preclinical safety assessments are conducted during only one of these sleep intervals (daytime sleeping) which excludes the sleeping pattern characteristic of the second typology (night-time sleeping). There is no clear parameter in the standard ICH-based safety pharmacology FOBs that would help to identify which of the two patterns of sleep should be included in the safety FOB that would be predictive to unmask the "silent adverse effects" of these sleep aids on the induction of somnambulism in the rat. In these cases, the standard FOBs may have been supplemented with other parameters that might have predicted the "silent" somnambulism induced by these drugs. The "biomarker" that may have identified the need to conduct Tier II sleep studies to address this "silent adverse event" were missed during the nonclinical programs of each of the three known drug substances that required the post-marketing addition of the "black box warnings."

8   Tailoring to the Test Article: Therapeutic Target and Mode of Action

It is the series of FOBs conducted by trained animal technicians that provide the evidence that is needed by regulatory approval agencies to establish relative safety, risk, and hazards to human patients prior to allowing access to the drug by the general population. Studies on the mode of action and/or effects of a substance in relation to its desired therapeutic target are primary pharmacodynamic studies. Studies on the mode of action and/or effects of a substance not related to its desired therapeutic target are secondary pharmacodynamic studies (these have sometimes been referred to as part of general pharmacology studies) ([8], Note 2, page 8).

It is the flexibility of choosing a set of parameters that make up the FOBs at each point of the drug development process (safety pharmacology vs toxicity studies) and the features of the tier-testing strategies that confer significant scientific and operational advantages of the current regulatory environment over the non-stratified toxicity testing of the last 5 decades of drug approval. The basic characteristic of the tier-testing system provides the most efficient use of animals and the most sensitive barometer of risk to the CNS readily interpretable risk assessment [17].

Screening of motor function [87] is critical in determining the safety of drugs to cortical spinal pathways. Laterality of function involving the motor and sensory cortex, cerebellum, basal ganglia (caudate, putamen, and globus pallidus) and spinal tracts are all targets of drug-induced toxicity that have the potential of compromising human performance and therapy compliance in the targeted patient population. Animals cannot tell us if an experimental drug is inadvertently inducing nausea, malaise or agitation, and yet we need to know this information before advancing the drug to human trials. There is little hope of retaining human subjects in clinical trials if the drug alters or adversely affects their well-being. As proposed by MacPhail, [29], while we certainly would agree that if any animal is feeling *sick* following drug administration, behavioral change seems reasonable. However, for drug regulatory approval we must frame the issue in objective or measurable terms. If the basis for describing an animal as "***sick***" is a change in behavior, it becomes a circular argument to explain this change in behavior on the belief that the *sickness* itself is that change in behavior [29]. The concept of *malaise* has also been misused in interpreting drug-induced changes in behavior [63]. As with the concept of *illness*, users of the term *malaise* express belief that behavioral changes can be interpreted as the consequence of non-specific actions of the drug, thus casting a cloud of doubt on the mechanisms of that behavioral change. For Evans [63], *malaise* is no more an acceptable account of animal behavior that it is a diagnosis of human adverse events in clinical trials. *Malaise* can be applied to all changes identified in the functional domains of the FOB (physiological, sensorimotor, autonomic, and neuromuscular) for which the controlling variables have not been identified but the use of this "explanation" only begs the question of CNS mechanisms involved.

Optional tests to add to an FOB to target a finer-grain analysis of normal stance, gait, balance, and equilibrium that would be sensitive to human patients' reports of subjective state changes such as feelings of malaise, disinterest, "unsteadiness," "light headedness" are shown in Table 8 below. For example, a simple horizontal ladder test can be added to the FOB that is a motor and coordination test for evaluating skilled walking and fore- and hindlimb coordination. By removing individual ladder rungs making the stride longer one can evaluate how the rodent targets individual rungs. Impairments on this test can also be seen for drugs that impair or affect the spinal motor system (peripheral nervous system) and sensory-motor feedback loops between spinal alpha motor neurons and distal muscle sensory controls (gamma motor reflexes).

Another elementary test easily setup in the lab is the balanced beam test—a simple wooden dowel of 30–40 inches in length placed on an angle (30–40 degrees), allowing the rat to transit up the beam is a predictive behavior that does not need training. The balance beam is sensitive to changes in the accuracy and speed of paw placement, balance, and equilibrium, predictive of human "unsteadiness," "light headedness," and malaise.

Table 8
Proposed behavioral endpoints to include in standard ICH S7A CNS safety assessment FOBs to better address typical motor "side effects" or "adverse events" reported by human volunteers in clinical trial studies

Parameter	Human Correlate	Reference
String test	Balance, coordination of fine paw movements and grasp, equilibrium	Barclay, Gibson and Blass, [88]
Balanced beam test	Agility, balance, grip strength and accuracy, vestibular function	Combs and D'Alecy [89] Modianos and Pfaff [90]
Horizontal ladder test	Skilled walking, forelimb and hindlimb placement, stepping, coordination	Metz and Whishaw [91]
Behavioral arrest (freezing)	Basal ganglia Motor pathway integration	Roseberry and Kreitzer [92]

Since these additional tests can be added to the current standard FOB, the test options must be simple, sensitive, of short duration and does not require training or acclimation to the novel situation. The full selected battery of tests should be limited to approximately 10 min per rat in order to complete the scheduled FOB in all animals in the same work day. The ability in walking on the horizontal ladder or ambulate up the inclined dowel are within the normal expression of "rat-like" innate behaviors that can be easily included in the "shot-gun" battery of tests.

The disparity between rat FOB data and "side effects" reported in patients in clinical trials does not mean that the assay cannot be predictive. This issue demonstrates that it is important to consider the ethological relevance of the behavior for the species in question (rat) and it is important to modify existing tests or design new ones, depending on the question being asked and the experimental manipulation used to target the most predictive behavioral change in the rat. It also illustrates that the importance of using ethologically relevant biomarkers does not imply that there are no general features and common molecular mechanisms underlying behavioral phenomena (vomiting, headache, etc.). The knowledge of the species-specific characteristics of rat behavior is essential for discovering these commonalities and generalizing the findings of animal research to humans. The challenge is to design appropriate behavioral tests that can be included in the standard FOB which tap into the rat's natural neural networks for ethologically relevant behaviors in the rat. By knowing the core ethology of the rat ensures these behavioral tests are sensitive enough to detect the subjective changes of the rat's internal milieu that is currently not used effectively by many laboratories. To use a common idiomatic expression, "there is no reason to throw the baby out with the bath water." The common-industry FOB can be easily modified to capture these discordant animal–human safety concerns.

9 Ethologically Relevant Parameters

Weaver and Valentin [93] have recently concluded that some nervous system adverse events can only be detected in early humans clinical trials such as headache, dizziness, nausea, anxiety because all these human adverse events lack nonclinical in vivo models. But is this true? Just because a rat does not vomit or cannot communicate to the technician the severity of a headache doesn't mean that critically relevant behavioral changes in the rat are not present at the time of the scheduled FOB that can be used as a predictive index for nausea and vomiting in humans. It is essential for the continued use of the FOB by drug regulators that the data have constructive and predictive validity with respect to risk and hazard assessments. Therefore, the human observers (technicians) of the laboratory rat must have the background, education, and "hands-on" experience with the regulatory-preferred animal to be able to identify the species-specific behavioral changes unique to rats that predict nausea, vomiting, and headaches in humans. Whishaw [4] has referred to these as "ethologically relevant" behaviors—observing the rat can inform us that the drug it was administered has induced subjective states of nausea that accurately predicts observations of vomiting in humans in clinical trials. Additionally, observing the rat can identify when the rat is experiencing a subjective state equivalent to the human in clinical trials reporting "headache."

To help address the unmet needs of the standard ICH S7A CNS safety assessment FOBs [8], an additional set of parameters for the FOB that may target more common features of concern in predicting "side effects" in humans are shown in Table 9 below. Human complaints of vomiting, nausea, malaise, lethargy, "light headedness" can be better predicted by observing the rat in the FOB.

Based on the observations of rats in the Irwin Screen, these FOB parameters can be easily scored during cage-side observations, handling observations, open-field assessments or manipulative tests (hindlimb splay, grip strength assessments, anti-nociception tests, etc.) already being conducted in standardized ICH S7A or EPA/FDA FOBs in most pharmaceutical development programs.

Published reports of these behaviors show them to be sensitive behavioral indicators that best represent the expected challenges to the nervous system following a drug dose administration in rats. Drugs and experimental conditions that produce self-reports of nausea and vomiting in humans produce pica and gaping in rats;

Table 9
Proposed behavioral endpoints to include in standard ICH S7A CNS safety assessment FOBs to better address typical subjective or interoceptive "side effects" or "adverse events" reported by human volunteers in clinical trial studies

Parameter	Human Correlate	Reference
Pica	Nausea, vomiting	Batra and Schrott [94]
Gaping	Nausea, vasovagal response	Parker and Brousseau [95] Parker and Mechoulam [96]
Facial grooming	Headache, trigeminal nucleus caudalis, migraines	Pyun, Son, Kwon [97]
Facial action coding (grimacing)	Pain, neuropathic pain	Sotocinal et al. [98] Langford et al. [99]
Skilled reaching, handedness	Sensorimotor integration, Attention	Sacrey and Whishaw [100]

by adding simple observational scores of pica and gaping in scheduled FOBs, a better prediction of these "side effects" can be identified prior to the first dose in man. Repetitive facial grooming in rats is induced by drugs and experimental conditions that induce self-reports of headache in humans, by identifying and scoring the qualitative and quantitative episodes of facial grooming or "partial grooming" will increase the likelihood that the new drug will engender similar subjective effects when administered in human patients screened for inclusion into Phase I Clinical Trials.

10   Knowing the Rat Is Critical

The Good Laboratory Practice Guidelines (GLPs) establish the scope and responsibilities of the Study Director (SD). The role of the SD is not just to coordinate study requirements. The preamble of the FDA GLP regulations clearly stated that the SD is charged with the "technical conduct of a study, including interpretation, analysis, documentation and reporting of results." The OECD and US FDA places "ultimate responsibility" over all aspects of the scientific conduct of the study and the interpretation the study data solely on the shoulders of the SD. Regulatory agencies appoint the SD as the single authority responsible for "drawing the final overall conclusions from the study" data [101]. There are no specific regulatory-based qualifications of a Study Director—the main interpreter of FOB data on GLP-based safety assessment protocols. How much does the SD have to know about physiology, pharmacology, or the functional organization of the nervous system? The current regulatory and industry standards do not require the SD to be an expert in every specialty field of the study, merely that they should have sufficient understanding to work with specialists to coordinate, integrate, and interpret those integrated results. SDs are not required to have even a bachelor's degree in science or be board certified in any specific discipline of science or medicine. All that is required under regulatory/administrative guidelines is test facility management assignment to the position and documentation of sufficient background, education, skills, and training to meet the standards of the institution.

For technical staff, training should be specific for the species to be tested. For example, experience with mice does not fully qualify one to evaluate rats; there should be separate training for FOBs conducted in mice, rats, NHPs, canines, and swine. Training should include both basic handling (i.e., how to hold the animal, injection techniques) as well as behavioral testing. Crucial to the demonstration of drug safety is the knowledge of the sensitivity of rats to drug insults. Observer (technician) and reviewer (Study Director) must be familiar with the diverse spectrum of behavioral and physiological changes that can be expressed by this test system, and the important environmental, social, and historical factors that can mediate changes in the test system's response to new drug administrations.

While training is of high priority within research organizations, the most critically important decisions for test facility management is the operational definitions of "competency" and "proficiency." Understanding the regulatory requirements for safety assessment is not simply a matter of training or the successful completion of some performance test. As an SD designing a safety study, selecting or modifying a standardized set of parameters for inclusion in the FOB protocol and interpreting the results for regulatory submission is established through documentation of both competency and proficiency.

A reproducible and valid FOB study protocol and its report must stand up to regulatory scrutiny. Both protocol development and the data interpretation of the report are critically dependent on the background, education, and experience of the SD in knowing the "instrument of measure"—the test system (i.e., the rat) as well as a proficiency in interpreting the neurological basis of behavioral change. Here are some examples of errors in the development and interpretation of rat FOB safety pharmacology assessment studies:

Example 1: *One standard measurement in most rat FOBs is the quantitative measurement of "grip strength." The grip strength parameter is an objective ratio scale quantified measure. In assessing neurotoxicity, it is a categorical error to presume that muscles contributing to gripping power is a measure of real grip strength. Grip assessment involves a process of rats being introduced to a bar or screen* *of which it grasps to hold. Once the technician observes the grasping of the bars the rat is pulled lateral to the device and a force meter measures the gripping power to "break the hold of the animal." This behavioral assessment is a performance measure not a strength measure.*

Grip strength assessments first require the motivation to grasp the bar by the animal. Although commonly assumed by many labs that the force meter reflects a measure of strength, it is not. The fingers and wrist flexors of the rat [102], monkey [103–105], and human are not corticospinal innervated and may not be compromised in the presence of a postmortem histopathological finding of corticospinal lesions [106–108]. Grip is a complex function with many different muscles involved and is not significantly affected by peripheral neuropathy either. In interpreting these data, it must be understood that (1) drugs affecting motivation (barbiturates, CNS depressants) may alter the volitional (frontal cortex) drive to initially grasp the bars, (2) the drug may induce an emotional subjective state (e.g., anxiety, relaxation; amygdala) which may increase or decrease the clasping power of the rat when approaching the device, (3) the drug may relax the skeletal muscles to a degree that the rat cannot physically hold on to the bars of the device (benzodiazepines) or increase the muscle tension such that releasing a hold on the bar (methamphetamine) is delayed.

Example 2: *There is a strong connection between the status of the intestinal environment and the function of the central nervous system (CNS). This so-called gut-brain axis incorporates bidirectional communication between the central and enteric nervous and endocrine systems as well as regulation of immune responses in the gut and brain [109]. One of the most well-characterized effects of intestinal inflammation on the CNS involve hyper-reactivity of the hypothalamic-pituitary-adrenal axis and imbalances in serotonergic activity [110, 111]. These changes have been associated with the manifestation of "sickness behaviors," as well as anxiety and depression, and these psychological conditions are frequently observed as comorbidities in individuals with diseases characterized by persistent intestinal inflammation, such as irritable bowel syndrome (IBS) and inflammatory bowel disease (IBD), [112–114]. These and other systemic effects of intestinal inflammation are almost certainly mediated by a host of immune factors, but at present, the cytokines interleukin-1β (IL-1β), interleukin-6 (IL-6), and tumor necrosis factor (TNF) have been most frequently implicated. In the case of protocol development for new drugs to treat these disorders what use is the rat, the regulatory-preferred species that does not vomit?*

The observation of pica in the rat, the eating of non-nutritive substances, is a predictive biomarker of vomiting and emesis in rats who innately do not vomit [94]. Gaping [95, 96] is a reliable predictor of drugs that induce nausea in human patients. Selecting these two parameters to include in the FOB in rats will provide a valid and reliable FOB for new drugs focussing on these therapeutic targets.

These two simple examples are meant to highlight that when designing FOB studies the SD must have a working understanding of the basic ethology of the test system, the rat, and use the regulatory-based strategy of selecting parameters based on the therapeutic target or known pharmacology of the new drug being tested and not some predetermined template used for both safety and toxicology programs.

11   Conclusion

Observational data are only as valid and reliable as the observers (technicians) who conduct the scoring and the SD that interprets their findings. Behavioral data is inherently variable. Behavioral changes following drug administration are not always toxic or adverse, and the determination that behavioral effects are minor or not physiologically meaningful is not an indictment of the FOB or the usefulness of the data in evaluating drug safety [82]. Dyer [115] proposed a complex system for classifying endpoints for regulatory review. He suggested that defining the parameters included in the FOB in terms of their similarity to known human health effects. For Dyer [115], the behavioral change in rats may be the same as, apparently the same as, a model of, or unrelated to a human health effect, whereas the adverse events reported by human patients in clinical trials are classified on the basis of their relationship to human health, their willingness to participate in the clinical trial, and their likelihood of taking the drug therapeutically, if approved by the FDA. Calabrese [27] has suggested that the reason the rat is used in predictive toxicology is because it is relatively hardy, available in large numbers, costs little, gives sufficient blood for analyses, and no other animal model, in a general sense, uniformly simulates the response of humans to drugs (p. 283). For Sette [82] and Dyer [115], the FOB shows how we can classify neurological effects in rats based on their "adverse" nature and a relationship to "adverse effects" in human patients. The protocol study design and selection of FOB parameters should be based on a comprehensive description of all available data, consideration of the severity of expected drug effects, and the duration of the adverse effects. The acute single dose FOB required prior to the first delivered dose to a human patient should not be a "cookie cutter" copy of the 3- or 9-month FOB included in general toxicology programs. These FOBs should be tailored to the therapeutic target and the known pharmacology of the new drug being tested.

References

1. Moser VC (2011) Functional assays for neurotoxicity testing. Toxicol Pathol 39:36–34

2. Kulig BM (1996) Comprehensive neurotoxicity assessment. Environ Health Perspect 104 (Suppl 2):317–322

3. Tilson HA (1990) Behavioral indices of neurotoxicity. Toxicol Pathol 18:96–104

4. Whishaw IQ, Haun F, Kolb B (1999) Analysis of behavior in laboratory rodents. In: Windhorst U, Johansson H (eds) Modern techniques in neuroscience. Springer-Verlag, Berlin, pp 1243–1275

5. Weiss B, O'Donoghue J.L. (eds). (1994) Neurobehavioral toxicity: analysis and interpretation. Raven Press, New York, NY

6. US Environmental Protection Agency (1979) Teratogenetic/reproductive health effects. Fed Register 44:44,087, July 26, 1979

7. National Academy of Sciences (NAS) (1975) Principles of evaluating chemicals in the environment. National Academy of Sciences Press, Washington, D.C.

8. International Conference on Harmonisation (ICH) (2000) Safety pharmacology studies for human pharmaceuticals. Presentation on S7A. http://www.ich.org/fileadmin/Public_Web_Site/ICH_Products/Guidelines/Safety/S7A/Presentation/S7A_Presentation.pdf. Accessed 19 Jan 2020

9. Mandella RC Applied neurotoxicology. In: Derelanko MJ, Hollinger MA (eds) Handbook of toxicology, 2nd edn. CRC Press, Boca Raton, FL, pp 371–400

10. Moser VC, Kallman MJ (2018) 6.22. Behavioral screening for toxicology and safety pharmacology. In: McQueen CA (ed) Comprehensive toxicology (3rd ed), volume 6: nervous system and behavioral toxicology. Elsevier, New York, NY, pp 409–423

11. Tilson HA, Mitchell CL (1984) Neurobehavioral techniques to assess the effects of chemicals on the nervous system. Ann Rev Pharmacol Toxicol 24:425–450

12. Tilson HA, Moser VC (1992) Comparison of screening approaches. Neurotoxicol 13:1–14

13. U.S. Environmental Protection Agency (1994) Final report: principles of neurotoxicity risk assessment; notice. Fed Regist 59:42,360–42,404

14. U.S. Environmental Protection Agency (1998) Guidelines for neurotoxicity risk assessment. Fed Regist 63 (933):26,926–26,954

15. US Food and Drug Administration. (2016). Good laboratory practice for nonclinical laboratory studies. Proposed rule change. https://www.fda.gov/about-fda/economic-impact-analyses-fda-regulations/good-laboratory-practice-nonclinical-laboratory-studies. Fed register, docket number. FDA-2010-N-0548; notice of proposed rulemaking. https://www.regulations.gov/document?D=FDA-2010-N-0548-0088. Accessed 13 Jan 2020

16. Lorenz MD, Coates JR, Kent M (2011) Handbook of veterinary neurology, 5th edn. Saunders (Elsevier), Philadelphia, PA

17. Lorenz MD, Kornegay JN (2004) Handbook of veterinary neurology, 4th edn. Saunders (Elsevier), Philadelphia, PA

18. Bagley RS (2005) Fundamentals of veterinary clinical neurology. Blackwell Publishing, Ltd., Oxford

19. U.S. Environmental Protection Agency. (1998) Guidelines for neurotoxicity risk assessment. Fed Register May 14, 1998. 63, 933; 26,926–26,954

20. Norton S (1978) Is behavior or morphology a more sensitive indicator of central nervous system toxicity? Environ Health Perspect 26:21–27

21. Tilson HA (1987) Behavioral indices of neurotoxicity: what can be measured? Neurotoxicol Teratol 9:427–443

22. Anger WK (1984) Neurobehavioral testing of chemicals: impact on recommended standards. Neurobehav Toxicol Teratol 6:147–153

23. International Conference on Harmonisation (ICH) (2009) Guidance on nonclinical safety studies for the conduct of human clinical trials and marketing authorization for pharmaceuticals M3(R2). https://database.ich.org/sites/default/files/M3_R2__Guideline.pdf. Accessed 19 Jan 2020

24. Whitmore E (2012) Development of FDA-regulated medical products: a translational approach, 2nd edn. Quality Press, Milwaukee, WI

25. Lineberry N, Berlin JA, Mansi B et al (2016) Recommendations to improve adverse event reporting in clinical trial publications: a joint pharmaceutical industry/journal editor perspective. Br Med J 355:i5078. https://doi.org/10.1136/bmj.i5078

26. Bettge K, Kahle M, Abd El Aziz MS et al (2017) Occurrence of nausea, vomiting and diarrhoea reported as adverse events in clinical trials studying glucagon-like peptide-1 receptor agonists: a systematic analysis of published clinical trials. Diabetes Obes Metab 19:336–347

27. Calabrese EJ (1991) Principles of animal extrapolation. Lewis Publishers, Inc., Chelsea, MI

28. Tilson HA, Cabe PA (1978) Strategy for the assessment of neurobehavioral consequences of environmental factors. Environ Health Perspect 26:287–299

29. MacPhail RC (1994) Behavioral analysis in neurotoxicology. In: Weiss B, O'Donoghue J (eds) Neurobehavioral toxicity: analysis and interpretation. Raven Press, New York, NY, pp 7–18

30. Evans HL (1989a) Behaviors in the home cage reveal toxicity: recent findings and proposals for the future. J Am Coll Toxicol 8:35–52

31. Evans HL (1989b) Quantitation of naturalistic behaviors. Toxicol Lett 43:345–359

32. Johnson RW (2002) The concept of sickness behavior: a brief chronological account of four key discoveries. Vet Immunol Immunopathol 87:443–450

33. Avitsur R, Kinsey SG, Bidor K et al (2007) Subordinate social status modulates the vulnerability to the immunological effects of social stress. Psychoneuroendocrinology 32:1097–1105

34. Abrams PA (2009) Adaptive changes in prey vulnerability shape the response of predator populations to mortality. J Theor Biol 261:294–304

35. Matzel LD, Kolata S, Light K, Sauce B (2016) The tendency for social submission predicts superior cognitive performance in previously isolated male mice. Behav Proc 134:12–21

36. Cooper MA, Seddighi S, Barnes AK et al (2017) Dominance status alters restraint-induced neural activity in brain regions controlling stress vulnerability. Physiol Behav 179:153–161

37. Tizard I (2008) Sickness behavior, its mechanisms and significance. Anim Health Res Rev 9:87–99

38. Kobrzycka A, Napora P, Pearson BL et al (2019) Peripheral and central compensatory mechanisms for impaired vagus nerve function during peripheral immune activation. J Neuroinflammation 16:150. https://doi.org/10.1186/s12974-019-1544-y

39. Irwin S (1968) Comprehensive observational assessment: 1a. A systematic quantitative procedure for assessing the behavioral and physiologic state of the mouse. Psychopharmacologia 13:222–257

40. Moser VC (1999) Neurobehavioral screening in rodents, unit 11.2. Curr Protocols Toxicol 11(2):1–11.2.16

41. Baird SJS, Catalano PJ, Ryan LM et al (1997) Evaluation of effect profiles: functional observational battery outcomes. Fund Appl Toxicol 40:37–51

42. Becker RA, Plunkett LM, Borzelleca JF et al (2007) Tiered toxicity testing: evaluation of toxicity-based decision triggers for human health hazard characterization. Food Chem Toxicol 45:2454–2469

43. Collier ZA, Gust KA, Gonzalez-Morales B et al (2006) A weight of evidence assessment approach for adverse outcome pathways. Regul Toxicol Pharmacol 75:46–57

44. Adeleye V, Andersen M, Clewell R et al (2015) Implementing toxicity testing in the 21st century (TT21C): making safety decisions using toxicity pathways, and progress in a prototype risk assessment. Toxicology 332:102–111

45. National Research Council (NRC)/National Academy of Sciences (NAS) (2007) Toxicity testing in the 21st century: a vision and a strategy. The National Academy Press, Washington, D.C.

46. Carmichael NG, Barton HA, Boobis AR et al (2006) Agricultural chemical safety assessment: a multisector approach to the modernization of human safety requirements. Crit Rev Toxicol 36:1–7

47. International Conference on Harmonisation (ICH) (2005) The non-clinical evaluation of the potential for delayed ventricular repolarization (QT interval prolongation) by human pharmaceuticals S7B. https://database.ich.org/sites/default/files/S7B_Guideline.pdf

48. Balls M, Amcoff P, Bremer S et al (2006) The principles of weight of evidence validation of test methods and testing strategies. The report and recommendations of ECVAM workshop 58. Altern Lab Anim 34:603–620

49. Suter GW II, Cormier SM (2011) Why and how to combine evidence in environmental assessments: weighing evidence and building cases. Sci Total Environ 409:1406–1417

50. Linkov I, Welle P, Loney D et al (2011) Use of multicriteria decision analysis to support weight of evidence evaluation. Risk Anal 31:1211–1225

51. Chapman PM, McDonald BG, Lawrence GS (2002) Weight-of-evidence issues and frameworks for sediment quality (and other) assessments. Hum Ecol Risk Assess 8:1489–1515

52. Menzie C, Henning MH, Cura J et al (1996) Special report of the Massachusetts weight-of-evidence workgroup: a weight-of-evidence approach for evaluating ecological risks. Hum Ecol Risk Assess 2:277–304

53. Kraft AD, Aschner M, Cory-Slechta DA et al (2016) Unmasking silent neurotoxicity following developmental exposure to environmental toxicants. Neurotoxicol Teratol 55:38–44

54. US Environmental Protection Agency. (1994) Office of prevention, pesticides and toxic substances. Health effects test guidelines. OPPTS 870.6200. Neurotoxicity screening battery. https://nepis.epa.gov/Exe/tiff2png.cgi/P100G6U5.PNG?-r+75+-g+7+D%3A%5CZYFILES%5CINDEX%20DATA%5C95THRU99%5CTIFF%5C00002709%5CP100G6U5.TIF. Accessed 20 Jan 2020

55. Parkinson C, McAuslane N, Lumley C et al (eds) (1994) CMR workshop: the timing of toxicological studies to support clinical trials. Kluwer, Boston, MA

56. Olson H, Betton G, Robinson D et al (2000) Concordance of the toxicity of pharmaceuticals in humans and in animals. Regul Toxicol Pharmacol 32:56–67

57. Claude J-R, Claude N (2004) Safety pharmacology: an essential interface of pharmacology and CMR workshop—animal toxicity studies: their relevance for man toxicology in the non-clinical assessment of new pharmaceuticals. Toxicol Lett 15:25–28

58. Fonck C, Easter A, Pietras MR et al (2015) CNS adverse effects: from functional observation battery/Irwin tests to electrophysiology. In: Pugsley MK, Curtis MJ (eds) Handbook of experimental pharmacology, Principles of safety pharmacology, vol 229. Springer-Verlag, Berlin, Heidelberg, pp 83–113

59. Mead AN, Amouzadeh H, Chapman K et al (2016) Assessing the predictive value of the rodent neurofunctional assessment for commonly reported adverse events in phase I clinical trials. Regul Toxicol Pharmacol 80:348–357

60. Gribkoff VK, Kaczmarek LK (2017) The need for new approaches in CNS drug discovery: why drugs have failed, and what can be done to improve outcomes. Neuropharmacology 120:11–19

61. Sette WF (1987) Complexity of neurotoxicological assessment. Neurotox Teratol 9:411–416

62. Moser VC (1990) Approaches for assessing the validity of a functional observational battery. Neurotox Teratol 12:483–488

63. Evans HL (1994) Neurotoxicity expressed in naturally occurring behavior. In: Weiss B, O'Donoghue J (eds) Neurobehavioral toxicity: analysis and interpretation. Raven Press, New York, NY, pp 111–135

64. Hartun T (2017) Food for thought … thresholds of toxicological concern—setting a threshold for testing below which there is little concern. ALTEX 34:331–351

65. Radermacher P, Haouzi P (2013) A mouse is not a rat is not a man: species-specific metabolic responses to sepsis—a nail in the coffin of murine models for critical care research? J Software Eng Res Dev 1:7. http://www.icm-experimental.com/content/1/1/7

66. Goldberg S (2017) The four-minute neurologic exam, 2nd edn. Medmaster, Inc., Miami, FL

67. Gauvin DV, Zimmermann ZJ (2019) FOB vs modified Irwin: what are we doing? J Pharmacol Toxicol Methods 97:24–28. https://doi.org/10.1016/j.vascn.2019.02.008

68. Ringblom J, Törnqvist E, Ove HS et al (2017) Assigning ethical weights to clinical signs observed during toxicity testing. ALTEX 34(148):156

69. Authier S, Arezzo J, Pouliot M et al (2019) EEG: characteristics of drug-induced seizures in rats, dogs and non-human primates. Pharmacol Toxicol Methods 97:52–58

70. Carstens E, Moberg GP (2000) Recognizing pain and distress in laboratory animals. ILAR J 41:62–71. https://doi.org/10.1093/ilar.41.2.62

71. McElwee KJ, Hoffmann R (2002) Alopecia areata—animal models. Clin Exp Dermatol 27:410–417

72. McElwee KJ, Freyschmidt-Paul P, Zöller M et al (2003) Alopecia areata susceptibility in rodent models. J Investig Dermatol Symp Proc 8:182–187

73. Romagnani S (1999) Th1/Th2 cells. Inflamm Bowel Dis 5:285–294

74. Mathieu M (2008) New drug development: a regulatory overview (8th Ed). Parexel International Corp, Waltham, MA

75. Weiss B, Reuhl K (1994) Delayed neurotoxicity: a silent toxicity. In: Chang LW

(ed) Principles of neurotoxicology. Marcel Dekker, Inc., New York, NY, pp 764–784

76. Reuhl KR (1991) Delayed expression of neurotoxicity: the problem of silent damage. Neurotoxicology 12:341–346

77. Giordano G, Costa LG (2012) Developmental neurotoxicity: some old and new issues. ISRN Toxicol 2012:814795. https://doi.org/10.5402/2012/814795

78. Adityanjee MKR, Thampy A (2005) The syndrome of irreversible lithium-effectuated neurotoxicity. Clin Neuropharmacol 28:38–49

79. Verdoux H, Bourgeois M (1991) Irreversible neurologic sequelae caused by lithium. Encéphale 17:221–224

80. Gauvin DV, Cheng EY, Holloway FA (1993) Recent developments in alcoholism: biobehavioral correlates. Recent Dev Alcohol 11:281–304

81. McKeon A, Frye MA, Delanty N (2008) The alcohol withdrawal syndrome. J Neurol Neurosurg Psychiatry 79:854–862

82. Payne LE, Gagnon DJ, Riker RR et al (2017) Cefepime-induced neurotoxicity: a systematic review. Crit Care 21:276. (Open Access)

83. Sioka C, Kyritsis AP (2009) Central and peripheral nervous system toxicity of common chemotherapeutic agents. Cancer Chemother Pharmacol 63:761–767

84. US Food and Drug Administration (2019) FDA adds boxed warning for risk of serious injuries caused by sleepwalking with certain prescription insomnia medicines: FDA drug safety communication. April 30, 2019. https://www.fda.gov/drugs/drug-safety-and-availability/fda-adds-boxed-warning-risk-serious-injuries-caused-sleepwalking-certain-prescription-insomnia

85. van Betteray JN, Vossen JM, Coenen AM (1991) Behavioural characteristics of sleep in rats under different light/dark conditions. Physiol Behav 50:79–82

86. Coenen AML, Van Hulzen ZJM, Van Luijtelaar ELJM (1983) Paradoxical sleep in the dark period of the rat: a dissociation between electrophysiological and behavioral characteristics. Behav Neural Biol 37:350–356

87. Gauvin DV, Zimmermann ZJ, Dalton JA, Baird TJ, Kallman MJ (2019) CNS safety screening under ICH S7A guidelines requires observations of multiple behavioral units to assess motor function. Int J Toxicol 38 (5):339–356. https://doi.org/10.1177/1091581819864836

88. Barclay LL, Gibson GE, Blass JP (1981) The string test: an early behavioral change in thiamine deficiency. Pharmacol Biochem Behav 14:153–157. https://doi.org/10.1016/0091-3057(81)90236-7

89. Combs DJ, D'Alecy LG (1987) Motor performance in rats exposed to severe forebrain ischemia: effect of fasting and 1,3-butanediol. Stroke 18:503–511. https://doi.org/10.1161/01.str.18.2.503

90. Modianos DT, Pfaff DW (1976) Brain stem and cerebellar lesions in female rats. I. Tests of posture and movement. Brain Res 106:31–46. https://doi.org/10.1016/0006-8993(76)90071-8

91. Metz G, Whishaw I (2002) Cortical and subcortical lesions impair skilled walking in the ladder rung walking test: a new task to evaluate fore- and hindlimb stepping, placing, and co-ordination. J Neurosci Meth 115:169–179

92. Roseberry T, Kreitzer A (2017) Neural circuitry for behavioural arrest. Philos Trans R Soc Lond B Biol Sci 372:20160197. https://doi.org/10.1098/rstb.2016.0197

93. Weaver RJ, Valentin J-P (2019) Today's challenges to de-risk and predict drug safety in human "mind-the-gap". Toxicol Sci 167:307–321

94. Batra VR, Schrott LM (2011) Acute oxycodone induces the pro-emetic pica response in rats. J Pharmacol Exp Ther 339:738–745. https://doi.org/10.1124/jpet.111.183343

95. Parker LA, Brosseau L (1990) Apomorphine-induced flavor-drug associations: a dose-response analysis by the taste reactivity test and the conditioned taste avoidance test. Pharmacol Biochem Behav 35:583–587. https://doi.org/10.1016/0091-3057(90)90294-r

96. Parker LA, Mechoulam R (2003) Cannabinoid agonists and antagonists modulate lithium-induced conditioned gaping in rats. Integr Physiol Behav Sci 38:133–145. https://doi.org/10.1007/BF02688831

97. Pyun K, Son JS, Kwon YB (2014) Chronic activation of sigma-1 receptor evokes nociceptive activation of trigeminal nucleus caudalis in rats. Pharmacol Biochem Behav 124:278–283. https://doi.org/10.1016/j.pbb.2014.06.023

98. Sotocinal SG, Sorge RE, Zaloum A et al (2011) The rat grimace scale: a partially automated method for quantifying pain in the laboratory rat via facial expressions. Mol Pain 7:55. https://doi.org/10.1186/1744-8069-7-55

99. Langford DJ, Bailey AL, Chanda ML et al (2010) Coding of facial expressions of pain in the laboratory mouse. Nat Methods

7:447–449. https://doi.org/10.1038/nmeth.1455

100. Sacrey LA, Whishaw IQ (2010) Development of collection precedes targeted reaching: resting shapes of the hands and digits in 1-6-month-old human infants. Behav Brain Res 214:125–129. https://doi.org/10.1016/j.bbr.2010.04.052

101. Cosenza ME (2014) Introduction to the study director. In: Brock WJ, Mounho B, Fu L (eds) The role of the study director in non-clinical studies: pharmaceuticals, chemicals, medical devises, and pesticides. Wiley, Hoboken, NJ, pp 1–6

102. Tosolini AP, Morris R (2012) Spatial characterization of the motor neuron columns supplying the rat forelimb. Neuroscience 200:19–30

103. Kuypers HG (1964) The descending pathways to the spinal cord, their anatomy and function. Prog Brain Res 11:178–200

104. Kuypers HGJM (1981) Anatomy of the descending pathways. In: Brookhart JM, Mountcastle VB (eds) Handbook of physiology, the nervous system, vol II. Williams and Wilkins, Baltimore, MD, pp 597–666

105. Porter R, Lemon RN (1993) Corticospinal function and voluntary movement. Clarendon Press, Oxford, NY

106. Iwaniuk AN, Whishaw IQ (2000) On the origin of skilled forelimb movements. Trends Neurosci 23:372–376

107. Cenci MA, Whishaw IQ, Schallert T (2002) Animal models of neurological deficits: how relevant is the rat? Nat Rev Neurosci 3:574–578

108. Campbell WW, Barohn RJ (2020) DeJong's the neurologic examination, 8th edn. Wolters Kluwer, Philadelphia, PA

109. Houser MC, Tansey MG (2017) The gut-brain axis: is intestinal inflammation a silent driver of Parkinson's disease pathogenesis? NPJ Parkinsons Dis 3:3. https://doi.org/10.1038/s41531-016-0002-0

110. El Aidy S, Dinan TG, Cryan JF (2014) Immune modulation of the brain-gut microbe axis. Front Microbiol 5:146. https://doi.org/10.3389/fmicb.2014.00146

111. Clarke G, Quigley EM, Cryan JF et al (2009) Irritable bowel syndrome: towards biomarker identification. Trends Mol Med 15:478–489

112. Dantzer R, O'Connor JC, Freund GG et al (2008) From inflammation to sickness and depression: when the immune system subjugates the brain. Nat Rev Neurosci 9:46–56

113. Dunn AJ (2006) Effects of cytokines and infections on brain neurochemistry. Clin Neurosci Res 6:52–68

114. Graff LA, Walker JR, Bernstein CN (2009) Depression and anxiety in inflammatory bowel disease: a review of comorbidity and management. Inflamm Bowel Dis 15:1105–1118

115. Dyer RS (1984) Cross species extrapolation and hazard identification in neurotoxicology. Neurobehav Toxicol Teratol 6:409–411

Chapter 9

Behavioral Assessment of Vestibular Dysfunction in Rats

Alberto F. Maroto, Erin A. Greguske, Meritxell Deulofeu, Pere Boadas-Vaello, and Jordi Llorens

Abstract

Rodents with toxic lesions in the vestibular system display abnormalities in spontaneous and reflex behaviors that are easily recognized by knowledgeable observers. However, a quantitative assessment of the functional deficit is necessary in addition to its simple identification. In this chapter, we describe a semi-quantitative behavioral test battery that our laboratory has successfully used for almost three decades to evaluate vestibular toxicity in rats. We also describe a method recently developed for the same purpose. Using high-speed video recording, the minimum angle formed by the nose, the back of the neck, and the base of the tail during the tail-lift reflex is obtained as a fully objective and quantitative measure of vestibular function. Data collected on the same animals show high correlation values between the test battery scores and the tail-lift angles.

Key words Ototoxicity, Vestibular system, Behavioral test battery, Tai-lift reflex, Video tracking, Rat

1 Introduction

1.1 The Motor Syndrome Caused by Vestibular Dysfunction in Laboratory Animals

The vestibular system comprises several end-organs located in the inner ear. It detects gravity and accelerations generated by movements of the head [1]. The organism constantly uses vestibular information to control gaze as well as body posture and movements. Vestibular input is also necessary for cognitive functions, including among others the sense of orientation in space. Individuals are largely unaware of their vestibular sense in healthy status, but they become dramatically disabled in case of vestibular loss. Many clinically relevant vestibular pathologies are unilateral, generating an imbalance of vestibular input that most often results in vertigo. In contrast, vestibular toxicity causes a symmetric bilateral loss of function. In humans, this may not cause vertigo but disequilibrium, oscillopsia, and a defective proprioception that can generate disembodiment feelings and loss of spatial orientation.

In rodents and other species, bilateral loss of vestibular function produces a syndrome of abnormalities in motor behavior. This

Jordi Llorens and Marta Barenys (eds.), *Experimental Neurotoxicology Methods*, Neuromethods, vol. 172,
https://doi.org/10.1007/978-1-0716-1637-6_9, © Springer Science+Business Media, LLC, part of Springer Nature 2021

syndrome was described in animals exposed to several compounds, such as 3,3′-iminodipropionitrile (IDPN) [2], some nitrogen mustard derivatives [3] and arsanilic acid [4] well before their vestibular toxicity was identified [5–11]. Rats and mice with vestibular deficiency caused by toxic exposure commonly show hyperactivity, stereotyped circling, backward displacement, and abnormal head movements (head bobbing). However, some variability exists among species and strains and, for instance, IDPN-induced vestibular loss results in hyperactivity in many strains of rats and mice, but not in guinea pigs [11] and 129S1/SvImJ mice [12]. This syndrome has been named the ECC (excitation with choreiform and circling movements) syndrome [13] or the waltzing syndrome [14]. The similarity of these chemically induced motor abnormalities with those shown by rodents with genetic vestibular dysfunction is now well established, so it comes to no surprise that several vestibular mutant strains have been named because for these behaviors, i.e., the waltzing rat [15], the waltzing guinea pig [16], the waltzer mouse [17], the dancer mouse [18], the shaker mouse [19], or the stargazer mouse [20], among others.

1.2 Assessing Loss of Vestibular Function in Laboratory Rodents

In comparison to other sensory systems, the vestibular system poses a greater difficulty for quantitative assessment of its dysfunction in laboratory animals. This is due to difficulties in both delivering appropriate stimuli, which are linear and angular accelerations, and recording specific and sensitive endpoints. Notwithstanding, several methods are available for this purpose, based on recording of eye movements, nervous system evoked potentials and myogenic evoked potentials (reviewed by [21]). Each of these methods has its strengths and weaknesses to evaluate one particular aspect of vestibular function. However, their technical complexity and their requirements of labor time make them suboptimal for addressing the most common problems in toxicity assessment: the easy and quick identification of agents with potential toxicity on the system, and the obtention of dose–response and time-course data of functional loss following the toxic exposure. While the behavioral syndrome caused by vestibular loss in rodents is dramatically apparent, clear criteria for its recognition are necessary and methods to quantify the loss are required. For almost three decades, our laboratory has been using a behavioral test battery to assess vestibular toxicity. This battery [9], slightly modified from the initial design [8], uses subjective assessment to obtain a semi-quantitative rating score that has been well documented [10, 11, 22]. Its rationale and implementation will be described in this chapter. While using this test battery, we noticed a particularly robust consistency for one of its items, the tail-lift reflex (see [23]). Recently, we have developed a method to obtain quantitative and objective data from this reflex and obtained good evidence of its promising interest [24, 25]. This chapter will therefore include a detailed description of the assessment of the tail-lift reflex.

2 Behavioral Test Battery for the Semi-Quantitative Assessment of Vestibular Loss in Rodents

2.1 Rationale and Overall Description

Initially, our test battery was inspired by that designed to evaluate the effects of IDPN on spontaneous motor behavior [26, 27]. These authors evaluated the IDPN-induced motor syndrome without recognition of its vestibular basis, which was discovered later [8]. The battery includes six items (Fig. 1), of which three are spontaneous behaviors and three are anti-gravity reflexes, as vestibular dysfunction results in abnormalities in both spontaneous and reflex activity. The observed spontaneous behaviors are: (1) circling, (2) backward walking, and (3) head bobbing; the items assessing reflex activity are: (4) the tail-lift reflex, (5) the contact-inhibition of the righting reflex, and (6) the air-righting reflex.

Stereotyped circling activity is the most obvious result of vestibular deficiency in rats and mice, as shown in several internet videos that can be found by searching "waltzing mouse." This can take the form of ample circles, of about 50 cm of diameter in rats, or smaller circles, as small as allowed by the animal's size, well described as "tail-chasing" activity (Fig. 1a). This can last for tens of seconds with an apparent difficulty of the animal to stop, and usually, but not always, associates with hyperactivity. Indeed, circling is also observed in vestibular-deficient rats and mice that do not show an increased level of activity [12]. While vestibular deficiency causes circling, circling activity alone can have other causes, such as an imbalance in dopamine systems sources [28].

Another consequence of vestibular deficiency is backward displacement or walking (Fig. 1b). This is more often seen in the initial days after the ototoxic treatment and tends to disappear or become less frequent at longer times. It usually presents as bursts of backward displacements that last a few seconds until the animal encounters a wall or obstacle. In some cases, the movement is quite coordinated on the four paws, but is more often caused by pedaling with the forepaws, as if animals were trying to stop slipping downhill.

The third item in the test battery is the abnormal head movements (Fig. 1c). Vestibular-deficient rats show exaggerated repetitive movements of the head or bobbing. These can be side to side ("no"), but more frequent are nodding ("yes") movements or backward hyperextension of the neck. The bobbing behavior has been used as a single measure to quantify the vestibulotoxic effect of several nitriles [29, 30]. When circling fast, vestibular-deficient rats do not show bobbing, but this movement may appear at the end of circling bursts or together with slower circling or non-circling ambulation. The bobbing and even more the backward hyperextension of the neck may combine with backward walking. Quite frequently, the rat stays for several seconds with

85
86
87
88
89
90
91
92
93
94
95
96
97
98
99
100
101
102
103
104
105
106
107
108
109
110
111
112
113
114
115
116
117
118
119
120
121
122
123
124
125
126
127
128

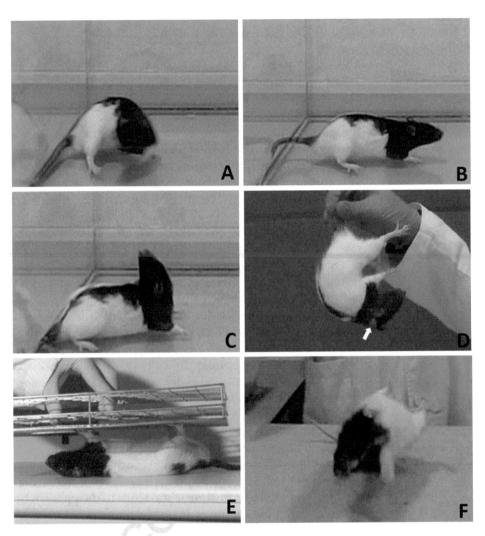

Fig. 1 Pictures illustrating the six items included in the behavioral test battery to obtain vestibular dysfunction ratings. (**a**) Circling. The position of this rat denotes a spinning activity, with a circle as small as allowed by the maximal possible lateral flexion of the body. (**b**) Backward walking. Note the ataxic opening of the hindlimbs. In this posture, the vestibular-deficient rat may push the floor with the hind and fore paws generating a backward displacement. (**c**) Characteristic hyperextension of the neck (star-gaze), usually alternating with repetitive bobbing movements. (**d**) Tail-lift reflex in a rat with vestibular deficiency. The arrow points to the marble used for the analysis of this reflex by high-speed video recording. (**e**) Contact-inhibition of the righting reflex in a vestibular-deficient rat. Rats with intact vestibular function immediately right when placed supine on the bench. (**f**) Back landing due to a defective air-righting reflex

the neck in exaggerated hyperextension, a behavior that has generated the name of a mutant strain of mice, the stargazer mouse [20].

The three reflexes included in the battery are anti-gravity reflexes. The first one is the tail-lift reflex [6]. A healthy rat submitted to a tail-lift maneuver shows an extension reflex of the trunk and paws, allowing it to land on its forepaws first, followed by the hind

paws, when lowered down. This landing reflex makes evident the 135
tonic anti-gravity input that the vestibular system exerts on the 136
spinal motor nuclei to control body posture [31, 32]. When the 137
vestibular input is absent, the rat bends ventrally and tends to 138
occipital or back landing (Fig. 1d) [6, 24, 33]. A remarkable 139
observation is that the bending response shows a gradation, from 140
just an incomplete extension response in slightly deficient rats to a 141
total flexion of the body, in which the forepaws or the nose touch 142
the experimenter's hand grasping the tail, in rats suffering a deep 143
vestibular deficiency [24]. The second reflex examined is the right- 144
ing reflex of the rat placed supine between two surfaces [7, 34]. The 145
aim of the test is to evaluate the loss of perception of the direction 146
of gravity. Healthy rats quickly right themselves without hesitation. 147
Rats with absent vestibular function will stay in supine position, 148
frequently walking and circling (Fig. 1e). In the absence of vestibu- 149
lar function, touch becomes the main sensory input signaling body 150
orientation, and the presence of a surface contacting the paws 151
suffices to inhibit the righting response. In the absence of this 152
surface, or if it is removed when the rat stays supine, the rat will 153
quickly right to the prone position. The same inhibition by contact 154
of the righting reflex is observed in vestibular-deficient mice and 155
guinea pigs, while no contact is necessary for the absence of right- 156
ing in frogs [11]. The third anti-gravity reflex included in the test 157
battery is the well-known air-righting reflex test [7], in which the 158
rat is dropped in supine position to fall on a foam cushion (Fig. 1e). 159
Healthy rats right themselves while in the air to land on their paws, 160
while vestibular-deficient animals fail to right. 161

A widely accepted test of vestibular dysfunction not included in 162
the present battery is swimming (e.g., [35]). Vestibular-deficient 163
rodents are unable to swim and show barrel rolling when placed in 164
water, so they need to be immediately rescued to avoid drowning. A 165
swimming test was included in our first version of the battery [8], 166
but eliminated later [9]. This was due to the evidence that the 167
swimming test did not add significant value to the battery while 168
being quite stressful for healthy animals and very stressful for the 169
vestibular-deficient ones. 170

171

2.2 Detailed Protocol for Rats (See Note 1)   Rats are first observed for 1 min for the abnormalities in spontane- 172
ous behavior in an adequate arena. We have used either a 173
100 × 100 × 50 cm white wood open field or a 174
50 × 50 × 40 cm glass cube placed on a bench. As in any observa- 175
tional measure of behavior, the experimenter must stay quiet or use 176
a remote video system for observation. The reflex tests are then 177
delivered, beginning with the tail-lift, followed by the contact- 178
righting, and finally testing the air-righting reflex. This sequence 179
goes from less to more stressing manipulations. Each reflex is 180
examined once but can be repeated a second or even a third time 181
if needed to decide a rating. To generate the tail-lift reflex, the rat is 182

grasped by the base of the tail and quickly but gently lifted about 30–40 cm, then lowered down. The animal must be lifted and immediately descended, not held in the air (*see* **Note 2**), so the maneuver is completed in less than a second. To evaluate the contact-righting response, the rat is grasped by the body, then quickly turned to the supine position, and placed on its back on a bench while a metal bar grid (we use a spare shelf from a fridge) is simultaneously placed on its paws. This grid must be held in contact with the paws, at a distance from the bench that allow the rat to be in a posture like its natural walking position, with the extremities in semiflexion. For the air-righting reflex, the experimenter cups the rat's body in supine position with the two hands at about 40 cm above a foam cushion and suddenly releases it to fall on the cushion (*see* **Note 3**).

Each of the six items in the battery are rated 0–4, and the values are then added to obtain a Vestibular Dysfunction Rating (VDR) with a maximum of 24 [9, 10]. The values are given according to the descriptions in Table 1, which respond to the following overall criteria: 0, normal behavior or response, as seen in healthy rats before any toxic treatment; 1, not sure if the behavior or response is normal or not; 2, slightly but undoubtedly abnormal behavior or response; 3, moderately abnormal behavior or response; and 4, extremely abnormal behavior or response (*see* **Note 4**).

2.3 Consistency of the Test Battery

In repeated testing, the test battery shows a remarkable stability in its measure of irreversible vestibular damage [8–10, 24] and a capacity to detect functional recovery in case of reversible damage, as occurs in the case of chronic IDPN exposure [24, 36]. Nevertheless, not all items show the same stability. With time, backward walking tends to become less frequent, while circling is more persistent (*see* **Note 5**). Also, some degree of learning seems to affect the contact-righting reflex, as some control rats may show a less immediate response when tested repeatedly. By contrast, the tail-lift reflex shows a remarkable stability.

In dose–response studies comparing with histological assessment, the test battery has proved to be quite sensitive, so only undamaged animals and animals with only small lesions located in the apical crista show baseline levels (VDRs in the 0–3 range) [8, 10, 23, 24].

After proper training, variability across experimenters in our laboratory is low. However, we have not carried out a systematic evaluation of this variability. Also, no data has been gathered to evaluate cross-laboratory variability.

2.4 Adaptation of the Vestibular Test Battery for Mice

The behavioral syndrome induced by vestibular toxicity in the mouse [1, 29, 30] is identical to that observed in the rat, so the same test battery has been successfully used in this species after minor adaptation [11, 22, 37]. We use a standard Macrolon cage

t.1
Table 1
Scores used to obtain Vestibular Dysfunction Ratings (VDRs) in the Behavioral Test Battery

Score Item	0 (Normal)	1 (Dubious)	2 (Mild)	3 (Overt)	4 (Extreme)
Circling	The rat ambulates normally and does not display stereotyped circling	The experimenter is unsure of whether the rat is moving in circles more than normal	The rat shows an unmistakable tendency to move in circles	The rat spends a long time moving in wide circles or shows bursts of rapid circling in a small area (tail-chasing behavior)	The rat spends most of the time moving in circles or displays prolonged bursts of fast tail-chasing activity, being apparently unable to stop circling
Backward walking	The rat ambulates normally and does not display backward displacement	The experimenter is unsure of whether the rat has a tendency to push to its back when not walking or immediately after being placed in the observational arena	The rat shows one or two short (up to one body length) backward displacements	The rat makes several (3 or more) and long (more than its body length) displacements backward	The rat frequently moves backward or keeps pushing backward against the wall of the open field after reaching it. The hind limbs may move up the wall until the rat falls on its side
Head bobbing	The rat ambulates normally and does not display abnormal head movements	The experimenter is unsure whether the movement of the head is normal or not	The rat displays mild but overtly abnormal nodding movements of its head	The rat does wide and quite frequent nodding movements, sometimes up to extreme hyperextension of the neck (stargazer behavior)	The rat displays very frequent wide nodding activity and frequently stays for several seconds in extreme hyperextension of the neck
Tail-lift reflex	The rat extends the fore and hind limbs and the trunk in a coordinated landing response	The experimenter is unsure whether the rat is showing an incomplete extension	The rat shows a mild ventral flexion, with a minimum angle that would cause nose or face landing	The rat shows an ample ventral flexion, with a minimum angle that would cause neck landing	The rat completely curls the body, reaching a posture that may cause landing on the middle to caudal part of its back

(continued)

t.2

t.3

t.4

t.5

t.6

Table 1
(continued)

Item	0 (Normal)	1 (Dubious)	2 (Mild)	3 (Overt)	4 (Extreme)
Contact-inhibition of the righting reflex	The rat immediately rights itself between the bench and the grid	The experimenter is unsure whether the rat hesitates before righting	The rat spends a noticeable time in supine position before righting	The rat stays in supine position for a long time (more than 10 s) and then rights hesitantly	The rat stays in supine position as long as the grid is maintained in contact with its paws. It usually walks and circles in this upside-down position. If the grid is lifted slowly, it grasps it to keep the contact
Air-righting reflex	The rat immediately rights itself in the air, and lands symmetrically with the four paws	The experimenter is unsure that the rat has landed symmetrically	The rat has incompletely righted when landing, the landing denotes disequilibrium	The rat shows an overtly defective righting response and lands on its side or back	The rat shown no righting response, it falls on the cushion in the same supine position it had when released

t.7

t.8

t.9

t.10

for rat housing ($215 \times 465 \times 145$ mm) as open field for the observation of spontaneous behavior. The correct evaluation of the spontaneous behavior requires adequate avoidance of any stress. Other differences between the rat and the mouse protocols are related to the righting reflexes. When testing the contact-inhibition of the righting reflex, it is important that a flat surface, such a A4-size plastic board, is used, not a grid. Placing the mouse in supine position on the bench and below the board becomes somehow tricky and requires some practice. To test the air-righting reflex, we use the following maneuver: the mouse is placed facing the experimenter with its paws on a small plastic board; the animal is then pulled by the base of the tail away and into the air with a rotating gesture, so it attains a supine position in the air and the experimenter releases the animal at 10–20 cm above a foam cushion.

3   The Tail-Lift Test for Rats

3.1   Rationale

As explained above, we observed that the tail-lift reflex was particularly robust as a measure of the degree of vestibular dysfunction. Thus, we observed that small but consistent changes in the tail-lift reflex were an early effect of vestibular toxicity and increasing degrees of vestibular damage caused an increase in the abnormal ventral flexion of the trunk. This led us to the hypothesis that the angle formed between the nose, the back of the neck, and the base of the tail (Fig. 2) could provide an objective measure of the degree of vestibular dysfunction. We therefore developed a protocol to

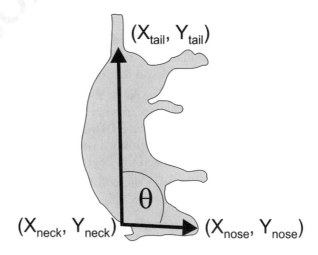

Fig. 2 Geometric interpretation of the neck-to-tail vector and neck-to-nose vector defining the tail-lift angle. The analysis of the video movies provides the three pairs of X and Y coordinates at a frequency of 240/s

obtain a good measure of this angle and evaluated its ability to assess irreversible and reversible vestibular dysfunction [24]. This method, described in more detail here, was found to offer a good measure of vestibular dysfunction.

3.2 Objective Measure of the Tail-Lift Reflex in Rats

Materials.

– Standard rat cage, without lid.
– Background for recording offering high contrast with the rats. For our hooded Long-Evans rats, we taped a red board on the wall.
– Small rubber bands with a white, blue, or green marble, to be used as collars on the neck of the rat. These colors offer good contrast with the brown fur of the rat and the red board on the back wall.
– Video camera capable of high-speed (240 frames per second) recording (*see* **Note 6**). We have used two different cameras, a Casio Exilim ZR700 and a GoPro 5.
– Small tripod to hold the camera.
– Video tracking software. We use the free software Kinovea (www.kinovea.org).
– Data analysis software. We use the free software environment R-Studio.

Method for video recording.

– Place the rat cage on the bench, with its long axis parallel to the high contrast background.
– Place the video camera on the tripod, oriented perpendicular to the cage and background. The movie must cover the entire up and down trajectory of the points to be used (nose, back of the neck, base of the tail) when above the cage limit (*see* **Note 7**).
– Place the rubber collar around the neck of the rat, with the marble on the back (Fig. 1d) (*see* **Note 8**).
– Place the rat on the cage, turn on the video recording at 240 fps, and proceed with the tail-lift maneuver as described above (*see* **Notes 9** and **10**). The movie must show the rat from its profile for a continuous view of the nose, the marble on the back of the neck, and the beginning of the tail.
– Remove the collar with the marble and get the rat back to its home cage.

Method for video analysis using Kinovea.

– Open Kinovea and select the movie to be analyzed.
– In the Motion menu, select "High speed camera" and enter the recording speed (240 fps).

- To calibrate pixels to physical lengths, select the line tool from the drawing tool bar and draw a line over a desired segment of known length. A right click on the line will open the menu to calibrate the measure, where to add the real length of the segment. Next, delete the line to not disturb the video analysis; the calibration remains as determined.

- In the Image menu, select "Coordinates system origin" to locate the origin of the axis that will be used to define point coordinates. Select a desired location to place the coordinate system. We use a corner of the cage as a standard point of origin.

- In the options menu, time marker formats can be selected. We use total milliseconds.

- To track the points of interest, select a cross-marker from the drawing tool bar and place three of them on the nose, neck (marble), and base of the tail, in this order (*see* **Note 11**). Then right click on the cross-marker and select Track path (*see* **Note 12**). Move the video forward by using the Play button. Alternatively, use the Next Frame button, or the Mouse Wheel for a frame by frame tracking. Check that tracking is correct and adjust if necessary (*see* **Note 13**). If desired, display options can be selected to change the drawing of the path.

- To finish tracking, right click and select End Path Edition.

- Examine the tracking paths and make further corrections if necessary (*see* **Note 14**).

- Save the video including only the frames of interest by selecting Start and End of a Working Zone. The video can be saved with the tracking data in the file, which will be suitable to be opened in Kinovea for further analysis or re-analysis. Alternatively, the video can be saved with permanent paint of the image data, generating a file that can be viewed using standard video players.

- From the main menu, export the path data by selecting File > Export to Spreadsheet > Trajectories to simple text.

- If a second rat is included in the same video, the tracked path of the first rat must be deleted after the data have been saved and before starting the new tracking.

- Open the exported data file and check coherence of the data, meaning that all setups were correctly adjusted, the file contains data from only one rat, and the tracking has been adequate (*see* **Notes 15** and **16**). There are three columns, showing left to right: time in milliseconds, X position, and Y position. Data for each marker are placed one below another, in order: nose, neck, and tail.

*Method for analysis of the coordinates data (see **Note 17**).*

- Use the XY coordinates of the nose, back of the neck (marble), and base of the tail to calculate the angle formed by these three points at each time point. The coordinates define two vectors "a" (neck to tail) and "b" (neck to nose) that define the angle θ (Fig. 2). The dot product of the two vectors is defined by: $a \cdot b = |a||b| cos\theta$, from which the angle can be obtained. Convert resulting radians into degrees.

– Obtain the minimum angle from the tail-lift angle series of each reflex.

4 Notes

1. This test battery does not usually require habituation, at least in young adult Long-Evans rats. We routinely evaluate the animals a few days before the exposure to the ototoxic treatment to obtain pre-test data. During these pre-test sessions, the animals have occasionally shown behaviors interfered by an excessive reactivity to manipulation, as movements driven against the hand grasping the tail during the tail-lift maneuver. In these cases, we have habituated the animals with a second pre-test session.

2. Grasping the tail near its base, not at the middle or tip segments is important to avoid hurting or stressing the rat, and to elicit the desired reflex. The up and down movement must be continuous and smooth. Adequate execution of the maneuver is made evident by the behavior of healthy rats. If not properly performed, the rat will show active responses, such as turning the body to one side and attempts to oppose the manipulation. The desired control response is that in which the rat behavior is only directed to smooth landing. Active (undesired) responses are more frequent the first time of testing, but quickly disappear with repeated testing.

3. For the air-righting reflex, the rat must be held firmly but gently, to decrease as much as possible its stress but limiting its freedom to move. The experimenter must take care in holding all animals similarly and at the same height. The release of the rat must be fast and must avoid giving it any acceleration. If the grip is decreased too slowly, the rat starts righting while in the hands of the experimenter and then gets a rotatory force when the experimenter moves them away.

4. The use of a 0–4 range of scores provides robustness to the ratings. For any observation, the experimenter may doubt between one of two scores (for instance, between 3 and 4), but not a third one (your choice will not be 2 if you doubt between 3 and 4). Therefore, choosing one or the other of the two possible ratings will have a small impact on the global result, so when all items for each animals are pooled, differences among groups (for instance, dose groups) become robust.

5. The spontaneous behaviors are partly competing and a decrease 389
in the frequency or intensity of backward walking or bobbing 390
frequently associates with increased circling, so on repeated 391
testing the composite score tends to be more stable than each 392
individual item. 393

6. High-speed video recording is available in many domestic cam- 394
eras and smartphones. The 240-fps speed is eight times the 395
speed of standard video recording (30 fps) and provides 396
enough precision to assess the tail-lift angle. 397

7. The optimal location will vary with different cameras and tri- 398
pods, but their placement respect to the cage must be constant. 399
We have it marked on the bench. 400

8. The diameter of the rubber collars must be adequate. If too 401
loose, they will not hold the marble in the desired position. If 402
too tight, they cause discomfort and the rats remove them. 403

9. The main difficulty is to obtain a movie that provides a constant 404
profile image of the rat, with a continuous view of the three 405
reference points, as this view is lost if the animal twists the body 406
to one side. This may be caused by a deficient performance of 407
the experimenter, but this problem is greatly reduced by prac- 408
tice. More often, the rat may twist its trunk as a reaction against 409
the passive lift, causing both interference with the vestibulo- 410
motor reflex and impairment to the quality of the movie. To 411
avoid this interference of the tail-lift reflex with spontaneous 412
motor behavior, the rat must be unstressed. We obtain better 413
results laying some wood shavings on the cage, and simulta- 414
neously placing on it the 2–4 rats that are housed together. We 415
then record these animals one after the other in the same 416
movie. We display a handmade sign to identify rats as we record 417
them for individual assessment. 418

10. The movie must contain a reference for distance calibration. 419
We use the dimensions of a post-it label that we stick on the 420
cage with the number of the group of rats being recorded. 421

11. Changing the order of the points will change the order of the 422
data in the output file. 423

12. Kinovea can provide angles as output. However, our protocol 424
obtains the raw coordinates of the three points and then calcu- 425
lates the angles from these coordinates. The coordinates can be 426
used for other analyses in addition of the angle analysis. 427

13. Tracking in Kinovea is a semi-automatic process. Point loca- 428
tions are automatically computed, but they can be adjusted if 429
needed. During tracking, misplaced points can be adjusted to 430
the correct positions dragging the outer rectangle until its 431
center cross-matches the correct location. 432

14. After tracking, a path can be edited again by going back to the previous frame, right clicking on it, and selecting Restart Path Edition.

15. The whole movement should be around 400–800 ms. If it shows oddly great numbers, like 1000–8000 ms, make sure that you have adjusted the time display to 240 fps.

16. Check that each of the three markers has been tracked the same amount of time, so the three lists of coordinates have the same length. Unequal series will generate an error when processing the data files.

17. A script in R programming to perform this data analysis has been published as supplementary material for the article Martins-Lopes et al. (2019) [24].

5   Results and Discussion

Figure 3 shows raw results from the tail-lift test from one of our experiments using young (2–4 months) adult male Long-Evans rats. Healthy (control) rats show tail-lift angles in the $130–170°$ range, while vestibular-deficient rats show reduced angles as the extension landing reflex is substituted by ventral curling. Together, the data collected in two different experiments [24] indicated that this method offers an objective and fully quantitative assessment of vestibular dysfunction. The minimum angle during the tail-lift maneuver was found to better correlate with the vestibular lesion than other measures of the reflex, such as the average angle during a frame of time around its peak.

One notable observation was that of the high correlation found between the tail-lift angle and the VDR values obtained with the behavioral test battery in the same rats [24]. We concluded that both measures provide a similar assessment of vestibular function. This validates the behavioral test battery with data from a fully objective and quantitative method. Both methods have minimal needs of equipment and expertise, making them suitable for easy implementation in many laboratories. Importantly, recent data [25] from our laboratory indicates that the tail-lift angle measured in rats, after completion of the ototoxic damage, mostly associates with the loss of type I hair cells, one of the two types of sensory cells in the vestibular epithelium. These cells are consistently more susceptible to toxic damage than the other type of vestibular sensory cells (named type II), so the evaluation of the tail-lift reflex has an undoubtable interest in vestibular toxicity assessment.

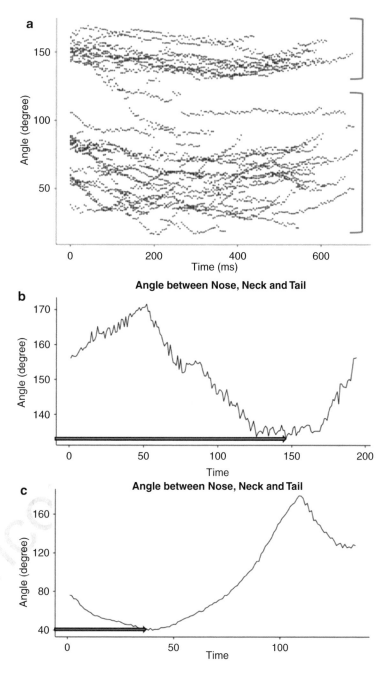

Fig. 3 Raw data of angle vs time. (**a**) All points from rats in two groups, control (upper bracket) and vestibular deficient (lower bracket) are shown with no connecting lines. The existence of two groups with different behaviors is evident. (**b**) Representative record from a control rat, with the data points connected by a line, displaying a minimum angle of about 130°. (**c**) In this representative record from a vestibular-deficient rat, the minimum angle is near 40°. Note the difference in scale between graphs in **b** and **c**

Acknowledgments

This study was supported by grants RTI2018-096452-B-I00 (Ministerio de Ciencia, Innovación y Universidades, Agencia Estatal de Investigación, Fondo Europeo de Desarrollo Regional, MCIU/AEI/FEDER, UE), and 2017 SGR 621 (Agència de Gestió d'Ajuts Universitaris i de Recerca, Generalitat de Catalunya). We thank former undergraduate and master students of the laboratory Sílvia Prades, Adrià Ricarte, Anna Bellmunt, and Vanessa Martins-Lopes for their contributions to the development of the tail-lift test.

References

1. Bronstein AM (2013) Oxford textbook of vertigo and imbalance. Oxford University Press, Oxford

2. Delay J, Pichot P, Thuillier J, Marquiset JP (1952) Action de l'amino-dipropionitrile sur le comportement moteur de la souris blanche. C R Sot Biol 146:533–534

3. Goldin A, Noe HA, Landing BH, Shapiro DM, Goldberg B (1948) A neurological syndrome induced by administration of some chlorinated tertiary amines. J Pharmacol Exp Therap 94:249–261

4. Oliver WT, Roe CK (1957) Arsanilic acid poisoning in swine. J Am Vet Med Assoc 130:177–178

5. Anniko M, Wersall J (1977) Experimentally (atoxyl) induced ampullar degeneration and damage to maculae-utriculi. Acta Otolaryngol 83:429–440

6. Hunt MA, Miller SW, Nielson HC, Horn KM (1987) Intratympanic injections of sodium arsanilate (atoxil) solution results in postural changes consistent with changes described for labyrinthectomized rats. Behav Neurosci 101:427–428

7. Ossenkopp KP, Prkacin A, Hargreaves EL (1990) Sodium arsanilate-induced vestibular dysfunction in rats: effects on open-field behavior and spontaneous activity in the automated digiscan monitoring system. Pharmacol Biochem Behav 36:875–881

8. Llorens J, Demêmes D, Sans A (1993) The behavioral syndrome caused by 3,3′-iminodipropionitrile and related nitriles in the rat is associated with degeneration of the vestibular sensory hair cells. Toxicol Appl Pharmacol 123:199–210

9. Llorens J, Rodríguez-Farré E (1997) Comparison of behavioral, vestibular, and axonal effects of subchronic IDPN in the rat. Neurotoxicol Teratol 19:117–127

10. Boadas-Vaello P, Riera J, Llorens J (2005) Behavioral and pathological effects in the rat define two groups of neurotoxic nitriles. Toxicol Sci 88:456–466

11. Soler-Martín C, Diez-Padrisa N, Boadas-Vaello P, Llorens J (2007) Behavioral disturbances and hair cell loss in the inner ear following nitrile exposure in mice, Guinea pigs, and frogs. Toxicol Sci 96:123–132

12. Boadas-Vaello P, Sedó-Cabezón L, Verdú E, Llorens J (2017) Strain and sex differences in the vestibular and systemic toxicity of 3,3′-iminodipropionitrile in mice. Toxicol Sci 156:109–122

13. Selye H (1957) Lathyrism. Rev Can Biol 16:1–82

14. Thuillier J, Burger A, Mouille P (1953) Contribution to the study of a motor syndrome induced in mice by amino-dipropionitrile (waltzing mice). C R Seances Soc Biol Fil 47:1052–1055

15. Rabbath G, Necchi D, de Waele C, Gasc JP, Josset P, Vidal PP (2001) Abnormal vestibular control of gaze and posture in a strain of a waltzing rat. Exp Brain Res 136:211–223

16. Sobin A, Anniko M (1986) Otoconial pathology in a strain of the waltzing Guinea pig. Am J Otol 7:449–453

17. Demêmes D, Sans A (1985) Pathological-changes during the development of the vestibular sensory and ganglion-cells of the bronx waltzer mouse—scanning and transmission electron-microscopy. Dev Brain Res 18:285–295

18. Wenngren BI, Anniko M (1989) Vestibular hair cell pathology in the dancer mouse mutant. Acta Otolaryngol 107:182–190

19. Anniko M, Sobin A, Wersäll J (1980) Vestibular hair cell pathology in the Shaker-2 mouse. Arch Otorhinolaryngol 226:45–50

20. Khan Z, Carey J, Park HJ, Lehar M, Lasker D, Jinnah HA (2004) Abnormal motor behavior and vestibular dysfunction in the stargazer mouse mutant. Neuroscience 127:785–796

21. Llorens J, Callejo A, Greguske EA, Maroto AF, Cutillas B, Martins-Lopes V (2018) Physiological assesment of vestibular function and toxicity in humans and animals. Neurotoxicology 66:204–212

22. Saldaña-Ruíz S, Boadas-Vaello P, Sedó-Cabezón L, Llorens J (2013) Reduced systemic toxicity and preserved vestibular toxicity following co-treatment with nitriles and CYP2E1 inhibitors: a mouse model for hair cell loss. J Assoc Res Otolaryngol 14:661–671

23. Balbuena E, Llorens J (2001) Behavioural disturbances and sensory pathology following allylnitrile exposure in rats. Brain Res 904:298–306

24. Martins-Lopes V, Bellmunt A, Greguske EA, Maroto AF, Boadas-Vaello P, Llorens J (2019) Quantitative assessment of anti-gravity reflexes to evaluate vestibular dysfunction in rats. J Assoc Res Otolaryngol 20:553–563

25. Maroto AF, Barrallo-Gimeno A, Llorens J. BioRxiv preprint. https://doi.org/10.1101/2020.12.21.423804

26. Diamond BI, Reyes MG, Borison R (1982) A new animal model for Tourette syndrome. Adv Neurol 35:221–225

27. Crofton KM, Knight T (1991) Auditory deficits and motor dysfunction following iminodipropionitrile administration in the rat. Neurotoxicol Teratol 13:575–581

28. Willis GL, Kennedy GA (2004) The implementation of acute versus chronic animal models for treatment discovery in Parkinson's disease. Rev Neurosci 15:75–87

29. Tanii H, Hayashi M, Hashimoto K (1989) Nitrile-induced behavioral abnormalities in mice. Neurotoxicology 10:157–166

30. Tanii H, Kurosaka Y, Hayashi M, Hashimoto K (1989) Allylnitrile: a compound which induces long-term dyskinesia in mice following a single administration. Exp Neurol 103:64–67

31. Wilson VJ, Yoshida M (1968) Vestibulospinal and reticulospinal effects on hindlimb, forelimb, and neck alpha motoneurons of the cat. Proc Natl Acad Sci U S A 60:836–840

32. Basaldella E, Takeoka A, Sigrist M, Arber S (2015) Multisensory signaling shapes vestibulo-motor circuit specificity. Cell 163:301–312

33. Pellis SM, Pellis VC, Teitelbaum P (1991) Labyrinthine and other supraspinal inhibitory controls over head-and-body ventroflexion. Behav Brain Res 46:99–102

34. Shoham S, Chen Y-C, Devietti TL, Teitelbaum P (1989) Deafferentation of the vestibular organ: effects on atropine-resistant EEG in rats. Psychobiology 17:307–314

35. Pau H, Hawker K, Fuchs H, De Angelis MH, Steel KP (2004) Characterization of a new mouse mutant, flouncer, with a balance defect and inner ear malformation. Otol Neurotol 25:707–713

36. Sedó-Cabezón L, Jedynak P, Boadas-Vaello P, Llorens J (2015) Transient alteration of the vestibular calyceal junction and synapse in response to chronic ototoxic insult in rats. Dis Model Mech 8:1323–1337

37. Boadas-Vaello P, Jover E, Saldaña-Ruíz S, Soler-Martín C, Chabbert C, Bayona JM, Llorens J (2009) Allylnitrile metabolism by CYP2E1 and other CYPs leads to distinct lethal and vestibulotoxic effects in the mouse. Toxicol Sci 107:461–472

Chapter 10

Assessment of Olfactory Toxicity in Rodents

David C. Dorman and Melanie L. Foster

Abstract

This chapter focuses on the main olfactory system of rodents that is used to detect small molecules. The olfactory system begins with sensory olfactory neurons found within the olfactory epithelium that lines part of the dorsal or dorsoposterior nasal cavity. These neurons play a critical role in olfaction and current tests of the integrity or function of the olfactory system in rodents often begins with a histologic assessment of this epithelium. Toxicant-induced olfactory epithelial lesions are often site-specific and dependent on the intranasal regional dose of the inhaled chemical and the inherent sensitivity of this epithelial tissue to the chemical of interest. In some cases, additional behavioral assessment of olfactory function may be performed in rodents. These behavioral tests can range from simple efforts to assess an animal's response to a novel odorant to more complicated tests evaluating olfactory discrimination or olfactory thresholds in exposed animals. These more complicated tests often require specialized equipment and extensive training of the animal to display an odor-mediated behavior. This chapter provides an update of our current understanding of chemically induced olfactory toxicity with a special emphasis on research of interest to neurotoxicologists. Despite advancements made to the study of olfaction in rodents, our knowledge of the olfactory toxicity of many chemicals remains incomplete. This is surprising given the known association between changes in olfactory ability in people following either occupational exposure to chemicals or with a variety of neurodegenerative diseases.

Key words Rodent olfaction, Nasal pathology, Olfactory discrimination testing, Olfactory threshold

1 Introduction

This chapter will focus on current methods used to assess olfactory toxicity in mice, rats, and other rodents. The chapter is not intended to provide an exhaustive review of the topic, but rather our goal was to introduce key concepts to the reader. Our discussion of this topic is not intended to be a complete survey of methods. For example, we have not addressed the role neurochemistry and other biochemical measures may play in assessing the olfactory system for toxicant-induced injury. Whenever possible the reader has been directed to recent reviews for more information and we have largely restricted our efforts to toxicology rather than neuroscience more broadly.

Jordi Llorens and Marta Barenys (eds.), *Experimental Neurotoxicology Methods*, Neuromethods, vol. 172,
https://doi.org/10.1007/978-1-0716-1637-6_10, © Springer Science+Business Media, LLC, part of Springer Nature 2021

1.1 The Rodent Olfactory System

In rodents, there are two major olfactory apparatuses in the nasal cavity: the main olfactory epithelium used to detect odorants (e.g., volatile organic molecules) and the vomeronasal organ which predominantly responds to pheromones. Although these structures are anatomically distinct, functional separation between these two systems may be incomplete. This chapter focuses entirely on the main olfactory system used to detect odorants.

The main olfactory system begins with sensory olfactory neurons found within the olfactory epithelium that lines part of the dorsal or dorsoposterior nasal cavity. This pseudostratified epithelium contains olfactory receptor neurons, glial-like sustentacular cells, and basal cells [1, 2]. The olfactory mucosa also contains Bowman's glands and olfactory axons that traverse the lamina propria [3, 4]. Bowman's glands have a secretory function including secretion of mucin and odorant-binding proteins [5]. These glands may have additional functions including transport of odorants and metabolism of odorants and other chemicals [6]. The olfactory mucosa is located in the dorsal aspect of the nasal cavity. Olfactory receptor neurons comprise the main cell type found in the olfactory epithelium. In rats and humans, approximately 50% and 3–5% of the nasal cavity is lined by olfactory epithelium, respectively [7]. The surface area dedicated to the olfactory mucosa has been used as a proxy for olfactory function. However, surface area measurements alone can underestimate the number of olfactory neurons [8]. In general, rodents and other animals with a large number of olfactory neurons have a well-developed sense of smell (i.e., macrosmatic). The rodent nasal cavity has both a dorsal meatus and an olfactory recess. These structures enhance odorant delivery to the olfactory epithelium and are often found in macrosmatic species with a well-developed sense of smell [9, 10]. Olfactory receptor neurons have a lifetime of 2–4 weeks and are continually replaced by replication from underlying stem cells found in the basal cell layer [11–14]. These stem cells may also give rise to sustentacular cells and olfactory ensheathing cells which envelope the olfactory nerve [13].

Olfactory sensory cells are bipolar neurons, with a short apical dendrite and a long, thin (~0.2 μm diameter) unmyelinated axon that passes into the underlying lamina propria to form prominent olfactory nerve bundles. The dendritic portions of these neurons extend above the epithelial surface and terminate in a protruding olfactory knob with 10–15 immotile cilia embedded in the mucus layer [15]. Olfactory cilia provide an extensive surface area for reception of odorants. Odorants in the air diffuse into this mucus sheet and subsequently bind to olfactory receptors located on the olfactory cilia. Mucosal odorant-binding proteins also facilitate odorant–receptor binding [16–18]. Expression patterns of olfactory receptors help identify several subzones of olfactory neurons [19].

Axons from sensory olfactory neurons form bundles that traverse the lamina propria and bony cribriform plate to synapse with the dendrites of mitral cells in the olfactory bulb. The olfactory bulb is the terminal nucleus of the olfactory nerve. Sensory inputs to the olfactory bulb are organized into olfactory glomeruli each comprised of synapses between ~25,000 receptor cell axons and 25 mitral cells [20, 21]. Axonal projections from these glomeruli contribute to formation of the olfactory tract to the cerebrum. The olfactory system also has an area of associated neocortex. A neural stem cell niche found in the adult subventricular zone also retains the ability to produce new neurons and glia to the adult olfactory bulb [22].

Odorants and many olfactory toxicants often reach the olfactory mucosa via the inspired air. Each olfactory receptor cell possesses one type of odorant receptor, and each odorant receptor subtype can detect a limited number of odorants [23]. Information from multiple olfactory receptors is ultimately combined in the olfactory bulb, forming an odor signature [24]. Rapid metabolism of many odorants occurs within the olfactory mucosa. Indeed, the metabolic capacity of the nasal cavity rivals that found in the liver. Detoxification of odorants and other xenobiotics via phase one (hydroxylation and other reactions increasing hydrophilicity) and phase two (conjugation) reactions occurs extensively in the nasal cavity [25, 26] Cytochrome P450 isoforms, including CYP2A3/5/10/13 and CYP2G1 found in the olfactory epithelium are thought to contribute to the clearance of odorants and other inhaled chemicals [27]. Local metabolism probably accounts for much of the cellular specificity of toxic responses to xenobiotics in the nose through bioactivation [28, 29]. Transport proteins including organic anion transporters and divalent metal transporters (DMT) are also expressed in the nasal epithelium [30, 31].

1.2 Olfactory Dysfunction: Why Animal Testing Is Important

Olfactory dysfunction in people takes many forms including decreased (i.e., hyposmic) or an absent (i.e., anosmic) ability to detect or correctly label odors. Anosmia can be partial or complete and may be either permanent or reversible. Partial anosmia in some individuals may be attributed to changes in the expression of one or more olfactory receptors. For example, polymorphisms in the odorant receptor (OR7D4) alter the ability of people to recognize androstenone in cooked pork [32]. Dysosmia can also occur and includes improper identification of an odor (i.e., parosmia) and olfactory hallucinations (i.e., phantosmia). Olfactory dysfunction in people has been associated with a wide range of syndromes including viral infections, idiopathic Parkinson's disease, Alzheimer-type dementia, and certain other neurodegenerative diseases as well as following chemical exposure [33–35]. Animal models of degenerative neurologic syndromes also demonstrate olfactory dysfunction [36, 37]. A diminished sense of smell is associated with decreased quality of life, depression, degraded abilities of workers to sense a hazardous chemical, and other consequences [38, 39].

The ability of clinicians to recognize these diverse syndromes often relies initially on patient self-reporting. Self-reported cases of anosmia following acute COVID-19 infection [40–43] illustrates just one such example. Additional assessment of olfactory function may be performed in some clinical cases. Specific tests to assess a patient's olfactory function may involve administration of an odor threshold or olfactory discrimination task [44]. Odor identification tests usually involve the patient identifying odors from odor-impregnated test strips (e.g., the University of Pennsylvania Smell Identification Test [UPSIT]), felt-tip pens (e.g., Sniffin' Sticks), or other odor sources. Of these the UPSIT has been widely used and age- and sex-based norms have been established [45]. This self-administered scratch-and-sniff test consists of 40 chemically micro-encapsulated odor patches that release an odor when scratched. The UPSIT evaluates an individual's ability to identify odorants at the suprathreshold level. Psychophysical tests of olfactory ability in people are generally more sensitive and reliable in detecting and quantifying chemosensory disturbances than electrophysiological tests [46]. Despite the availability of the UPSIT and other psychophysical tests of olfaction, self-reporting all too often remains the mainstay for diagnosis of olfactory dysfunction in people.

2   Materials

One of the more common odorants used in rodent studies is vanillin (4-hydroxy-3-methoxybenzaldehyde), the main chemical component of natural vanilla. Sources of this chemical range from analytical grade material to commercial preparations used for cooking and other applications. The concentration and purity of many of these sources of vanillin is unknown which could contribute to a decreased ability of an investigator to replicate results. Vanillin is considered a "pure odor," i.e., excites olfactory neurons only [47, 48]. Other odorants (e.g., mustard oil, eucalyptol) predominantly evoke trigeminal responses [48]. For many odorants, it is unknown whether they act via olfactory or trigeminal pathways or a combination of the two.

There are few standardized test apparatuses developed specifically for evaluating olfaction in rodents. Therefore, instrumentation, test apparatuses, and materials can vary between laboratories. In our laboratory, several of our tests of olfaction have been created by ourselves. One of our simple tests of olfactory discrimination (Foster ML, unpublished) uses a novel open field created from a black rectangular polyethylene storage box (approximately $60 \times 50 \times 13$ cm). Holes were drilled into each corner of the

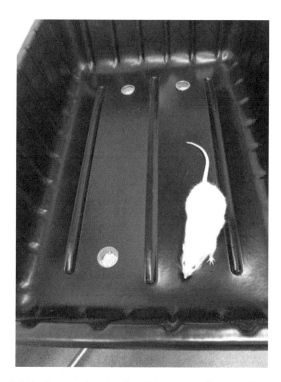

Fig. 1 Open field test apparatus developed to assess olfactory discrimination and memory. This simple test uses odor cues that are placed into disposable plastic cups. In some cases, the odor cue is paired with a food reward and in other cases the odor cue is used as a distractor. Odor cues and food rewards are buried using a neutral substrate (clean bedding). The number of correct and incorrect trials to find the correct odor is determined during each 5-min trial

box to accommodate 2 ounce (59 mL) disposable plastic condiment cups (Fig. 1). These cups were used to hold food treats, clean bedding, and odorants. Presentation of the odorant could occur using variable volumes (e.g., 1–10 μL) of the odorant onto filter paper, cotton balls, or another "odorless" substrate. These materials could be used directly or placed into containers to prevent the animals from having direct contact with the odorant (e.g., cassettes used for histology are helpful).

In contrast, we also developed an olfactometer (Fig. 2) that was designed to present an olfactory cue (acetaldehyde) to an animal performing an odor-cued active avoidance task [49]. In this case, a commercially available shuttle box (Coulbourn Instruments; Allentown, PA) was extensively modified to have two exterior 0.64-cm polycarbonate (Lexan, GE Structured Products Department, Mt. Vernon, IN) walls with inlets for the odorant. The center divider was modified to have a guillotine door with an air current that helped remove odor from the animal's fur as it crossed from one side of the shuttle box to the other. Each chamber received fresh air or odorant diluted with air through Teflon inlet lines

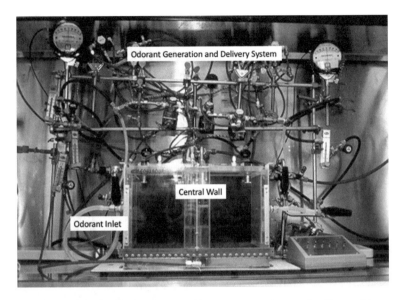

Fig. 2 Acetaldehyde olfactometer used to assess an odor-cued active avoidance test. The modified shuttle box has three main components: Odorant Generation and Delivery System (1) Odorant Generation and Delivery System, (2) Central Wall, and (3) Odorant Inlet – numerous solenoids were used to allow deliver either fresh air or odorant containing air to be to either side of the shuttle box and exhaust air through the Central Wall. The Central Wall also contained an air curtain and a guillotine door. The third major component (Odorant Inlet) delivered fresh air or odorant containing air to either side of the shuttle box. A schematic drawing showing additional details is available [49]

(0.95 cm od). A stainless-steel exhaust panel was also incorporated into the central panel containing the air current and it provided a flow of ~3.00 L/min to each chamber side. Acetaldehyde atmospheres were generated in Tedlar bags (SKC Co., Eighty Four, PA) by the injection of acetaldehyde at −20 °C into fresh air. Daily determinations of acetaldehyde concentrations in the Tedlar bags and at the inlet to each chamber during constant delivery were made with an infrared spectrophotometer (MIRAN 1A, Foxboro Co., East Bridgewater, MA). Experiments were conducted at a chamber inlet concentration of 5 ppm, which was verified by infrared spectrometry. In this case, some animals were trained using an auditory cue to confirm that toxicant exposure did not impair the animal's ability to perform the task.

A third example for a Go/No-Go test of olfactory discrimination used a commercially available operant chamber equipped with a stimulus light, response lever, nose-poke port, and odorant delivery system (Habitest System, Coulbourn Instruments, Holliston, MA) [50]. All tubing connections were located outside of the operant chamber. The odorant delivery system was constructed with a small air pump (Fluval, Baie d'Urfé, Québec Canada), Tygon polyethylene (PE) tubing (VWR International, Radnor,

PA), glass jars for odorants, flowmeters (Cole-Parmer, Bunker, CT), and solenoid-operated valves controlled by a computer running Habitest System software (Graphic State 3). The software system controlled the function of several solenoids that controlled delivery of fresh air or fresh air with odorant to the nose-poke port. Airflow through the clean air stream was never interrupted during the session. A house-light turned on, which signaled to the rat that the trial could begin. Then, a nose-poke by the rat would trigger the solenoid, which would allow the odorized stream to flow into the odor port. While the rat kept its nose in the odor port (i.e., while the photobeam was interrupted), the solenoid remained open for 1.5 s. When an odorant's solenoid valve was open, airflow to the odorant jar was 1.5 L/min and clean airflow was 1 L/min.

3 Methods

3.1 Olfactory Toxicity Testing in Rodents: An Overview

The assessment of olfactory toxicity in rodents generally rests on two main approaches, namely nasal pathology and behavioral tests of olfaction. Histopathology evaluates the structure of the nasal cavity, olfactory bulb, and other brain structures and inferences regarding olfactory function can be made from these findings. Behavioral tests assess the function of the olfactory system. Some neurophysiologic and brain imaging techniques including odor-evoked electroolfactograms (EOG) and olfactory electroencephalograms (EEG) also provide information about the function of the olfactory system—however, these approaches are rarely used in toxicology. Indeed, assessment of olfactory toxicity in rodents remains all too limited in many studies. This general lack of screening for olfactory toxicity in rodent studies should not be presumed to suggest that adverse effects of chemicals on olfaction lack importance. Indeed, the human olfactory system has been reported to be adversely affected by approximately 200 compounds of toxicological interest [1, 35].

3.2 Nasal Pathology

Nasal pathology is used to assess the integrity of the olfactory, respiratory, squamous, transitional, and lymphoepithelial epithelium that lines the nasal passage [1, 2, 51]. Sensory olfactory neurons are directly exposed to the external environment, which increases their vulnerability to certain chemical agents, especially following inhalation. Unlike most other neurons, sensory olfactory neurons age and undergo biochemical and morphologic characteristics of apoptotic cell death. These factors may contribute to the frequency with which olfactory epithelial lesions are observed in rodent toxicity studies.

Toxicant-induced epithelial lesions in the nasal passages are often site-specific and dependent on the intranasal regional dose of the inhaled chemical and the sensitivity of the nasal epithelial

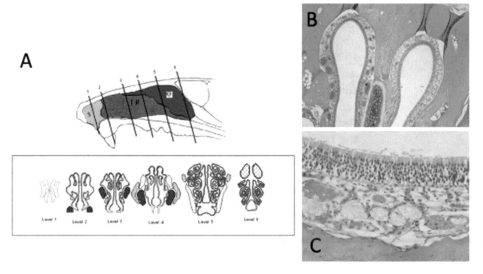

Fig. 3 (**a**) Sagittal view of the rat nose with corresponding cross-sections showing the distribution of the major epithelial types: squamous (S), transitional/respiratory (T/R), and olfactory (O). (**b**) Lower magnification of the dorsal recess from transverse section taken at Level 4 in (**a**). (**c**) Higher magnification of the olfactory epithelium lining the dorsal recess of the rat nasal cavity

tissue to the specific chemical. The location of olfactory epithelial lesions is attributable to local dose of the chemical, tissue susceptibility, or, more frequently, a combination of these factors [52]. Nasal lesions can arise from either inhalation or systemic delivery of a chemical and several reviews of this topic are available [1, 2, 53, 54]. Because of the complex anatomy of the rodent nasal cavity, systematic approaches for collecting and examining rodent nasal tissue have been developed [2, 55, 56]. These approaches result in the collection of multiple transverse nasal sections which result in representative sections of all epithelial subtypes being available for evaluation (Fig. 3). Recommendations to collect multiple sections of the nasal cavity have been included in some regulatory inhalation testing requirements [57]. For example, the Organisation for Economic Co-operation and Development guidelines for inhalation toxicity studies include recommendations for the evaluation of multiple transverse nasal sections [58]. Standard procedures are generally used to prepare tissue sections, with the exception that decalcification is often required prior to sectioning. Routine evaluation of nasal sections with hematoxylin and eosin is typically performed on multiple cross-sections. Additional immunohistochemical markers including olfactory marker protein (OMP) for the identification of mature sensory olfactory neurons can also be used to characterize the extent of the injury [18, 23].

Damage to the olfactory epithelium is a common histologic finding seen in rodents following the inhalation of some chemicals (Fig. 4). Chemical exposure can result in apoptosis, loss of olfactory

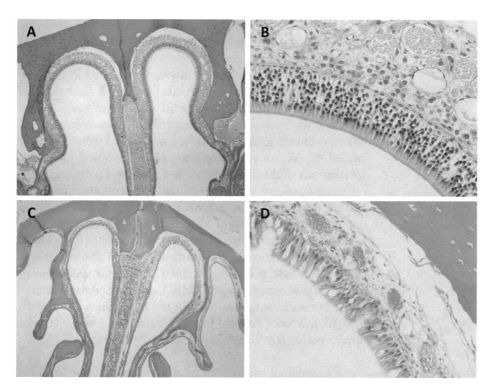

Fig. 4 (**a**) Normal olfactory epithelium from the rat dorsal medial meatus. (**b**) Higher magnification of normal olfactory epithelium. (**c**) Olfactory neuronal loss following subchronic exposure to hydrogen sulfide. (**d**) Higher magnification of lesion shown in (**c**) illustrating thinning of the epithelium and loss of olfactory neurons [57]

neurons, epithelial degeneration, and thinning of the olfactory epithelium. Chemical-induced olfactory neuron loss often exhibits a concentration or dose–response relationship and often presents as progressive degeneration with or without an inflammatory response. These initial changes may be followed by regeneration of the olfactory epithelium. In other cases, adaptive responses such as replacement of the olfactory epithelium by respiratory epithelium (i.e., metaplasia) may be seen. Loss of olfactory sensory neurons may occur with or without injury to the adjoining sustentacular cells and in some chronic studies the sustentacular cells can take on an engorged appearance with eosinophilic proteinaceous deposits of unknown toxicological significance [59, 60]. Neoplastic lesions in the olfactory epithelium are also recognized. Nasal carcinogens include formaldehyde, acrolein, ethylene oxide, among others [6]. As mentioned previously, stem cells found in the olfactory epithelium may partially or completely lead to epithelial repair. It is largely unknown whether olfactory function normalizes in these cases. Many rodent inhalation toxicity studies have satellite groups of animals to evaluate possible lesion recovery at 30–90 days after the end of the exposure. Regeneration appears to decrease with age [61].

Lesions to the rodent olfactory epithelium and nasal cavity can also be appreciated using magnetic resonance imaging (MRI) techniques. For example, the time course of thinning of the olfactory epithelium has been followed in vivo using MRI in rats following exposure to zinc oxide nanoparticles [62] and 3-methylindole [63].

Collectively olfactory epithelial and other nasal lesions have provided the basis for some chemical risk assessments [57]. These risk assessments need to consider a number of factors including the animal model, route of exposure, physiochemical properties, and metabolism of the chemical, airflow (rate and patterns), and dose–response relationships when evaluating the relevance of an olfactory epithelial lesion seen in rodents to humans. Anatomically accurate computational models of the nasal airways have been used to support quantitative interspecies extrapolations [64]. These computer models are generally constructed from sequential cross-sections of the nasal passages, obtained from tissue specimens or from magnetic resonance imaging (MRI) scans of the nasal cavity. These models provide broad support for the hypotheses that the distribution of inhaled chemical-induced olfactory lesions is related to uptake patterns in the nasal/olfactory epithelium [65, 66].

3.3 Behavioral Approaches to Assess Olfactory Ability of Rodents

Neurobehavioral approaches used to assess olfaction in rodents can be categorized into suprathreshold and subthreshold tests. Suprathreshold tests include evaluation of olfactory discrimination and use odorant concentrations well above the threshold needed to elicit a response. Subthreshold tests are used to estimate an olfactory threshold for an odorant that is the lowest air concentration of the odorant that can be reliably detected by an animal. These tests can be challenging to perform since odor detection thresholds in rodents are often reported to occur in the picomolar range [67, 68]. Another way to categorize olfactory testing in rodents is to consider whether the test method can be performed on an untrained animal that may also be naïve to the odor.

One of the simplest tests to assess olfaction in rodents involves the presentation of a cotton swab containing an odorant to the animal, followed by evaluation for orientation toward and sniffing of the swab. This test requires no training of the animal and is analogous to assessing hearing in rodents with the use of a clicker, both of which can easily be incorporated into a functional observational battery [69]. Multiple odorants could be used to assess olfaction in this simple way. Simple tests of olfactory discrimination are commonly performed in developmental neurotoxicity studies and depending upon the experimental methods can be performed on untrained naïve animals. These tests are generally performed on neonatal rats after post-natal day 8 [70]. Rat pups are typically placed in the middle of a T- or Y-shaped maze in which one arm is baited with clean shavings while the other contains soiled home cage bedding. A positive response is indicated by the animal orienting and moving toward the side containing the home cage bedding. These tests simultaneously assess olfaction and motor activity.

Innate rodent responses to predator odors can also be used as a simple test of olfaction. For example, mice will avoid or "freeze" near filter paper scented with 2,4,5-trimethylthiazoline, a constituent of fox urine and feces [71]. Similar results are seen in rats exposed to cat odors [72].

Another test that has been used in toxicology is the buried food pellet test. In this test, food is withheld from untrained animals overnight (12–14 h) and placed in a test cage in which a food pellet has been placed 1–2 cm below the surface of a familiar bedding substrate [73]. The increases in the latency to eat food in the buried food test suggest an impairment of olfaction is present. The assumption is that food-restricted mice which fail to use odor cues to locate the food within a 10–15-min period likely have an olfactory deficit. The animal's motivation to eat the pellet is also tested in control trials in which the food item is placed on the surface of the substrate. Another approach is to reduce food supplied to the animals over several days to attain a ~10% reduction in body weight. This simple test has been used with a number of toxic agents including zinc gluconate, rotenone, and methyl bromide [74–77]. Detailed protocols for the buried food pellet test in rodents are provided in several publications [73, 78].

The olfactory habituation/dishabituation test is another simple test of rodent olfaction [73, 79]. This test relies on the tendency of rodents to investigate novel smells. A commonly used sequence for odor presentation is water, two non-social odors, and two social odors. Each odor (or water) is presented in three consecutive trials for a duration of 2 min with an inter-trial interval of 1 min, Habituation is defined by a progressive decrease in olfactory investigation (sniffing) toward a repeated presentation of the same odor stimulus. Dishabituation is defined by a reinstatement of sniffing when a novel odor is presented using an odor-impregnated applicator. This test has been recently used to show decreased olfactory performance in male rats exposed to phencyclidine, valproic acid, and chlorpyrifos [80–82]. Detailed protocols for the olfactory habituation/dishabituation test are provided in several publications [73, 78].

Simple tests of olfactory discrimination can be performed. For example, animals can be quickly trained over several days to find a food reward that is paired with an odor cue. In these studies, the food reward and odorant are buried in small containers with a substrate (e.g., bedding, sand) that requires the animal to dig for a reward. The test arena can have several null containers (no odor) or other odorants as distracting odors. More elaborate tests including go-no-go olfactory discrimination tasks have been developed for use in operant conditioning. In these tests, animals are trained to either lever press ("go") in response to a positive conditioned stimulus (e.g., vanillin) or to do nothing ("no go") when a negative conditioned stimulus (e.g., amyl acetate) is presented. Animals can be subsequently tested for their ability to reverse the associated stimuli (odor reversal learning). Training of animals can often require multiple sessions over several weeks. These tests have been used to identify functional deficits in rodents exposed to cadmium, N-methyl-D-aspartate (NMDA), Δ^9-tetrahydrocannabinol, and manganese [50, 83–85]. Examples of protocols for the olfactory discrimination tests are provided in several publications [86, 87].

Tests of olfactory discrimination can also be modified to measure olfactory thresholds in rodents. In these studies, odors are presented at varying concentrations. In some studies, they are presented sequentially from low to high concentrations or high to low concentrations [67]. In other cases, a staircase approach that uses trials with ascending, descending and random orders of concentration can also be used [88, 89]. Threshold is determined when the animal can no longer reach the specified criterion for detection of the target odor when compared with a vehicle (usually correct response rates >75%). Olfactory thresholds for 2-propanol, D-limonene, and ethyl acetoacetate were largely unaltered in rats treated with methyl bromide despite extensive (>95%) damage to the olfactory epithelium [90].

3.4 Neurophysiologic and Brain Imaging Approaches to Assess Olfactory Function

The electroolfactogram (EOG) assesses the collective response of groups of olfactory neurons following odorant exposure. The EOG has been primarily used to study spatial and temporal responses and help map the distribution of olfactory receptor genes in zones in the rodent olfactory epithelium [91, 92]. The EOG has been used both in vivo and in vitro using organotypic cell cultures [93, 94]. To date, application of the EOG to rodent toxicity studies has been limited [95, 96]. A useful video demonstration of the EOG is available [93].

The EOG directly measures olfactory neuron responses to an odorant. An alternative approach to assess the integrity of the olfactory system would be to assess odor-induced changes in the electroencephalogram (EEG). In such studies, EEG signals are recorded during odorant presentation [97, 98]. A related method, magnetoencephalography, has also been used to evaluate human olfaction [99]. In humans, surface electrodes are used to collect the EEG signal and studies using a combination of both EEG and functional MRI have been reported [100, 101]. In rodents electrodes implanted in the olfactory bulb are typically used [102, 103]. As is the case with the EOG, odor-induced EEGs have been infrequently used to assess olfaction in rodents following toxicant exposure.

Brain imaging has received increased use in the evaluation of smell disorders in people. These methods have been used to identify cortical structures associated with olfactory stimulation and have also been introduced into clinical practice. Both positron emission tomography [104–106] and functional MRI [107–109] have been applied to the study of olfaction in people. Many of these approaches have been used in rodents as well [110–115]. Other more specialized approaches including optogenetics and two-photon imaging have been used to study rodent olfaction [116–118]. As these methods become more widely available, we anticipate that they may be increasingly used in future rodent toxicity studies.

4 Notes

Several technical issues need to be considered when assessing olfaction in rodents. Training of staff to detect cues associated with olfaction is critical. Inter-individual variability in assessment of olfaction should be minimized. One approach is to use the same individual to assess olfaction in animals. Blinding of the observer to animal treatment groups can also reduce bias.

Species selection often depends on the research objectives of the toxicity study. Regulatory toxicology studies often rely on rats whereas more investigative studies often favor mice due to availability of knockout mice and other gene targeted models. Sex- and strain-differences in rodent olfaction are recognized and these factors may affect results in toxicity studies. For example, aging in mice is associated with decreased numbers of sensory olfactory neurons, reduced regenerative capacity of the olfactory epithelium, and decreased olfactory ability [119, 120]. It is plausible that the aged olfactory system may be at greater risk to toxicant exposure. Female mammals generally have a better sense of smell than males [119–121], and this can be further influenced by the reproductive status of the animal [122, 123]. The method of chemical exposure, inhalation, intranasal instillation, or systemic delivery, depends upon the toxicant of interest and the technical capabilities of the laboratory performing the work.

In mice, vanillin is neutral with respect to odor preference—mice show neither attraction nor revulsion to the odor [124]. Short-term (20 min) exposure of mice to vanillin has minimal impacts on their performance in an elevated plus maze, open field, Y-maze, among other behavioral tests. Short-term exposure to vanillin was associated with both decreased hot plate response and reduced forelimb grip strength [124]. The impact of other odorants on neurobehavioral test performance by rodents remains incompletely understood. Concerns might be raised regarding whether or not the use of a limited number of odorants to study rodent olfaction is sufficient given the large number of odors that exist in the environment. Studies performed in humans have shown that as few as three odors, namely cinnamon, banana, and fish odor may be needed to quickly identify an individual as having normal olfactory function [125].

Stimulus control can be a daunting technical challenge since unlike sound or light, odor cues cannot be easily generated or turned on or off. Extraneous odors from personal care products, cleaning supplies, and other materials should be avoided. Likewise, maintaining a constant odorant concentration is difficult since the odor stimulus can fluctuate over time [126]. Odor cues in some suprathreshold tests of olfaction are presented using cotton balls or other media that are impregnated with solutions that contain the odorant of interest. Other tests of rodent olfaction rely on the generation of an atmosphere containing the odorant of interest. These atmospheres are often produced by placing the odorant of interest in water or another liquid vehicle and then relying on vaporization of the odorant to produce the atmosphere. Saturated solutions of the odorant are often used as the odor source resulting in suprathreshold exposures. Air odorant concentration derived from solutions of an odorant depends upon the concentration of the odorant in the vehicle, temperature, and the odorant's vapor pressure. In other cases, Tedlar bags or other inert materials containing known amounts of the odorant of interest mixed with air or nitrogen can be used to generate the odor source [49]. Air odorant concentrations delivered to the animal or test system can be modified by mixing the odor source air with a clean (no odorant) air source producing subthreshold concentrations useful for the measurement of an olfactory threshold. The odorant generation and delivery system should use nonreactive components that do not contribute to the odor signature (e.g., Teflon). Separate odor generation systems may be needed to reduce cross-contamination when multiple odors are being used. Nominal air concentrations of the odorant can be confirmed using gas chromatography and mass spectrometry or other analytical chemistry methods.

Once the odorant containing atmosphere has been generated it needs to be presented to the breathing zone of the animal. Commercially available microprocessor controlled "odor ports" are available that can present the odorant to an animal performing a behavior in an operant chamber. Odorant can also be presented in other ways including infusing a test chamber with the odorant of interest. Residual presence of an odorant in a test apparatus or an animal room may confound results—therefore, some effort to reduce residual odor (e.g., using an exhaust system) may be helpful. Using your sense of smell to determine when an odor is no longer present is unreliable since olfactory thresholds for many odorants are appreciably lower in rodents when compared with people. Likewise, test apparatuses should be cleaned with distilled water or other unscented cleaning solutions to reduce distracting odors in a test system.

It can also be challenging to recognize when an animal responds to the odorant. Active sniffing in small rodents with high respiratory rates is difficult to perceive visually. In many cases, tests of rodent olfaction rely on the expression of a second behavior (e.g., uncovering of a buried food pellet, lever pressing) that is presumed to indicate that the animal has detected the odor. Animals that can no longer perform these secondary behaviors may be misclassified as having a decreased olfactory ability. It is therefore critical that controls be incorporated into the study to confirm that the animal's ability to perform the task is unaffected by toxicant exposure. In addition, some commonly used test methods rely on the animal responding to an odor cue in order to receive a food reward. In these cases, animals are often maintained at 80 to 90% of their normal body weight or held off food overnight prior to testing. Chemicals that decrease motivation may result in spurious results on these assays.

5   Conclusions

In closing there are several "take home" messages that we want to impart to the reader. Evaluation of olfactory function in rodents occurs widely in neuroscience; however, approaches used by neuroscientists have rarely been adopted by toxicologists. There is an important opportunity for future application of these methods to neurotoxicology. For this and other reasons, olfaction and toxic effects on the rodent olfactory system remains underappreciated. Most studies used to evaluate the rodent olfactory system rely on histologic evaluation of multiple sections of the nasal cavity and relatively simple tests of olfaction. To date many of these behavioral tests have been unable to demonstrate a lack of olfactory function even in animals with significant pathology in the olfactory system. For example, extensive injury and loss of sensory olfactory neurons as assessed using histopathology is often required before a functional deficit can be observed [49, 90, 127]. Other studies have shown that near total ablation of the rat olfactory bulb did not impair an animal's ability to detect and discriminate between odors [128]. These findings suggest that a tremendous "reserve capacity" in the rodent for the sense of smell likely exists. In humans, we rarely have information regarding the structural integrity of the olfactory system; instead, clinicians more commonly rely on behavioral tests of olfaction. It is possible that, for people, behavioral effects may prove to be more sensitive than nasal pathology in detecting toxicant-induced injury to the olfactory system. This dichotomy of responses between the species may reflect the relative importance of the sense of smell.

References

1. Dorman DC (2018) Olfactory system. In: McQueen CA (ed) Comprehensive toxicology, 3rd edn. Elsevier, New York, NY, pp 361–375

2. Harkema JR, Carey SA, Wagner JG (2006) The nose revisited: a brief review of the comparative structure, function, and toxicologic pathology of the nasal epithelium. Toxicol Pathol 34:252–269

3. Solbu TT, Holen T (2012) Aquaporin pathways and mucin secretion of Bowman's glands might protect the olfactory mucosa. Chem Senses 37(1):35–46

4. Sarnat HB, Flores-Sarnat L (2017) Olfactory development, part 2: neuroanatomic maturation and dysgeneses. J Child Neurol 32:579–593

5. Badonnel K, Durieux D, Monnerie R, Grébert D, Salesse R, Caillol M, Baly B (2009) Leptin-sensitive OBP-expressing mucous cells in rat olfactory epithelium: a novel target for olfaction-nutrition crosstalk? Cell Tissue Res 338:53–66

6. Jeffrey AM, Iatropoulos MJ, Williams GM (2006) Nasal cytotoxic and carcinogenic activities of systemically distributed organic chemicals. Toxicol Pathol 34:827–852

7. Gross EA, Swenberg JA, Fields S, Popp JA (1982) Comparative morphometry of the nasal cavity in rats and mice. J Anat 135:83–88

8. Smith TD, Bhatnagar KP, Tuladhar P, Burrows AM (2004) Distribution of olfactory epithelium in the primate nasal cavity: are microsmia and macrosmia valid morphological concepts? Anat Rec A Discov Mol Cell Evol Biol 281:1173–1181

9. Craven BA, Paterson EG, Settles GS (2010) The fluid dynamics of canine olfaction: unique nasal airflow patterns as an explanation of macrosmia. J R Soc Interface 7:933–943

10. Rygg AD, Van Valkenburgh B, Craven BA (2017) The influence of sniffing on airflow and odorant deposition in the canine nasal cavity. Chem Senses 42:683–698

11. Beites CL, Kawauchi S, Crocker CE, Calof AL (2005) Identification and molecular regulation of neural stem cells in the olfactory epithelium. Exp Cell Res 306:309–316

12. Mackay-Sim A (2010) Stem cells and their niche in the adult olfactory mucosa. Arch Ital Biol 148:47–58

13. Schwob JE, Jang W, Holbrook EH, Lin B, Herrick DB, Peterson JN, Hewitt Coleman J (2017) Stem and progenitor cells of the mammalian olfactory epithelium: taking poietic license. J Comp Neurol 525:1034–1054

14. Yu CR, Wu Y (2017) Regeneration and rewiring of rodent olfactory sensory neurons. Exp Neurol 287:395–408

15. McEwen DP, Jenkins PM, Martens JR (2008) Olfactory cilia: our direct neuronal connection to the external world. Curr Top Dev Biol 85:333–370

16. Pelosi P, Zhu J, Knoll W (2018) Odorant-binding proteins as sensing elements for odour monitoring. Sensors (Basel) 18(10): pii: E3248

17. Sun JS, Xiao S, Carlson JR (2018) The diverse small proteins called odorant-binding proteins. Open Biol 8:180208

18. Baker H, Grillo M, Margolis FL (1989) Biochemical and immunocytochemical characterization of olfactory marker protein in the rodent central nervous system. J Comp Neurol 285:246–261

19. Ressler KJ, Sullivan SL, Buck LB (1993) A zonal organization of odorant receptor gene expression in the olfactory epithelium. Cell 73:597–609

20. Sarnat HB, Flores-Sarnat L (2019) Development of the human olfactory system. Handb Clin Neurol 164:29–45

21. Zou DJ, Chesler A, Firestein S (2009) How the olfactory bulb got its glomeruli: a just so story? Nat Rev Neurosci 10:611–618

22. Lim DA, Alvarez-Buylla A (2016) The adult ventricular-subventricular zone (V-SVZ) and olfactory bulb (OB) neurogenesis. Cold Spring Harb Perspect Biol 8(5):pii: a018820

23. Heydel JM, Coelho A, Thiebaud N, Legendre A, Le Bon AM, Faure P, Neiers F, Artur Y, Golebiowski J, Briand L (2013) Odorant-binding proteins and xenobiotic metabolizing enzymes: implications in olfactory perireceptor events. Anat Rec (Hoboken) 296:1333–1345

24. Grabe V, Sachse S (2018) Fundamental principles of the olfactory code. Biosystems 164:94–101

25. Genter MB (2004) Update on olfactory mucosal metabolic enzymes: age-related changes and N-acetyltransferase activities. J Biochem Mol Toxicol 18:239–244

26. Watelet JB, Strolin-Benedetti M, Whomsley R (2009) Defence mechanisms of olfactory neuro-epithelium: mucosa regeneration, metabolising enzymes and transporters. B-ENT Suppl 13:21–37

27. Ling G, Gu J, Genter MB, Zhuo X, Ding X (2004) Regulation of cytochrome P450 gene expression in the olfactory mucosa. Chem Biol Interact 147:247–258

28. Dahl AR, Hadley WM (1991) Nasal cavity enzymes involved in xenobiotic metabolism: effects on the toxicity of inhalants. Crit Rev Toxicol 21:345–372

29. Reed CJ (1993) Drug metabolism in the nasal cavity: relevance to toxicology. Drug Metab Rev 25:173–205

30. Burckhardt G (2012) Drug transport by organic anion transporters (OATs). Pharmacol Ther 136:106–130

31. Thiebaud N, Menetrier F, Belloir C, Minn AL, Neiers F, Artur Y, Le Bon AM, Heydel JM (2011) Expression and differential localization of xenobiotic transporters in the rat olfactory neuro-epithelium. Neurosci Lett 505:180–185

32. Lunde K, Egelandsdal B, Skuterud E, Mainland JD, Lea T, Hersleth M, Matsunami H (2012) Genetic variation of an odorant receptor OR7D4 and sensory perception of cooked meat containing androstenone. PLoS One 7 (5):e35259

33. Boesveldt S, Postma EM, Boak D, Welge-Luessen A, Schöpf V, Mainland JD, Martens J, Ngai J, Duffy VB (2017) Anosmia—a clinical review. Chem Senses 42:513–523

34. Doty RL (2018) Age-related deficits in taste and smell. Otolaryngol Clin N Am 51:815–825

35. Genter MB, Doty RL (2019) Toxic exposures and the senses of taste and smell. Handb Clin Neurol 164:389–408

36. Coronas-Sámano G, Portillo W, Beltrán Campos V, Medina-Aguirre GI, Paredes RG, Diaz-Cintra S (2014) Deficits in odor-guided behaviors in the transgenic 3xTg-AD female mouse model of Alzheimer's disease. Brain Res 1572:18–25

37. Bermúdez ML, Seroogy KB, Genter MB (2019) Evaluation of carnosine intervention in the Thy1-aSyn mouse model of Parkinson's disease. Neuroscience 411:270–278

38. Croy I, Nordin S, Hummel T (2014) Olfactory disorders and quality of life—an updated review. Chem Senses 39:185–194

39. Gobba F (2006) Olfactory toxicity: long-term effects of occupational exposures. Int Arch Occup Environ Health 79:322–331

40. Heidari F, Karimi E, Firouzifar M, Khamushian P, Ansari R, Mohammadi Ardehali M, Heidari F (2020) Anosmia as a prominent symptom of COVID-19 infection. Rhinology 58:302–303

41. Hopkins C, Surda P, Kumar N (2020) Presentation of new onset anosmia during the COVID-19 pandemic. Rhinology 58:295–298

42. Kaye R, Chang CWD, Kazahaya K, Brereton J, Denneny JC 3rd (2020) COVID-19 anosmia reporting tool: initial findings. Otolaryngol Head Neck Surg 163:132–134

43. Klopfenstein T, Kadiane-Oussou NJ, Toko L, Royer PY, Lepiller Q, Gendrin V, Zayet S (2020) Features of anosmia in COVID-19. Med Mal Infect 50(5):436–439

44. Wrobel BB, Leopold DA (2004) Clinical assessment of patients with smell and taste disorders. Otolaryngol Clin N Am 37:1127–1142

45. Doty RL, Shaman P, Kimmelman CP, Dann MS (1984) University of Pennsylvania Smell Identification Test: a rapid quantitative olfactory function test for the clinic. Laryngoscope 94:176–178

46. Doty RL (2018) Measurement of chemosensory function. World J Otorhinolaryngol Head Neck Surg 4:11–28

47. Radil T, Wysocki CJ (1998) Spatiotemporal masking in pure olfaction. Ann N Y Acad Sci 855:641–644

48. Tremblay C, Frasnelli J (2018) Olfactory and trigeminal systems interact in the periphery. Chem Senses 43:611–616

49. Owens JG, James RA, Moss OR, Morgan KT, Bowman JR, Struve MF, Dorman DC (1996) Design and evaluation of an olfactometer for the assessment of 3-methylindole-induced hyposmia. Fundam Appl Toxicol 33:60–70

50. Foster ML, Rao DB, Francher T, Traver S, Dorman DC (2018) Olfactory toxicity in rats following manganese chloride nasal instillation: a pilot study. Neurotoxicology 64:284–290

51. Chamanza R, Wright JA (2015) A review of the comparative anatomy, histology, physiology and pathology of the nasal cavity of rats, mice, dogs and non-human primates. Relevance to inhalation toxicology and human health risk assessment. J Comp Pathol 153:287–314

52. Morgan KT, Monticello TM (1990) Airflow, gas deposition, and lesion distribution in the nasal passages. Environ Health Perspect 85:209–218

53. Maronpot RR (1990) Pathology Working Group review of selected upper respiratory

tract lesions in rats and mice. Environ Health Perspect 85:331–352

54. Monticello TM, Morgan KT, Uraih L (1990) Nonneoplastic nasal lesions in rats and mice. Environ Health Perspect 85:249–274

55. Mery S, Gross EA, Joyner DR, Godo M, Morgan KT (1994) Nasal diagrams: a tool for recording the distribution of nasal lesions in rats and mice. Toxicol Pathol 22:353–372

56. Morgan KT (1991) Approaches to the identification and recording of nasal lesions in toxicology studies. Toxicol Pathol 19:337–351

57. Dorman DC (2019) Use of nasal pathology in the derivation of inhalation toxicity values for hydrogen sulfide. Toxicol Pathol 47:1043–1048

58. Organization for Economic Co-operation and Development (2009). Draft OECD guidance document on histopathology for inhalation toxicity studies, supporting TG 412 (subacute inhalation toxicity: 28-Day) and TG 413 (subchronic inhalation toxicity: 90-Day). http://www.oecd.org.prox.lib. ncsu.edu/chemicalsafety/testing/ 43062801.pdf. Accessed 4 Sep 2020

59. Gross EA, Patterson DL, Morgan KT (1987) Effects of acute and chronic dimethylamine exposure on the nasal mucociliary apparatus of F-344 rats. Toxicol Appl Pharmacol 90:359–376

60. Wolf DC, Morgan KT, Gross EA, Barrow C, Moss OR, James RA, Popp JA (1995) Two-year inhalation exposure of female and male B6C3F1 mice and F344 rats to chlorine gas induces lesions confined to the nose. Fundam Appl Toxicol 24:111–131

61. Morrison EE, Costanzo RM (1992) Morphology of olfactory epithelium in humans and other vertebrates. Microsc Res Tech 23:49–61

62. Gao L, Yang ST, Li S, Meng Y, Wang H, Lei H (2013) Acute toxicity of zinc oxide nanoparticles to the rat olfactory system after intranasal instillation. J Appl Toxicol 33:1079–1088

63. Wiethoff AJ, Harkema JR, Koretsky AP, Brown WE (2001) Identification of mucosal injury in the murine nasal airways by magnetic resonance imaging: site-specific lesions induced by 3-methylindole. Toxicol Appl Pharmacol 175:68–75

64. Kimbell JS, Subramaniam RP (2001) Use of computational fluid dynamics models for dosimetry of inhaled gases in the nasal passages. Inhal Toxicol 13:325–334

65. Moulin FJ, Brenneman KA, Kimbell JS, Dorman DC (2002) Predicted regional flux of

hydrogen sulfide correlates with distribution of nasal olfactory lesions in rats. Toxicol Sci 66:7–15

66. Schroeter JD, Kimbell JS, Gross EA, Willson GA, Dorman DC, Tan YM, Clewell HJ 3rd (2008) Application of physiological computational fluid dynamics models to predict interspecies nasal dosimetry of inhaled acrolein. Inhal Toxicol 20:227–243

67. Schellinck H (2018) Measuring olfactory processes in *Mus musculus*. Behav Process 155:19–25

68. Wackermannová M, Pinc L, Jebavý L (2016) Olfactory sensitivity in mammalian species. Physiol Res 65:369–390

69. Moser VC (2011) Functional assays for neurotoxicity testing. Toxicol Pathol 39:36–45

70. Gregory EH, Pfaff DW (1971) Development of olfactory-guided behavior in infant rats. Physiol Behav 6:573–576

71. Kaneko-Goto T, Sato Y, Katada S, Kinameri E, Yoshihara S, Nishiyori A, Kimura M, Fujita H, Touhara K, Reed RR, Yoshihara Y (2013) Goofy coordinates the acuity of olfactory signaling. J Neurosci 33:12,987–12,996a

72. Staples LG, McGregor IS, Apfelbach R, Hunt GE (2008) Cat odor, but not trimethylthiazoline (fox odor), activates accessory olfactory and defense-related brain regions in rats. Neuroscience 151:937–947

73. Yang M, Crawley JN (2009) Simple behavioral assessment of mouse olfaction. Curr Protoc Neurosci Chapter 8:Unit 8.24

74. Hao S, Yu F, Yan A, Zhang Y, Han J, Jiang X (2012) In utero and lactational lanthanum exposure induces olfactory dysfunction associated with downregulation of βIII-tubulin and olfactory marker protein in young rats. Biol Trace Elem Res 148:383–391

75. Hsieh H, Horwath MC, Genter MB (2017) Zinc gluconate toxicity in wild-type vs. MT1/ 2-deficient mice. Neurotoxicology 58:130–136

76. Hurtt ME, Thomas DA, Working PK, Monticello TM, Morgan KT (1988) Degeneration and regeneration of the olfactory epithelium following inhalation exposure to methyl bromide: pathology, cell kinetics, and olfactory function. Toxicol Appl Pharmacol 94:311–328

77. Liu Y, Sun JD, Song LK, Li J, Chu S, Yuan Y, Chen N (2015) Environment-contact administration of rotenone: a new rodent model of Parkinson's disease. Behav Brain Res 294:149–161

78. Lehmkuhl AM, Dirr ER, Fleming SM (2014) Olfactory assays for mouse models of neurodegenerative disease. J Vis Exp 90:e51804

79. Oummadi A, Meyer-Dilhet G, Béry A, Aubert A, Barone P, Mortaud S, Guillemin GJ, Menuet A, Laugeray A (2020) 3Rs-based optimization of mice behavioral testing: the habituation/dishabituation olfactory test. J Neurosci Methods 332:108550

80. Campolongo M, Kazlauskas N, Falasco G, Urrutia L, Salgueiro N, Höcht C, Depino AM (2018) Sociability deficits after prenatal exposure to valproic acid are rescued by early social enrichment. Mol Autism 9:36

81. De Felice A, Venerosi A, Ricceri L, Sabbioni M, Scattoni ML, Chiarotti F, Calamandrei G (2014) Sex-dimorphic effects of gestational exposure to the organophosphate insecticide chlorpyrifos on social investigation in mice. Neurotoxicol Teratol 46:32–39

82. Tarland E, Brosda J (2018) Male rats treated with subchronic PCP show intact olfaction and enhanced interest for a social odour in the olfactory habituation/dishabituation test. Behav Brain Res 345:13–20

83. Czarnecki LA, Moberly AH, Rubinstein T, Turkel DJ, Pottackal J, McGann JP (2011) In vivo visualization of olfactory pathophysiology induced by intranasal cadmium instillation in mice. Neurotoxicology 32:441–449

84. Ferry AT, Lu XC, Price JL (2000) Effects of excitotoxic lesions in the ventral striatopallidal--thalamocortical pathway on odor reversal learning: inability to extinguish an incorrect response. Exp Brain Res 131:320–335

85. Sokolic L, Long LE, Hunt GE, Arnold JC, McGregor IS (2011) Disruptive effects of the prototypical cannabinoid Δ^9-tetrahydrocannabinol and the fatty acid amide inhibitor URB-597 on Go/No-Go auditory discrimination performance and olfactory reversal learning in rats. Behav Pharmacol 22:191–202

86. Arbuckle EP, Smith GD, Gomez MC, Lugo JN (2015) Testing for odor discrimination and habituation in mice. J Vis Exp 99:e52615

87. Liu G, Patel JM, Tepe B, McClard CK, Swanson J, Quast KB, Arenkiel BR (2018) An objective and reproducible test of olfactory learning and discrimination in mice. J Vis Exp 133:e57142

88. Clevenger AC, Restrepo D (2006) Evaluation of the validity of a maximum likelihood adaptive staircase procedure for measurement of olfactory detection threshold in mice. Chem Senses 31:9–26

89. Walker JC, O'Connell RJ (1986) Computerized odor psychophysical testing in mice. Chem Senses 11(4):439–453

90. Youngentob SL, Schwob JE, Sheehe PR, Youngentob LM (1997) Odorant threshold following methyl bromide-induced lesions of the olfactory epithelium. Physiol Behav 62:1241–1252

91. Chen X, Xia Z, Storm DR (2013) Electroolfactogram (EOG) recording in the mouse main olfactory epithelium. Bio Protoc 3(11): pii: e789

92. Coppola DM, Waggener CT, Radwani SM, Brooks DA (2013) An electroolfactogram study of odor response patterns from the mouse olfactory epithelium with reference to receptor zones and odor sorptiveness. J Neurophysiol 109:2179–2191

93. Cygnar KD, Stephan AB, Zhao H (2010) Analyzing responses of mouse olfactory sensory neurons using the air-phase electroolfactogram recording. J Vis Exp 37:pii: 1850

94. Pinato G, Rievaj J, Pifferi S, Dibattista M, Masten L, Menini A (2008) Electroolfactogram responses from organotypic cultures of the olfactory epithelium from postnatal mice. Chem Senses 33:397–404

95. Faure F, Da Silva SV, Jakob I, Pasquis B, Sicard G (2010) Peripheral olfactory sensitivity in rodents after treatment with docetaxel. Laryngoscope 120:690–697

96. Thiebaud N, Veloso Da Silva S, Jakob I, Sicard G, Chevalier J, Ménétrier F, Berdeaux O, Artur Y, Heydel JM, Le Bon AM (2013) Odorant metabolism catalyzed by olfactory mucosal enzymes influences peripheral olfactory responses in rats. PLoS One 8(3):e59547

97. Aydemir O (2017) Olfactory recognition based on EEG gamma-band activity. Neural Comput 29:1667–1680

98. Lorig TS (2000) The application of electroencephalographic techniques to the study of human olfaction: a review and tutorial. Int J Psychophysiol 36:91–104

99. Walla P, Hufnagl B, Lehrner J, Mayer D, Lindinger G, Imhof H, Deecke L, Lang W (2003) Olfaction and face encoding in humans: a magnetoencephalographic study. Brain Res Cogn Brain Res 15:105–115

100. Gudziol H, Guntinas-Lichius O (2019) Electrophysiologic assessment of olfactory and gustatory function. Handb Clin Neurol 164:247–262

101. Masaoka Y, Harding IH, Koiwa N, Yoshida M, Harrison BJ, Lorenzetti V, Ida M, Izumizaki M, Pantelis C, Homma I (2014) The neural cascade of olfactory processing: a combined fMRI-EEG study. Respir Physiol Neurobiol 204:71–77

102. Gervais R, Buonviso N, Martin C, Ravel N (2007) What do electrophysiological studies tell us about processing at the olfactory bulb level? J Physiol Paris 101:40–45

103. Zhang B, Zhuang L, Qin Z, Wei X, Yuan Q, Qin C, Wang P (2018) A wearable system for olfactory electrophysiological recording and animal motion control. J Neurosci Methods 307:221–229

104. Benveniste H, Lazebnik Y, Volkow ND (2017) Seeing how we smell. J Clin Invest 127:447–449

105. Kim YK, Hong SL, Yoon EJ, Kim SE, Kim JW (2012) Central presentation of postviral olfactory loss evaluated by positron emission tomography scan: a pilot study. Am J Rhinol Allergy 26:204–208

106. Risacher SL, Tallman EF, West JD, Yoder KK, Hutchins GD, Fletcher JW, Gao S, Kareken DA, Farlow MR, Apostolova LG, Saykin AJ (2017) Olfactory identification in subjective cognitive decline and mild cognitive impairment: association with tau but not amyloid positron emission tomography. Alzheimers Dement (Amst) 9:57–66

107. Han P, Zang Y, Akshita J, Hummel T (2019) Magnetic resonance imaging of human olfactory dysfunction. Brain Topogr 32:987–997

108. Levy LM, Henkin RI, Hutter A, Lin CS, Martins D, Schellinger D (1997) Functional MRI of human olfaction. J Comput Assist Tomogr 21:849–856

109. Moon WJ, Park M, Hwang M, Kim JK (2018) Functional MRI as an objective measure of olfaction deficit in patients with traumatic anosmia. AJNR Am J Neuroradiol 39:2320–2325

110. Chuang KH, Lee JH, Silva AC, Belluscio L, Koretsky AP (2009) Manganese enhanced MRI reveals functional circuitry in response to odorant stimuli. NeuroImage 44:363–372

111. Han Z, Chen W, Chen X, Zhang K, Tong C, Zhang X, Li CT, Liang Z (2019) Awake and behaving mouse fMRI during Go/No-Go task. NeuroImage 188:733–742

112. Lehallier B, Coureaud G, Maurin Y, Bonny JM (2012) Effects of manganese injected into rat nostrils: implications for in vivo functional study of olfaction using MEMRI. Magn Reson Imaging 30:62–69

113. Litaudon P, Bouillot C, Zimmer L, Costes N, Ravel N (2017) Activity in the rat olfactory cortex is correlated with behavioral response to odor: a microPET study. Brain Struct Funct 222:577–586

114. Muir ER, Biju KC, Cong L, Rogers WE, Torres Hernandez E, Duong TQ, Clark RA (2019) Functional MRI of the mouse olfactory system. Neurosci Lett 704:57–61

115. Van de Bittner GC, Riley MM, Cao L, Ehses J, Herrick SP, Ricq EL, Wey HY, O'Neill MJ, Ahmed Z, Murray TK, Smith JE, Wang C, Schroeder FA, Albers MW, Hooker JM (2017) Nasal neuron PET imaging quantifies neuron generation and degeneration. J Clin Invest 127:681–694

116. Courtiol E, Neiman M, Fleming G, Teixeira CM, Wilson DA (2019) A specific olfactory cortico-thalamic pathway contributing to sampling performance during odor reversal learning. Brain Struct Funct 224:961–971

117. Ogg MC, Ross JM, Bendahmane M, Fletcher ML (2018) Olfactory bulb acetylcholine release dishabituates odor responses and reinstates odor investigation. Nat Commun 9:1868

118. Wilson CD, Serrano GO, Koulakov AA, Rinberg D (2017) A primacy code for odor identity. Nat Commun 8(1):1477

119. Kass MD, Czarnecki LA, McGann JP (2018) Stable olfactory sensory neuron in vivo physiology during normal aging. Neurobiol Aging 69:33–37

120. Lee AC, Tian H, Grosmaitre X, Ma M (2009) Expression patterns of odorant receptors and response properties of olfactory sensory neurons in aged mice. Chem Senses 34:695–703

121. Kunkhyen T, Perez E, Bass M, Coyne A, Baum MJ, Cherry JA (2018) Gonadal hormones, but not sex, affect the acquisition and maintenance of a Go/No-Go odor discrimination task in mice. Horm Behav 100:12–19

122. Kass MD, Czarnecki LA, Moberly AH, McGann JP (2017) Differences in peripheral sensory input to the olfactory bulb between male and female mice. Sci Rep 7:45851

123. van der Linden C, Jakob S, Gupta P, Dulac C, Santoro SW (2018) Sex separation induces differences in the olfactory sensory receptor repertoires of male and female mice. Nat Commun 9:5081

124. Ueno H, Shimada A, Suemitsu S, Murakami S, Kitamura N, Wani K, Takahashi Y, Matsumoto Y, Okamoto M, Fujiwara Y, Ishihara T (2019) Comprehensive

behavioral study of the effects of vanillin inhalation in mice. Biomed Pharmacother 115:108879

125. Lötsch J, Ultsch A, Hummel T (2016) How many and which odor identification items are needed to establish normal olfactory function? Chem Senses 41:339–344

126. Persaud KC (2013) Engineering aspects of olfaction. In: Persaud KC, Marco S, Gutiérrez-Gálvez A (eds) Neuromorphic olfaction. CRC Press/Taylor & Francis, Boca Raton, FL; Chapter 1. Frontiers in neuroengineering

127. Evans JE, Miller ML, Andringa A, Hastings L (1995) Behavioral, histological, and neurochemical effects of nickel (II) on the rat olfactory system. Toxicol Appl Pharmacol 130:209–220

128. Lu XC, Slotnick BM (1998) Olfaction in rats with extensive lesions of the olfactory bulbs: implications for odor coding. Neuroscience 84:849–866

Chapter 11

Assessment of Neurotoxicant-Induced Changes in Behavior: Issues Related to Interpretation of Outcomes and Experimental Design

Deborah A. Cory-Slechta, Katherine Harvey, and Marissa Sobolewski

Abstract

Behavioral assessment is a critical component of neurotoxicological research as the consequences of neurotoxicant exposure frequently include changes in behavior. Behavior encompasses multiple domains, and numerous behavioral testing paradigms, ranging from simple to complex, are available for assessment within each such domain and described in the literature. Many laboratories adapt the simplest procedures, based on the assumption that these procedures would be both simple to implement and to interpret. Such assumptions fail to recognize that behavior is actually quite complex. For example, it is critical to consider potential confounding effects (stress, nutritional state, motivation, motor endurance, fear, pain sensitivity) inherent to common behavioral tests, i.e., to understand behavioral mechanisms of an observed behavioral change, because they may shift the interpretation of a change in behavior, particularly for cognitive/executive functions, such as learning. Additionally, this chapter addresses issues related to the use and misuse of test batteries and the influence of the behavioral history resulting from use of a test battery that may interact with the toxicant treatment. It also emphasizes the critical need to recognize that behavioral testing itself modifies the brain; hence, attempting to define biological mechanisms of a neurotoxicant from brains of organisms that have behavioral experience may prove misleading, as brain changes will reflect the effects of both the toxicant exposure and the behavioral experience. Finally, the chapter emphasizes the advantages of using the same behavioral paradigms across species, with appropriate confounds measured and ethologically relevant modifications made, to enhance translation.

Key words Behavioral experience, Behavioral history, Behavioral mechanisms, Neurotransmitters, Lead, Prenatal stress, Translation

1 Introduction and Purpose

The brain is comprised of the dynamic and highly interactive activity of interdependent cells, networks and systems, and an organism's behavior reflects the functional output of these systems (Fig. 1). Additionally, the brain interacts with and controls functions of peripheral organs and systems. It is critical to recognize that behavior does not originate from molecular events within the central nervous system, but is ultimately a response to environmental

Jordi Llorens and Marta Barenys (eds.), *Experimental Neurotoxicology Methods*, Neuromethods, vol. 172, https://doi.org/10.1007/978-1-0716-1637-6_11, © Springer Science+Business Media, LLC, part of Springer Nature 2021

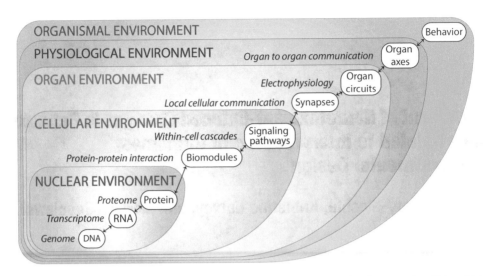

Fig. 1 Behavior originates in the environment while physiology creates the framework within which the environment influences behavior. Consequences of behavior can feed back to modify the physiological framework

stimuli, i.e., *behavior is context-dependent.* Different molecular environments can change the probability that an organism will respond in a particular way to a particular environmental stimulus. But it is simplistic to consider that gene x environment interactions are constants, e.g., that simply activating gene A or hormone B will invariably invoke behavior Y across all environmental contexts in all individuals [1]. For example, an evolving understanding demonstrates that the relationship between increased testosterone and aggressive behavior is not linear or invariant, but, rather, is dependent on the type of aggression, on genetic sex, on parental status, on age, and more [2, 3]. However, providing environmental context improves our understanding of when such relationships will exist and when they do not.

A common adage has been that genes + environment = phenotype. However, to define it more completely, organismal physiology (and its related subcomponents) + organismal environment (with its behavioral history and context) = behavior (Fig. 1). Associations of biochemical effects with behavior can be altered based on variations in nuclear, cellular, organ, and physiological environments, and these environments are molded and shaped by previous rewards, behavioral histories, and social interactions in a highly interactive and dynamic capacity across time in behaving organisms. Within that framework, interpretations of behavioral toxicity can only be derived from an understanding of an organism's previous behavioral history, social rank/relationships, stimulus conditioning history, context-dependency, and/or the deprivation of these experiences. Further, many behavioral test methods designed to evaluate behavioral domains such as learning, particularly those based on simple assays, actually rely on a behavioral response that encompasses other behavioral domains, i.e., motor and sensory function, which if altered by the toxicant, will itself alter learning. Without an appreciation of this bidirectional influence, we risk missing the potential confounding physiological and environmental factors in our outcomes, and our predictive validity and ability to reproduce associations will be limited and potentially misleading, as this chapter seeks to further elaborate through discussions of: (1) the importance of confounding environmental and/or physiological influences on "behavioral mechanisms" of action and (2) the influences of prior behavioral experience/history on neurotoxicology studies.

In addition to understanding the context of the origins of an organism's behavior, consideration of the dynamic temporal relationship between a biochemical response and a behavioral response is critical. Behavioral responses provoked by a stimulus can occur extremely rapidly, i.e., sometimes within milliseconds of a stimulus. Consider, e.g., the rapidity of removing one's hand from a sudden fire. Fire is clearly the environmental context for the response, and such rapidity and the plasticity of behavior are critical in this example to prevent burns. This response reflects the interaction between environment and physiological functions, such as neuronal myelination, synaptic connectivity, and activation of endocrine stress responses. In contrast, biological changes, such as in gene expression and protein synthesis, require a far greater time frame to process and would, if such factors actually controlled behavior, already have resulted in burns. Instead the association between transcriptional shifts and/or epigenetic reprogramming and an environmental stimulus is often temporally delayed. In fact, the behavioral response to a stimulus alters an organism's physiology through these mechanisms to provide a "memory" of this event that then defines the parameters of future responses, including rapidity and plasticity, by molding cellular signaling and neuronal function.

As this implies, genes and the nuclear environment, with their influences on physiology and morphology, act to create a framework or parameters within which the environment acts to shape the behavior of an individual. Behavioral responses can feed back throughout the organism to alter this biological framework at multiple levels. As characterized by Skinner: "We shall know the precise neurological conditions which immediately precede, say, the response, "No, thank you." These events in turn will be found to be preceded by other neurological events, and these in turn by others. This series will lead us back to events outside the nervous system and, eventually, outside the organism" [4]. This is the framework requisite for interpretation of behavioral toxicity, particularly for ostensible changes in executive functions. Behavior is a critical component both of the understanding of the neurotoxic consequences of environmental chemical exposures, as well as to advancing our understanding of brain function, as it is often one of the most sensitive and translational endpoints. Both simple and complex methods exist to assess virtually all domains of behavioral function. Reported descriptions and instructions for implementation of a sizeable array of specific behavioral procedures for carrying out such assessments are already widely available for almost every behavioral domain [5] and include parametric considerations related to the species being tested.

Probably more so than in any other area, the scientific community appears to consider it relatively simple to implement behavioral paradigms into laboratories even without any background or expertise. This tendency was presaged by Skinner [4] who stated that "Actually there is no subject matter with which we could be better acquainted, for we are always in the presence of at least one behaving organism." However, as he also went on to note: "But this familiarity is something of a disadvantage, for it means that we have probably jumped to conclusions which will not be supported by the cautious methods of science." This quote seems particularly relevant today given the tendency of many laboratories to utilize the simplest behavioral methods in lieu of more complex techniques, likely given the greater rapidity and lowered costs with which they can be implemented.

However, as has become clear through years of the experimental analysis of behavior, the less complicated the procedure, the more difficult to interpret the nature of an observed behavioral change and whether the observed change actually or specifically reflects the behavioral domain ostensibly being tested in a given behavioral procedure, or whether alternate mechanisms are operative [6]. For example, in the use of a supposed "learning" paradigm, a behavioral change in response to a toxicant may not reflect a learning deficit at all, but instead a change in motivation and/or motor function and/or sensory capacities, all of which are components of the designated response/procedure being used to evaluate learning. Thus, it is inappropriate to presume that because a behavioral procedure is stated to measure learning, that any observed behavioral change in that procedure in response to a toxicant exposure is necessarily a learning deficit.

While the scientific community prefers simplicity of explanation and unqualified, broad generalizations, this chapter focuses on the potential problems that can arise when behavioral context is ignored as it relates to: the interpretation of outcomes from various behavioral procedures, to appropriate experimental designs, as well as to reliability and reproducibility of effects. After discussing the origins of behavior, it endeavors to underscore the importance of recognizing how different physiological and environmental contexts alter behavioral mechanisms of action, and how these alterations influence behavior, particularly in relation to the least complicated behavioral assays that are most widely used. Secondly, it discusses the potential limitations and confounds to be considered when using a series of behavioral tests, i.e., a behavioral battery, in the same subjects. Thirdly, it re-introduces the associated topic of "behavioral history," i.e., the influence of behavioral experience itself on subsequent behavior, drug effects, etc. In addition, it emphasizes the potential limitations and confounds introduced by the fact that behavioral assessment itself introduces changes in brain, indeed, across all scales, as this has significant consequences for experimental design and interpretation, particularly as related to biological mechanisms of action of a neurotoxicant. Finally, it emphasizes the utility of the use of the same behavioral paradigms across species to minimize difficulties of extrapolation to human subjects.

2 Origins of Behavior: Evolution, Phylogeny, Development, and Individual Experience

The behavioral repertoire ranges from unlearned to learned behaviors. The overlaps across species in characteristic behavioral repertoires are a function of shared phylogenetic ancestry and evolutionary adaptive significance [7, 3]. Unlearned behaviors include unconditioned reflexes (URs) and some fixed action patterns. However, some unlearned behaviors, particularly some fixed action patterns, actually have the capacity to be modified [8]. The bounds of learned behavioral flexibility reflect cumulative developmental and individual behavioral history. Additionally, explanations for the origins of learned behaviors overlap, as throughout evolution, primates (and rodents) have been evolutionarily selected for increased behavioral and phenotypic plasticity, increasing the role of our developmental environments and individual learning experiences to produce wide behavioral complexity.

Two processes underlie the origins of learned behavior across species, specifically Pavlovian conditioning and operant conditioning. Pavlovian conditioning (Fig. 2) builds upon unconditioned reflexes (UR) built into the organism, which are elicited by an unconditioned environmental stimulus, e.g., a light shining into the eye (US; unconditioned stimulus) elicits pupillary dilation (unconditioned response, UR) which can then serve as the basis for the establishment of new conditioned reflexes. In that second-order conditioning process, a new, initially neutral (from the organism's perspective) stimulus is repeatedly paired with the unconditioned stimulus, with the neutral stimulus preceding the unconditioned stimulus. With multiple pairings, the neutral stimulus itself acquires conditioning properties, i.e., becomes a conditioned stimulus (CS) such that it alone can now elicit the

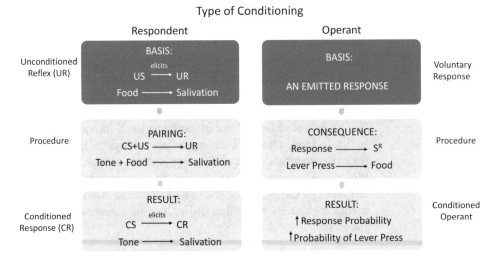

Fig. 2 Schematic depiction of respondent (left column) and operant (right column) conditioning paradigms. *CS* Conditioned stimulus, *US* Unconditioned stimulus; *S^R* Reinforcer

unconditioned response, now defined as a conditioned reflex or conditioned response (CR). The classic example has been the repeated pairing of a tone with food powder such that the tone itself comes to elicit a conditioned salivation response in dogs. More complex and higher order Pavlovian conditioning can also occur, such as when another neutral stimulus is paired with a conditioned stimulus eliciting a conditioned response; with repeated pairing that neutral stimulus will also acquire conditioned stimulus properties. In the absence of intermittent re-pairing, conditioned stimuli lose their ability to elicit conditioned responses.

The basis of operant conditioning (Fig. 2) is voluntary, not elicited, behavior. In operant conditioning, a voluntary response is followed by some environmental stimulus (consequence). If that stimulus is rewarding (reinforcing), the frequency of the response it followed will increase. A reinforcing stimulus may be a positive reinforcer (its presentation is reinforcing, e.g., delivery of a food reward following completion of a designated response) or a negative reinforcer (its removal is reinforcing, e.g., termination of an ongoing shock following completion of a designated response). If the reinforcing stimulus is discontinued, the frequency of the response will decline, a process known as extinction. If the environmental stimulus that follows a response is punishing (aversive stimulus), the frequency of the response it follows will decline. Initially, neutral stimuli can acquire reinforcing or aversive properties through pairing with other unconditioned (e.g., primary reinforcers) or with conditioned (secondary reinforcers) stimuli that have acquired such properties. Indeed, some such stimuli, i.e., generalized reinforcers, become associated with multiple other reinforcing stimuli through such pairings, such as in the case of money or attention. As an extensive experimental literature has documented over the years, these procedures and principles are operative across species.

3 Recognizing the Importance of Behavioral Mechanisms of Action

As alluded to above, behavior is a reflection of the environment and of experience. While we often consider biological mechanisms of action, behavioral mechanisms of action and their potential influence on behavior, are generally not fully considered despite their critical role in interpretation of behavioral changes. These interdependent mechanisms include the antecedent stimulus environment, which can include interoceptive and/or exteroceptive stimuli, the nature of the designated response, and the ultimate consequences of the response, all of which ultimately influence outcome, as shown in Fig. 3.

Antecedent conditions are critical, with environmental stimuli providing significant and crucial information as to the probability of how or whether a response will be reinforced; such stimuli are deemed discriminative stimuli. For example, if one sees that there are no cookies inside of a glass cookie jar, the odds that reaching into the jar would be reinforced are zero, likely then precluding emission of the response. Antecedent conditions likewise influence motivation, i.e., if you haven't eaten for 24 h, food is highly likely to serve as a reinforcer, whereas for someone with a stomach flu (interoceptive stimulus events), this would not be the case. A red traffic light signals that braking behavior will be reinforced, while continuing to drive through it could be punished. Thus, environmental antecedents are critical in influencing the likelihood of a response. Notably, behavior under very strong control by the antecedent stimuli can also be resistant to effects of drugs and toxicants. For example, the effects of d-amphetamine were found to be significantly diminished in a context where a specific stimulus was presented when pigeons had reached the designated number of key pecking responses required such that a response on a different key would now produce food reward relative to the same behavioral procedure but without the signaling stimulus [9, 10].

A study that directly compared a nose-poke vs. a bar press response in mice found that administration of pentobarbital or methocarbamol (a muscle relaxant) prior to the session resulted in greater reduction of lever pressing responses than of nose-pokes [11], underscoring the importance of the topography of the chosen response. In another such report, acquisition of avoidance behavior of rats was found to be faster and to exhibit sharper temporal discrimination in a shuttle-box apparatus requiring a jumping response rather than in a Skinner box in which lever pressing served

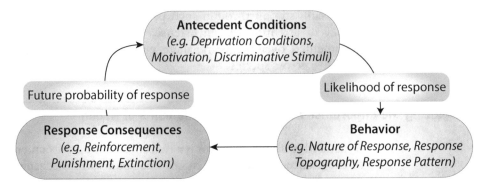

Fig. 3 Behavioral mechanisms of action: Behavior occurs in a context. The likelihood of a response occurring depends upon antecedent conditions, e.g., motiviation, stimuli signaling likelihood of reward. Response topography and its relevance to the species being studied can influence the rate at which a designated response may occur, while consequences of the response determine its future probability of reoccurrence

as the operant response [12]. Hence, ease and species appropriateness of the nature and parameters of the designated response can increase reward densities and learning opportunities.

The third component of this behavioral sequence is the nature of the consequence that follows the response. A consequence that follows a response is only defined as a reinforcer if it increases the probability of the response that it follows, or only a punisher if it decreases the frequency of behavior. As a corollary, it is important to recognize that the reinforcing or aversive properties of an environmental stimulus are not invariant, they are context-dependent. For example, as alluded to above, food may not serve as a reinforcer for an organism that is already sated, i.e., the behavioral mechanism is related to the establishing operation of pre-session feeding or restriction of food. In addition to the examples cited above, a counterintuitive example is that shock delivery itself, normally considered an inherently aversive stimulus, has been shown under certain environmental and behavioral history conditions to serve as a positive reinforcer [13]. Importantly, the timing of reward delivery is critical. In fact, delay of reward has long been known to hinder learning, and it can end up rewarding the wrong behavior, resulting in the increased frequency of superstitious behaviors [14–16].

4   Consideration of Behavioral Mechanisms in Interpreting the Specificity of the Behavioral Produced for Measuring a Particular Behavioral Function

Recognition of the role of the three-term contingency shown in Fig. 3 has particular significance for interpretation of the outcome of neurotoxicant exposure on behavior. As noted, behavior has been characterized into different domains, such as motor function,

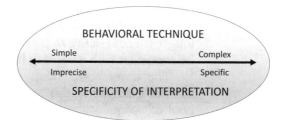

Fig. 4 A wide array of techniques, ranging from very simple to far more complex are available for measurement of virtually every behavioral domain. However, the simpler the technique, the more difficult it is to determine the specificity of the observed behavior

sensory behavior, and executive responses as well as social behaviors. Multiple behavioral procedures have been developed for assessment of each of these domains and, within each such domain, behavioral testing procedures can range from very uncomplicated assays to more complex approaches in terms of, e.g., equipment, behavioral training requirements, ease of implementation, etc. The trend in behavioral toxicological research, and in neuroscience in general, has predominantly been to utilize uncomplicated behavioral assays as they are easier to implement, provide data more rapidly and are not perceived to require significant behavioral expertise within the laboratory.

In general, however, the simpler the behavioral assay used, the more difficult it can be to interpret the outcome (Fig. 4). Consider simple paradigms used to measure "learning." In a technique such as a Y maze or a T maze, the subject learns to turn to the side of the maze that contains a presumed reinforcing stimulus, and the number of trials and or latency to reach the reward is measured as outcomes. One potential finding in a behavioral toxicology study could be that subjects exposed to the chemical (i.e., treated group) required a greater number of trials and/or a longer latency to learn to turn to the correct side of the maze (i.e., the side in which the reinforcer is placed). While this might be interpreted as a learning deficit, numerous different behavioral mechanisms could actually be influencing the outcome that would also need to be considered. For example, the chemical exposure itself may impair motor function such that it takes treated subjects a longer time to traverse the maze and thus delays the time to reward (alters antecedent conditions) and/or increases the required effort for a reward (alters response topography). Alternatively, the treatment may impair sensory capacities of the subject, for example, olfactory or visual capabilities, thus impairing the ability of the subject to use visual or odor cues in guiding its direction (antecedent conditions). Finally, the stimulus being used to reward behavior may not in fact be reinforcing or not be sufficiently reinforcing (response consequences as a behavioral mechanism), particularly if the chemical treatment itself induces any sickness effects. All of these impairments would lead to the outcome of greater number of trials or longer latency relative to non-treated controls, but they would not be consistent with a learning impairment.

The water maze, often used for testing rodents, serves as another example. In this maze, the subject is placed in a circular pool of opaque water in which an escape platform is submerged beneath the surface; finding the escape platform serves as a negative reinforcer in providing escape from water, a non-preferred environment for rodents. Here too, motor and sensory impairments, among other factors, can influence outcome. Swimming is a highly effortful response, and thus any motor impairment produced by treatment could influence the ability to sustain a swimming response, i.e., reduce motor endurance. It has become accepted that a measure of swimming speed per se is a control for this problem; however, this does not measure changes in physical endurance which is actually requisite to the correct response. Similarly, impaired sensory capacity will alter ability of the subject to utilize environmental cues to learn the placement of the submerged platform. Olfactory cues left by rodents in these test environments have also been shown to influence outcome [17, 18].

Another physiological factor at play in water maze paradigms is hypothermia [19]. Studies have reported this effect to underlie the putative age-related learning deficits in rats reported in water maze performance, as preventing hypothermia by warming 23 months old rats between trials in a water maze task was shown to significantly improve their performance levels to that of young animals. As noted by the authors, these results suggest that age-related deficits in the water maze during aging are not due to the loss of visual acuity, which did not influence performance; but, as a specific measure of cognitive function, performance in the water maze can be confounded by the loss of thermoregulatory control [20]. Furthermore, water maze utilizes an escape response, i.e., negative reinforcement, which studies suggest may result in differential behavioral effects than occurs under conditions of positive reinforcement [21].

More complex behavioral techniques, albeit more expensive and time-consuming and requiring more behavioral expertise, have the capacity to address questions about specificity of the effect observed, and to concomitantly rule out potential confounding behavioral mechanisms. One such example for learning is the multiple schedule of repeated learning and performance. Initially developed for human subjects by Boren, it has since been widely used to study the effects of drugs on learning [22, 23] and to be useful across numerous species including human, monkeys, rats, and mice [24–31]; performance on this behavioral baseline also shows correlation with IQ levels in humans [29]. One version of this behavioral paradigm requires the organism to execute response sequences, i.e., a designated chain of responses, for reward. In the repeated learning component (i.e., the RL component) of the session, the correct sequence changes at each successive experimental session. The RL component alternates with another component during the session, the performance (P) component. In contrast to the RL component, the response sequence reinforced during the performance component remains constant across sessions, so the organism is simply performing an already learned or rote response sequence. By alternating the repeated learning and performance components within each experimental session, e.g., after every tenth reinforcer delivery, with component changes signaled by a change in, e.g., illumination or an auditory cue, drug- or toxicant-induced changes in learning can be differentiated from nonspecific changes in motor function or motivation, i.e., confounding alternative behavioral mechanisms can be considered and/or eliminated. This is based on the premise that if a compound selectively affects learning, then decreases in accuracy should only be seen during the repeated learning components of the schedule, as no learning at all is required in the performance components. However, if decrements in accuracy also occur in the performance components, it would suggest that nonspecific influences are occurring, as intact motor, sensory, and motivational functions are required in both the repeated learning and performance components in the execution of the response and impairments of these functions would result in decreases in accuracy in both the repeated learning and performance components.

Our laboratory used this paradigm to examine whether effects of developmental lead exposure on learning represented selective learning impairments [32]. Figure 5 compares cumulative records of the behavior of a typical control rat (top) to that of a rat exposed to 250 ppm lead acetate from weaning. The session began with a performance (P1) component, followed by the repeated learning component (RL1), a return to the performance component (P2), and a final presentation of the repeated learning component (RL2); components alternated after 25 reinforcer deliveries and were accompanied by a change in the lights illuminated within the chamber. Reinforcement followed each completion of a correct three-response sequence. Following extensive training on the schedule, the control rat (top record) had high levels of accuracy in P1, generating a high rate of reinforcement delivery and a relatively low rate of errors, as expected. The onset of RL1, signaled to the subject by a change in illumination within the operant chamber, required the rat to now learn a new three-response sequence that differed from the correct RL sequence from the prior session. As expected, the error rate was initially high, producing a lower rate of reinforcement delivery, but by the end of RL1, the rate of errors had begun to decline, and the rat was earning food deliveries at a faster rate as it gradually began to learn the correct

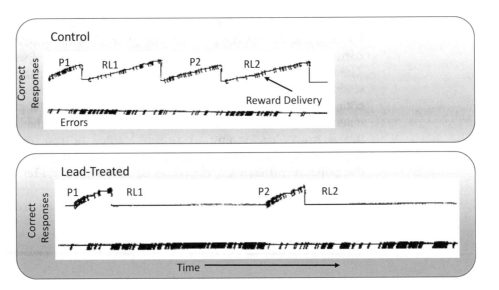

Fig. 5 Cumulative records of performance of a control rat (top) and a rat exposed from weaning to 250 ppm lead acetate in drinking water (bottom). Correct responses cumulate vertically, while errors are depicted on the bottom line of each record. *RL* Repeated learning; *P* Performance

sequence. The learning process was even more pronounced during RL2, as the rat continued to learn the correct RL sequence for this particular session. The bottom record shows a dramatic impact of lead exposure that was selective for the RL components of the schedule. Specifically, the lead-exposed rat earned virtually no food deliveries during either the RL1 or RL2 components, despite emitting hundreds of incorrect responses over the course of these components; in contrast, it exhibited high accuracy levels during both the P1 and P2 components, rapidly earning all 25 available reinforcers and eliminating potential confounding mechanisms as the explanation for the deficits in the RL components.

In the case of methylmercury exposure of mice, behavioral changes in the repeated acquisition procedure were found instead to be due to motoric effects which was only possible to evaluate with the inclusion of a performance component [33]. Thus, without experimental control over behavior, such as in uncomplicated behavioral procedures, it is not possible to rule out these other potential determinants.

Such alternative explanations for ostensible learning changes can apply to other negative reinforcement-based techniques as well, potentially confounding interpretations of behavioral outcome and/or effects of the treatment of interest. For example, studies have demonstrated significant increases in peripheral corticosterone levels during performance of a shuttle-box avoidance task [34–36]. In another study in Wistar rats, swimming in a water maze caused a significantly increased release of other hormones, namely, vasopressin (AVP) within the paraventricular nucleus of the hypothalamus and of oxytocin (OXT) within the supraoptic nuclei on each of the three test sessions. Furthermore, plasma ACTH and corticosterone were found to be similarly elevated in response to water maze testing during each of the test sessions [37]. In another study [35], corticosterone concentrations increased when extinction of a previously rewarded response was imposed, but not when positive reinforcement continued. Activation of stress or endocrine systems modify the physiological environment and thus the context under which learning is being assayed, which can have significant consequences if these changes vary by treatment, i.e., interact with the neurotoxicant, and/or further differ by sex. If, for example, the corticosterone release is requisite to learning of the avoidance task, but is reduced by the neurotoxicant, then the framework for learning has been altered. In relation to sex differences, a study in Long-Evans rats found that corticosterone was highly elevated during and after water maze training, with females showing greater increases than males, and strong inverse correlations were observed between corticosterone and measures of water maze performance in females [38]. While such negative reinforcement-based fear conditioning behavior is typically acquired more rapidly than positive reinforcement-based behaviors, it is important to remember that the former are not in fact models of voluntary behavior [21].

Fear conditioning has been a widely used technique based on aversive stimulus presentation to study learning/memory. Like water maze or forced swimming procedures, fear conditioning also evokes significant corticosterone release. For example, a study comparing the role of shock intensity in fear conditioning found that fear conditioning resulted in corticosterone release regardless of the shock intensity used [39]. Similarly, foot-shocked rats showed higher corticosterone levels than controls that were not footshock conditioned [40]. Negative reinforcement-based shock avoidance techniques have similar consequences [41]. An additional factor that could influence learning/memory in shock-based techniques that have long been known to depend upon shock intensity [42], is sensitivity to pain; should the chemical treatment alter pain sensitivity, it would decrease learning in this paradigm. Given this, it becomes necessary to measure toxicant-induced changes in baseline shock sensitivity which is almost invariably never considered.

Thus, testing in a negative reinforcement paradigm where the reinforcer consists of removal of an aversive stimulus, or shock presentation paradigms, can be associated not only with a significant activation of the hypothalamic–pituitary–adrenal axis but also with an intrahypothalamic release of AVP and OXT. Such increases in stress physiology have the potential to interact with the independent variable, particularly in the case of a chemical exposure. This could mean that the outcomes of the study do not reflect the effects of the chemical exposure per se, but the interaction of the treatment and stress hormones if one is examining effects in behaviorally tested subjects.

In summary, any behavioral assessment requires that other behavioral mechanisms of effect need to be considered. This may be particularly true for negative reinforcement and shock presentation-based behavioral paradigms and for simpler behavioral assays. But even with positive reinforcement, a stimulus usually rewarding may not increase the frequency of a response in a subject that is physically ill from a chemical exposure. In fact, if the subject is sick, lethargic, having seizures, etc. this precludes any cognitive, motor, or social testing. Ruling out alternative explanations is critical for interpretation of the behavioral changes observed.

It should also be noted that many of the uncomplicated behavioral procedures are one-time snapshots taken without providing the sort of behavioral history that can reduce variability and the influence of other extraneous variables like the handler, time of day, last feeding, etc.

5   Behavioral History and Behavioral Testing Effects

It is frequently the case in studies of behavioral toxicology that animals from a given exposure cohort are subjected to multiple behavioral tests within an experiment. This approach is often adopted as it results in an efficiency as it relates to costs of such experiments, as well as, to the time required. However, as has been described in numerous prior reports, experience of an organism in a given behavioral test paradigm can have significant "carry-over" effects and influence performance not only in subsequent behavioral tests, but can even alter the dose-effect curves for drug effects on subsequent behavioral tests.

For example, a comparison of performances on various behavioral procedures in mice tested on a battery of tests as compared to use of mice that were naïve for each of the behavioral procedures used, showed multiple differences, particularly for behavioral paradigms such as open field, rotarod, and the hot plate test [43]. This study likewise found order effects when the sequence of behavioral tests was rearranged. In another such study assessing affective behavior in rats, performance in a forced swim test was found to be dependent upon the order in which it occurred in the test battery. Furthermore, repeated testing in the open field or in the forced swim paradigm resulted in significant behavioral changes relative to earlier performances in these paradigms [44] showing the influence of repeated testing. Such changes are quite apparent in comparison to video recordings of behavior in such tests across sessions; it underscores the importance of watching what behavior is occurring in a test. A study comparing performance of C57Bl/6J mice that were experienced vs. comparable aged mice that were naïve to a test battery comprised of the elevated plus maze test, the dark-light test, an open field test, and a novel cage test found that prior experience with the battery significantly reduced exploratory behavior and open arm-related measures. Moreover, these effects of prior experience were significantly more pronounced in older (13 weeks) as compared to younger (9 week) mice [45].

Such behavioral experience can then interact with the chemical exposure. In support of such possibilities, we found evidence for effects of prior behavioral experiences on subsequent complex learning. Rats were given the experience of either a single restraint stress followed by a single forced swim test (R + FS), or the behavioral experience of receiving food reward for lever pressing on a fixed interval (FI) schedule of reinforcement. Learning was then subsequently assessed in all groups using a repeated learning paradigm that required acquisition of a new two-response chain in each successive behavioral test session. Figure 6 depicts three outcome measures across six sessions for these offspring. As they show, accuracy tended to be lower in groups with R + FS experience than with FI experience. In addition, under those conditions, marked reductions in percent correct learning were found in both males and females subjected to the forced swim test in the two-response chain repeated learning paradigm, particularly in the early test sessions, which was accompanied by marked increases in the first response of the sequence being incorrect. Thus, an early powerful stressor had a greater impact than did early experience with positively reinforced behavior in terms of subsequent acquisition of a behavioral chain.

Prior behavioral experience can also markedly alter the dose response curves for drug effects in subsequent behavioral tests. For example, a comparison of the effects of d- amphetamine on punished responding was measured in squirrel monkeys in which food-maintained responding was suppressed by the presentation of electric shock (punishment). Two monkeys were experimentally naive and two had a prior history of responding under conditions of both shock postponement (avoidance behavior) and shock presentation (punishment) schedules. D-amphetamine did not increase punished responding by naive monkeys, but markedly increased rates in the behaviorally experienced group [46]. Other examples include studies reporting that prior experience in a light/dark exploration test resulted in chlordiazepoxide-induced increases in exploration in the light compartment in naïve subjects, whereas the drug had no effects in test-experienced mice [47]. A recent study found that testing in the light, as compared to the dark phase, increased locomotor activity as well as the response to d-amphetamine, while reducing social approach behavior [48].

Such findings collectively suggest that in a study design in which multiple behavioral tests are administered, the effects of a toxicant upon behavior may well depend upon the specific

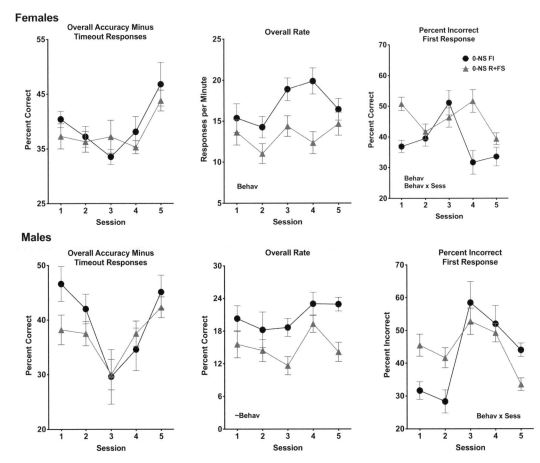

Fig. 6 Changes in overall accuracy (left column), overall response rates (middle column and percent of first responses in a two-response behavioral chain that were incorrect (right column) in females (top row) and male (bottom row) rats that had prior experience of either food reward on a fixed interval schedule of reinforcement (FI) or a single exposure to restraint stress followed by forced swim (R + FS). The correct two-response changed with every session. * = significant difference between FI and R + FS groups at the indicated session. Cory-Slechta et al. unpublished data

behavioral paradigms that are used, as well as the order in which the testing occurs.

As a further problem, behavioral history can interact with toxicant treatment. For example, consider an experimental design in which pain is tested on a hot plate followed by a fear conditioning procedure. Assume the toxicant modifies pain sensitivity and thus hot plate performance. This modified behavioral history is now carried forward into the fear conditioning paradigm. If differences between toxicant-treated and control subjects are found in the fear conditioning paradigm, it may not be possible to ascertain whether these reflect the effect of the toxicant itself, the altered behavioral history, or perhaps an interaction of the two. Such effects may be particularly pronounced with fear conditioning. Prior studies in rats have shown that even a single footshock can produce a persistent hypoactivity in exposure to subsequent unknown environments [49] which in a learning test could slow the time to reward. For such reasons, it is equally important to recognize that the trajectory of behavioral responses and behavioral experience is dynamic and life-long.

6 Behavioral Testing Itself Modifies the Brain

A frequent experimental design in behavioral neurotoxicology involves chemical exposure followed by behavioral testing and subsequent collection of brains from behaviorally tested subjects to be examined, for example, for potential mechanisms of the observed behavioral deficits or perhaps an intervention to mitigate the adverse behavioral effects. This approach is typical, as it may be considered both more efficient and economical in terms of animal costs.

Unfortunately, this approach is confounded by the fact that, as has long been known but subsequently forgotten, behavioral experience itself modifies the brain [50–52], which would mean that changes in the brain of organisms with a behavioral history reflect the interaction of the chemical exposure with the behavioral experience, rather than simply being the result of the chemical exposure alone. The plasticity of the brain in response to behavioral experience was demonstrated more than 50 years ago. For example, early studies in this area showed systematic differences in cortical acetylcholinesterase enzyme activity not only in rats tested in different behavioral apparati, but also in comparison to rats not subjected to behavioral testing at all [53]. Furthermore, increases in rates of protein synthesis and protein levels and in amounts of RNA were found in rats from an enriched environment, and maze training itself increased cortical RNA:DNA ratios and such differences were subsequently shown to be induced across ages, although slower to emerge in later ages. Moreover, these changes can occur rapidly, being reported after only 4 days of differential housing in one study [54] and after only 40 min of experience in another report [55]. Such plasticity has been reported across mammalian species, as well as in birds, fish, fruit flies, and spiders [56]. Subsequent control experiments have confirmed that these effects did not appear to be attributable to differences in handling, stress, accelerated maturation, differential locomotion, or hormonal mediation [57].

Given this potential confound, we have implemented the inclusion of a non-behavioral control group in our experiments, as examination of brains in this group following chemical treatments should reflect actual potential sites and mechanisms of its toxicity with both groups having equivalent housing conditions. Moreover, given that we have also found behavioral experience to mitigate the effects of exposures in our studies of neurodevelopmental toxicity of lead [58], this approach permits assessment of specific behavioral experiences that may accelerate such reversals. Of course, this also raises the issue of what constitutes a "non-behavioral" control group for such purposes and what experience best translates to human conditions.

In further support of such differences, we have found evidence for effects of prior behavioral experiences on subsequent neurotransmitter changes that further interacted with prior developmental exposures to lead and/or prenatal stress. We exposed mice to 0 or 100 ppm lead (0 vs. 100) in drinking water from 2 to 3 mos prior to breeding through lactation, with half of the dams in each lead group also receiving prenatal stress (NS = no stress; PS = prenatal stress). Subsequently, offspring from each of these groups were given either a series of four forced swim (FS) experiences, or the behavioral experience of receiving food reward for lever pressing on a fixed interval (FI) schedule of reinforcement, leading to a total of eight groups of male and of female offspring (0-NS-FI, 0-NS-FS, 0-PS-FI, 0-PS-FS, 100-NS-FI, 100-NS-FS, 100-PS-FI, 100-PS-FS) [59]. Figure 7 depicts changes in brain dopamine concentrations in four different brain regions of male mice following developmental lead exposure and/or prenatal stress and subsequent FI vs. FS experience. Here, behavioral history interacted with developmental lead and/or prenatal stress to markedly modify brain dopamine relative to 0-NS control values. This included marked increases in frontal cortex brain dopamine after FI testing following lead exposure, but particularly in organisms that had been exposed to both lead and prenatal stress, whereas increases in dopamine concentrations were seen in response to FS experience in groups receiving either prenatal stress only or Pb only in striatum, midbrain, and hippocampus.

In another such study from our laboratory in which rats were maternally exposed to lead (0, 50, or 150 ppm in drinking water) and/or prenatal stress, offspring were given experience on the FI schedule of food reward, as previously described, or no behavioral experience (NFI) and subsequent brain neurochemical alterations were measured [58]. As can be seen in Fig. 8, a similar interaction of behavioral history occurred with developmental insults, with effects in both sexes. For example, increases in nucleus accumbens norepinephrine concentrations occurred in females with developmental exposures to lead and to lead + prenatal stress only if they also had an FI behavioral history. Frontal cortex dopamine turnover concentrations were markedly increased only in mice whose dams received 50 ppm lead plus prenatal stress, or the 150 ppm exposure level and/or prenatal stress followed by no behavioral experience. In striatum, dopamine concentrations of males with prior FI experience only were elevated if they had maternal exposures to 50 ppm

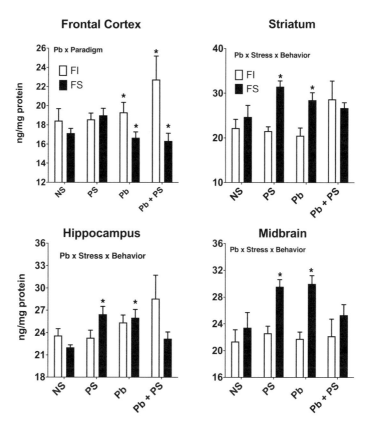

Fig. 7 Dopamine concentrations in indicated brain regions of male mice exposed to maternal lead exposure (0 or 100 ppm in drinking water and/or prenatal stress followed by behavioral experience on either a fixed interval schedule of food reward (FI) or a series of four forced swim experiences (FS). Modified from Cory-Slechta et al. (2013). * = significantly different from corresponding treatments in the other behavioral experience condition

lead plus prenatal stress, or the 150 ppm exposure level and/or prenatal stress, whereas dopamine turnover was markedly increased by developmental lead and/or prenatal stress in males that also had no behavioral experience.

Unintended behavioral histories may also be produced by variations in housing conditions for laboratory animals, particularly rodents. A recent emphasis over the past decade has been the implementation of "enriched" housing as a mechanism to improve animal welfare within vivarium housing. Such enrichment has ranged from marked differences in cage sizes and numbers of rodents within a cage, to provisions of nesting materials, different bedding, and various physical structures. However, such "enrichment" may have very unintended and unrecognized consequences for studies of behavioral toxicity, as such types of enrichment can produce changes in both physiology and behavior [60, 61], and as noted above, such effects may interact with the neurotoxicant being

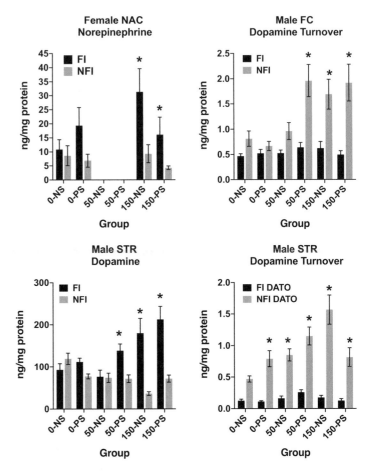

Fig. 8 Concentrations of indicated neurotransmitters in brain regions and by sex of rats that were maternally exposed to lead in drinking water (0, 50, or 150 ppm) and/or prenatal stress followed by experience on the FI schedule or with no FI experience (NFI). * = significantly different from corresponding treatments in the other behavioral experience condition. Modified from Cory-Slechta et al. (2009)

studied. Such interactive effects may also significantly differ by sex [62]. The use of "enriched housing" has also been reported to increase variability among animals in outcome measures [60]. Here too such differences can significantly contribute to differences across studies in reliability and reproducibility [63].

7   Use of the Same Behavioral Test Paradigm across Species

One of the goals of behavioral toxicology is to ascertain the extent to which chemical exposures may impact the human brain and behavior, i.e., a translational objective. For such reasons, an approach which could significantly enhance the translational nature of behavioral toxicology experiments derived from animal models is the use of the same behavioral paradigms across species. Such an approach incorporates both forward translation and backward translational capabilities into our understanding. For example, findings in experimental animal studies can be forward translated into subsequent epidemiological studies in populations exposed to the chemical of interest. Furthermore, information from human behavioral testing can be pursued in animal models to elaborate mechanisms of effect, biomarkers, and potential behavioral intervention or therapeutic strategies. Importantly, the use of the same paradigm, rather than different tests stated to measure the same domain, eliminates many of the questions that arise about extrapolation across different behavioral paradigms. One such example was the use of repeated acquisition paradigm across species cited above [22–31]. A wide of cross-species behavioral procedures are available and have been described in prior reports [64–66].

8   Summary and Conclusions

In conclusion, studies that report alterations in learning following toxicant exposure, drug administration, genetic modification, etc. need to consider that behavior is the product of an organism's physiology and environment. Further, the equation is not static, as developmental and behavioral histories feedback on one another across the life span. Context matters in discussions of predictability and reproducibility of behavioral results. Given that behavioral experience can modify the effects of treatment, researchers must describe behavioral histories for appropriate comparisons. Additionally, the effects of a treatment on the brain following behavioral experience represent the effects of both the treatment and behavioral experience. Furthermore, understanding the behavioral mechanisms of learning is critical in ensuring that confounds, such as alterations in motivation (deprivation/satiation), motor function, pain sensitivity, and sex differences in stress responsivity, are controlled for to rule out these effects as primary contributors to toxicity, and not learning. In these cases, there is no substitute for a performance component or additional behavioral assays that ensure other behavioral domains are not altered. To this end, there is also no substitute for watching your test subjects. Understanding the evolutionary history of the species being utilized will help improve learning assays by increasing reward frequencies through more naturalistic behavioral responses such as nose-pokes or jumping. With these considerations included in experimental designs, behavior, and learning paradigms specifically remain a critical and strong endpoint for translation across species for neurotoxicology.

References

1. Dick DM (2011) Gene-environment interaction in psychological traits and disorders. Annu Rev Clin Psychol 7:383–409. https://doi.org/10.1146/annurev-clinpsy-032210-104518

2. Wingfield JC, Hegner RE, Dufty AM, Ball GF (1990) The "challenge hypothesis": theoretical implications for patterns of testosterone secretion, mating systems, and breeding strategies. Am Nat 136(6):829–846

3. Bateson P, Laland KN (2013) Tinbergen's four questions: an appreciation and an update. Trends Ecol Evol 28(12):712–718. https://doi.org/10.1016/j.tree.2013.09.013

4. Skinner BF (1953) Science and human behavior. Simon & Schuster Inc., New York

5. Buccafusco JJ (2009) Methods of behavior analysis in neuroscience, 2nd edn. CRC Press/Taylor & Francis, Boca Raton, FL

6. Nestler EJ, Hyman SE (2010) Animal models of neuropsychiatric disorders. Nat Neurosci 13 (10):1161–1169. https://doi.org/10.1038/nn.2647

7. Tinbergen N (1963) On aims and methods of ethology. Z Tierpsychol 20(4):410–433. https://doi.org/10.1111/j.1439-0310.1963.tb01161.x

8. Kenward B, Rutz C, Weir AAS, Kacelnik A (2006) Development of tool use in new Caledonian crows: inherited action patterns and social influences. Anim Behav 72 (6):1329–1343. https://doi.org/10.1016/j.anbehav.2006.04.007

9. Laties VG (1975) The role of discriminative stimuli in modulating drug action. Fed Proc 34(9):1880–1888

10. Wood RW, Rees DC, Laties VG (1983) Behavioral effects of toluene are modulated by stimulus control. Toxicol Appl Pharmacol 68 (3):462–472. https://doi.org/10.1016/0041-008x(83)90291-0

11. Gerhardt S, Liebman JM (1981) Differential effects of drug treatments on nose-poke and bar-press self-stimulation. Pharmacol Biochem Behav 15(5):767–771. https://doi.org/10.1016/0091-3057(81)90020-4

12. Riess D (1971) Shuttleboxes, Skinner boxes, and Sidman avoidance in rats: acquisition and terminal performance as a function of response topography. Psychon Sci 25(5):283–286. https://doi.org/10.3758/BF03335878

13. Barrett JE, Valentine JO, Katz JL (1981) Effects of chlordiazepoxide and d-amphetamine on responding of squirrel monkeys maintained under concurrent or second-order schedules of response-produced food or electric shock presentation. J Pharmacol Exp Ther 219(1):199

14. Lett BT (1975) Long delay learning in the T-maze. Learn Motiv 6(1):80–90. https://doi.org/10.1016/0023-9690(75)90036-3

15. Messing RB, Kleven MS, Sparber SB (1986) Delaying reinforcement in an autoshaping task generates adjunctive and superstitious behaviors. Behav Process 13(4):327–338. https://doi.org/10.1016/0376-6357(86)90028-8

16. Saltzman IJ (1951) Delay of reward and human verbal learning. J Exp Psychol 41(6):437–439

17. Roullet P, Lassalle JM, Jegat R (1993) A study of behavioral and sensorial bases of radial maze learning in mice. Behav Neural Biol 59 (3):173–179. https://doi.org/10.1016/0163-1047(93)90926-9

18. Means LW, Alexander SR, O'Neal MF (1992) Those cheating rats: male and female rats use odor trails in a water-escape "working memory" task. Behav Neural Biol 58(2):144–151

19. Panakhova E, Buresova O, Bures J (1984) The effect of hypothermia on the rat's spatial memory in the water tank task. Behav Neural Biol 42(2):191–196. https://doi.org/10.1016/s0163-1047(84)91059-8

20. Lindner MD, Gribkoff VK (1991) Relationship between performance in the Morris water task, visual acuity, and thermoregulatory function in aged F-344 rats. Behav Brain Res 45 (1):45–55

21. Magoon MA, Critchfield TS, Merrill D, Newland MC, Schneider WJ (2017) Are positive and negative reinforcement "different"? Insights from a free-operant differential outcomes effect. J Exp Anal Behav 107 (1):39–64. https://doi.org/10.1002/jeab.243

22. Boren JJ (1963) Repeated acquisition of new behavioral chains. Am Psychol 17:421

23. Boren JJ, Devine DD (1968) The repeated acquisition of behavioral chains. J Exp Anal Behav 11:651–660

24. Brooks AI, Cory-Slechta DA, Murg SL, Federoff HJ (2000) Repeated acquisition and performance chamber for mice: a paradigm for assessment of spatial learning and memory. Neurobiol Learn Mem 74(3):241–258

25. Brooks AI, Stein CS, Hughes SM, Heth J, McCray PM Jr, Sauter SL, Johnston JC, Cory-Slechta DA, Federoff HJ, Davidson BL (2002) Functional correction of established central nervous system deficits in an animal model of lysosomal storage disease with feline

immunodeficiency virus-based vectors. Proc Natl Acad Sci U S A 99(9):6216–6221

26. Arnold MA, Newland MC (2018) Variable behavior and repeated learning in two mouse strains: developmental and genetic contributions. Behav Process 157:509–518. https://doi.org/10.1016/j.beproc.2018.06.007

27. Johnson JM, Bailey JM, Johnson JE, Newland MC (2010) Performance of BALB/c and C57BL/6 mice under an incremental repeated acquisition of behavioral chains procedure. Behav Process 84(3):705–714. https://doi.org/10.1016/j.beproc.2010.04.008

28. Shen AN, Pope DA, Hutsell BA, Newland MC (2015) Spatial discrimination reversal and incremental repeated acquisition in adolescent and adult BALB/c mice. Behav Process 118:59–70. https://doi.org/10.1016/j.beproc.2015.06.005

29. Baldwin RL, Chelonis JJ, Prunty PK, Paule MG (2012) The use of an incremental repeated acquisition task to assess learning in children. Behav Process 91(1):103–114. https://doi.org/10.1016/j.beproc.2012.06.004

30. Ferguson SA, Gopee NV, Paule MG, Howard PC (2009) Female mini-pig performance of temporal response differentiation, incremental repeated acquisition, and progressive ratio operant tasks. Behav Process 80(1):28–34. https://doi.org/10.1016/j.beproc.2008.08.006

31. Garey J, Paule MG (2010) Effects of chronic oral acrylamide exposure on incremental repeated acquisition (learning) task performance in Fischer 344 rats. Neurotoxicol Teratol 32(2):220–225. https://doi.org/10.1016/j.ntt.2009.10.001

32. Cohn J, Cox C, Cory-Slechta DA (1993) The effects of lead exposure on learning in a multiple repeated acquisition and performance schedule. Neurotoxicology 14:329–346

33. Bailey JM, Hutsell BA, Newland MC (2013) Dietary nimodipine delays the onset of methylmercury neurotoxicity in mice. Neurotoxicology 37:108–117. https://doi.org/10.1016/j.neuro.2013.03.011

34. Coover GD, Hart RP, Frey MJ (1986) Corticosterone, free fatty acid and glucose responses of rats to footshock, fear, novel stimuli and instrumental reinforcement. Psychoneuroendocrinology 11(3):373–388

35. de Boer SF, de Beun R, Slangen JL, van der Gugten J (1990) Dynamics of plasma catecholamine and corticosterone concentrations during reinforced and extinguished operant behavior in rats. Physiol Behav 47 (4):691–698. https://doi.org/10.1016/0031-9384(90)90079-J

36. Port RL, Sisak ME, Finamore TL, Soltrick ML, Seybold KS (1998) Role of corticosterone in extinction of an appetitive instrumental response. Int J Neurosci 96(1–2):13–21. https://doi.org/10.3109/00207459808986454

37. Engelmann M, Ebner K, Landgraf R, Wotjak CT (2006) Effects of Morris water maze testing on the neuroendocrine stress response and intrahypothalamic release of vasopressin and oxytocin in the rat. Horm Behav 50 (3):496–501. https://doi.org/10.1016/j.yhbeh.2006.04.009

38. Beiko J, Lander R, Hampson E, Boon F, Cain DP (2004) Contribution of sex differences in the acute stress response to sex differences in water maze performance in the rat. Behav Brain Res 151(1):239–253. https://doi.org/10.1016/j.bbr.2003.08.019

39. Pietersen CY, Bosker FJ, Postema F, den Boer JA (2006) Fear conditioning and shock intensity: the choice between minimizing the stress induced and reducing the number of animals used. Lab Anim 40(2):180–185. https://doi.org/10.1258/002367706776319006

40. Finn DP, Jhaveri MD, Beckett SR, Madjd A, Kendall DA, Marsden CA, Chapman V (2006) Behavioral, central monoaminergic and hypothalamo-pituitary-adrenal axis correlates of fear-conditioned analgesia in rats. Neuroscience 138(4):1309–1317. https://doi.org/10.1016/j.neuroscience.2005.11.063

41. Wade S (1984) Corticosterone availability in male rats at rest and during shock avoidance: differences among individuals, temporal patterns and effects of dexamethasone. J Endocrinol 103(2):187–194. https://doi.org/10.1677/joe.0.1030187

42. Davis M, Astrachan DI (1978) Conditioned fear and startle magnitude: effects of different footshock or backshock intensities used in training. J Exp Psychol Anim Behav Process 4 (2):95–103. https://doi.org/10.1037/0097-7403.4.2.95

43. McIlwain KL, Merriweather MY, Yuva-Paylor LA, Paylor R (2001) The use of behavioral test batteries: effects of training history. Physiol Behav 73(5):705–717

44. Blokland A, Ten Oever S, van Gorp D, van Draanen M, Schmidt T, Nguyen E, Krugliak A, Napoletano A, Keuter S, Klinkenberg I (2012) The use of a test battery assessing affective behavior in rats: order effects. Behav Brain Res 228(1):16–21. https://doi.org/10.1016/j.bbr.2011.11.042

45. von Kortzfleisch VT, Kastner N, Prange L, Kaiser S, Sachser N, Richter SH (2019) Have I been here before? Complex interactions of age and test experience modulate the results of behavioural tests. Behav Brain Res 367:143–148. https://doi.org/10.1016/j.bbr.2019.03.042

46. Barrett JE (1977) Behavioral history as a determinant of the effects of d-amphetamine on punished behavior. Science 198(4312):67. https://doi.org/10.1126/science.408925

47. Holmes A, Iles JP, Mayell SJ, Rodgers RJ (2001) Prior test experience compromises the anxiolytic efficacy of chlordiazepoxide in the mouse light/dark exploration test. Behav Brain Res 122(2):159–167

48. Richetto J, Polesel M, Weber-Stadlbauer U (2019) Effects of light and dark phase testing on the investigation of behavioural paradigms in mice: relevance for behavioural neuroscience. Pharmacol Biochem Behav 178:19–29. https://doi.org/10.1016/j.pbb.2018.05.011

49. Daviu N, Fuentes S, Nadal R, Armario A (2010) A single footshock causes long-lasting hypoactivity in unknown environments that is dependent on the development of contextual fear conditioning. Neurobiol Learn Mem 94 (2):183–190. https://doi.org/10.1016/j.nlm.2010.05.005

50. Rosenzweig MR (2007) Frontiers in neuroscience. Modification of brain circuits through experience. In: Bermudez-Rattoni F (ed) Neural plasticity and memory: from genes to brain imaging. CRC Press/Taylor & Francis Group, LLC., Boca Raton, FL

51. Bennett EL, Diamond MC, Krech D, Rosenzweig MR (1964) Chemical and anatomical plasticity of brain. Science 146(3644):610. https://doi.org/10.1126/science.146.3644.610

52. Kozorovitskiy Y, Gross CG, Kopil C, Battaglia L, McBreen M, Stranahan AM, Gould E (2005) Experience induces structural and biochemical changes in the adult primate brain. Proc Natl Acad Sci U S A 102 (48):17,478–17,482. https://doi.org/10.1073/pnas.0508817102

53. Rosenzweig MR, Krech D, Bennett EL, Diamond MC (1962) Effects of environmental complexity and training on brain chemistry and anatomy: a replication and extension. J Comp Physiol Psychol 55(4):429–437. https://doi.org/10.1037/h0041137

54. Wallace CS, Kilman VL, Withers GS, Greenough WT (1992) Increases in dendritic length in occipital cortex after 4 days of differential housing in weanling rats. Behav Neural Biol 58(1):64–68. https://doi.org/10.1016/0163-1047(92)90937-y

55. Ferchmin PA, Eterovic VA (1986) Forty minutes of experience increase the weight and RNA content of cerebral cortex in periadolescent rats. Dev Psychobiol 19(6):511–519. https://doi.org/10.1002/dev.420190604

56. Mohammed AH, Zhu SW, Darmopil S, Hjerling-Leffler J, Ernfors P, Winblad B, Diamond MC, Eriksson PS, Bogdanovic N (2002) Environmental enrichment and the brain. Prog Brain Res 138:109–133. https://doi.org/10.1016/s0079-6123(02)38074-9

57. Lim R (1970) Grant Newton and Seymour Levine (Eds.). Early experience and behavior (the psychobiology of development). Springfield, III: Charles C Thomas, 1968. Behav Sci 15(4):367–369. https://doi.org/10.1002/bs.3830150411

58. Cory-Slechta DA, Virgolini MB, Rossi-George A, Weston D, Thiruchelvam M (2009) Experimental manipulations blunt time-induced changes in brain monoamine levels and completely reverse stress, but not Pb+/−stress-related modifications to these trajectories. Behav Brain Res 205(1):76–87

59. Cory-Slechta DA, Merchant-Borna K, Allen J, Liu S, Weston D, Conrad K (2012) Variations in the nature of behavioral experience can differentially alter the consequences of developmental exposures to Lead. Prenatal Stress Combination Toxicol Sci. https://doi.org/10.1093/toxsci/kfs260

60. Benefiel AC, Dong WK, Greenough WT (2005) Mandatory "enriched" housing of laboratory animals: the need for evidence-based evaluation. ILAR J 46(2):95–105. https://doi.org/10.1093/ilar.46.2.95

61. Bayne K (2005) Potential for unintended consequences of environmental enrichment for laboratory animals and research results. ILAR J 46(2):129–139. https://doi.org/10.1093/ilar.46.2.129

62. Juraska JM (1998) Neural plasticity and the development of sex differences. Annu Rev Sex Res 9:20–38

63. Crabbe JC, Wahlsten D, Dudek BC (1999) Genetics of mouse behavior: interactions with laboratory environment. Science 284 (5420):1670–1672. https://doi.org/10.1126/science.284.5420.1670

64. Wallace TL, Ballard TM, Glavis-Bloom C (2015) Animal paradigms to assess cognition with translation to humans. Handb Exp Pharmacol 228:27–57. https://doi.org/10.1007/978-3-319-16522-6_2

65. Brown VJ, Tait DS (2016) Attentional set-shifting across species. Curr Top Behav Neurosci 28:363–395. https://doi.org/10.1007/7854_2015_5002

66. Barnett JH, Blackwell AD, Sahakian BJ, Robbins TW (2016) The paired associates learning (PAL) test: 30 years of CANTAB translational neuroscience from laboratory to bedside in dementia research. Curr Top Behav Neurosci 28:449–474. https://doi.org/10.1007/7854_2015_5001

Part IV

Molecular Methods

Chapter 12

Assessment of Neurofilament Light Protein as a Serum Biomarker in Rodent Models of Toxic-Induced Peripheral Neuropathy

Giulia Fumagalli, Guido Cavaletti, Henrik Zetterberg, and Cristina Meregalli

Abstract

Chemotherapy-Induced Peripheral Neurotoxicity (CIPN) is a side effect frequently caused by common antitumor drugs, which may induce a severe and persistent limitation of the quality of life of cancer patients. Today, nerve conduction studies are considered the most objective indicators for CIPN diagnosis. Unfortunately, they are not easily available at most oncology centers. Therefore, a noninvasive and highly sensitive method is required to confirm nerve damage.

Increased evidence supports the potential utility of fluid-based biomarkers to predict tissue damage and to monitor neurotoxicity due to drug administration or the efficacy of disease-modifying treatments. Neurofilaments, the major intermediate filaments in neurons that are specifically expressed in axons, have been investigated as potential biomarker candidates that might be used for this purpose. Neurofilament light chain (NfL) protein is increasingly proposed as a blood biomarker in several neurological diseases mainly affecting the central nervous system. In addition, analysis of serum NfL was evaluated also in peripheral neuropathies including Guillain–Barré syndrome, chronic inflammatory demyelinating and vasculitic neuropathies, and Charcot–Marie–Tooth.

This chapter aims to provide an overview of the methods that allow NfL quantification in serum, focusing on the most recent ultrasensitive single molecule array (Simoa) assay. This technique is likely to be the best method for NfL dosage in CIPN models to predict the onset of large caliber neuronal dysfunction. Since blood sampling is an easily accessible technique, serum NfL may provide important help to monitor neuroaxonal damage after chemotherapy treatment, and might represent promising tools to follow CIPN progress.

Key words Neurofilament light, Axonal injury, Chemotherapy-induced peripheral neurotoxicity, Serum biomarker, Neurotoxicity

1 Introduction

The intermediate filaments are specific neuronal proteins composing the cellular cytoskeleton, and they exert a critical role in supporting axon and dendrite outgrowth, stabilization, and function. Among intermediate filaments, neurofilaments are particularly

Jordi Llorens and Marta Barenys (eds.), *Experimental Neurotoxicology Methods*, Neuromethods, vol. 172,
https://doi.org/10.1007/978-1-0716-1637-6_12, © Springer Science+Business Media, LLC, part of Springer Nature 2021

abundant and they are important for the increase of axonal caliber and consequently for the speed of electrical impulse transmission and structural stability [1, 2]. Neurofilament subunits are classified on the basis of their molecular weight as light, medium, and heavy [1].

During the last few years, several studies have indicated the quantification of serum neurofilament light chain (NfL) as an important biomarker of damage in different diseases of the central nervous system such as Alzheimer's, Huntington's, and Parkinson's diseases, amyotrophic lateral sclerosis, and multiple sclerosis [2–4]. Recently, it has been demonstrated that the use of NfL concentration can be used to detect peripheral nervous system diseases, including inherited [5, 6] and acquired peripheral neuropathies [7] and vasculitic neuropathy [8]. This is due to the fact that upon axonal injury, irrespective of its cause, NfLs are released into the interstitial fluid, CSF, and blood; the magnitude of the increase correlates with the intensity and/or severity of the axonal injury process [4].

To date, grading neuroaxonal damage in a particular peripheral neurotoxicity induced by toxic agents remains an unmet clinical need. Chemotherapy-induced peripheral neurotoxicity (CIPN) is a serious side effect in patients undergoing anticancer therapies commonly used for breast, colorectal, head and neck, lung, prostate, ovarian, hematological, and testicular cancers [9]. The common anticancer drugs known to cause CIPN include platinum derivatives (e.g., cisplatin), vinka-alkaloids (e.g., vincristine), taxanes (e.g., paclitaxel), and proteasome inhibitors (e.g., bortezomib) [10].

Neurotoxicity may significantly affect patient daily activities and quality of life, leading to anticancer treatment modification or even withdrawal [11, 12]. In fact, neuroaxonal damage and loss of fibers can result in severe and/or permanent disability. For this reason, it is important to assess CIPN in a simple and reproducible manner.

Currently, nerve conduction studies are the conventional method employed to objectively detect and monitor CIPN, in combination with grading scales based on subjective evaluations [13, 14]. However, this approach is not completely satisfactory in evaluating CIPN because on the one hand nerve conduction studies routinely performed deal in the detection of only large myelinated fibers whereas they are not able to detect any alteration in small fibers. In fact, techniques for assessing the involvement of small fibers are limited, and they are not employed in the standard practice [15, 16]. On the other hand, the oncological scales (e.g., the National Cancer Institute Common Toxicity Criteria, NCI CTC) usually underestimate the severity and the frequency of symptoms, and they are not always unambiguously described, leading to a variable interpretation [13]. Moreover, to allow a clinical feasibility of an early and reliable detection of CIPN aimed at avoiding an irreversible and permanent damage, the method needs to be simple and reproducible.

The use of biomarkers released into blood from injured axons would fit this aim, and additionally it would not add an invasive procedure to the standard patients' workout as blood sampling is routinely performed in patients undergoing chemotherapy. Moreover, each subject could be examined before, during, and after chemotherapy in order to observe any change from the baseline value.

In preclinical murine models of CIPN-induced by cisplatin, vincristine, and paclitaxel, we have demonstrated the possible role of serum NfL concentration as a potential biomarker for axonal damage severity [17, 18]. In particular, the observed increase in NfL concentration correlated with the severity of axonal damage demonstrated by neurophysiology and morphological investigations and with the temporal course of the pathology. Moreover, at least in some of these models, NfL levels were higher in treated animals even before overt neurophysiological and morphological changes.

1.1 Methods for NfL Detection

In the last three decades, several studies were conducted in order to develop a sensitive immunoassay for NfL detection. The methods for neurofilament protein detection could be divided in four groups, depending on assay sensitivity [4]. The first-generation immunoassays were semi-quantitative methods based on electrophoretic protein separation followed by immunoblotting. These techniques have never been employed in clinical laboratory practice due to their semi-quantitative nature and moreover they are quite laborious [19]. The sandwich enzyme-linked immunosorbent assay (ELISA) belongs to the second-generation assays. It provides quantitative data and a higher analytical accuracy of neurofilament detection. However, while this approach can quantify neurofilaments in cerebrospinal fluid, its analytical sensitivity is not good enough for blood [4, 20]. The electrochemiluminiscence (ECL) technology is a third-generation assay that is approximately 10-fold more sensitive than ELISA [21]. ECL-based measurement of NfL can quantify pathological levels of the biomarker but does not cover the normal range of values, which could be a target values in treatment studies.

Our analyses were performed using an innovative technique called single molecule array (Simoa) [22]. Simoa-based measurement of NfL is basically a bead-based sandwich ELISA where the detection reaction is compartmentalized in small micro-wells (each well-fitting one single antibody-conjugated bead), which permits single molecule counting of NfL [23]. Simoa-based NfL measurement belongs to the fourth-generation NfL quantification technology, and it is 120-fold and 25-fold more sensitive than ELISA and ECL assays, respectively.

2 Materials

2.1 CIPN Animal Models

All the procedures were approved by Animal Care and Use Committee of the University of Milano-Bicocca. The experiments were performed in conformity with the institutional and governmental guidelines for human treatment of laboratory animals set forth in the Guide for the Care and Use of Laboratory Animals (Office of Laboratory Animal Welfare) as well as with the Italian D.L.vo n.26/2014 and the Europeans Union directive 2010/63/UE.

Adult female Wistar rats (175–200 g, Envigo, Udine, Italy) were used in order to study CIPN. Animals were housed in a certified and limited access animal facility under constant temperature (21 °C ± 2) and humidity (50% ± 20). Artificial lighting provided a 12 h light/12 h dark (7 a.m.–7 p.m.) with food and water ad libitum. Animal health condition was monitored daily and rats were sacrificed under deep anesthesia with CO_2 in order to collect samples.

In order to assess CIPN, three different models were obtained using repeated injections of the drugs in rats: vincristine (intravenous (i.v.) administration, 0.2 mg/kg, q7dx4 ws, TEVA Pharma B. V., Mijdrecht, The Netherlands), cisplatin (intraperitoneal (i.p.) administration, 2 mg/kg, 2qwx4 ws, Accord Healthcare Limited, Middlesex, UK) and paclitaxel (i.v. administration, 10 mg/kg, q7dx4 ws, LC laboratories, Woburn, MA, USA).

2.2 CIPN Animal Assessment

Despite the large diffusion of peripheral neurotoxicity associated with antineoplastic drugs, the molecular mechanism of the onset of this side effect remains largely unknown. Here, we report several highly reliable and reproducible rat models that effectively allow to use multimodal approaches (neurophysiological and morphological investigations) to identify the features of the peripheral neurotoxicity. In addition, our models are able to reflect multiple aspects of human peripheral sensory neurotoxicity disease since they manifested axonal degeneration, as well as neurophysiological deficits and loss of intraepidermal nerve fibers, as clinically observed. All the treatments induced CIPN of different severity depending on the chemotherapy drugs employed and their mechanism of action on peripheral nerve system.

In these preclinical models, the axonal damage was evaluated by several techniques. In particular, neurophysiological analyses were performed on peripheral nerves. Briefly, conduction studies (potential amplitude and conduction velocity) of caudal nerves and hind limb digital nerves were performed through an orthodromic stimulation. During all the registration period, the animal was kept under deep isoflurane anesthesia and body temperature was maintained constant (37 °C ± 0.5) using a heating pad. Myto II EMG apparatus (EBN Neuro, Florence, Italy) and subdermal needle electrodes (Ambu Neuroline, Ambu, Ballerup, Denmark) were employed for all recordings [17]. The severity of axonal damage detected by neurophysiological investigations was confirmed by morphological study on collected nerves; otherwise, the evaluation of intraepidermal nerve fibers revealed a decrease of small unmyelinated fiber density as previously described [18].

3 Methods

3.1 Blood Specimen Collection and Processing

The aim of this section is to describe the procedure required to collect suitable serum sample in order to perform the Simoa assay.

1. Blood samples were collected by venipuncture 2 days after drug administration. The animals were kept in a plexiglas restrain cage and the tail was dipped in water at 37 °C for few minutes in order to obtain a dilatation of the vein.

2. Blood collection was performed using a deltaven T 22G (MEDVET srl, Taranto, Italy) and then the sample was withdrawn by gravity and collected in serum-separating tubes (APTACA Spa, Canelli (AT), Italy). Almost 500 μL of blood was collected for each animal.

3. Serum was obtained through centrifugation at 4 °C, 3500 g for 15 min and then aliquoted and stored at −80 ° C until NfL quantification.

3.2 Single-Molecule Array (Simoa) Technology

Rat serum NfL concentration was measured using an in house-developed Simoa assay, in which the same monoclonal antibodies (UD1 and UD2) and calibrator as in the NF-light ELISA for CSF NfL (UmanDiagnostics) were transferred onto the Simoa platform (Quanterix, Billerica, MA, USA), as previously described in detail [24]. Human and rodent NfL sequences are completely conserved and the assay works on both rat and mouse samples.

1. Paramagnetic carboxylated beads (Quanterix Corp, Boston, MA, USA) were coated with an anti-neurofilament light antibody (UD1, UmanDiagnostics, Umeå, Sweden) and incubated 35 min with sample and a biotinylated anti-neurofilament light antibody (UD2, UmanDiagnostics) in a Simoa HD-1 instrument (Quanterix). The bead-conjugated immunocomplex was thoroughly washed before incubation with streptavidin-conjugated β-galactosidase (Quanterix).

2. After additional washes, resorufin β-D-galactopyranoside (Quanterix) was added and the immunocomplex was applied to a multi-well array designed to enable imaging of every single bead to capture light emission and quantify it in digital (average number of captured enzymes per bead [AEB]) mode.

3. The AEB of samples was interpolated onto the calibrator curve constructed by AEB measurements on bovine NfL (Uman-Diagnostics) serially diluted in assay diluent.

4. Samples were analyzed using one batch of reagents and animal treatment information was blinded to the one performing the analysis. The average repeatability of the assay was assessed by measurements of quality control samples and the coefficient of variation was 6.2% for a sample with a mean NfL concentration of 50.7 pg/mL, and 12.3% for a sample with a mean NfL concentration of 22.6 pg/mL.

This assay set up forms the basis for the commercially available NF-Light Simoa assay (Quanterix).

4   Conclusions

Over the last decades, increasing data have been supporting the use of NfL as a reliable biofluid-based biomarker for neuroaxonal injury in central nervous system diseases. NfL is a structural protein also in peripheral nerves, but the literature data regarding its potential role as a peripheral nerve injury marker is scarce and more studies on the topic are required. Conventional ELISA and ECL are the most common tests available for measuring NfL concentration in biofluids, but blood levels are too low to allow for reliable quantification, at least in the normal range [25]. In 2015, the first ultrasensitive NfL assay that allowed for reliable quantification of NfL in blood samples was published [23]. This assay is based on Simoa technology that permits single molecule counting of NfL and subfemtomolar quantification of the protein. Therefore, we used this innovative approach in order to study the onset and progression of CIPN in different preclinical models. In particular, we demonstrated the utility of Simoa for the detection and grading of axonal damage severity in a preclinical model with vincristine [17], cisplatin and paclitaxel [18], as showed in Fig. 1. Therefore, although using the same reagents as in the ELISA kit, the Simoa system provides the best analytical sensitivity across all currently available immunoassays, representing an ideal tool for implementation in clinical setting.

The development of ultrasensitive blood-based detection approach will provide additional information on axon pathology by monitoring longitudinally neuronal degeneration, in order to pursue an early detection of CIPN and avoid irreversible nerve damage. Moreover, further investigations may be extended to other antineoplastic drugs and a potential use of this technique would be evaluated for monitoring the responses to therapy in trials.

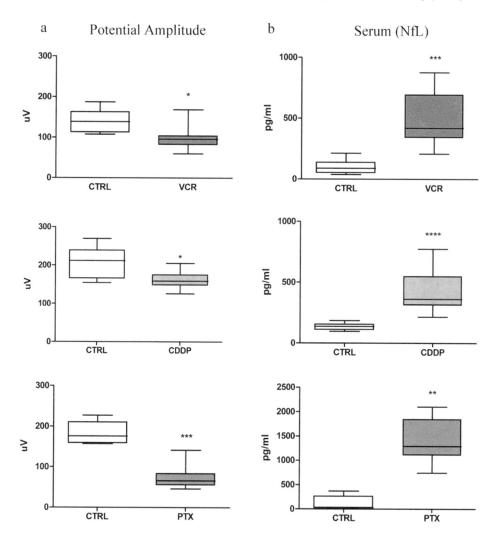

Fig. 1 The graphs show the study of the axonal damage in preclinical models of neurotoxicity induced by vincristine (VCR), cisplatin (CDDP), and paclitaxel (PTX). The most significant neurophysiological results (a) and NfL dosage (b) obtained at the end of treatment are reported. $*p < 0.05$; $**p < 0.01$; $***p < 0.001$; $****p < 0.0001$ vs control (CTRL) (2-sided Mann–Whitney U test). Full original data of VCR model were published on Experimental Neurology [17], whereas results regarding CDDP and PTX models were published on Archives of Toxicology [18]

Acknowledgments

GC is supported by the Italian PRIN grant (#2017ZFJCS3) and CM is supported by Fondazione Cariplo grant (#2019-1482) . HZ is a Wallenberg Scholar supported by grants from the Swedish Research Council (#2018-02532), the European Research Council (#681712), Swedish State Support for Clinical Research (#ALFGBG-720931) and the UK Dementia Research Institute at UCL.

Conflicts of Interest

HZ has served at scientific advisory boards for Roche Diagnostics, Wave, Samumed and CogRx, has given lectures in symposia sponsored by Alzecure and Biogen, and is a co-founder of Brain Biomarker Solutions in Gothenburg AB, a GU Ventures-based platform company at the University of Gothenburg. The other authors declare that they have no conflict of interest.

References

1. Perrot R, Berges R, Bocquet A et al (2008) Review of the multiple aspects of neurofilament functions, and their possible contribution to neurodegeneration. Mol Neurobiol 38(1):27–65

2. Perrot R, Eyer J (2009) Neuronal intermediate filaments and neurodegenerative disorders. Brain Res Bull 80(4–5):282–295

3. Novakova L, Zetterberg H, Sundström P et al (2017) Monitoring disease activity in multiple sclerosis using serum neurofilament light protein. Neurology 89(22):2230–2237

4. Khalil M, Teunissen CE, Otto M et al (2018) Neurofilaments as biomarkers in neurological disorders. Nat Rev Neurol 14(10):577–589

5. Sandelius Å, Zetterberg H, Blennow K et al (2018) Plasma neurofilament light chain concentration in the inherited peripheral neuropathies. Neurology 90(6):e518–e524

6. Kapoor M, Foiani M, Heslegrave A et al (2019) Plasma neurofilament light chain concentration is increased and correlates with the severity of neuropathy in hereditary transthyretin amyloidosis. J Peripher Nerv Syst 24(4):314–319

7. Mariotto S, Farinazzo A, Magliozzi R et al (2018) Serum and cerebrospinal neurofilament light chain levels in patients with acquired peripheral neuropathies. J Peripher Nerv Syst 23(3):174–177

8. Bischof A, Manigold T, Barro C et al (2018) Serum neurofilament light chain: a biomarker of neuronal injury in vasculitic neuropathy. Ann Rheum Dis 77(7):1093–1094

9. Park SB, Goldstein D, Krishnan AV et al (2013) Chemotherapy-induced peripheral neurotoxicity: a critical analysis. CA Cancer J Clin 63(6):419–437

10. Wolf S, Barton D, Kottschade L et al (2008) Chemotherapy-induced peripheral neuropathy: prevention and treatment strategies. Eur J Cancer 44(11):1507–1515

11. Carozzi VA, Canta A, Chiorazzi A (2015) Chemotherapy-induced peripheral neuropathy: what do we know about mechanisms? Neurosci Lett 596:90–107

12. Cavaletti G, Marmiroli P (2015) Chemotherapy-induced peripheral neurotoxicity. Curr Opin Neurol 28(5):500–507

13. Cavaletti G, Frigeni B, Lanzani F et al (2010) Chemotherapy-induced peripheral neurotoxicity assessment: a critical revision of the currently available tools. Eur J Cancer 46(3):479–494

14. Cornblath DR, Chaudhry V, Carter K et al (1999) Total neuropathy score: validation and reliability study. Neurology 53(8):1660–1664

15. Themistocleous AC, Ramirez JD, Serra J et al (2014) The clinical approach to small fibre neuropathy and painful channelopathy. Pract Neurol 14:368–379

16. Svilpauskaite J, Truffert A, Vaiciene N et al (2006) Electrophysiology of small peripheral nerve fibers in man. A study using the cutaneous silent period. Medicina (Kaunas) 42(4):300–313

17. Meregalli C, Fumagalli G, Alberti P et al (2018) Neurofilament light chain as disease biomarker in a rodent model of chemotherapy induced peripheral neuropathy. Exp Neurol 307:129–132

18. Meregalli C, Fumagalli G, Alberti P et al (2020) Neurofilament light chain: a specific serum biomarker of axonal damage severity in rat models of chemotherapy-induced peripheral neurotoxicity. Arch Toxicol. https://doi.org/10.1007/s00204-020-02755-w

19. Petzold A (2005) Neurofilament phosphoforms: surrogate markers for axonal injury, degeneration and loss. J Neurol Sci 233(1–2):183–198

20. Norgren N, Rosengren L, Stigbrand T (2003) Elevated neurofilament levels in neurological diseases. Brain Res 987(1):25–31

21. Kuhle J, Nourbakhsh B, Grant D et al (2017) Serum neurofilament is associated with

progression of brain atrophy and disability in early MS. Neurology 88(9):826–831

22. Rissin DM, Kan CW, Campbell TG et al (2010) Single-molecule enzyme-linked immunosorbent assay detects serum proteins at subfemtomolar concentrations. Nat Biotechnol 28 (6):595–599

23. Gisslén M, Price RW, Andreasson U et al (2015) Plasma concentration of the neurofilament light protein (NFL) is a biomarker of CNS injury in HIV infection: a cross-sectional study. EBioMedicine 3:135–140

24. Rohrer JD, Woollacott IO, Dick KM et al (2016) Serum neurofilament light chain protein is a measure of disease intensity in frontotemporal dementia. Neurology 87 (13):1329–1336

25. Kuhle J, Barro C, Andreasson U et al (2016) Comparison of three analytical platforms for quantification of the neurofilament light chain in blood samples: ELISA, electrochemiluminescence immunoassay and Simoa. Clin Chem Lab Med 54(10):1655–1661

Chapter 13

Assessing Neurotoxicant-Induced Inflammation in the Central Nervous System: Cytokine mRNA with Immunostaining of Microglia Morphology

Christopher A. McPherson and G. Jean Harry

Abstract

Inflammation occurs as a normal response of the organism to harmful stimuli such as microbial pathogens, irritants, or toxic cellular components that result from injury and trauma. It serves as a balanced process of pro- and anti-inflammatory responses to maintain normal tissue. The increased role of inflammation in the manifestation of neurotoxicity, whether directly induced or the result of pathological changes, has led to assessments of inflammatory factors within models of environmental exposures. Within the brain, an inflammatory response can be elicited from the resident central nervous system (CNS) glia (microglia and astrocytes) but can also be influenced by endothelial cells and peripherally derived immune cells depending on the nature of the insult, chemical-induced insults. There is a complex and dynamic response in the brain to regulate the inflammatory process. This chapter outlines methods to assess occurrence of a neuroinflammatory response with examination of mRNA levels for pro-inflammatory cytokines and receptors by qRT-PCR, combined with immunocytochemical staining for resident microglia immune cells and their morphological assessment. Analysis of the endpoints described in this chapter provides a framework to assess chemical-induced inflammation in the CNS.

Key words Microglia, Pro-inflammatory cytokines, TNF, IL-6, Neuroinflammation

1 Introduction

Inflammation occurs as a normal response of the organism to harmful stimuli such as microbial pathogens, irritants, or toxic cellular components that result from injury and trauma. Within the brain, an inflammatory response can be elicited from the resident CNS glia (microglia and astrocytes), endothelial cells, and peripherally derived immune cells. These cells mediate pro- and anti-inflammatory responses and are major sources of soluble molecules, cytokines, chemokines, hormones, and neuropeptides. These small molecules and membrane receptors provide cells with the tools to sense, process, and relay physiological signals beyond their canonical roles. A complex series of immune-like reactions

Jordi Llorens and Marta Barenys (eds.), *Experimental Neurotoxicology Methods*, Neuromethods, vol. 172,
https://doi.org/10.1007/978-1-0716-1637-6_13, © Springer Science+Business Media, LLC, part of Springer Nature 2021

are initiated to neutralize invading pathogens, repair injured tissues, and promote wound healing. In all cases, inflammation represents a coordinated process to serve a specific goal to restore tissue homeostasis. Such reactions include the clearance of cellular debris, secretion of neurotrophic factors, secretion of cytokines, and activation of proteases for matrix remodeling. In this framework, inflammation can be viewed as a complicated series of local immune responses to deal with a threat to the microenvironment. The appropriate regulation of the initial cellular response to tissue damage facilitates recovery while uncontrolled neuroinflammation can induce secondary injury. Within the area of neurotoxicity assessment, neuroinflammation can represent an underlying stress placed upon the system, a cellular process associated with cell injury and repair, as well as a possible mode of action to initiate cell death or exacerbate/prolong injury. This chapter is not intended to serve as a review of the components of neuroinflammation for which there is an extensive literature rather, it will present points of consideration when designing experiments to evaluate exposure-related neuroinflammation and provide methods for examining in vivo morphological responses of microglia and quantification of mRNA expression of inflammatory markers. Many of the general concepts and approaches can be applied to cells in culture.

1.1 Cytokines

Neuroinflammation is closely linked with the induction and presence of inflammatory factors including pro-inflammatory cytokines such as tumor necrosis factor (TNF), interleukin 1 (IL-1), and IL-6 [1]. Elevations in such factors can represent a response to adjacent cellular damage, direct action on immune cells, cell interactions, and infiltration of peripheral immune cells [2]. Upon activation of transcription factors or inhibition of negative transcription regulators, mRNA synthesis is initiated for the eventual production and secretion of pro-inflammatory cytokines. Increased expression of cytokines occurs transcriptionally, post-transcriptionally, translationally, and through the conversion of latent precursors to biologically active protein [3–9]. Translational control offers a strategic advantage to these cells, allowing the use of pre-existing mRNAs to bypass the lengthy nuclear control mechanisms (e.g., transcription, splicing, and transport); additionally, it provides for reversibility through modifications of the regulatory intermediates. Combined, these features allow rapid activation or termination of synthesis of a specific protein or group of proteins required for the inflammatory process.

Cytokines can be secreted upon complete intracellular processing while others are stored intracellularly and require an additional stimulus to trigger protein secretion. There is additional evidence demonstrating that, while most cytokines are secreted from the cell in a biologically active form for some, the latent form can be secreted, requiring additional extracellular processing for biological activity. A biologically active cytokine can bind to a specific membrane receptor, a carrier protein, or a soluble receptor. This allows for induction and regulation of the inflammatory response often in an autocrine/paracrine fashion. For the pro-inflammatory cytokine, tumor necrosis factor (TNF) signaling, upon release of mature biologically active protein, TNF-α interacts with two different receptors, designated TNFR1 and TNFR2, which are differentially expressed on cells and tissues and initiate both independent and overlapping signal transduction pathways, leading to multiple cellular responses [10]. In addition to membrane-associated receptors, soluble forms of some cytokine receptors can be generated in response to cell activation such as soluble TNF receptors that bind released TNF and block its biological activity. The interleukin-1 (IL-1) family of cytokines and receptors broadly affects a broad range of inflammatory responses [11]. IL-1α and IL-1β are of primary interest in the nervous system. The family of IL-1 receptors contains pro- and anti-inflammatory receptors including IL-1R1 that binds IL-1α, IL-1β, and IL-1 receptor antagonist (IL-1Ra) [12]. The association between cytokine production and the concurrent upregulation of receptors allows for the opportunity to use mRNA levels to examine biological activity of a specific cytokine by measuring both the cytokine and the cytokine receptor. This would then allow for examination of downstream signaling events as a result of changes observed in cytokine mRNA levels.

1.2 Consideration of the Source of Inflammatory Factors

Under healthy conditions, microglia are the only immune cell type present in the CNS parenchyma yet there are various immune regulatory cells such as macrophages and dendritic cells present in the CNS-adjoining tissues [13, 14]. The CNS is mechanically separated from the circulation by the blood–brain barrier (BBB). This separation influences immune responses by excluding many peripherally derived innate and adaptive immune cells and inflammatory molecules. However, infiltrating cells significantly contribute to any neuroinflammatory response following disruption of the BBB, as can occur with physical injury or high levels of inflammation. While a predominant interest is in the response of resident nervous system immune cells, microglia, the possible contribution from infiltrating blood-borne cells or circulating factors requires consideration. Induction of a peripheral inflammatory response and elevations of inflammatory factors in the blood can influence either a localized response along the vascular wall or actually infiltrate into the brain parenchyma. Thus, the contribution from peripheral immune cells must be considered with any experimental design and interpretation of the resulting data. The responses of microglia versus peripheral macrophages display distinct properties under polarization conditions and may play different roles in the inflamed CNS [15, 16]. Figure 1 shows an example of how the blood compartment can significantly influence the data outcome.

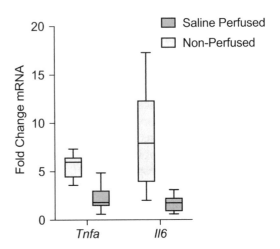

Fig. 1 Adult male mice were exposed to Chemical X for 3 months. At 3 months following exposure, *Tnfa* and *Il6* mRNA expression was analyzed from the hippocampus of saline perfused vs non-perfused animals. mRNA expression for *Tnfa* and *Il6* was significantly higher in the non-perfused animals compared to saline perfused animals ($^*p < 0.05$). These data demonstrate the importance of considering the contribution of peripheral blood compartment

1.2.1 Peripheral

The BBB mechanically separates the CNS from the circulation by the presence of specialized endothelial cells tightly attached to each other via tight junctions and adherens junctions. As such, it excludes plasma proteins and many of the peripherally derived innate and adaptive immune cells and associated inflammatory molecules [17, 18]. Thus, it can be considered that the BBB actively contributes to the immune response of the CNS not only by regulating entry of peripheral factors but also with signaling upon stimulation of inflammatory cells along the vascular wall and perivascular microglia. With physical injury to the BBB or highly inflamed CNS parenchyma, blood-borne monocytes can enter the brain parenchyma and contribute to the inflammatory response and pathogenic process [19, 20].

1.2.2 Central Nervous System

In the CNS, microglia serve as the primary resident innate immune cell population [21]. Depending on the anatomical region, microglia account for between 0.5 and 16% of the total cell population in the human brain [22] and 5–12% in the mouse brain [23]. Microglia are unique from peripheral monocytes in that they develop from myeloid progenitors from the embryonic yolk sac [24, 25]. Microglia migrate into the embryonic CNS where they proliferate to populate the brain without further contribution of peripheral progenitors to the resident microglia pool [26, 27]. This developmental process establishes a life-long population of resident microglial cells which is maintained by a limited self-renewal [28–30]. This also makes the early establishment of the microglia population a critical event in development that can have long-term consequences. Thus, one can consider that this low turnover rate renders microglia susceptible to the effects of environmental exposure, age, injury, stress, or any combination. Microglia display morphological heterogeneity and differential density across brain regions [23, 31]. They dynamically and actively sense their microenvironment and perform crucial physiological functions for tissue development, architecture refinement, and tissue remodeling/repair [32, 33]. In their quiescent surveillance phenotype, microglia display fine mobile processes extending into the surrounding microenvironment [32]. The initial concept of a stereotypic response of microglia with limited response variability has been replaced by data showing that microglia respond with a variety of different morphological changes as a result of integrating multifarious inputs and represent a diverse spectrum of responses that do not fall within a dichotomy of surveying or activated [33–35]. When microglia sense pathogens, dying cells, debris, or aberrant proteins, they adjust to their environment which often manifests as a phenotypic change including morphological changes and increase in phagocytic capability. As damage is resolved, the correlated pro-inflammatory response diminishes.

1.2.3 Distinguishing Cell Origin

While it is enticing to consider any specific marker as unique for either resident microglia or peripheral monocytes/macrophages, such confirmation of the cellular source requires additional approaches. One approach relies on flow cytometry to discriminate between microglia [CD11b+ and CD45 low] and peripheral macrophages [CD11b+ and CD45 high] [36]. A more confirmative method used to identify infiltrating blood-borne monocytes and discriminate from resident microglia relies on bone marrow chimera animals to track fluorescently tagged bone marrow-derived cells [37]. While microglial gene expression shares homology with tissue-resident macrophages, it maintains a unique gene expression signature [38, 39].

1.3 Microglia Receptors

The distinct features associated with a neuroinflammatory response will impact the design and interpretation of experimental models to assess effects within a neurotoxicology framework. In a sterile inflammatory response, immune cells are activated in the absence of microbial compounds by endogenous molecules called danger-associated molecular patterns (DAMPs) [40]. These damage signals can activate immune cells through pattern recognition receptors (PPR) including: Toll-like receptors (TLRs), nucleotide-binding domain leucine-rich repeat containing proteins (also known as NOD-like receptors, NLRs), and Rig1-like receptors (RLRs). The engagement of these receptors results in the induction of specific pathways and release of cytokines that contribute to resolving injury. Molecules that function as DAMPs include nucleic acids, lipids, proteins that immune cells do not normally see until they are released or unmasked with cells death occurring due to tissue injury. It is thought that receptors on microglia can act as molecular switches to control microglial responses [33, 41]. Microglia express age and sex-dependent P1 receptors for nucleosides and monophosphate containing nucleotides (adenosine and AMP) and most of the receptors of the P2 family for di- and tri-phosphate containing nucleotides (e.g., ADP, ATP) that can regulate both pro-inflammatory and anti-inflammatory functions [42].

1.4 Cell Polarization

In general, the macrophage inflammatory response is characterized based on the nature of the activating stimulus and the resulting production of factors which can span a vast spectrum [34, 43–45]. A diverse array of inflammatory stimuli has been used alone or in combinations to generate polarized macrophage populations in vitro and in vivo. While a classification of M1 and M2 for a dichotomy of responses has taken hold in the neuroinflammation literature, this is considered an inappropriate and simplified terminology. Rather, it is recommended that the stimulus inducing the response should be identified [45]. Examples include M[LPS], m[IFNγ], M[LPS + IFNγ] to elicit release of pro-inflammatory factors and m[IL-4], M[Il-10], and M[IL-13] to elicit the release of anti-inflammatory factors. Stimulation of these pathways lead to production of pro-inflammatory cytokines (e.g., interferon gamma (IFNγ), IL-12, TNF-α, IL-6, and IL-1β), chemokines (CCL2, CXCL10, and CXCL11), and antigen presentation molecules, such as major histocompatibility complex (MHC) or anti-inflammatory cytokines (e.g., IL-4, IL-10, IL-13, and transforming growth factor beta, (TGF-β)) as well as arginase-1 (Arg-1), CD206, and Chitinase-3-like-3 (Ym-1 in rodents). As research on polarization advances, it is becoming clear that reliance on a limited number of factors for either polarization state is not recommended or reflective of the biology.

1.5 Age

Microglia originate from a primitive monocyte population derived from the yolk sac during a defined window of time before vascularization or definitive hematopoiesis and in rodents, they are present as early as embryonic day 8 [28, 29]. The cells demonstrate a structural maturation that starts from a more rounded morphology in the early stages of development followed by a progressive complexity of cell processes [46]. Many of the pro-inflammatory and anti-inflammatory factors also show a pattern over the course of development followed by a shift with aging [47]. Thus, when designing a study or interpreting data from a developmental or aging study, reliance upon literature derived from the adult animal is not recommended given the multiple and shifting roles for cytokine signaling. Also, with the dynamic nature of brain development, including the high level of cell death and remodeling, any elevation in phagocytic action of microglia may reflect the normal high demands placed on the cells for clearance of cellular material. With aging, a decreased phagocytic capability would be considered detrimental. Over the life span, microglia continue to change. Morphologically, microglia in the aged brain display cytoplasmic structures showing excessive beading and spheroid swellings that may be reflective of dystrophy and senescence [48]. The observed decrease in process motility may compromise surveillance features and responsivity [49, 50]. In addition, a slight elevation in the basal level of pro-inflammatory cytokines and a diminished ability to recruit an anti-inflammatory response is observed. Thus, as these limited examples demonstrate, it is critical that investigators ensure an understanding of the related biological processes as they apply to the age under study.

1.6 Time Interval for Assessment

An acute or short-term exposure regimen requires consideration of when to assess for an associated neuroinflammatory response. The sequence of biological events required to initiate a neuroinflammatory or microglia response would require time for the effect to occur and reach an adequate level of detection. For example, an acute response might rely on chemoattractant signaling molecules such as the chemokines, transcription factor modifications allowing for cytokine production, and receptor activation for a downstream biological effect. Many pro-inflammatory cytokines display an autocrine and paracrine signaling and thus, self-regulate expression. In addition, many cytokines share an ATTTA sequence in their 3' end that allows for rapid mRNA degradation [51]. Thus, depending on the time of assessment, relative to the pro-inflammatory response, different stages of the response may be required.

1.7 Quantitative Real-Time PCR for mRNA Levels of Inflammatory Markers In Vivo

Quantitative reverse transcription-polymerase chain reaction, qRT-PCR, uses reporter dyes to detect PCR products in real time. The assay relies on measuring the increase in fluorescent signal, which is proportional to the amount of DNA produced during each PCR cycle. Individual reactions are characterized by the PCR cycle at which the fluorescence first rises above a defined background level, a parameter known as the threshold cycle (C_t). The more gene target there is in the starting material the lower the C_t. This correlation between fluorescence and the amount of amplified product allows for a quantitated assessment of target mRNA expression. Detection of qRT-PCR products can be accomplished with either SYBR™ Green dye or TaqMan® probes. SYBR™ Green dye fluoresces when bound to the double-stranded PCR product amplicons from investigator-designed forward and reverse PCR primers. The primers attach to the anti-sense and sense strand of the target DNA of interest. As the PCR progresses, more amplicons of the target sequence are created which in turn bind more SYBR™ Green, resulting in increased fluorescent intensity that is directly

Table 1
Advantages and limitations of SYBR™Green and TaqMan® probes

Detection method	Advantages	Limitations
SYBR™Green	• Primers can be designed for any double-stranded DNA sequence • Does not require probes, lower assay costs	• SYBR Green binds to all double-stranded DNA generated in the reaction and may generate false positives • Sensitivity varies depending on primer design • Time-consuming primer optimization is required for assays • A dissociation curve analysis PCR product is required at the end of each assay to ensure specificity of primer binding, lengthening the assay time
TaqMan®	• Specific hybridization between target and probe is required to generate fluorescent signal • High sensitivity for low copy targets • Dissociation curves are not necessary • Predesigned and validated probes are commercially available	• Probes cost more than SYBR Green Primers • A different probe needs to be generated for each target sequence of interest

Adapted from: http://tools.thermofisher.com/content/sfs/manuals/cms_083618.pdf

proportional to the amount of target gene in the sample. Important optimization steps have been described [52, 53] and must be undertaken when SYBR™ Green probes are used. All of these optimization steps are done to ensure sensitivity, specificity, reproducibility, and a large dynamic range of the PCR reaction. They are briefly outlined in Subheadings 3.5.4.1 and 3.5.4.2.

TaqMan® probes contain a reporter dye (FAM or TAMRA) linked to the 5′ end of the probe, and a non-fluorescent quencher (NFQ) at the 3′ end of the probe. While intact the TaqMan® fluorescent reporter on the 5′ end is quenched by the NFQ on the 3′ end. When the probe binds to the target sequence, it is cleaved by Taq DNA polymerase during primer extension. This cleavage separates the quencher resulting in an increase in the reporter dye. With each PCR cycle probes bind and are cleaved, increasing the fluorescent signal intensity. The resulting increased fluorescent intensity is directly proportional to the amount of target sequence in the sample. The advantages and disadvantages of each detection chemistry (Table 1) should be considered when designing experiments.

1.7.1 Target Gene Primers

The assay should be specifically designed for each target gene to identify all transcripts of interest including splice variants while discriminating between closely related family members. This protocol outlines the use of "off the shelf" TaqMan® assays that are confirmed to be specific to the genes of interest allowing for high specificity detection of low copy number inflammatory molecules in the CNS. It is also important to ensure that the assay has an amplification efficiency close to 100% in the tested system. A less efficient assay can result in reduced sensitivity and reduced dynamic range, limiting the ability to detect low abundance cytokine transcripts.

The quantitation method should be determined. The procedures outlined in this chapter have been validated for use of either the comparative C_t $(2^{-\Delta\Delta C_t})$ or the relative standard curve method. The $2^{-\Delta\Delta C_t}$ method is a relative quantitation method used to analyze fold changes of gene expression in a given sample relative to a reference sample (i.e., untreated control), using the mathematical model first described by Livak and Schmittgen [52]. The $2^{-\Delta\Delta C_t}$ method is useful to compare levels in treated vs untreated samples within a study. The relative standard curve method offers more direct analysis options when a more complex experimental design is needed such as measurement expression levels in samples treated with a compound under different experimental conditions. This is especially valuable under conditions where any prior manipulation or exposure alters the baseline expression level of a gene prior to any secondary manipulation. Using the standard curve method allows for a more refined experimental design and statistical analysis. The relative standard curve method uses a dilution series created from a positive control sample run with both the target and endogenous control gene. The standard curve can be generated from any known biological source of the gene but needs to generate a curve for which the experimental samples fall within the linear portion. For all experimental samples, a relative quantity is determined from this dilution series [53].

1.7.2 Selection of Housekeeping Gene

Regardless of the quantitation method used, selection of a valid endogenous control to normalize for RNA sampling is critical to avoid misinterpretation of results. Suitable endogenous controls include housekeeping genes such as GAPDH, β-actin, or 18S ribosomal RNA (18S rRNA). It is imperative that expression of the endogenous control gene used in the qRT-PCR reaction does not change with the experimental manipulation. Subtle changes in the level of a high-copy housekeeping gene that may not appear significant can significantly influence interrogating low copy number cytokine mRNA transcripts. The fold change of endogenous housekeeping genes following treatment can be calculated using the 2^{-C_t} method (Table 2) [53]. It is suggested that, for any new experimental manipulation, selection of the housekeeping gene includes comparing of more than one gene as well as the impact of any subtle changes on target gene expression level is evaluated.

Table 2
Validation of endogenous housekeeping gene

GAPDH		β-actin		18S rRNA	
Chemical X	Vehicle	Chemical X	Vehicle	Chemical X	Vehicle
$2^{(-16.6)} = 8.01E-06$	$2^{(-17.5)} = 5.36E-06$	$2^{(-19.3)} = 1.54E-06$	$2^{(-19.2)} = 1.76E-06$	$2^{(-19.6)} = 1.22E-06$	$2^{(-20.7)} = 5.68E-06$
$2^{(-16.6)} = 9.78E-06$	$2^{(-17.4)} = 5.64E-06$	$2^{(-20.6)} = 6.12E-06$	$2^{(-19.3)} = 1.51E-06$	$2^{(-20.9)} = 5.28E-06$	$2^{(-21.1)} = 4.42E-06$
$2^{(-17.8)} = 4.25E-06$	$2^{(-17.6)} = 4.92E-06$	$2^{(-20.6)} = 6.12E-06$	$2^{(-19.3)} = 1.51E-06$	$2^{(-20.8)} = 5.36E-06$	$2^{(-20.9)} = 5.25E-06$
$2^{(-18.0)} = 3.95E-06$	$2^{(-17.9)} = 4.18E-06$	$2^{(-20.7)} = 5.97E-06$	$2^{(-19.7)} = 1.21E-06$	$2^{(-21.1)} = 4.39E-06$	$2^{(-21.4)} = 3.71E-06$
$2^{(-17.7)} = 4.59E-06$	$2^{(-16.9)} = 8.01E-06$	$2^{(-20.5)} = 6.63E-06$	$2^{(-18.9)} = 2.09E-06$	$2^{(-20.8)} = 5.64E-06$	$2^{(-20.5)} = 6.67E-06$
Mean = 5.23E-06	Mean = 5.63E-06	Mean = 8.06E-07	Mean = 1.06E-06	Mean = 5.23E-06	Mean = 5.63E-06

Gene expression of endogenous controls can be determined using the $2^{-\alpha}$ method. First calculate the mean $2^{-\alpha}$ for each housekeeping gene
Next calculate the fold change of each housekeeping gene
GAPDH: 5.23E−06/5.63E−06 = 9.29E−01; −1/9.29E−01 = −1.08 fold change
β-actin: 8.06E−07/1.06E−06 = 5.03E−01; −1/5.03E−01 = −2.0 fold change
18s rRNA: 5.23E−06/5.63E−06 = 1.27 fold change
It is recommended not to use a housekeeping gene that changes over 1.4-fold under experimental conditions [43, 44]

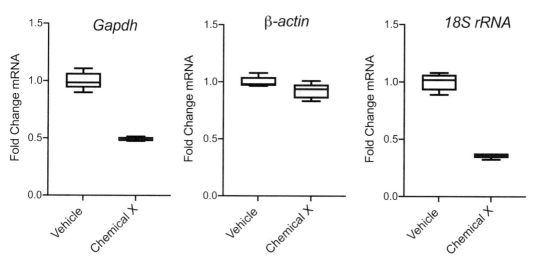

Fig. 2 Three housekeeping genes (*Gapdh, b-actin, and 18 s rrna*) were examined for fold change following exposure to Chemical X, using the $2^{-\alpha}$ method (Table 2). The fold change for *Gapdh, 18 s rrna, and b-actin* were −1.0, 1.2, and 2.0, respectively. When *Gapdh* and *18 s rrna* were each used as the endogenous controls in the $2^{-\Delta\Delta\alpha}$ calculation to compare differences with vehicle controls, a significant decrease in *Tnfa* mRNA level was observed following exposure to Chemical X ($^*p < 0.05$). In contrast, when *b-actin* was used, no changes in *Tnfα* mRNA levels were observed. *Chemical X* reduced *b-actin* mRNA levels by 2.0-fold and was not considered a suitable internal control gene

An example of how changing housekeeping gene expression may affect interpretation of the pro-inflammatory cytokine TNF-α mRNA levels as determined by the $2^{-\Delta\Delta Ct}$ method is provided (Fig. 2).

1.8 Considerations for an Experimental Approach

It is highly recommended that a spectrum of markers is employed to determine neuroinflammation. Of additional interest are to determine if any change is a result of chemical exposure, if it occur in the absence of neuropathology, and if it is sufficient to initiate a biologically active inflammatory response. These markers should reflect not only the initial stage but also confirm a biological downstream activation. This could be a combination of mRNA levels for pro-inflammatory cytokines, associated receptors, receptor antagonists, and anti-inflammatory cytokines. The selection of genes may reflect whether or not the system is capable of responding to the inflammatory process for down-regulation. Depending on the model system, one may want to consider additional markers that reflect resolution and repair.

Combining neuroanatomical assessments with molecular or biochemical assessments lends support for any interpretation. A change in microglia morphology as a result of damage in the surrounding environment is characterized by increased soma size and the retraction of elongated fine processes to shorter, coarser cytoplasmic processes displaying a bushy appearance. This

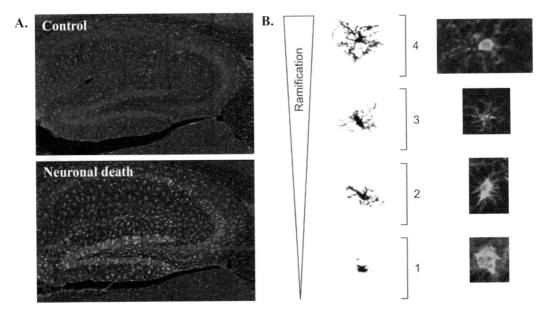

Fig. 3 Representative images of microglial response to chemical-induced neuronal cell death in the hippocampus of adult mice. Adult male CD-1 mice received a single intraperitoneal injection of trimethyltin hydroxide (TMT; 2.3 mg/kg), (**a**) Iba-1+ microglia cells (green), and nuclear DAPI (blue) in controls and with neuronal death at 72 h post-injection. In the control hippocampus, Iba-1+ cells displayed thin elongated processes with distal ramifications. With neuronal death, Iba-1+ cells differentiated into a rounded amoeboid morphology suggesting a high level of phagocytic activity to remove dead neurons in the dentate granule region and in the CA pyramidal region cells with enlarged bushy cells with retracted processes were observed in areas not necessarily associated with neuronal death. Cells are also present that do not appear to shift significantly from controls. (**b**) Representative rating scale that can be used for discriminating microglia morphology based on cell soma shape, and orientation, density, and complexity of processes

morphology can further shift to fully ameboid morphology depending on the nature and severity of the insult (Fig. 3). The response in a developing animal requires additional considerations given the maturation of resident microglia with regard not only to morphology but also to cytokine production and the functions of such cytokines. It is recommended that the morphology of the cell be evaluated rather than to either count number of cells or simply assign a classification to the cell response A number of quantitative and semi-quantitative approaches have been developed for this purpose [54, 55].

2   Materials

The protocol is focused on the use of rodents (mice, rats). For any use of animal, procedures should be conducted in compliance with all requirements of institutional committees regarding use of animals in research.

2.1 Whole-Body Saline Perfusion Equipment and Materials

1. Balance to weigh animal.
2. Perfusion pump or 50 mL syringe.
3. Restraint board to restrain body and limbs.
4. Small dissecting scissors.
5. Hemostat to secure perfusion needle.
6. Fatal-Plus® Solution (Vortech Pharmaceuticals, Ltd., 9373).
7. 25 g x ¾″ Winged perfusion needle with 12″ tubing.
8. Sterile 0.9% Saline solution.

2.2 Excision of Brain

1. Scissors for decapitation.
2. Small surgical scissors for cutting scalp.
3. Small animal rongeurs or sturdy forceps to remove skull.
4. Flat edge weighing spatula to remove brain from skull.
5. Single edge razor blade to block brain.

2.3 Tissue Collection

2.3.1 Frozen Tissue

1. Flat surface on dry ice (*see* **Note 1**).
2. Labeled RNAse-free microcentrifuge tubes (1.5–1.8 mL) for frozen tissue storage.
3. RNaseZap® RNase decontamination solution (Thermo Fisher, AM9780).

2.3.2 Fixed Tissue

1. Labeled flat bottom containers.
2. 0.1 M phosphate buffer (PB) pH 7.4
   (a) 77.4 mL 1 M Na_2HPO_4 (Sigma, S5136).
   (b) 22.6 mL 1 M NaH_2PO_4 (Sigma, S5011).
   (c) 900 mL dH_2O.
3. 4% weight/volume (w/v) paraformaldehyde solution (*see* **Note 2**)
   (a) 0.4 g Paraformaldehyde, EM grade (Ted Pella, 18501).
   (b) 0.1 M phosphate buffer (bring to 10 mL using a volumetric flask).
4. 15% (w/v) sucrose solution
   (a) 150 g sucrose.
   (b) 0.1 M PB (bring to 1 L using a volumetric flask).
5. 30% (w/v) sucrose solution
   (a) 300 g sucrose.
   (b) 0.1 M PB (bring to 1 L using a volumetric flask).
6. Tissue-Tek® O.C.T. Compound (Sakura, 4583).
7. Tissue-Tek® Cryomold (Andwin Scientific, 4557).

2.4 Cryosectioning

1. Cryostat (*see* **Note 3**).
2. Disposable cyrostat blades (Thermo Fisher, 3051835).
3. Fine artist paintbrush (#3) for handling sections to minimize damage to sections.
4. 2.0 mL cryovials (Nunc, 368632)
5. 6-well tissue culture plate or equivalent
6. 0.1 M PB
7. FD Section Storage Solution® (FD NeuroTechnologies, Inc., PC101) (*see* **Note 4**).

**2.5
Immunofluorescent
Staining**

1. 24-well tissue culture plate or equivalent.
2. Fisherbrand™ Superfrost™ Plus Microscope Slides (Fisher, 22-037-246).
3. Coverslips.
4. Fine paint brush.
5. Clear nail polish.
6. Fluorescent microscope.
7. Blocking Solution
   (a) 0.1 M PB.
   (b) 10% Normal Goat Serum (Gemini Bio, 100-109).
   (c) 0.3% Triton X-100.
8. Anti Iba1, Rabbit polyclonal (Wako 019-19741) (*see* **Note 5**).
9. Goat anti-rabbit IgG Alexa Fluor488 conjugate (Invitrogen, A-11008).
10. 4′,6-diamidino-2-phenylindole, 5 mg/mL stock solution (DAPI; Invitrogen D1306) (*see* **Note 6**)
11. Dissolve the contents of one vial of DAPI (10 mg) in 2 mL of dH$_2$O.
12. ProlongGold™ Antifade Mountant (Invitrogen, P36930).

2.6 qRT-PCR

1. Latex gloves.
2. Hand-held tissue homogenizer with polytron probe (*see* **Note 7**).
3. Round bottom 2 mL microcentrifuge tubes (*see* **Note 8**).
4. RNase-free Microcentrifuge tubes (1.5 mL).
5. Vortex.
6. PCR machine.
7. Spectrophotometer.
8. Pipette and tips.

9. Optical 96-well or 384-well PCR plate (Applied Biosystems, 4326813 or 4309849).

10. Optical adhesive film (Applied Biosystems, 4360954).

11. Multichannel pipet 0.5–10 µL and 5–100 µL.

12. TRIzol™ (Invitrogen, 15596026) (*see* **Note 9**).

13. 70% ethanol

14. 100% Isopropanol

15. Fresh chloroform (*see* **Note 10**).

16. RNAse-Free H_2O.

17. Superscript II Reverse Transcriptase (Invitrogen, 18064014).

18. Oligo dT primer (Applied Biosystems, N8080128).

19. Oligonucleotides, 10 mM DNTP mix (Invitrogen, 18420788).

20. TaqMan® Universal PCR master mix (Applied Biosytstems, 4305719).

21. TaqMan® Gene Expression Assays FAM (Applied Biosystems, 4331182).

22. Power SYBR™ Green master mix (Applied Biosystems, 4368706).

3 Methods

3.1 Whole-Body Saline Perfusion

1. Weigh animal.

2. Deeply anesthetize animal with a single intraperitoneal injection 100 mg/kg body weight Fatal-Plus® Solution.

3. Once anesthesia has taken effect secure the animal on a dissection stage with each limb immobilized.

4. Using a scalpel make an incision along the underside of the animal to expose the chest cavity.

5. Using scalpel or small surgical scissors open the chest cavity without injuring the lungs or heart.

6. Insert the perfusion needle (winged or blunt edge needle shaft) into the left ventricle of the heart and stabilize then perform a small incision on the right atrium.

7. Perfuse the mouse with a minimum of 10 mL (mouse) or 30 mL (rat) of cold 0.9% saline using a syringe or diastolic perfusion pump attached to the needle inserted into the left ventricle of the heart. Saline should be perfused at a rate of 1 mL per minute.

8. Following perfusion, decapitate the head and carefully extract the brain from the skull.

3.2   Tissue Collection

*3.2.1   Histology and mRNA from Each Brain to Allow for Assessments in Each Animal (See **Note 11**)*

1. Place the brain on a flat surface with cortical side up and cut with a single or double edge razor blade along the midline of the brain hemispheres to bisect the brain into two hemispheres. Prefer a straight edge of a razor blade to a scalpel blade to facilitate straight even cut.

2. To fix tissue, place one hemisphere in flat bottom container (e.g., specimen jar) containing fresh 4% paraformaldehyde/ PBS (approximately 50 mL/g tissue weight).

3. Ensure that the cut face of the tissue lies on the flat surface to maintain accurate cytoarchitecture.

4. Allow brains to fix in 4% paraformaldehyde/PBS for 18 h at 4 °C.

5. If additional blocking of brain is required for specific regional sectioning, this is conducted only after tissue fixation.

6. Proceed to Cryopreservation Subheading 3.3.1.

7. From the non-fixed contralateral extracted hemisphere, sub-dissect any region of interest (e.g., freehand, micro-punch),

8. Immediately place tissue on a pre-chilled flat metal surface on a bed of dry ice and allow tissues to freeze.

9. Store froze tissue in microcentrifuge tubes at −80 °C. Frozen tissue will remain loose in the tube during storage and if adhered to the tube wall that will indicate a thaw/freeze during storage and degradation.

3.3   Cryosectioning

3.3.1   Cryopreservation

1. Transfer brain to a flat bottom specimen jar containing 15% w/v sucrose solution in 0.1 M PB (approx. 30 mL/g tissue weight), incubate at 4 °C until the brain sinks, approximately 3 days.

2. Transfer brain to specimen jar containing equivalent volume of 30% (w/v) sucrose solution in 0.1 M PB, incubate at 4 °C until the brain sinks.

3. Remove cryo-preserved brains from the sucrose.

4. Place tissue with the cutting face down into a cryomold, add OCT to each tissue cassette sufficient to completely cover the brain.

5. Immediately freeze brain in cryomold on dry ice, ensuring maintaining correct architecture, store at −20 °C.

3.3.2   Cryosectioning

1. Cool cryostat to −20 °C before beginning.

2. Remove brain from cryomold and place in cryostat for 15 min to equilibrate to temperature.

3. Mount brain on chuck with OCT compound and allow to freeze for 15 min.

4. Mount chuck into the specimen holder, orient specimen to desired position using adjustable screws on the holder.

5. Insert disposable blade into the knife holder adjust knife angle to 5–10°.

6. Set section thickness to 10 μm.

7. Trim the block to expose the brain region of interest. This is done by carefully taking 10 μm sections through the block. Initial sections will be OCT only and eventually will contain brain samples. If the trimming is not collecting a full uniform section of the sample, adjust the specimen holder until a full section is obtained.

8. Once the block is trimmed set the section thickness to 30 μm.

9. Section and collect tissues using a fine artist paintbrush (#3) into a 6-well plate containing 0.1 M PB (*see* **Note 12**).

10. When sectioning is completed, sections can be either processed immediately or gently transferred to cryovial containing section storage solution and stored at −20 °C until processing.

3.4
Immunofluorescent
Staining

3.4.1 Day 1

1. Add 800 μL 0.1 M PB to the required number of wells of a 24-well tissue culture plate or equivalent.

2. Remove required brain sections from the section storage solution using a fine artist paintbrush and place one section into each well of the 24-well plate containing 0.1 M PB.

3. Allow sections to equilibrate to room temperature (RT) for 15 min.

4. Gently remove fluid and replace with 800 μL 0.1 M PB.

5. Block nonspecific IgG binding in sections by applying 200 μL Blocking Solution with gentle orbital shaking for 2 h at RT.

6. Prepare primary Iba-1 antibody solution by diluting in Blocking Solution. For Wako Chemicals Iba-1 antibody, the recommended dilution is 1:200, this may require optimization dependent on the brain region of interest. Prepare sufficient antibody solution to incubate each section with 200 μL. Store on ice while sections are undergoing blocking (**step 5**).

7. Aspirate blocking solution using pipet and add 200 μL of primary antibody solution to each section. Allow to incubate at 4 °C for 48 h using gentle orbital shaking.

3.4.2 Day 2

1. Gently remove primary antibody solution from the wells.

2. Wash sections 3× with 0.1 M PB, 5 min each.

3. Prepare secondary antibody detection solution by diluting goat anti-rabbit IgG Alexa Fluor488 conjugate 1:5000 in blocking solution.

4. Add 200 µL secondary antibody solution to each section. Allow to incubate for 2 h at RT with gentle orbital shaking.

5. Gently aspirate secondary antibody solution from wells.

6. Wash sections 3× with 0.1 M PB, 5 min.

7. Prepare DAPI solution by diluting DAPI stock 1:25000 in 0.1 M PB.

8. Add 200 µL DAPI solution to each section allow to incubate at RT for 5 min.

9. Wash sections 3× with 0.1 M PB, 5 min.

10. Using a fine paint brush, gently lift section from well and place on glass microscope slide with adjustments for flat and correct orientation. Remove excess liquid from the slide using a dry laboratory wipe.

11. Apply 1–2 drops of ProLong™ Antifade Mountant to the mounted section.

12. Cover the sections with a coverslip and allow to dry for 24 h at RT in the dark. Drying for 24 h allows for optimal performance of the antifade reagent.

13. Seal stained slides by applying a thin layer of clear nail polish around the outer perimeter of the coverslips; don't apply sealant over the section. Sealed sections are ready for viewing or storage at −20 °C.

3.5 qRT-PCR

3.5.1 RNA Extraction Using TRIzol™ Reagent

1. Maintain your experimental design within the isolation and RNA analysis (*see* **Note 13**).

2. Isolate 6 or maximum of 8 samples at a time.

3. Add 1 mL of TRIzol™ Reagent per 50–100 mg to sample in microcentrifuge tube (2.0 mL) (*see* **Note 14**).

4. Disrupt subdissected tissue using a polytron tissue homogenizer for 30–40 s.

5. After each tissue, rinse homogenizer probe using 80% ethanol/RNase-free H_2O followed by a rinse with RNase-free H_2O. Ensure no residual ethanol on the probe.

6. Centrifuge at $12,000 \times g$ at 4 °C for 5 min.

7. Transfer the cleared supernatant to a new 1.5 mL microcentrifuge tube.

8. Incubate the homogenized sample for 5 min at room temperature to permit complete dissociation of the nucleoprotein complex.

9. Add 0.2 mL of chloroform per 1 mL of TRIzol™ Reagent used for homogenization. Cap the tube securely.

10. Vortex tube for 15 s.

11. Incubate for 2–3 min at room temperature.

12. Centrifuge the sample at 12,000 × g for 15 min at 4 °C. Following centrifugation, the mixture separates into a lower red phenol-chloroform phase, an interphase, and a colorless upper aqueous phase. RNA remains in the aqueous phase.

13. Remove the colorless upper aqueous phase of the sample by angling the tube and pipetting the solution out. Avoid drawing any of the interphase or organic layer when removing the aqueous phase and place into a new tube. If material from the interphase or lower phase is drawn up, allow the liquid to partition in the pipette tip and gently push out to discard the unwanted phase and collect the upper phase.

14. Add 0.5 mL of 100% isopropanol to the aqueous phase per 1 mL of TRIzol™. Reagent used for homogenization and incubate at room temperature for 10 min.

15. Place tubes into the centrifuge with a standard orientation to identify where the pellet would form (e.g., plate the hinge of the cap outward and the pellet will form on that side of the tube).

16. Centrifuge the sample at 12,000 × g for 10 min at 4 °C. Following centrifugation, the RNA should form a gel-like pellet at the bottom of the tube.

17. Gently remove supernatant from the tube, leaving only the RNA pellet.

18. Gently wash the pellet with 1 mL of 75% ethanol.

19. Vortex the samples briefly, then centrifuge the tube at 7500 × g for 5 min at 4 °C.

20. Air dry the RNA pellet for 5–10 min. It is critical that all ethanol is evaporated from the pellet but if the pellet dries to hard it may be difficult to resuspend and may require a series of freeze/boil to resuspend.

21. With samples of approximately 80–100 mg starting material, resuspend RNA pellet in 30 μL RNAse-free water. For samples derived from starting material <80 mg one may consider resuspending in a smaller volume (20 μL) for determination of RNA yield.

22. Place in a heating block or hybridization oven at 60 °C for 10 min.

23. Determine RNA concentration by measuring the absorbance at 260 and 280 nm in a spectrophotometer.

24. Calculate RNA concentration using the formula A260 × dilution × 40 = μg RNA/mL. Calculate the A260/A280 ratio, with expectations of a ratio of ~2. Ratios below 1.7 or above

2 may be contaminated with DNA, solvents, or protein and should not be used (*see* **Note 15**).

25. Aliquot RNA into 5–10 µg total RNA for use in qPCR experiments. It is recommended that samples be aliquoted for each follow-up assay. Multiple freeze thaws of RNA should be avoided given the degradation with each cycle following the criteria of a maximum of three with recalculation of concentration with each cycle.

3.5.2 Perform Reverse Transcription Using SuperScript™ II Reverse Transcriptase

1. Program PCR machine to execute the following steps to prepare cDNA by reverse transcription

Primer anneal step	Hold	65 °C	10 min
	Hold	4 °C	10 min
Reverse transcriptase activation step	Hold	42 °C	2 min
Reverse transcription step	Hold	42 °C	50 min
Inactivation step	Hold	70 °C	15 min
	Hold	4 °C	Indefinite

2. Add the following components to a nuclease-free PCR tube:

Oligo(dT)$_{12-18}$ (500 µg/mL) primers	1 µL
1.5 µg total RNA	x µL
1 µL dNTP mix (10 mM each)	1 µL
Sterile, distilled water	To 12 µL

3. Run the primer anneal step on the PCR machine (Subheading 3.5.2, **step 1**) then add:

5× First-Strand Buffer	4 µL
0.1 M DTT	2 µL

4. Mix contents of tube by gently pipetting up and down five times, then run the reverse transcriptase activation step on the PCR machine (Subheading 3.5.2, **step 1**).

5. Add 1 µL (200 units) of SuperScript™ II RT and mix by pipetting gently up and down two times.

6. Run the PCR reverse transcription and heat inactivation steps (Subheading 3.5.2, **step 1**).

7. Hold reaction at 4 °C for 10 min then add 60 μL dH$_2$O. Samples can be used immediately or stored at −20 °C.

8. For TaqMan gene expression analysis, proceed to Subheading 3.5.3. For SYBR Green gene expression analysis proceed to Subheading 3.5.4.

3.5.3 TaqMan® Gene Expression Analysis

1. Set up plate and experiment using the PCR system software. Use the following thermal cycling parameters.

Polymerase activation	Hold	95 °C	10 min
PCR (40 cycles)	Denature	95 °C	15 s
	Anneal	60 °C	1 min

2. Determine the number of reactions to perform for each assay, include three replicates of each sample for each reaction. If using the relative standard curve method, prepare a tenfold serial dilution of positive control cDNA. Add 10% more reactions than calculated to account for pipetting error.

3. Prepare the 20 μL reaction mix for each replicate using the components listed below.

TaqMan®Universal Master Mix (2×)	10 μL
TaqMan® gene expression assay (20×)	1 μL
cDNA template (1–100 ng) + RNase-free H$_2$O	9 μL
Total volume	20 μL

4. Seal plate MicroAmp™ Optical adhesive film.

5. Centrifuge plate briefly to spin down the contents and eliminate air bubbles in the sample.

6. Open experiment file in the PCR system that corresponds to the plate.

7. Load plate into the PCR system and start the run.

8. Proceed to data analysis in (Subheading 3.5.5).

3.5.4 SYBR™ Green Gene Expression Analysis
(See Note 16)

1. Set up plate and experiment using the PCR system software. Use the following thermal cycling parameters.

Optimization of SYBR™ Green Primer Concentrations

Polymerase activation	Hold	95 °C	10 min
PCR (40 cycles)	Denature	95 °C	15 s
	Anneal	60 °C	1 min

Add dissociation curve protocol to the end of the run

2. Dilute stock forward and reverse SYBR green primers to the following concentrations: 50, 300, and 900 nM.

3. A plate should be run to test each of the nine forward and reverse primer conditions [FWD/REV] nM: [50/50], [300/50], [900/50], [50/300], [300/300], [900/300], [50/900], [300/900], and [900/900] in quadruplicate.

4. Prepare the 50 μL reaction master mix for each replicate using the components listed below:

Power SYBR™Green PCR Master Mix (2×)	20 μL
Forward primer	Variable
Reverse primer	Variable
cDNA template	5 μL
dH$_2$0	Variable
Total volume	50 μL

5. Seal plate with MicroAmp Optical adhesive film.

6. Centrifuge late briefly to spin down the contents and eliminate bubbles in the sample.

7. Open experiment file in the PCR system corresponding to the plate.

8. Load plate into the PCR system and start the run.

9. Following the run determine the minimum primer concentration that gave the lowest Ct value for the target gene of interest with the least nonspecific binding following melt curve analysis.

SYBR™ Green PCR Reaction Optimization

1. Set up plate and experiment using the PCR system software. Use the following thermal cycling parameters.

Polymerase activation	Hold	95 °C	10 min	
PCR (40 cycles)	Denature	95 °C	15 s	
	Anneal		60 °C	1 min

Add dissociation curve protocol to the end of the run

2. Make tenfold serial dilution of a positive control sample cDNA. Prepare enough to amplify the target primer sequence as well as the housekeeping gene of interest in quadruplicate.

3. Prepare 50 μL reaction master mix for each replicate using the components listed below and use the optimized primer concentrations empirically determined in **step 3** of Subheading 3.5.4.

Power SYBR Green PCR Master Mix (2×)	20 µL
Forward primer	Variable
Reverse primer	Variable
cDNA template	Variable
dH₂0	Variable
Total volume	50 µL

4. Seal plate with MicroAmp Optical adhesive film.

5. Centrifuge late briefly to spin down the contents and eliminate bubbles in the sample.

6. Open experiment file in the PCR system corresponding to the plate.

7. Load plate into the PCR system and start the run.

8. Following the run, plot the Ct value (y-axis) versus the log cDNA dilution (x-axis), then determine the slope of the line using the equation $y = mx + b$.

9. Calculate the PCR efficiency using the equation $m = -(1/\log E)$, where m = slope and E = PCR efficiency.

10. Efficiency of the target gene of interest and endogenous housekeeping gene should fall with 10% of each other.

11. If 10% difference in efficiency is not obtained, try further optimization of primer concentrations; design of new primers is required if optimal PCR efficiency is not achieved.

Preparation of SYBR™ Green Master Mix for Gene Expression Analysis

1. Set up plate and experiment using the PCR system software. Use the following thermal cycling parameters.

Polymerase activation	Hold	95 °C	10 min
PCR (40 cycles)	Denature	95 °C	15 s
	Anneal	60 °C	1 min

Add dissociation curve protocol to the end of the run

2. Determine the number of reactions needed for the experimental design. If conducting relative standard curve quantitation, make tenfold serial dilution of a positive control sample cDNA. Prepare enough to amplify the target primer sequence as well as the housekeeping gene of interest in quadruplicate.

3. Prepare 50 µL reaction master mix for each replicate using the components listed below, use the optimized primer concentrations empirically determined in **step 3** of Subheading 3.5.4.

Power SYBR Green PCR Master Mix (2×)	20 μL
Forward primer	Variable
Reverse primer	Variable
cDNA template	Variable
dH$_2$0	Variable
Total volume	50 μL

4. Seal plate with MicroAmp Optical adhesive film.

5. Centrifuge late briefly to spin down the contents and eliminate bubbles in the ample.

6. Open experiment file in the PCR system corresponding to the plate.

7. Load plate into the PCR system and start the run.

8. Proceed with data analysis (Subheading 3.5.5).

3.5.5 Data Analysis

1. Examine the amplification plots of each sample and identify abnormal plots. For example, note samples that record a C_t value with no amplification plot.

2. Remove any abnormal plots and allow software to assign C_t values to samples.

3. Ensure the C_t has been assigned correctly to each well. This is achieved by ensuring the threshold has been set in the exponential phase of the curve. If the C_t is set above or below, the exponential phase manual adjust the cycle threshold to fall within the exponential phase.

4. Check controls for expected results.

5. Examine C_t values of all replicates. C_t values of replicates should fall within 0.5 C_t of each other. C_t values above 35 will have greater variability, making quantitation unreliable.

6. Use the relative standard curve method or the $2^{-\Delta\Delta Ct}$ method to analyze the data (*see* **Note 17**).

4 Notes

1. A thick metal plate cleaned thoroughly with RNase decontamination solution on dry ice is a suitable surface for freezing brain regions for mRNA analysis. Brain regions may also be placed in cryovials and snap frozen in a Dewar's flask containing liquid nitrogen.

2. Heat solution to 68 °C in a water bath or hybridization oven until paraformaldehyde is dissolved. Adjust pH to 7.4 after the solution cools. Paraformaldehyde is hazardous; avoid contact with skin, eyes, or inhalation of powder or fumes.

3. The methods outlined in this chapter are optimized for immunofluorescent staining in 40 μm fixed free-floating sections obtained with a Leica CM3050-S cryostat (Leica Biosystems).

4. FD Section Storage Solution is ideal for the prolonged storage of free-floating fixed tissue sections while preserving morphology and antigenicity of the tissues. Sections kept in this solution may be stored at −20 °C for 5 years before processing for immunofluorescence.

5. The anti Iba-1 polyclonal antibody is stored at −20 °C, avoid multiple freeze thaw cycles of the antibody by immediately aliquoting the antibody upon receipt from the vendor.

6. For long-term storage, stock solution can be aliquoted and stored at ≤−20 °C, protected from light. DAPI is a known mutagen and should be handled with care.

7. A hand-held tissue homogenizer such as the Omni International Tissue homogenized (TH115) with a 5 mm × 75 mm rotor-strator fine generator probe (G5-75) works well for disrupting brain tissues. The rotor-strator probe allows for flow through of the tissue through the probe allowing for quick and efficient shearing of tissue and efficient disruption of tissue. If a larger tissue mass is to be used for isolation, one can consider pulverization of the frozen tissue prior to homogenization.

8. Do not use a tube with a conical-shaped bottom for homogenization, a round or flat bottom tube allows for a better material flow though the homogenizer probe. Microcentrifuge tubes of 1.8 mL volume usually have a straight wall configuration that can be used.

9. Other isolation methods may be used for the isolation of high-quality total RNA from brain samples such as column-based systems. For brain tissues, a column that can accommodate the high lipid content is required to prevent diminished total RNA yields or possible interference with DNA contamination. In addition, one-step quick isolations that do not include a partitioning step may not be appropriate for tissue with high lipid content.

10. Chloroform can absorb H_2O from the air especially in humid climates. Ensure that the chloroform is well sealed when not in use and change to fresh chloroform no longer than every 6 months. This will prevent H_2O contamination in the RNA extraction process and loss of sample.

11. This protocol is designed for examination of histology in sagittal sections of one hemisphere and then examining mRNA expression in subdissected regions from the contralateral hemisphere.

12. If quantitation of microglia in the region of interest is required, then serial 40 μM sections of the region should be collected and immunostained for analysis following unbiased stereological sampling requirements [56].

13. A large source of experimental variance is the RNA isolation; take care to maintain experimental design during the RNA isolation. Limit the number of samples extracted at any one time to maintain consistency across samples and ensure successful isolation. Recommended limit is 6–8 samples per isolation.

14. Volumes of TRIzol™ and all other reagents are for samples weighing between 50 and 100 mg. Samples <50 mg can be isolated with 500 μL TRIzol™ with corresponding vol adjustments made for chloroform partitioning. If sample size is <10 mg, add 5–10 μg RNAse-free glycogen as a carrier during the aqueous phase as recommended by the TRIzol™ protocol. (http://tools.thermofisher.com/content/sfs/manuals/trizol_reagent.pdf).

15. The traditional way to determine RNA quantity and purity is to measure UV absorption of the sample using a spectrophotometer. RNA has a maximum absorption at 260 nm. Concentration is estimated with an A260 OD of 1.0 is equivalent to 40 μg/mL of RNA. Measurements should also be taken at 280 nm for protein contamination and 230 nm for other contaminants such as guanidine salts and phenol. The ideal A260/A230 ratio is greater than 1.5.

    UV absorbance can change depending on pH of the RNA solution; thus, best results are obtained when RNA is solubilized in TE buffer. Always blank the spectrophotometer with the vehicle. Readings of RNA concentrations below 20 μg/mL may not be reliable. RNA integrity is critical for any analysis and can be determined either by running on an agarose gel or using a Bioanalyzer to measure the size of the rRNA bands and determine RNA integrity number (RIN).

16. Primer optimization and PCR reaction efficiency steps are necessary to ensure sensitivity, specificity, reproducibility, and a large dynamic range of the PCR reaction.

17. Most qRT-PCR systems and software make calculations for the user; see these references for detailed descriptions of the relative standard curve method [53] or the $2^{-\Delta\Delta Ct}$ method [52].

References

1. Becher B, Spath S, Goverman J (2017) Cytokine networks in neuroinflammation. Nat Rev Immunol 17:49–59

2. Kempuraj D, Thangavel R, Selvakumar GP et al (2017) Brain and peripheral atypical inflammatory mediators potentiate neuroinflammation and neurodegeneration. Front Cell Neurosci 11:216. https://doi.org/10.3389/fncel.2017.00216

3. Falvo JV, Tsytsykova AV, Goldfeld AE (2010) Transcriptional control of the TNF gene. Curr Dir Autoimmun 11:27–60

4. Mazumder B, Li X, Barik S (2010) Translation control: a multifaceted regulator of inflammatory response. J Immunol 184(7):3311–3319

5. Casanova JL, Abel L, Quintana-Murci L (2011) Human TLRs and IL-1Rs in host defense: natural insights from evolutionary, epidemiological, and clinical genetics. Annu Rev Immunol 29:447–491

6. Rattenbacher B, Bohjanen PR (2012) Evaluating posttranscriptional regulation of cytokine genes. Methods Mol Biol 820:71–89

7. Khabar KSA (2014) Post-transcriptional control of cytokine gene expression in health and disease. J Interf Cytokine Res 34(4):215–219

8. Mino T, Takeuchi O (2018) Post-transcriptional regulation of immune responses by RNA binding proteins. Proc Jpn Acad Ser B Phys Biol Sci 94(6):248–258

9. Kany S, Vollrath JT, Relja B (2019) Cytokines in inflammatory disease. Int J Mol Sci 20(23):6008. https://doi.org/10.3390/ijms20236008

10. Brenner D, Blaser H, Mak T (2015) Regulation of tumour necrosis factor signalling: live or let die. Nat Rev Immunol 15:362–374

11. Dinarello CA (2018) Overview of the IL-1 family in innate inflammation and acquired immunity. Immunol Rev 281(1):8–27

12. Garlanda C, Riva F, Bonavita E et al (2013) Negative regulatory receptors of the IL-1 family. Semin Immunol 25(6):408–415

13. Mrdjen D, Pavlovic A, Hartmann FJ et al (2018) High-dimensional single-cell mapping of central nervous system immune cells reveals distinct myeloid subsets in health, aging, and disease. Immunity 48(2):380–395

14. Norris GT, Kipnis J (2019) Immune cells and CNS physiology: microglia and beyond. J Exp Med 216(1):60–70

15. Durafourt BA, Moore CS, Zammit DA et al (2012) Comparison of polarization properties of human adult microglia and blood-derived macrophages. Glia 60(5):717–727

16. Yamasaki R, Lu H, Butovsky O et al (2014) Differential roles of microglia and monocytes in the inflamed central nervous system. J Exp Med 211:1533–1549

17. Hawkins BT, Davis TP (2005) The blood-brain barrier/neurovascular unit in health and disease. Pharmacol Rev 57:173–185

18. Muldoon LL, Alvarez JI, Begley DJ et al (2013) Immunologic privilege in the central nervous system and the blood-brain barrier. J Cereb Blood Flow Metab 33:13–21

19. King IL, Dickendesher TL, Segal BM (2009) Circulating Ly-6C+ myeloid precursors migrate to the CNS and play a pathogenic role during autoimmune demyelinating disease. Blood 113:3190–3197

20. Brendecke SM, Prinz M (2015) Do not judge a cell by its cover—diversity of CNS resident, adjoining and infiltrating myeloid cells in inflammation. Semin Immunopathol 37:591–605

21. Michelucci A, Mittelbronn M, Gomez-Nicola D (2018) Microglia in health and disease: a unique immune cell population. Front Immunol 9:1779

22. Mittelbronn M, Dietz K, Schluesener HJ et al (2001) Local distribution of microglia in the normal adult human central nervous system differs by up to one order of magnitude. Acta Neuropathol 101(3):249–255

23. Lawson LJ, Perry VH, Gordon S (1992) Turnover of resident microglia in the normal adult mouse brain. Neuroscience 48(2):405–415

24. Alliot F, Godin I, Pessac B (1999) Microglia derive from progenitors, originating from the yolk sac, and which proliferate in the brain. Brain Res Dev Brain Res 117:145–152

25. Gomez Perdiguero E, Klapproth K, Schultz C et al (2015) Tissue-resident macrophages originate from yolk-sac-derived erythro-myeloid progenitors. Nature 518:547–551

26. Ajami B, Bennett JL, Krieger C et al (2011) Infiltrating monocytes trigger EAE progression, but do not contribute to the resident microglia pool. Nat Neurosci 14:1142–1149

27. Mildner A, Schmidt H, Nitsche M et al (2007) Microglia in the adult brain arise from Ly-6ChiCCR2+ monocytes only under defined host conditions. Nat Neurosci 10:1544–1553

28. Ginhoux F, Jung S (2014) Monocytes and macrophages: developmental pathways and tissue homeostasis. Nat Rev Immunol 14:392–404

29. Ginhoux F, Greter M, Leoeuf M et al (2010) Fate mapping analysis reveals that adult microglia derive from primitive macrophages. Science 330:841–845

30. Ajami B, Bennett JL, Krieger C et al (2007) Local self-renewal can sustain CNS microglia maintenance and function throughout adult life. Nat Neurosci 10:1538–1543

31. Harry GJ, Kraft AD (2012) Microglia in the developing brain: a potential target with lifetime effects. Neurotoxicology 33:191–206

32. Nimmerjahn A, Kirchhoff F, Helmchen F (2005) Resting microglial cells are highly dynamic surveillants of brain parenchyma in vivo. Science 308:1314–1318

33. Kettenmann H, Hanisch UK, Noda M et al (2011) Physiology of microglia. Physiol Rev 91:461–553

34. Hanisch UK, Kettenmann H (2007) Microglia: active sensor and versatile effector cells in the normal and pathologic brain. Nat Neurosci 10:1387–1394

35. Kierdorf K, Prinz M (2013) Factors regulating microglia activation. Front Cell Neurosci 7:44

36. Carson MJ, Reilly CR, Sutcliffe JG et al (1998) Mature microglia resemble immature antigen-presenting cells. Glia 22:72–85

37. Lassmann H, Hickey WF (1993) Radiation bone marrow chimeras as a tool to study microglia turnover in normal brain and inflammation. Clin Neuropathol 12:284–285

38. Gautier EL, Shaty T, Miller J et al (2012) Gene-expression profiles and transcriptional regulatory pathways that underlie the identity and diversity of mouse tissue macrophages. Nat Immunol 13:1118–1128

39. Larochelle A, Bellavance MA, Michaud JP, Rivest S (2016) Bone marrow-derived macrophages and the CNS: an update on the use of experimental chimeric mouse models and bone marrow transplantation in neurological disorders. Biochim Biophys Acta 1862(3):310–322

40. Kono H, Rock KL (2008) How dying cells alert the immune system to danger. Nat Rev Immunol 8:279–289

41. Pocock JM, Kettenmann H (2007) Neurotransmitter receptors on microglia. Trends Neurosci 30:527–535

42. Crain JM, Nikodemova M, Watters JJ (2009) Expression of P2 nucleotide receptors varies with age and sex in murine brain microglia. J Neuroinflammation 6:24

43. Mosser DM, Edwards JP (2008) Exploring the full spectrum of macrophage activation. Nat Rev Immunol 8:958–969

44. Martinez FO, Gordon S, Locati M et al (2006) Transcriptional profiling of the human monocyte-to-macrophage differentiation and polarization: new molecules and patterns of gene expression. J Immunol 177:7303–7311

45. Murray PJ, Allen JE, Biswas SK et al (2014) Macrophage activation and polarization: nomenclature and experimental guidelines. Immunity 41:14–20

46. Lopez-Atalaya JP, Askew KE, Sierra A et al (2018) Development and maintenance of the brain's immune toolkit: microglia and non-parenchymal brain macrophages. Dev Neurobiol 78:561–579

47. Hefendehl JK, Neher JJ, Suhs RB et al (2014) Homeostatic and injury-induced microglia behavior in the aging brain. Aging Cell 13:60–69

48. Streit WJ, Miller KR, Lopez KO et al (2008) Microglial degeneration in the aging brain—bad news for neurons? Front Biosci 13:3423–3438

49. Tremblay ME, Zette ML, Ison JR et al (2012) Effects of aging and sensory loss on glial cells in mouse visual and auditory cortices. Glia 60:541–558

50. Damani MR, Zhao L, Fontainhas AM et al (2011) Age-related alterations in the dynamic behavior of microglia. Aging Cell 10:263–276

51. Caput D, Beutler B, Hartog K et al (1986) Identification of a common nucleotide sequence in the 3′-untranslated region of mRNA molecules specifying inflammatory mediators. Proc Natl Acad Sci U S A 83:1670–1674

52. Livak KJ, Schmittgen TD (2001) Analysis of relative gene expression data using real-time quantitative PCR and the 2(-Delta Delta C (T)) method. Methods 25:402–408

53. Schmittgen TD, Livak KJ (2008) Analyzing real-time PCR data by the comparative C (T) method. Nat Protoc 3:1101–1108

54. Davis BM, Salinas-Navarro M, Cordeiro MF et al (2017) Characterizing microglia activation: a spatial statistics approach to maximize information extraction. Sci Rep 7:1576

55. Paasila PJ, Davies DS, Kril JJ et al (2019) The relationship between the morphological subtypes of microglia and Alzheimer's disease neuropathology. Brain Pathol 29:726–740

56. Phoulady HA, Goldgof D, Hall LO, Mouton PR (2019) Automatic ground truth for deep learning stereology of immunostained neurons and microglia in mouse neocortex. J Chem Neuroanatomy 98:1–7

Chapter 14

Mitochondrial Stress Assay and Glycolytic Rate Assay in Microglia Using Agilent Seahorse Extracellular Flux Analyzers

Gabrielle Childers and G. Jean Harry

Abstract

Regulation of cellular processes to meet the energy demands for multiple cellular processes has long been known to be a critical endpoint for assessing cellular health and functional capabilities. The more recent evidence that indicates mitochondria as a critical central component of immunity and inflammatory responses has prompted research into mitochondrial functions in immune cells. Mitochondrial biogenesis, fusion, and fission have contributory roles in aspects of immune-cell activation. The association of neuroinflammation across essentially all neurodevelopmental and neurodegenerative disorders has led to a research inquiry on the functional aspects of the primary brain immune cell, the microglia. While mitochondria may serve as a platform for innate immune signaling it is likely that, in the central nervous system (CNS), resident microglia mitochondria functions are critically linked to the various task of the cells such as providing signals and functions for neurodevelopment, synapse formation and remodeling, protective responses to promote repair following injury, and clearance of aberrant proteins and debris from the CNS. Given the multiple functions and tasks of microglia, they have become a cell of interest in the area of neurotoxicology, no longer as a simple indicator of brain injury/pathology but rather as a critical team player for maintaining the normal homeostatic state of the nervous system and contributing to the structural and functional integration of the system. In this regard, the various and often rapid functions of microglia require sufficient and appropriate detection of signals in the environment and the energy to respond. Thus, gaining a better understanding of the microglia mitochondria powerhouse will allow for the evaluation of subtle impacts of environmental chemicals, pharmacological agents, genetic alterations, and disease states on the ability of the nervous system to develop and function properly. The following chapter provides a protocol built upon information available on the Agilent website [https://community.agilent.com/docs/DOC-8069-collection-of-cell-analysis-resources] but focuses on specific experiments conducted with the murine BV-2 microglial cell line and rodent primary microglia cultures.

Key words Microglia, Mitochondria, Bioenergetics, Glycolysis, Immune response

1 Introduction

Under normal oxygen conditions, cells obtain energy via two different mechanisms. In the first, glucose is converted to pyruvate via glycolysis, entering the mitochondrial tricarboxylic acid cycle

Jordi Llorens and Marta Barenys (eds.), *Experimental Neurotoxicology Methods*, Neuromethods, vol. 172,
https://doi.org/10.1007/978-1-0716-1637-6_14, © Springer Science+Business Media, LLC, part of Springer Nature 2021

(TCA) to produce adenosine triphosphate (ATP) through oxidative phosphorylation (OXPHOS). When oxygen is present, the major energy-producing pathway in cells is OXPHOS. Nutrients derived from different metabolic pathways are oxidized to produce cellular energy in the form of ATP and macromolecules for cellular function. The mitochondrion is the central organelle where these processes occur. Independent of oxygen, cells also generate ATP through glycolysis, the conversion of glucose to lactate. The measurement of lactic acid produced is indirect via the protons released into the extracellular medium surrounding the cells.

Increasing evidence suggests a role of metabolic reprogramming in the regulation of the innate inflammatory response [1]. The specific modulation of glycolytic energy flux is critical to macrophage activation and is likely a critical factor to define cell polarization [2, 3]. Similar to other immune cells, when stimulated with toll-like receptor (TLR) agonists (e.g., lipopolysaccharide; LPS), microglia, which serve as the primary brain immune cells, switch from the preferred oxidative metabolism towards glycolytic metabolism [4]. Under this condition, cells preferentially use glycolysis rather than catabolic mitochondrial pathways to conserve and generate metabolic resources for multiple cellular functions. This metabolic switch is designed to support macrophages during the demands of an immune challenge [2]. In classically activated macrophages and dendritic cells, metabolism is shifted towards glycolysis with nitric oxide and citrulline production. This switch increases glucose uptake and lactate production and decreases mitochondrial oxygen consumption. Activation of the pentose phosphate pathway produces nitric oxide and citrulline. Additional evidence supports a link between inflammation and metabolic changes in microglia similar to what has been reported in monocyte-derived macrophages [5–8]. In the anti-inflammatory state induced by interleukin (IL) 4 (IL4) and IL13, oxidative respiration and increased fatty acid oxidation are observed [9]. Under this state, cultured microglia, BV-2 murine cell line and primary murine microglia, show decreased glucose consumption and lactate production that is potentially associated with a reduced need for anabolic reaction. IL-4 exposure also increases oxygen consumption rate, basal respiration, and ATP production in microglia [10].

The ability to interrogate the real-time mitochondrial-related responses of immune cells to stimulation has provided significant information on the heterogeneity of microglial reaction and alterations in respiratory function. Of the currently available commercial systems, mitochondrial respiration can be measured using the Oxygraph-2k high-resolution respirometer or the Seahorse Extracellular Flux (XF) Analyzer. These two technologies propose two different methods of analysis but provide similar information. The following protocol details methods for examining BV-2 and primary microglia for mitochondrial bioenergetic changes using the Agilent Seahorse XF analyzers. By using optical sensors, the analyzers simultaneously measure proton and oxygen levels in a small volume of media directly above a monolayer of cultured cells. More recently, the unit has been used to examine zebrafish, rodent embryos, and tissue slices (*see* https://www.agilent.com/cell-reference-database/). While specifically describing methods for the XFe systems, the methods are applicable for the XFp with specific plate size adaptations.

1.1 Mitochondrial Stress

The Agilent Seahorse XF Cell Mitochondria Stress Test measures key parameters of mitochondrial function by directly measuring the oxygen consumption rate (OCR) of cells using Seahorse XF analyzers. This assay is a plate-based live-cell assay that monitors OCR in real time. As a first measure, basal OCR levels are established. Built-in injection ports on XF sensor cartridges then allow for the addition of chemical modulators of respiration directly into wells for real-time measures during the assay. This sequence of responses examines key parameters of mitochondrial function. Within the mitochondrial stress assay, modulators include oligomycin, carbonyl cyanide-4 (trifluoromethoxy) phenylhydrazone (FCCP), rotenone, and antimycin (Fig. 1). Basal respiration is defined as oxygen consumption used to meet cellular ATP demand resulting from mitochondrial proton leak and represents the energetic demands of the cell under basal conditions. Basal measurements are followed by the addition of oligomycin to inhibit ATP synthase (complex V). It alters or decreases electron flow through the electron transport chain (ETC), resulting a reduction in mitochondrial respiration or OCR. The decrease in oxygen consumption rate upon injection of the ATP synthase inhibitor, oligomycin, represents the portion of basal respiration that was being used to drive ATP production. Remaining basal respiration not coupled to ATP production can represent proton leak and a sign of mitochondria damage. It can also be a mechanism used by the cell to regulate mitochondrial ATP production. This is followed by carbonyl cyanide-4 (trifluoromethoxy) phenylhydrazone (FCCP), an uncoupling agent that collapses the proton gradient and disrupts the mitochondrial membrane potential. As a result, electron flow through the ETC is uninhibited and oxygen consumption by complex IV reaches maximum capacity and allows for the determination of the maximal oxygen consumption rate. A physiological "energy demand" is mimicked by stimulating the respiratory chain to operate at maximum capacity. This causes rapid oxidation of various substrates such as sugars, fats, or amino acids, in order to meet the metabolic challenge. FCCP-stimulated OCR can be used to calculate spare respiratory capacity, defined as the difference between

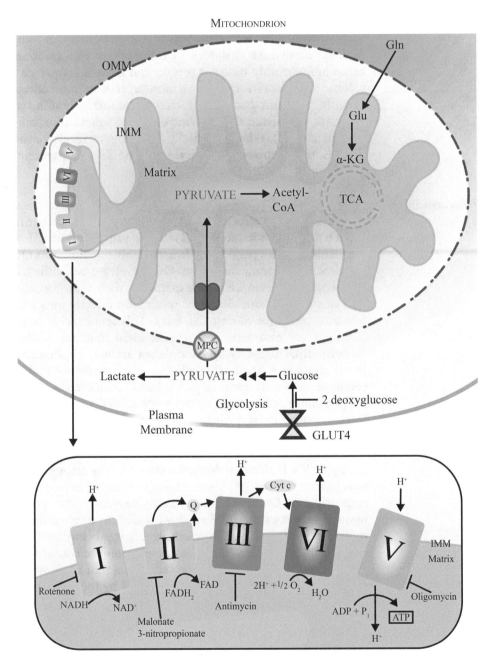

Fig. 1 Schematic diagram of a mitochondrion illustrating metabolism associated-cellular components. Abbreviations: *MPC* mitochondrial pyruvate carrier, *OMM* outer mitochondrial membrane, *IMM* inter-mitochondiral membrane, *Acetyl-CoA* acetyl coenzyme A, *Gln* glutamine, *Glu* glucose, *GLUT4* a glucose transporter, α-*KG* alpha ketoglutarate dehydrogenase, *TCA* tricarboxylic acid cycle. Representation of Complex I, II, III, IV, and V in the electron transport chain. The electron flow from nicotinamide adenine dinucleotide (NADH) binding to complex I or from succinate/flavin adenine dinucleotide (FAD) hydroquinone (FADH$_2$) binding to Complex II initiate electron flow to produce adenosine triphosphate (ATP) in Complex V. Rotenone is a complex I inhibitor; malonate or 3-nitropropionate are complex II inhibitors; antimycin is a complex III inhibitor; oligomycin is a complex V inhibitor. Abbreviations: *NAD+* nicotanamide adenine dinucleotide, *ADP* adenosine diphosphate, *H+* hydrogen, *O$_2$* oxygen, *Pi* phosphate, *Cyt c* cytochrome C

maximal respiration and basal respiration. Spare respiratory capacity is a measure of the ability of the cell to respond to increased energy demand or under stress and therefore, be an indicator of cell fitness or flexibility. Decreased spare respiratory capacity has been linked to aging effects on immune-cell function and plasticity [11, 12] as well as neurodegeneration [13]. The final injection is of rotenone, a complex I inhibitor, or a combination of rotenone and antimycin A, a complex III inhibitor. This shuts down mitochondrial respiration and the oxygen consumption that persists is due to cellular enzymes that continue to consume oxygen, such as those found in the peroxisomes [14], represents nonmitochondrial respiration.

1.2 Glycolytic Function

Most cells possess the ability to shift between two energy-producing pathways, glycolysis and OXPHOS. This ability allows cells to adapt to their environment. With glycolysis, glucose is converted to pyruvate followed by conversion to lactate in the cytoplasm or carbon dioxide and water in the mitochondria. This process results in the production and release of protons into the extracellular space (Fig. 1). The Seahorse XF Glycolytic Rate Assay is a method to assess key parameters of glycolytic function of cells. Acidification of the medium occurs as the result of extrusion of protons from the cells. The Seahorse XF instrument directly measures the acidification rate and reports this as ECAR. The calculated rate, Proton Efflux Rate (PER), is a measure of extracellular acidification that is considered to be more accurate than the traditional extracellular acidification rate (ECAR) measurements. PER considers the contribution of carbon dioxide to extracellular acidification due to mitochondrial/TCA cycle activity, it also accounts for media buffer factor and well volume. The glycolytic PER rate has been shown to be directly comparable to lactate accumulation [15]. Real-time measurements for changes in glycolysis rates requires a sequence of basal measurements followed by an injection of rotenone and/or antimycin A to inhibit oxidative phosphorylation. Inhibition of the OXPHOS pathway results in an adaptive shift of the energy profile in the cell to a predominantly glycolytic metabolism. This shows as an increased PER and is considered to represent compensatory glycolysis with mitochondrial inhibition as a measurement of the reserve glycolytic capacity to meet energy demands. The injection of 2-deoxy-glucose (2-DG) inhibits glycolysis through competitive binding to hexokinase and shows as a decrease in PER. This represents acidification that was produced by glycolysis and the remaining PER representing glycolysis-independent acidification. The reader is referred to the *Agilent Seahorse XF Protocol User Guides* for further details on assay specifics.

2 Materials

2.1 Agilent Seahorse Equipment and Materials

1. Agilent Seahorse XFe96 or 24 Analyzer Agilent Technologies.
2. Agilent Seahorse XF Base Medium™ Agilent Technologies.
3. Agilent Seahorse XFe96 or 24 FluxPaks™.
4. Agilent Seahorse XFe96 or 24 Cell Culture Microplates™.
5. Agilent Seahorse XF Calibrant ™ (500 mL).
6. Agilent Seahorse XF Glycolytic Rate Assay Kit™.

2.2 Equipment

1. Ambient air 37 °C incubator.
2. Water bath at 37 °C.
3. Calibrated pH meter.
4. 37 °C tissue culture incubator preferably a tri-gas incubator to closely/rapidly regulate oxygen levels.
5. Multichannel 200 µL pipet.
6. Vortex.

2.3 Supplies

1. 0.2 µm Sterile Filter (syringe).
2. Narrow pipet tips (e.g., Fisherbrand™ SureOne™ Micropoint Pipet Tips).
3. 50 mL conical tubes.
4. Cell culture grade sterile water.

2.4 Reagents

1. 1 M Sodium hydroxide (NaOH).
2. DMEM cell culture medium: (high glucose, no sodium pyruvate; *see* Subheading 4.1).
3. Streptomycin.
4. Penicillin.
5. Fetal bovine serum (<0.25 endotoxin units/mL).
6. Dimethyl sulfoxide (DMSO).
7. Glucose Stock concentration 1.0 M Glucose solution.
8. L-Glutamine solution stock concentration 200 mM (*See* Subheading 4.2).
9. Oligomycin, make 2.5 mM stocks in DMSO, store at −20 °C.
10. Antimycin A, make 2.5 mM stocks in DMSO, store at −20 °C.
11. Carbonyl cyanide 4-(trifluoromethoxy)phenylhydrazone [FCCP] make 2.5 mM stocks in DMSO, store at −20 °C.
12. 2-deoxy-D-glucose (2-DG) stock concentration 100 µm.
13. Sodium pyruvate 100 mM solution, final concentration 1 mM.

14. Rotenone-make 2.5 mM stocks in DMSO, store at −20 °C 15. Lipopolysaccharide (LPS)—Stocks of 1 endotoxin unit (EU)/ng E. coli 055:B5 sonicated in nuclease-free sterile water, stored at −20 °C. Do not reconstitute in media.

15. Stock compounds can be stored at −20 °C in aliquots for use, unless stated otherwise. It is recommended that a new aliquot be used for each assay.

3   Methods

3.1   Cell Density Optimization

1. Seeding Adherent Cells:

   - Visit the Cell Reference Database (https://www.agilent.com/cell-reference-database/) to search for any recent publications using microglia cells.

   - Optimal cell number is usually 5000–40,000 cells/well. While a minimum of 50% confluency is needed to generate enough metabolic rate for the instrument to read properly, it is best to aim for 80–90% confluency.

   - The cell surface area of the 96-well microplate is 0.106 cm^2/well. If you do not find your cell type in the reference database, you can determine a starting point based on the cells/cm^2 usually plated for high confluency. Do not allow the cells to overcrowd.

   - Microglia cell lines and primary microglia are maintained in DMEM (high glucose, no sodium pyruvate) supplemented with 10% fetal bovine serum (FBS; <0.25 EU/mL), 100 U/mL penicillin and streptomycin. The low endotoxin level of the FBS helps to prevent the activation state artifact that can occur in cultured microglia.

   - Microglia cell lines—expand in T75 tissue culture flasks to confluency. Gently trypsinize and resuspend cells in normal growth medium to final concentration to seed at 7.2×10^4 cells/cm^2 to achieve 70% confluency within 24 h.

   - Primary rodent microglia—obtain microglia from either postnatal day 1–3 rodent pups by tissue dissociation and filtration or from mature brain tissue using density gradient separation, magnetic-activated cell sorting, or flow cytometry. A number of detailed methods are available in the literature [16–22]. Seed cells at a density of approximately 8×10^4 cells/cm^2.

   - Cells are allowed to adhere to the wells for 24 h prior to running assay.

Table 1
Mitochondrial stress assay. Recommended drug concentrations for optimization

	10× stock concentration (μM)	Stock volume added (μL)	Final well concentration (μM)
Port A— Oligomycin	5	20	0.5
	15	20	1.0
	25	20	2.0
Port B—FCCP	1.25	22	0.125
	2.5	22	0.25
	5	22	0.50
	10	22	1.0
	20	22	2.0
Port C— Rotenone[a]	10	25	2.0

[a]Rotenone dose is sufficiently high and does not normally require optimization

Table 2
Glycolytic rate assay. Recommended drug concentration range

	10× stock concentration	Stock volume added (μL)	Final well concentration (μM)
Port A—Rotenone	10 μM	20	1.0 μM
Rotenone/Antimycin A	5 μM	20	0.5 μM
Port B—2-DG	500 mM	22	50 mM

These compounds do not interfere the function of each other. They can be administered at large dose levels to produce equivalent results across cell types

3.2 Optimizing Drug Concentrations

Optimization of drug concentrations is critical for accurate determination of OCR. For the Mitochondrial Stress Assay, optimization of FCCP concentration is critical and optimization of oligomycin is strongly encouraged for individual cell types. The rotenone dose is sufficiently high to not require optimization. Dose recommendations are provided in Tables 1 and 2.

1. BV-2 microglia cell lines.
   - Glycolytic Rate Assay—sensor cartridges are hydrated and loaded to deliver a final concentration of 50 mM 2-deoxyglucose and 0.5 μM rotenone.
   - Mitochondrial Stress Assay—sensor cartridges are hydrated and loaded to deliver a final concentration of 0.9 μM oligomycin, 0.75 μM carbonyl cyanide-4-(trifluoromethoxy)-phenylhydrazone (FCCP), and 1 μM rotenone (option of using combination with antimycin).

2. Primary rodent microglia.

- Glycolytic Rate Assay—sensor cartridges are hydrated and loaded to deliver a final concentration of 50 mM 2-deoxyglucose and 0.5 µM rotenone.

- Mitochondrial Stress Assay—sensor cartridges are hydrated and loaded to deliver a final concentration of 0.75 µM oligomycin, 0.75 µM FCCP, and 1 µM rotenone (option of using combination with antimycin).

3.3 Assay Optimization Design

Optimization of the cell system can be accomplished using 1–2 plates.

1. The Mitochondria Stress Test requires the most optimization.

   (a) Plate 1: Cell number and oligomycin optimization.

   - As a first step, search the Seahorse Cell Reference Database for new information on microglia.

   - BV-2 cells are plated at concentrations bracketing 7.2×10^4 cells/cm^2 with a 1000-cell interval.

   - Oligomycin is loaded into Port A at various doses (0.5, 1.0, 2.0 µM) even though 1.0 µM has been shown to work for most 2D cell cultures.

   (b) Plate 2: FCCP optimization.

   - Conduct after optimizing for plating density and oligomycin concentrations.

   - Add FCCP at varying concentrations into Port B (0.0, 0.125, 0.25, 0.5, 1.0, 2.0 µM).

   - Add rotenone into Port C to collect control data for optimized conditions.

2. Criteria for an Optimized Mitochondrial Stress Assay.

   - Cell density of 80–90% confluency at time of assay.

   - Oligomycin injection is at a concentration producing the greatest drop in OCR.

   - FCCP dose results in the highest peak response.

   - Rotenone dose lowers OCR to slightly below the Oligomycin measurement (Fig. 3).

3. The Glycolytic Rate Assay.

   - Cell number optimization (as described for the Mitochondria Stress Test) is required.

   - Drug levels high enough to ensure equal inhibition across all cell types.

4   Criteria for an Optimized Glycolytic Rate Assay

- Cell density of 80–90% confluency at time of assay.
- Rotenone injection at a concentration to sufficiently lower OCR and increase PER.
- 2-DG injection at a concentration that sufficiently lowers PER to below baseline, reaching plateau after 3 or 4 measurement cycles (Fig. 2).

4.1   Loading of Plates

1. Mark out the four corner wells of the plate (e.g., 96 well: A1, A12, H1, H12) as background correction wells that will contain only medium and no cells. The software has this configuration as default plate map for background and will use as such in the data analysis; however, if you are concerned with the difference between external versus internal wells, you can modify the coordinates of the background wells or add/remove background wells under "Assign groups manually."

2. Using sterile cell culture technique, load cells into a room temperature microplate by holding the pipette tip at an angle approximately 50% down the side of each inner well.

3. Leave plate under tissue culture hood at room temperature for 1 h. This has been found to minimize edge effect by allowing cells to settle and distribute evenly.

4. Check distribution and attachment of cells under a microscope.

5. Place in incubator (preferably a tri-gas incubator to maintain oxygen level).

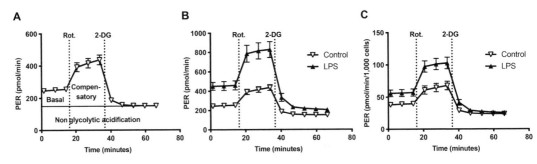

Fig. 2 (**a**) Representative profile of a Glycolytic Rate Assay. (**b**, **c**) Calculation of the proton efflux rate (PER) from the Glycolytic Rate Assay is significantly affected by cell plating density. Representative differences in data (**b**) without and (**c**) with optimizing to cell number under control and LPS conditions. *Rot* rotenone, *2-DG* 2 deoxyglucose

4.2 Hydrate the Sensor Cartridge (Day Before Assay)

1. Under the tissue culture hood, open the sensor cartridge pack. This pack will contain two well templates, a green upper insert (sensor cartridge) and a clear lower plate identical to the microplate (utility plate).

2. Place the sensor cartridge upside down next to the utility plate.

3. Using a multichannel pipette, add 200 µL of sterile water to each well of the utility plate and replace the sensor cartridge onto the utility plate, submerging the sensors in the water. Verify the water level is high enough to keep the sensors submerged.

4. Ensure there are no bubbles under the sensors.

5. Place assembled sensor cartridge and utility plate overnight in humidified ambient air 37 °C incubator.

6. Fill a 50 mL Falcon tube with approximately 30 mL Seahorse Calibrant solution and place into a 37 °C ambient air incubator overnight.

7. Hydrated cartridges can be covered with parafilm and maintained in a 37 °C ambient air incubator for up to 72 h.

4.3 Load Sensor Cartridge with Calibrant Solution (Day of Assay)

1. Remove assembled sensor cartridge with utility plate from the incubator.

2. Warm calibrant solution in 37 °C oven or water bath.

3. Remove the sensor cartridge and place upside down next to the utility plate.

4. Replace the sterile water with 200 µL of pre-warmed calibrant.

5. Place the sensor cartridge onto the utility plate ensuring that sensors are submerged in calibrant and that the cartridge and plate are in correct orientation.

6. Place reassembled sensor cartridge with utility plate in a humidified ambient air 37 °C incubator for approximately 45–60 min prior to running the assay.

4.4 Making Assay Media (Day of Assay)

1. Growth media is replaced with assay medium without bicarbonate buffer and low/no phenol red.

2. Supplement the Seahorse base medium (Agilent™):
   - Mitochondrial Stress Test: Low phenol red base medium, 4.5 g/L glucose (10 mM), 1 mM sodium pyruvate, and 2 mM L-glutamine (*see* Subheading 4.2).
   - Glycolytic Rate Assay: No phenol red base medium, 4.5 g/L glucose, 1 mM sodium pyruvate, 2 mM L-glutamine (*see* Subheading 4.2), and 5 mM HEPES buffer.

3. Adjust pH of the media to 7.4 using 1 M NaOH and sterile filter.

316 Gabrielle Childers and G. Jean Harry

4. The open cell microplate will be exposed to air several times over the course of the assay so full tissue culture sterility of the media is not a requirement but should be considered.

5. Place in a 37 °C ambient air humidified incubator or water/bead bath until use. Maximum 4 h.

4.5 Instrument Setup

1. Turn on the instrument by switching the black switch on the right side of the instrument.

2. Turn on the instrument computer by pressing the button below the screen.

3. Open Wave software.

4. Select the default template of the assay you will be running or create a custom template.

5. First tab will prompt you to fill in information on experiment conditions.
   • For the Glycolytic Rate Assay, you must enter the media conditions. If you do not identify that you are using HEPES-buffered media, the software will not calculate the proton efflux rate (PER).

6. The next tab will be the plate map. To complete this map, click on the group you want bulleted to the left then click and drag across the wells of the microplate diagram to the right. Add groups by clicking the "Add Group" button above the list.

7. The next tab will outline the instrument protocol. This protocol will outline measurements taken as a baseline prior to injection and then measurements taken following each port injection. The default protocol requires three baseline measurements and three measurements after each port injection. A minimum of three measurements is sufficient for microglia cells. These measurement cycles can be further broken down into a "mix," "wait," and "measure" phase. The mix phase is the time in which the sensor is removed from the culture plate and agitated to reintroduce oxygen into the media after forming the microchamber for measurement. The wait phase is the time in which the instrument delays in taking measurement. Finally, the measure phase is the time in which the sensors will measure within the well. Default measurements include a 3 min "mix," 0 min "wait," and 3 min "measure" for the XFe96 instrument and 3 min "mix," 2 min "wait," and 3 min "measure for the XFe24. All of these features can be customized; however, the default is sufficient for microglia. The time the assay will take will appear on the top right corner.

8. Wait until you have completed the sensor cartridge filling before hitting the "Run Assay" button.

4.6 Washing Cell Microplate

1. Confirm status of plated cells and confluency under microscope.

2. Confirm that there are no cells in the background wells.

3. Using a multichannel pipette, remove all 200 μL of medium from each well.

4. Gently replenish cells with 180 μL of Assay Media.

5. Confirm status of cells under microscope to ensure they were not disturbed or washed away.

6. Place plate in ambient air 37 °C humidified oven for 45–60 min to de-gas the plate and calibrate plate to instrument conditions.

7. Reconfirm health of cells by microscopy before running assay.

4.7 Fill Sensor Cartridge

1. While the microplate is de-gassing, fill sensor cartridge.

2. Remove sensor cartridge/utility plate unit from the oven and place under the tissue culture hood.

3. Using remaining assay media, dilute drug inhibitors to 10× their optimized concentrations (*see* optimization section for details).

4. pH drug inhibitors at 7.35–7.4 at 37 °C and sterile filter prior to loading.

5. Orient the assay cartridge with row labels (letters A-H) to the left. This will result in the notch of the plate in the bottom left-hand corner.

6. The four drug ports per well used for sequential injections of the drugs are labeled A–D. Ensure loading of ports in correct order. If not, you need to modify order in software. Do Not Pipette In the Center Hole as that will break the optic sensors.

7. When filling ports, the sensor cartridge must remain in the utility plate and placed flat on work surface. Minimize movement and do not lift or tilt the plate/cartridge.

8. Place the well template (A/D loading guide) flat on top of the assay cartridge. Hold in place while loading. Remove the template before filling.

9. It is recommended that a constant concentration of solution be added however, a constant volume can be added if concentrations are adjusted for each compound (Tables 1 and 2).

10. It is critical that each series of ports contains an equal volume.

11. Gently fill the ports starting with A guides progressing to D. Insert a 10–100 μL multichannel pipette ¾ of the way down the port at a slight angle and ejecting down the side of the port. Do not force the tips completely into the ports. Switch the guides and repeat for each of the ports.

12. Avoid creating any air bubbles.

13. Ensure that the drug solution is contained at the bottom of the well. If the drug solution is on the side of the well, it will not eject. **DO NOT TAP OR SHAKE THE PLATE**. It is very easy to prematurely eject the drugs. Try to fill the cartridge as close to the seahorse machine as possible to minimize any disturbance when transferring to machine. When moving the cartridge, hold the base of the utility plate.

14. Fill all ports for every well of the plate regardless of the presence of cells. This ensures accurate delivery of drug.

15. Very gently, visually confirm equal loading for ports. Make sure all liquid is in the port with no residual on top of the cartridge.

4.8 Mitochondria Stress Test

1. For the standard assay, load the ports in sequence of oligomycin, FCCP, then Rotenone or Rotenone/Antimycin. For a standard assay the following volumes per port for constant concentration: $10\times$ drug solutions: A: 20 μL, B: 22 μL, C: 25 μL (Table 1).

2. The additional port allows for a modification of the assay to examine a response of basal respiration to exposure to an experimental compound. In this model, Port A is loaded with an equal volume of a test compound concentration followed by the standard assay sequence of inhibitors in the remaining three ports.

4.9 Glycolytic Rate Assay

For the standard assay, load Port A with Rotenone/Antimycin A, Port B with 2 DG (Table 2).

For a modified assay, load Port A with optional compound, Port B with Rotenone/Antimycin A, and Port C with 2 DG.

4.10 Loading the Instrument and Running Assay

1. After reviewing the group definitions, plate map orientation, and protocol press the "Start Run" button in Wave.

2. Confirm the location for saving your results file.

3. The software will prompt for insertion of the sensor cartridge and utility plate unit onto the tray.

4. Remove all loading guides and plate lids before inserting the cartridge into the analyzer.

5. Very carefully transfer the plate to the instrument.

6. Make sure sensor cartridge fits properly on the utility plate.

7. Align the plate with A1 to the left as you are facing the instrument.

8. Make sure there are no lids or templates on top of the sensor cartridge before hitting "I'm Ready" button to load the plate into the instrument. Failure to do so will **DAMAGE THE INSTRUMENT**.

9. The instrument will take the sensor cartridge through a calibration cycle, which will take about 15 min for assays at 37 °C.

10. When calibration finishes, hit the "Open Tray" button to eject the utility plate from the instrument. A message will then prompt you to insert the Cell Plate. The sensor cartridge remains inside the analyzer for this step. Replace the utility plate with the cell microplate.

11. **REMOVE THE LID OF THE MICROPLATE** before hitting "Load Cell Plate". Failure to remove the lid can damage the instrument.

12. Confirm orientation of the Cell Plate on the tray.

13. Hit the Load Cell Plate button to initiate equilibration. Once completed the assay will automatically begin acquiring baseline measurements per instrument protocol.

14. Upon completion of final measurement, the Unload Sensor Cartridge dialog will display.

15. Hit Eject and set aside cell plate for later analysis.

16. The Assay Complete dialog will appear. Hit View Results to open assay results file.

17. Or, hit Wave Home to return to the templates view and start another assay.

18. Save or transfer results for analysis using Wave Desktop software (PC).

19. See Agilent Wave Users Guide for additional instructions https://www.agilent.com/cs/library/usermanuals/public/ S789410000_Rev_C_Wave_2_6_User_Guide.pdf

4.11 Normalization for Cell Number

One of the more critical normalization features of the Seahorse Assays is the need to verify cell number within each well at the initial plating and at time of the assay (Fig. 2). There are several ways to normalize for cell number (nuclear cell staining, mitochondrial DNA, or protein content) and may depend on the experimental manipulation. One recommended method is to obtain an estimate of cells using a nuclear cell stain such as Hoechst 33342. This stain can be read using a plate reader (*see* Agilent Seahorse Protocol) to provide a general read of the amount of stain within the well. The following method outlines a combination of Hoeschst staining and microscopic cell imaging to provide a determination of cell number.

1. After the Seahorse run is complete, save the cell microplate and dispose of the sensor cartridge.

2. Add Hoechst 33342 dye to a concentration of 200 μg/mL in each well and incubate for 5 min.

3. Remove media and replace with 200 μL of PBS.

4. Read the intensity of dye staining using a plate reader.

5. Using a fluorescent microscope with a blue/cyan filter and a 20× objective, capture four defined equal sized regions of interest for representation of the well.
    - In the Seahorse plate, the well bottom includes three bubble-like structures that can impede cell counting.

6. Use image analysis software (i.e., FIJI https://fiji.sc) to determine cell count.

7. Calculate normalized data (PER or OCR) as pmol/min/1000 cells.

    While staining of cells is a recommended method of normalization, it is not appropriate for all experimental approaches. If the experimental approach may alter nuclear number, an alternative method of normalization, such as measurement of total protein content per well using a biochemical protein assay can be used. Data would then be calculated as pmol/min/ug total protein.

4.12   Data

Array result files are generated by the Analyzers after completion of assay. The default analysis view is "Quick View".

1. The 4 analysis views in Wave software:
    - Quick View—displays a kinetic graph of OCR vs Time, ECAR vs Time, and a scatter plot of OCR vs ECAR.
    - Overview—display a kinetic graph or rate (OCR, ECAR, PER, or PPR) vs time.
    - OCR vs ECAR.
    - Data.

2. Normalized data can be generated once that data has been manually entered to the result file.

3. Agilent provides Excel Macros sheets that can assist in calculations.

4. For Glycolytic Rate Assay.
    - For each individual well, calculate the mean for all three basal measurements (mean basal), mean response after rotenone, and mean of last three of the five measurements recorded following 2-DG injection (mean rotenone, mean 2-DG).
    - Basal Glycolysis: difference between the mean basal and mean 2-DG for each well.
    - Compensatory Glycolysis: difference between the mean rotenone and mean basal for each well.

5. For Mitochondrial Stress Test.

- Calculate mean for all three basal measurements (mean basal) as well as measurements after each drug injection for each individual well (mean oligomycin, mean FCCP, and mean rotenone).

- Basal Respiration: the difference of the mean basal and mean rotenone for each well.

- ATP-linked Respiration: the difference of the mean basal and mean oligomycin for each well.

- Maximum Respiration: the difference of the mean FCCP and mean oligomycin for each well.

- Reserve Capacity: the difference between the mean FCCP and mean basal for each well.

6. Show representative data as the line graph of mean responses for each time point.

7. Show OCR and PER data as a bar graph of the mean response including a scatter graph of the individual samples.

8. Generate line plot of the normalized data including mean and measure of variance.

5 Notes

5.1 Culture Medium

For cell culture medium, one has the option of obtaining DMEM base and supplementing with glucose and glutamine/Glutamax. The media does not contain sodium pyruvate or carbonate. DMEM/F12 medium has also been reported in the literature as a medium for microglia cultures.

5.2 Glutamine Source

Do not use Glutamax as a glutamine source in the assay media. While Glutamax has greater stability, the substrate is not as readily available to the cells, significantly decreasing maximum respiration (Fig. 3).

5.3 Experimental Design

1. Experimental biological samples are run in triplicate.

2. Recommend two experimental replications.

3. While triplicates per biological sample is recommended an alternative is to run duplicates and then increase the number of biological samples of each experimental condition.

4. Calculate results based upon individual plates due to assay-to-assay variability.

5. The design of any experiment to assess effects of chemicals or drugs depends on the question.

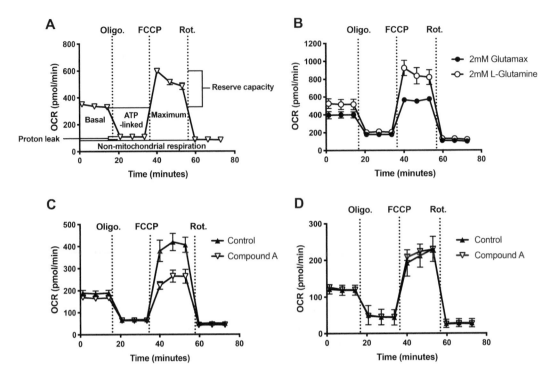

Fig. 3 (**a**) Representative profile of a Seahorse Mitochondria Stress Assay. (**b**) Representative. Images of the differences in response when using glutamax versus L-glutamine. Glutamax in assay media severely affects oxygen consumption rate (OCR) readings, especially the peak response. The lack of readily available glutamine can result in diminished peak response and actually result in a "failed" run. (**c**, **d**) Example of the effect of optimization of detecting subtle changes in OCR. (**c**) OCR profile under optimized conditions. (**d**) OCR profile under non-optimized conditions

- Cells can be exposed for a specific time period prior to running the assay.
- Effects of exposure on basal respiration can be measured by direct port delivery of the chemical/drug of interest.
- Exposure levels of a chemical/drug of interest should always be under those that produce cell death or stimulate cell proliferation. Remaining within a 20% range of control is recommended.

5.4 Troubleshooting

1. The drugs didn't deploy from the sensor cartridge ports.
   - When filling drug ports ensure that all of the solution is at the bottom of the port. Any drug solutions along the edges will not deploy.
   - Confirm that each port of the sensor cartridge is filled with equal amounts of injectable (i.e., If you are filling port A, B, and C, all three ports need to be filled for every well of the plate). The drugs will not deploy if not entered into ports for all wells. Seahorse media can be substituted for drugs to cover all wells.

2. The Mitochondria Stress Test curve peak is low.

- Reconfirm optimization of FCCP, oligomycin, and cell density.

- Ensure cell health and confluency prior to assay. Cell health and density is crucial for robust and accurate OCR measurements.

- Reconfirm the use of L-glutamine in your assay media and not Glutamax.

3. Large well-to-well variability within experimental groups.

- Drugs may have prematurely ejected during loading.

- Ensure even seeding of cell microplate and normalize data to cell number.

References

1. Mills EL, Kelly B, O'Neill LA (2017) Mitochondria are the powerhouses of immunity. Nat Immunol 18:488–498

2. Galvan-Pena S, O'Neill LA (2014) Metabolic reprograming in macrophage polarization. Front Immunol 5:420

3. Rodriguez-Prados JC et al (2010) Substrate fate in activated macrophages: a comparison between innate, classic, and alternative activation. J Immunol 185:605–614

4. Ghosh S et al (2018) Bioenergetic regulation of microglia. Glia 66:1200–1212

5. Gimeno-Bayon J et al (2014) Glucose pathways adaptation supports acquisition of activated microglia phenotype. J Neurosci Res 92:723–731

6. Orihuela R, McPherson CA, Harry GJ (2016) Microglial M1/M2 polarization and metabolic states. Br J Pharmacol 173:649–665

7. Rubio-Araiz A, Finucane OM, Keogh S, Lynch MA (2018) Anti-TLR2 antibody triggers oxidative phosphorylation in microglia and increases phagocytosis of beta-amyloid. J Neuroinflammation 15:247

8. Wang L et al (2019) Glucose transporter 1 critically controls microglial activation through facilitating glycolysis. Mol Neurodegener 14:2

9. Mills EL, O'Neill LA (2016) Reprogramming mitochondrial metabolism in macrophages as an anti-inflammatory signal. Eur J Immunol 46:13–21

10. Holland R et al (2018) Inflammatory microglia are glycolytic and iron retentive and typify the microglia in APP/PS1 mice. Brain Behav Immun 68:183–196

11. Chougnet CA et al (2015) Loss of phagocytic and antigen cross-presenting capacity in aging dendritic cells is associated with mitochondrial dysfunction. J Immunol 195:2624–2632

12. Pence BD, Yarbro JR (2018) Aging impairs mitochondrial respiratory capacity in classical monocytes. Exp Gerontol 108:112–117

13. Choi SW, Gerencser AA, Nicholls DG (2009) Bioenergetic analysis of isolated cerebrocortical nerve terminals on a microgram scale: spare respiratory capacity and stochastic mitochondrial failure. J Neurochem 109:1179–1191

14. Wanders RJA, Waterham HR (2006) Biochemistry of mammalian peroxisomes revisited. Annu Rev Biochem 75:295–332

15. Romero N, Swain P, Neilson A, Dranka BP. White paper: Improving Quantification of Cellular Glycolytic Rate Using Agilent Seahorse XF Technology. http://seahorseinfo. agilent.com/acton/fs/blocks/ showLandingPage/a/10967/p/p-00ca/t/ page/fm/1

16. Giulian D, Baker TJ (1986) Characterization of ameobid microglia isolated from developing mammalian brain. J Neurosci 6:2163–2178

17. Cardona AE et al (2006) Isolation of murine microglial cells for RNA analysis or flow cytometry. Nat Protoc 1:1947–1951. https://doi. org/10.1038/nprot.2006.327

18. Gordon R et al (2011) A simple magnetic separation method for high-yield isolation of pure-primary microglia. J Neurosci Methods 194:287–296

19. Tamashiro TT, Salgard CL, Byrnes KR (2012) Primary microglia isolation from mixed glial

cell cultures of neonatal rat brain tissue. J Vis Exp 66:e3814. https://doi.org/10.3791/3814

20. Lee J-K, Tansey MG (2013) Microglia isolation from adult mouse brain. Method Mol Biol 1041:17–23

21. Bohlen CJ, Bennett FC, Bennett ML (2018) Isolation and culture of microglia. Curr Protoc Immunol 125:e70. https://doi.org/10.1002/cpim.70

22. Stark JC et al (2018) Characterization and isolation of mouse primary microglia by density gradient centrifugation. J Vi s Exp 132: e57065. https://doi.org/10.3791/57065

Part V

Alternative Model Organism-Based Methods

Chapter 15

Assessment of Larval Zebrafish Locomotor Activity for Developmental Neurotoxicity Screening

Bridgett N. Hill, Kayla D. Coldsnow, Deborah L. Hunter, Joan M. Hedge, David Korest, Kimberly A. Jarema, and Stephanie Padilla

Abstract

Animal behavior has long been recognized as an informative endpoint for assessing effects of chemicals on the developing nervous system. Previous laboratory animal tests to screen and prioritize chemicals for developmental neurotoxicity have not met growing risk assessment demands, necessitating higher through-put testing strategies. Zebrafish, a small freshwater vertebrate species, have been proposed as one alternative vertebrate model. Here, we describe methodology for rapid screening of chemical libraries for developmental neurotoxicity potential utilizing larval zebrafish locomotor activity as an endpoint. The goal of this chapter is to provide guidance for applying this approach, as well as to provide a discussion of the advantages and limitations of various aspects of this method.

Key words *Danio rerio*, Zebrafish, Locomotor assay, Behavior, Developmental neurotoxicology, Alternative testing approach

1 Introduction

Nervous system development is a highly complex process requiring cellular proliferation, differentiation, migration, synaptogenesis, apoptosis, and myelination for the establishment of multifaceted cellular networks. These complex nervous system networks are crucial for many bodily functions and behaviors. Because of the importance of the nervous system for survival, irregularities in the nervous system are a large area of concern and research. Evidence suggests that early life stages can be more susceptible to damage from chemical exposure, with consequential effects that may persist into adulthood [1]. Of particular concern is developmental neurotoxicity, defined as alterations to the development of the nervous system caused by exposure to toxic substances. Reports of the increasing incidence of environmental chemicals detected in human umbilical cord blood and placental samples [2], increased diagnosis of neurodevelopmental disorders (e.g., autism spectrum,

Jordi Llorens and Marta Barenys (eds.), *Experimental Neurotoxicology Methods*, Neuromethods, vol. 172, https://doi.org/10.1007/978-1-0716-1637-6_15, © Springer Science+Business Media, LLC, part of Springer Nature 2021

attention deficit hyperactivity disorders) [3–5], and emergence of neurodegenerative diseases both in elderly and middle-aged populations [6, 7] all contribute to the need to understand causes and mechanisms underlying developmental neurotoxicity.

There are tens of thousands of chemicals currently in use in the United States, many of which lack the detailed toxicological information required for risk assessment. In fact, only a very small proportion of chemicals in use today has been assessed for their potential to cause developmental neurotoxicity [8]. Both the U.S. Environmental Protection Agency (EPA) and the Organization for Economic Co-operation and Development (OECD) have developmental neurotoxicity testing guidelines using rodents [9–11]. These guidelines assess a variety of motor and sensory tests, cognitive function, and neurohistopathology. Unfortunately, these types of rodent studies are often time and resource intensive, which reduces the feasibility of using the rodent model for screening thousands of chemicals. Furthermore, the U.S. EPA and the National Research Council in their visions for new toxicity testing [12–14] have increased the call for decreased use of mammals in toxicity testing, adaptation of alternative models, and development and application of new approach methods [15].

Zebrafish (*Danio rerio*), a small freshwater fish native to South Asia, have been utilized successfully in a wide variety of pharmacological and toxicological applications [16–19]. Zebrafish are also an intermediate model bridging the gap between traditional developmental neurotoxicity mammalian assays described above and high throughput in vitro cellular assays, which can lack important biological processes such as metabolic activation, and attendant hormonal and growth factor input necessary for brain development [20, 21]. Compared to other model organisms, zebrafish are easily maintained, have high fecundity, develop rapidly, and are transparent during early life stages, allowing for the observation of development using basic microscopy. Due to their smaller size, smaller quantities of test chemical are required, and experiments can be conducted in microtiter plates, analogous to in vitro cell assays, making assays amenable to the use of robotics, where applicable. Major brain subdivisions form quickly [by 24 h post fertilization (hpf)] and neurotransmitter-expressing neurons can be identified as early as 2–3 days post fertilization (dpf) [22]. All of this allows for the collection of multiple endpoints within a short time (embryonic and larval assays are usually 5–6 days in length). Lastly, zebrafish brain organization is comparable to mammals, with neurodevelopmental similarities [22–25]. Therefore, zebrafish larvae are a scientifically sound alternative model for developmental neurotoxicity research [20, 25] and for neuroscience research in general [26–29].

One method for measuring developmental neurotoxicity is through behavior. Behavior is a response to internal and external environmental cues and is controlled by both the central and peripheral nervous systems. Therefore, impairments in nervous system development will alter behavior, and ultimately influence an organism's health and survival. Traditional measures of behavior include motor, sensory, and cognitive endpoints. These endpoints are integrated across several levels of biological organization and although they indicate that a disruption has occurred, they are not usually specific enough to pinpoint the mechanisms underlying these responses. For example, changes in behavior could be caused by neurotransmitter inhibition or functional alterations to neurons. Despite this, behavior remains an optimal, first tier test for assessing the outcomes of an environmental chemical exposure.

In zebrafish, the effects of neurotoxicants on behavior can be assessed very early in development. Zebrafish exhibit spontaneous movements as early as 17 hpf. More complex behaviors begin at 4 dpf, which is comparable, on a functional level, to human behavior as shown by multiple neuroactive drug screens utilizing larval zebrafish behavior [30–33]. When exposed to neurotoxicants, behavior alterations usually occur at chemical concentrations below those inducing mortality or malformations. Some examples of multifaceted behavioral assays using the zebrafish model include: light/dark locomotor activity, circadian rhythm mapping, touch elicited response, thigmotaxis, prey-capture, and rest/wake [34]. These assays often assess alterations by determining the response to a stimulus. In terms of testing guidelines used to examine developmental neurotoxicity in zebrafish, there are currently no formal regulatory protocols. This has resulted in cross-laboratory variability in assay selection and experimental design within the research community.

Despite differences in research protocols, neurotoxicant-induced responses in zebrafish are in moderate to high concordance with mammalian data. For example, ethanol is known to cause mammalian developmental neurotoxicity resulting in cognitive and behavioral impairments. Zebrafish embryos exposed to ethanol elicited concentration-dependent behavioral changes [35], impaired learning performance in adults [36], and other neurotoxic effects including apoptotic cell death [37] and decreased retinotectal projection [38]. Similarly, lead (Pb) exposures in humans result in childhood cognitive decline, reduced IQ scores, and learning disabilities [39], while in zebrafish, Pb exposure during development causes altered behavior in zebrafish embryos and larvae and impaired learning and memory in adults [40].

Development of new approach methods for developmental neurotoxicity is an active area of research and thus various ways to determine a chemical's potential for developmental neurotoxicity have been employed using a zebrafish model. Here, we present our methods for screening the developmental neurotoxicity potential of chemicals by utilizing an early life stage zebrafish locomotor assay. Our procedures and experiment flow are summarized in the schematic presented in Fig. 1.

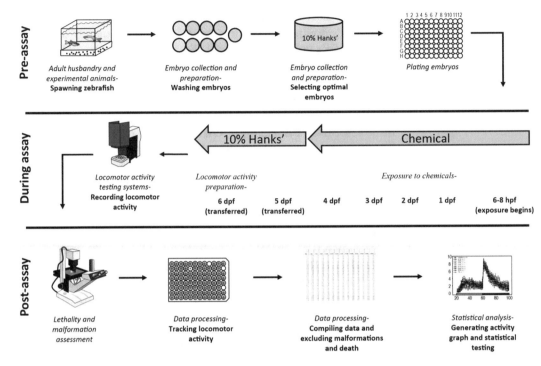

Fig. 1 Overview of larval zebrafish locomotor activity method for developmental neurotoxicity screening. The method is categorized into three sections: Pre-assay (*see* Subheadings 2.1–2.3) focusing on embryo collection and preparation, during assay (*see* Subheadings 2.4–2.8) concentrating on chemical exposure and larvae locomotor activity light/dark assay, and post-assay (*see* Subheadings 2.9–2.11) centering on larvae assessment, data processing, and statistical analysis

2   Materials and Methods

2.1   Pre-Assay: Adult Husbandry and Experimental Animals

All of our research and breeding colony procedures are reviewed and approved by the Office of Research and Development's Health Laboratory Institutional Animal Care and Use Committee (IACUC) at the U.S. EPA in Research Triangle Park, NC. Our animal facility is also Association for Assessment and Accreditation of Laboratory Animal Care (AAALAC) International accredited. In our facility, we house zebrafish on recirculating zebrafish housing systems (Tecniplast USA, West Chester, PA or Aquaneering Inc., San Diego, CA) in 3.5 or 6 L polycarbonate tanks at a density of approximately 8 adult zebrafish/liter. The housing rooms have a 14:10 h light/dark photoperiod cycle with lights on at 08:00 h. Our system water is composed of Durham, NC city tap water that is purified via reverse osmosis and buffered with sea salt (Instant Ocean, Spectrum Brands, Blacksburg, VA) and sodium bicarbonate (Church & Dwight Co., Ewing, NJ). This water is maintained at 28 °C, pH 7.4, conductivity of 1000 µS/cm, with negligible ammonia and nitrate/nitrite present. The turnover of rack water (system water circulating through 48–50 tanks) ranges from 10 to

15% of the total rack volume per day. We assess the rack water quality daily for temperature, pH, and conductivity, and weekly (or more often if needed) for ammonia, nitrite, nitrate, dissolved oxygen, hardness, and alkalinity.

Our breeding colony consists of wild-type adult zebrafish (*Danio rerio*), which are descended from an undefined, outbred stock of wild-type AB strain background zebrafish created by combining stock originally obtained from two suppliers approximately 15 years ago (Aquatic Research Organisms, Hampton, NH and EkkWill Waterlife Resources, Ruskin, FL). Zebrafish are set up to spawn starting at 3–4 months of age and this continues until they are approximately 15 months old. The breeding stock is then replaced with new zebrafish reared from the colony. The outbred nature has been maintained yearly by mixing in at least one other strain of zebrafish obtained from the Zebrafish International Research Center (ZIRC) as fertilized embryos, grown up in the colony, and allowed to spawn with our existing zebrafish. We feed these adult zebrafish a diet of decapsulated brine shrimp twice daily (E-Z Egg, Brine Shrimp Direct, Ogden, UT) and Gemma Micro 300 formulated diet twice daily (Skretting, Westbrook, ME) for a total of four feedings per day. We also monitor colony health daily by tank level observation and twice a year by histological and polymerase chain reaction (PCR) analysis for a panel of common zebrafish diseases including *Pseudoloma* and *Mycobacterium* using sentinel animals, environmental samples, and feed samples.

2.2 Pre-Assay: Embryo Collection and Preparation

The night before zebrafish spawning, we combine 2–3 tanks of adult zebrafish (approximately 75–100 total fish) in a spawning chamber with a mesh bottom. In the morning, when the lights come on, adults spawn and the eggs fall through the mesh. The mesh protects the eggs and fertilized embryos from predation by the adult zebrafish. Thirty minutes later, we collect the fertilized embryos, along with the rack water in which they are fertilized, into 500 mL beakers. Once collected, we keep the beakers in a water bath at 28 °C until we are ready to wash the embryos.

We wash embryos before plating (*see* Subheading 2.3), preferably within 2 h after spawning. The beakers often have waste material from the rack water, which is removed during washing and leads to healthier embryos. To obtain a random sampling of our fish population, we combine multiple beakers or random samples (1–2 mL) of all beakers. Our washing procedure begins with preparing a dilute 0.06% bleach solution consisting of 60 μL of 100% bleach to 100 mL of 10% Hanks' Balanced Salt Solution (hereafter referred to as 10% Hanks') [41]. 10% Hanks' contains the following salts—13.7 mM NaCl, 540 μM KCl, 25 μM Na_2HPO_4, 44 μM, KH_2PO_4, 130 μM $CaCl_2$, 100 μM $MgSO_4$, and 420 μM $NaHCO_3$ (all salts obtained from Sigma-Aldrich, St. Louis, MO). We find that Hanks' provides a more controlled and consistent media option than Instant Ocean or other general salts, and the 10% Hanks' dilution we use provides an optimal salinity for embryonic zebrafish. The dilute bleach solution is potent enough to eliminate biological contaminants, without posing a significant hazard to the embryos during their brief exposure time. We set a slide warmer (Premiere XH-2002, Manassas, VA) to 26 °C as this is our experimental temperature selected to slow down development of zebrafish and elongate observation periods during critical time periods. After the bleach solution is prepared, and the slide warmer is 26 °C, we place two petri dishes filled with 50 mL of the 0.06% bleach solution and seven additional petri dishes filled with 50 mL of 10% Hanks' on the slide warmer. Six of these dishes are used for rinsing embryos after bleaching and one as the final collection dish.

After setting up the slide warmer with the nine petri dishes, we remove our desired embryos from the water bath and gently pour the rack water containing the embryos through a metal sieve (~122 µM mesh size) to concentrate the embryos. We then submerge the sieve with the embryos into the first bleach dish for 5 min. During these 5 min, we slowly swirl the sieve around the dish and use a pipette to remove large pieces of debris and to apply additional bleach solution from that petri dish over the embryos to increase the amount of debris removed. After 5 min, we move the sieve to the first 10% Hanks' wash dish, and gently swirl it for 5 s. We repeat this with two more 10% Hanks' dishes. Next, we move the sieve to the second bleach dish, repeating the above process. After the final rinse, we collect embryos into the final 10% Hanks' dish by inverting the sieve and tapping it gently on the side of a petri dish.

Once the embryos are placed into the final dish, we select optimal embryos for plating. Separating viable (noncoagulated, clear, and defined cells) embryos from the dead embryos and from any remaining debris helps improve embryo quality and will expedite the plating process. At this point [2–3 h post fertilization (hpf)], the embryos rest on the slide warmer at 26 °C. We usually select embryos at the high blastula to oblong stage, which will have reached the dome to 50% epiboly stage (5–6 hpf) when we plate [42]. Since early life stage (embryo/larvae) zebrafish are transparent, we use a dissecting microscope (Olympus SZH10 Research Stereo, Center Valley, PA) for preselection, plating, and assessments. The use of a dissecting microscope allows plenty of working room between the lens and the stage for manipulating and selecting the embryos.

After the embryos are washed, some laboratories remove the chorion using mechanical or chemical techniques [43–45]. Our laboratory, however, does not dechorionate embryos because of the potential for negative consequences which should be considered. For example, dechorionating can affect survival though this may be dependent on the technique used and the timing of dechorionation after fertilization [43, 44]. It may also influence locomotor activity as it has been noted that dechorionated embryos may start to swim earlier in development [46]. Additionally, the act of dechorionating can be intrusive and some evidence suggests that the chorion may act as a barrier for certain substances leading to differences in chemical exposure compared to embryos with an intact chorion [43]. Lastly, dechorionating adds time, costs, and hurdles that can affect a laboratory's ability to use zebrafish, the experimental design, or the desired endpoints [43].

2.3 Pre-Assay: Plating Embryos

When placing embryos into experimental plates (plating), the developmental stage and experimental venue must be considered. The developmental stage selected might depend on the experimental design, the endpoint, and the chorion status [43]. We place our embryos into microtiter plates between 5 and 6 hpf, which is when embryos are between the dome and 50% epiboly stage when held at 26 °C [42]. This is an optimal development stage for selecting embryos because it reduces the chance of picking embryos with abnormalities since they are farther developed, but it still allows for early-development exposures [43]. For experimental venues, common arrangements include crystallization dishes or 24-, 48-, 96-, and 384-well microtiter plates [43, 47]. We have found that embryos move more in 24-well plates compared to 48- and 96-well plates [48]. This is likely from the larger circumference since larva commonly exhibit thigmotaxis, which is the tendency to avoid the center of the well and move closer to the wall, often called "wall-hugging" [46]. On the other hand, 384-well plates do not provide enough room for older, larger larva and therefore limit the experiment to 72 hpf or younger [43]. We often use 96-well mesh microtiter plates (Multiscreen™, MilliporeSigma, Burlington, MA) because the wells allow enough space for 6 dpf larva, provide many experimental units to assess several chemicals and/or concentrations simultaneously, and streamline plate renewals (see Subheading 2.5). The physiochemical properties of the chemical (s) of interest must be considered when choosing the type of 96-well plates (e.g., glass vial plates for volatile chemicals instead of these mesh plates). For the 96-well mesh microtiter plates, we add 1 embryo per well. In plates with larger wells, multiple embryos can be exposed together and then separated prior to the locomotor endpoint assay. Group rearing, however, can affect locomotor activity as peers can be stimulating, leading to higher activity levels [47]. Additionally, transferring from one venue to another may increase the likelihood of damaging larvae. When setting up multiple plates at once, we rotate our larval addition among the plates by loading two columns or two rows at a time to ensure that all plates are similar. Once plating is complete, we check over each plate for damaged chorions, missing embryos, or extra embryos, and replace the embryos accordingly.

2.4 During Assay: Experimental Plate Design

When designing the experimental plate arrangement, it is essential to include all experimental conditions on each plate. Another important aspect of experimental design is randomization and replication. In terms of randomization, we alternate chemicals/concentrations on each plate, but our main randomization pertains to our handling of embryos. As mentioned in Subheadings 2.2 and 2.3, we collect a random sample of embryos from multiple beakers and alternate plating across each plate. For replications, we have identified that at least 16 individual larvae per independent treatment group are needed to reduce variability and to accurately assess locomotor activity. Each experiment consists of multiple experimental plates that are chemically treated with the same stock plate. For example, sometimes an experiment includes four different chemicals, each with seven concentrations. This leads to an experimental plate containing roughly 3 embryos per chemical concentration, leaving about 12 embryos to serve as a negative or solvent control. With 6 plates, this leads to a maximum number of 18 total individual larvae per independent treatment group and 72 total control larvae. Determining the number of larvae per treatment group on each plate is essential to creating the layout of each plate and volumes of chemical required.

Inclusion of negative and positive controls are an extremely important part of our experimental design. We include negative controls on each plate and also test a positive control plate every few months. Negative controls allow us to understand the basal locomotor activity of unexposed larvae and monitor control activity consistency over time. Activity changes in the negative control can indicate chemical contamination or changes in our fish population. For the negative control, we commonly utilize the solvent dimethyl sulfoxide (DMSO) for our chemical preparations. While higher concentrations of DMSO have been shown to affect morphological features and locomotor activity [49–53], many experiments in our laboratory have shown that the final concentration we utilize (0.4%) does not affect locomotor activity or cause morphological changes. Additionally, our laboratory group has never observed any significant differences between our embryo media, 10% Hanks', and this solvent (0.4% DMSO in 10% Hanks') [54]. On the other hand, positive controls allow for verification of our techniques, ensuring we can detect locomotor activity changes. For our positive control, we primarily use a chemical that has been widely studied, is stable, and has produced consistent locomotor activity patterns.

When the positive and negative controls have been selected, we determine our test chemical(s) of interest and must decide on the maximum concentration(s). We are testing locomotor activity, and therefore need to select chemical concentrations that will not induce significant mortality or morphological abnormalities in larvae. This can be done by reviewing the literature on developmental and neurodevelopmental toxicity data if extensive information on the chemical(s) of interest is available. Alternatively, conducting a preliminary concentration range finding experiment will help in the selection of sublethal concentrations that are not expected to induce morphological abnormalities.

Once the maximum concentration for each chemical of interest has been determined, we prepare chemical stock plates that contain solvent controls and the test chemical solutions across a dilution series, most commonly in half-log steps. These stock plates are 250 times more concentrated than the final diluted concentration on the experimental plate. The highest concentration on the experimental plate is considered the highest environmentally and physiologically relevant concentration, so we prepare the highest stock concentration first. Our highest stock plate concentration never exceeds 30 mM with a final concentration of 120 μM on the experimental plate. Serial dilutions of each chemical are then subsequently prepared across the entire stock plate using a multichannel pipettor. While we do not analytically verify these concentrations, typically the highest concentration we select elicits some morphological abnormality, confirming that the chemical is present within each well. After preparation, these stock plates are sealed and stored at 4 or −80 °C (depending on chemical properties). We use this process to minimize the volumes handled, waste generated, and streamline the preparation of our experimental plates.

2.5 During Assay: Exposure to Chemicals

With the stock plate prepared (see Subheading 2.4) and the selected embryos placed into experimental plates with 10% Hanks' (see Subheading 2.3), we are ready to administer the chemical(s). We first prepare the dilutions for our experimental plate by loading 150 μL of 10% Hanks' and 1 μL of chemical from the stock plate, using a 96-channel Liquidator (Rainin, Mettler Toledo Intl., Columbus, OH), into a new bottom portion of the microtiter plate. Next, we lift the upper mesh insert of the microtiter plate (containing the plated embryos), carefully but quickly blot the bottom of the mesh insert once on filter paper [Whatman GF/B paper (fired), Brandel, Gaithersburg, MD] to remove any excess 10% Hanks' and place it in the new bottom plate containing the desired chemical dilutions in 10% Hanks'. After we transfer the mesh insert containing the embryos, 100 μL of 10% Hanks' is added to reach a final volume of 250 μL per well. Due to how the mesh insert rests in the bottom tray, the embryos are only exposed to the top 100 μL. Thus, during set up, the embryos are not exposed to the more concentrated chemical concentration prior to adding the final 100 μL a few moments later. After dosing, we seal the top of each plate with a non-adhesive Microfilm A film (Type A, BioRad, Hercules, CA), cover with the plate lid, and wrap the sides of the plate in Parafilm™ to secure the lid to the bottom plate. Once sealed we incubate plates at 26 °C on a standard 14:10 h light/dark cycle. We have chosen to provide the same light cycle as our animal facility because light only and dark only incubations have been shown to influence the development of zebrafish that could ultimately confound our behavioral results [55–63].

In addition to dosing the embryos on day 0 (plate set up day), we might also renew the solution at other time points throughout the next 4 days following the same method as described above. Solution renewals are determined based on the question being asked or the target developmental windows. For example, renewing the solution only once at 3 dpf minimizes the transfer of embryos while still targeting specific developmental stages (later brain development). On 5 dpf, we transfer the plates into 10% Hanks' only (no chemical) to remove the chemical(s) prior to locomotor activity testing. We complete this step to distinguish acute and developmental (morphological) from neurodevelopmental effects. We have found that overt toxicity leading to abnormal morphological features and larvae with uninflated swim bladders can cause locomotor activity changes compared to healthy morphologically normal control larvae (see Subheading 2.10) [47]. Unpublished data from our group has found that changing the plates additional times is correlated with higher rates of hatching and swim bladder inflation in unexposed larvae. We believe this is likely due to additional stimulation to the embryos and mild disruption of the exterior of the chorion from the plate changing process. It is important to note that while these plate changes are occurring, the embryo/larva are in the same well for the entire duration of the experiment. This is opposed to transferring larvae using pipettes to an entirely new plate, which can be more disruptive to larvae while also being much more labor intensive for researchers.

2.6 During Assay: Locomotor Activity Preparation

On the morning of day 6, we again transfer the larvae (as a whole plate) into fresh 10% Hanks' solution as a secondary measure to remove any residual chemical prior to testing. After plate changes, and approximately 2 h prior to locomotor activity testing, we take the plates to a darkened locomotor activity testing room set to 26 °C. We allow the larvae to acclimate in the dark prior to locomotor activity testing to reduce any disturbances that may be caused by plate changing, handling, and movement. While locomotor activity in response to temperature changes has not been thoroughly investigated by our group, other groups have found that a 3-degree temperature decrease resulted in an increased startle response [64]. Alternative methods to ensure that the temperature during testing does not change include the addition of temperature control devices or pumping water around the well plate. During the acclimation period, we warm up the visible and infrared (IR) lights to prevent any light fluctuations. As our locomotor activity protocol utilizes both light and dark photoperiods, IR light is required to record locomotor activity in the dark. Hartmann et al. (2018) observed that larvae exhibited negative phototaxis to 860 nm of IR light, suggesting that larvae are not in true "dark" conditions at wavelengths near 860 nm. While the wavelength of IR utilized in locomotor assays is seldomly reported, it has been found that 850 nm is the most common wavelength in older experimental devices and 850 nm is the IR wavelength for our present methodology [65]. Some testing chambers, including Viewpoint (Viewpoint, Lyon, France) and Noldus (Noldus Information Technology, Leesburg, VA), are either equipped with or offer newer systems at 950 nm. Whether this wavelength alters basal larvae locomotor activity and/or responses to chemical exposures in a light/dark assay requires future investigation. We have recently found that some testing systems, in addition to the IR wavelength mentioned, have a 10-lux background illumination during the "dark" setting, which may further affect results.

To begin the assay, we load the first plate onto the recording platform 4.5 h after the light has come on in the incubator (i.e., incubator used for entire duration of experiment) with the last plate concluding 4 h before the lights turn off. This allows for a window of approximately 6 h to test the locomotor activity. We selected this time frame due to stable and less variable activity patterns that have been observed for unexposed larvae (Fig. 2) [35]. Similar stable activity patterns have also been observed across multiple zebrafish larval ages [66]. If we are testing multiple plates in 1 day, we load

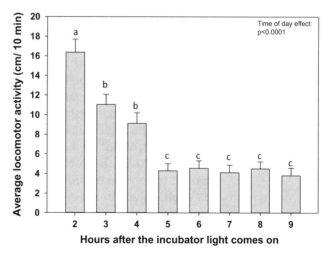

Fig. 2 The influence of time of day, represented as hours after the light comes on in the incubator, on control (unexposed) larvae locomotor activity in the dark. Results are based on averaged activity (cm) ± SEM over the entire 10 min recording of 6 dpf larvae ($n = 93$) from the same embryo brood. Statistical differences were determined using a One-Way ANOVA ($p < 0.0001$ overall time effect). Different letters represent statistical differences among groups as determined by a Fisher's PLSD

each plate onto the platform immediately following the conclusion of the previous plate. We keep sound, light, and movement to a minimum while loading the plate onto the platform to minimize any disturbance to the larvae, and no one enters the testing room during the recording.

We focus our assay on 6 dpf larvae but various zebrafish larval ages are utilized for locomotor activity screening by other laboratories [67]. These approaches primarily focus on larvae aged around 5 dpf or younger due to how regulation of fish have been defined in several countries [68]. While these ages could be applied to our protocol, one must be aware that the dpf is not always associated with the same developmental window. For example, temperature will influence the developmental rate [42]. Our protocol utilizes 6 dpf larvae reared at 26 °C which are roughly equivalent, developmentally, to 5 dpf larvae reared at 28 °C [42]. Baseline activity, both in terms of distance moved and pattern of activity, varies with these ages with higher activity typically observed for older larvae in light photoperiods [48, 69]. Similarly, the degree with which a toxicant alters locomotor activity varies among ages with different sensitivities based on the type of chemical (e.g., [70, 71]).

2.7 During Assay: Locomotor Activity Testing Systems

In our laboratory, we utilize commercially available larval zebrafish activity testing systems. These systems are devised to allow the simultaneous tracking of up to 96 zebrafish larvae in a multi-well plate. Each system consists of a video camera and a light box, which acts as a platform for the experimental plate. The camera and light box are enclosed to prevent extraneous light from interfering with the test procedure. As mentioned above, the light box is illuminated by both visible and IR light so that the larvae can be tested in both light and dark conditions. Specialized software controls both the lighting conditions and records a video of the testing session. Simultaneous tracking of the larval movements can be performed live during the session although we prefer to record the video and track later in order to simplify the protocol on testing day (*see* Subheading 2.9). Even if one opts to perform live tracking, we recommend always recording a video for proper recordkeeping and to safeguard against the event of a technical error occurring during live tracking.

One of the more useful aspects of utilizing these testing systems is the ease with which adjustments in lighting levels are accomplished. Larvae can be exposed to total darkness or to varying levels of visible light. This is important because our laboratory, as well as others, have determined that alternating light and dark photoperiods produce consistent patterns of locomotion [35, 72, 73]. Typically, when larvae in the dark are exposed to visible light, they initially stop moving, then slowly increase activity. When larvae are suddenly returned to dark conditions after exposure to visible light, the larval activity dramatically increases then slowly tapers off. When using varying photoperiods for developmental neurotoxicity locomotor activity studies, one must be aware of the effects due to the duration of the light and dark photoperiods [35]. The length of the dark photoperiods does not appear to affect the level of activity in the following light photoperiod, nor does it affect the activity level when the larvae are subsequently returned to dark. The length of the light photoperiod, on the other hand, does affect activity levels in a following dark photoperiod; the longer the light photoperiod, the higher the level of activity when the larvae are returned to dark [35].

In addition to the duration of light, locomotor activity levels can also be affected by the intensity of the light photoperiod. The general dark-to-light pattern of ceasing movement then slowly increasing activity occurs regardless of the light intensity but higher intensities of light will slow down the activity increase after ceasing movement, taking slightly longer for larvae to reach stable activity level with a higher final activity (Fig. 3). Light intensity also affects the activity levels when larvae are returned to total darkness. The brighter the light, the higher the activity level when abruptly returned to a dark photoperiod (Fig. 3). Armed with the

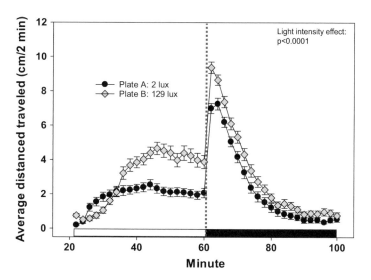

Fig. 3 Light intensity effect on control (unexposed) larvae locomotor activity. *Left side*: Plate **a** was exposed to 2 lux light intensity and Plate **b** exposed to 129 lux light intensity for 40 min. *Right side*: Both plates were then transitioned to the same level of darkness (0.5 lux) for 40 min. Results are represented as average activity (cm) $\pm$ SEM per 2 min of 6 dpf larvae from the same embryo brood (Plate A, $n = 93$; Plate B, $n = 89$). The maroon dotted vertical line represents the transition from light to dark. Light bar represents light photoperiod and black bar represents dark photoperiod. Overall light intensity effect ($p < 0.0001$) determined by repeated-measures ANOVA

knowledge of how larval zebrafish respond to varying photoperiods, one can look for variations in these patterns when testing the effects of chemical exposure during development.

As previously mentioned, there is no standardized zebrafish locomotor activity protocol. Among the lighting-related factors that vary among laboratories are different lighting protocols (e.g., multiple changes from light/dark photoperiod cycles over shorter times frames), acclimation phases (duration and lighting), and light intensity, which has been noted as a potential source of variability among different testing strategies [74]. Light intensity for both photoperiods is seldom reported in the literature. Although the influence of varying light intensities on locomotor activity following chemical exposure has not been examined, avoiding deviations in light intensity is recommended. This typical pattern, observed under alternating photoperiods, may be disrupted after exposure to chemicals. For example, Fig. 4 depicts depressed larvae locomotor activity in the light and dark photoperiods from developmental exposure to one of our laboratory's positive controls. Developmental neurotoxicants may affect larvae in one or both photoperiods, as shown in Fig. 4.

Our common schedule for assessing locomotor activity consists of an initial 20 min of dark (acclimation phase) to reduce perturbation from the transfer onto the behavioral platform, followed by

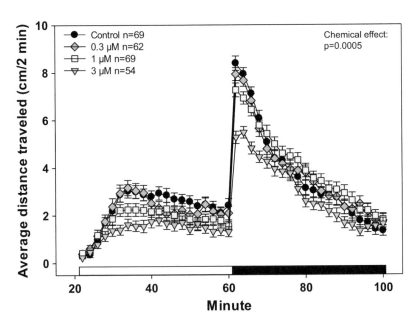

Fig. 4 Effect of a positive control on 6 dpf larval locomotor activity. Larvae were developmentally exposed to three concentrations of the chemical (0.3, 1, 3μM) and also vehicle control (0.4% DMSO) for 5 dpf and rinsed with 10% Hanks' solution for 24 h prior to locomotor testing on 6 dpf. Results are represented as average locomotor activity (cm ± SEM) per 2 min in either light (white bar) or dark (black bar) photoperiods. Overall concentration effect ($p = 0.0005$) determined by a repeated-measures ANOVA

40 min of light and 40 min of dark. We selected these time periods in order to assess the point at which locomotor activity is stable and reaches an asymptote. As seen in Fig. 4, it takes approximately 10–20 min to observe a significant deviation from control in the light and dark photoperiods. We select a light intensity of 18-lux, periodically checking every few months to make sure that there are no light fluctuations or that the light intensity has not changed throughout a study. As mentioned previously, it is important to warm up the testing system to avoid lighting fluctuations.

2.8 During Assay: Locomotor Alternative Strategies

We have developed the current testing protocol to yield high-throughput testing data, but there are of course alternative assays and/or methodologies that have been utilized for developmental neurotoxicity screening. Depending on the testing system, loco-motor activity can be automatically distributed into speed threshold categories that can be defined by the user. These speed categories have been used to distinguish and classify locomotor responses such as hypo- or hyperactivity. We choose to define locomotor activity as total distance traveled per 2 min. We primarily focus on this end-point because all other locomotor activity endpoints our laboratory uses depend on this measurement. There are other locomotor end-points that have been examined to determine developmental neu-rotoxicity; for example, some researchers study startle response, which is defined as the immediate reaction of larvae to a sudden change in lighting conditions [20, 71, 75]. In addition to the light/dark locomotor responses, testing systems can be adapted for other behavioral endpoints. Circadian, touch response, thigmo-taxis, prey-capture, and rest/wake are all examples of multifaceted behavioral assays that have been applied with the zebrafish model to investigate responses to stimuli [34]. The incorporation of multiple different behavioral assays may produce robust assessment [74]. Whatever protocol decisions are made, it is crucial for repro-ducibility and comparability for all experimental details to be reported in detail, as the multitude of variations outlined above may influence locomotor activity.

2.9 Post-Assay: Lethality and Malformation Assessment

We assess larval condition immediately after locomotor activity recording is complete. While we inform the assessor about the presence of the specific chemical(s), they are not aware of the chemical concentration in each well. These blinded assessments are important to ensure that we are only using data from live, hatched, and morphologically normal larvae. We exclude mal-formed larvae from our behavioral analysis because they swim differently than morphologically normal larvae, most notably they have lower activity in both dark and light photoperiods [48]. Addi-tionally, larvae with uninflated swim bladders that otherwise look morphologically normal have a lower activity level in the light and higher activity in the dark photoperiods compared to

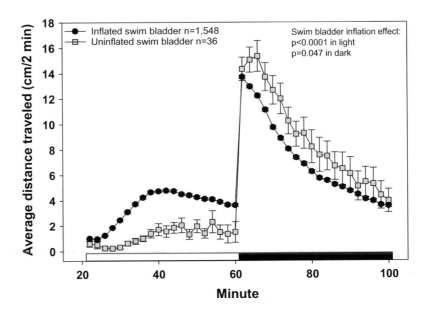

Fig. 5 Effect of swim bladder inflation status on baseline control (unexposed) 6 dpf larvae. Larvae with inflated swim bladders (black circles) have a markedly higher activity in the light compared to larvae with uninflated swim bladders (grey square). Only larvae that were otherwise morphologically normal were included in analysis. Activity is represented as average distance (cm ± SEM) per 2 min in the light (white bar) and dark (black bar) photoperiods. Swim bladder status effect determined by Mann-Whitney U test ($p < 0.0001$ for light and $p = 0.047$ for dark photoperiods)

morphologically normal larvae with inflated swim bladders (Fig. 5). While the goal of the assay is to determine locomotor differences to study nervous system disruptions, morphological abnormalities can affect locomotor activity [48].

We complete assessments using a dissecting microscope (Olympus SZH10 Research Stereo, Center Valley, PA, at the magnification 17.5×) viewing the dorsal position of the larva. We evaluate various morphological features including craniofacial (abnormal eyes, head, or otoliths), spinal (stunted, curved, or kinked tail), abdominal (edema or emaciation), thoracal (distention or heart malformations), and swim bladder inflation, as well as the position in the water column (floating or lying on side). Depending on the amount and degree of malformations, these assessments take a trained assessor approximately 5–10 min per 96-well plate. As stressed above, it is important to identify morphologically normal larvae. An example of our assessment is depicted in Fig. 6, which compares a normal and severely abnormal larva. The normal larva has a straight spine, well-developed eyes and head, and an inflated swim bladder. In contrast, the abnormal larva has an uninflated swim bladder, craniofacial deformity, curved spine, and shortened length. To accept and proceed with analyzing locomotor data, most (~85%) of the controls must be alive, hatched, and morphologically

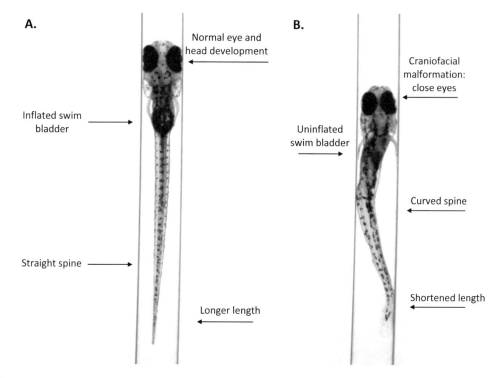

Fig. 6 Descriptions of common morphological features of (**a**) a morphologically normal larva and (**b**) morphologically abnormal larva at 6 dpf. Photos are representative of the ventral view of larva which is assessed in 96-well plates using a dissecting microscope. Morphological abnormalities have been shown to influence baseline activity and are therefore excluded from our neurotoxicity experiments [7]. Photos were taken using the Vertebrate Automated Screening Technology (Union Biometrica, Holliston, MA)

normal. Additionally, similar inclusion/exclusion data criteria (~70% must be alive, hatched, and morphologically normal) are used to determine whether a specific chemical concentration is included in our locomotor activity data.

2.10 Post-Assay: Data Processing

The resulting locomotor activity videos are analyzed using commercially available software. Here, we will refer to tracking locomotion using Ethovision software (Version 13, Noldus Corp.). Ethovision, with proper parameters, locates a single larval zebrafish in each well and tracks the movement of that larva's body center point. The software then uses these data to determine several possible endpoints and can be utilized with videos recorded using other systems (e.g., Viewpoint videos can be tracked using Ethovision software). Our protocol for tracking our recorded videos is outlined in Table 1.

For tracking from a video, the length of the video (or part of the video) is entered; this is the amount of time that the software will track the larvae. In our locomotor activity examples (Figs. 3, 4, and 5), this duration would be 100 min (20 min of acclimation, 40 min of light followed by 40 min of dark). Although we do not

Table 1
Data processing overview for locomotor activity using Noldus Ethovision 13 (Noldus Corp.) tracking software for videos that were recorded and not live tracked. Bolded text indicates Experiment Explorer options with sequential window options denoted by >

Step #	Experiment component	Purpose	Notes
	SETUP	Set parameters for tracking	
1	**Experiment settings** >Video source >Arenas >Tracked feature >Units	Set tracking from video or live Specifies # of wells on plate Select center-point detection Select unit of distance	We prefer tracking from videos Data reported here is in cm
2	**Arena settings**	Select video for tracking Draw arenas for each well Draw scale for calibration	
3	**Trial control settings**	Select duration of trial	Length of video to be tracked
4	**Detection settings** >Video >Detection	Select sample rate Select method for detection Set reference image Enter parameters for detection and verify that software accurately tracks movements	Number of frames per sec to be used We use dynamic subtraction We use a dynamic reference image We use subject color that is darker than background We use dark setting 8 to 105 and frame weight 10
	ACQUISITION	Tracks movement of larvae	
5	**Acquisition** >Playback control >Video window	Tells software to start tracking Shows tracking of larvae	DDS (detection determines speed) should be checked Verify that larvae movements are being tracked
6	**Track smoothing parameters**	Removes sources of noise during tracking	We use MDM (**minimal distance moved**) set t o 0.135 cm
	ANALYSIS	Reports results of tracking	
7	**Data profiles**	Select trials to report and time bins	Data reported here in 2-minute bins
8	**Analysis profiles**	Endpoints to be calculated	Data reported here is **distance moved**
9	**Results** >Statistics and charts >Track visualization	Calculates endpoints Exports data to excel file Tracks (movements) are displayed	

statistically analyze our acclimation period, we still include this as part of our protocol settings and recording. Ethovision allows for the tracking of up to 100 wells (i.e., arenas) with one larva per well. We select all wells (i.e., 96) for tracking and then once tracking is complete, we remove data collected for wells containing dead, unhatched, and abnormal larvae that were identified in Subheading 2.10. Arenas can be subdivided into zones: in zone, distance to zone, and distance to point (as in, distance to the center) can be calculated. Creating zones can be useful when using plates with larger sized wells, but 96-well plates do not provide enough area per well for zoning to be useful. A mandatory calibration scale must be drawn in the arena settings window. This tells the software how to translate pixels into actual measurements.

We use the dynamic subtraction method for our detection settings, which compares the subject color to the background (in our case, the subject is darker than the background). We also incorporate a minimal distance moved threshold that we determined by observing videos of unexposed larvae. This feature filters out the small inconsequential movements of the larvae that are not recognized by a trained observer to be locomotion (i.e., movement from one place to another). If a minimum distance moved threshold is not incorporated, the software will detect slight movements such as head bobbing as locomotor activity. We are only interested in distance moved (e.g., moving from one point to another point in the well) in our results, and we report these distances moved in time bins of 2-min intervals.

An important feature of Ethovision software is the ability to calculate a host of endpoints with relative ease. Once acquisition is complete (movements are tracked for each larva) the user can select different endpoints from the list of Dependent Variables provided in the Analysis Profiles section of the software. As mentioned above, we typically focus on distance moved, but there are other movement endpoints, such as velocity, movement (duration for which the animal is changing location, calculated as moving and not moving), acceleration, and acceleration state (high acceleration and low acceleration). Additional variables available include various endpoints for path, direction, and mobility. Each user will need to learn about these endpoints to determine which variable(s) will answer the question(s) being raised in the experiment. When selecting a variable, the software provides a definition of that feature and allows the selection of trial statistics and group statistics for that variable. Once endpoints are selected, the Results section is utilized to calculate the values for dependent variables. Independent variables can be selected from the drop-down menu using the Show/ Hide feature which includes information about start time, video name, and detection settings used, among many others. When the final results with desired endpoints are calculated, we choose to export our data to a Microsoft Excel (Office 365) file. In our example, these data are reported as distance moved in centimeters in the time bins selected. Because the results are reported per larva in chronological order per 2 min intervals/time bins (0–2 min interval, 2–4 min interval, etc.), we use an Excel macro to transpose the data so that the results per time bin are displayed across the sheet, utilizing one row for each larva. We find that transposing these data helps with data manipulation (removing any undesired larvae) and analysis (see Subheading 2.11).

2.11   Post-Assay: Statistical Analysis

Deciding whether to label a chemical as having an effect (i.e., a "hit") is not straightforward. The literature is replete with many distinct approaches to assessing locomotor activity for differences between the treated and control larvae, as well as differences among the various doses of each chemical (e.g., [73, 76–78]). There are probably as many different ways of analyzing zebrafish locomotor activity data as there are laboratories engaged in testing. To date there has not been a rigorous comparison among all the methods, so it is difficult to know the strengths and weaknesses of each.

In our laboratory, we have tried many different methods for our analyses. In almost all cases, we are trying to assess if there was an effect of chemical treatment and which concentration groups were different from the control. Two facts that need to be kept in mind when devising a statistical approach: the data are not necessarily normally distributed, and data are being collected on the same larvae over time. One can compare the entire time course of activity, which would mean comparing one curve to another, or one can compare the activity at one portion during the testing. For example, the total activity during the light photoperiod or during the dark photoperiod or both. Using parametric statistics is easiest for comparison because by doing an overall repeated-measure Analysis of Variance (ANOVA) and then determining when interactions are significant, we are automatically correcting for multiple comparisons. Another approach would be to use the ANOVA (parametric) analysis to compare the total activity in the light or dark photoperiods to determine if they are different. If the ANOVA tells us that there is an overall effect of concentration, we usually follow up with a Fisher's PLSD post hoc test to determine which concentrations are different from control.

As we often have many more individuals in our control groups than in our treated groups, and because of the aforementioned lack of normal distribution of the locomotor activity data, we also use a non-parametric approach for assessment of the light or dark activity. Usually, this consists of a Kruskal Wallis test of the control and all concentration groups to determine if there is a significant concentration response. If there is a significant concentration response, we follow up with a Mann-Whitney U test to determine which concentration groups are different from the control. In these comparisons, we also use a correction procedure for multiple comparisons, such as a type of Bonferroni correction. Often, we will analyze the locomotor activity data using both parametric and non-parametric methods to assess data consistency.

Recent criticism of the use of stringent statistical cut-offs (e.g., p-value <0.05) have raised questions about ubiquitous use for all scientific applications [79–81]. Of the many different approaches being proffered, the one which we plan to adopt is to calculate the degree of change in locomotor activity, (i.e., percent decrease or increase in activity compared to the control mean or median), and present that data with 95% confidence intervals which allows the reader to determine the level of change in the chemically treated larvae that provokes concern. For future analyses, we are planning to present our analysis of these activity data by including both traditional statistical analyses and degree of effect with 95% confidence intervals with no statistical analyses.

3   Conclusion

Relative to other developmental neurotoxicitymethods, evaluating behavior provides a way to examine the nervous system without using labor-intensive techniques. Zebrafish larval locomotor activity is an example of applying this connection to investigate disruption of the developing nervous system. Here, we present methodology for screening chemicals for locomotor activity changes in larval zebrafish with the ability to generate large amounts of data in a relatively short period of time. The integration of this rapidly derived in vivo endpoint with in vitro cellular assays for developmental neurotoxicity will create a more complete picture of the developmental neurotoxicity potential of a given library of chemicals.

We selected a light/dark locomotor assay because larvae produce consistent patterns in the light (decreased activity) and dark (higher activity) photoperiods. Neurotoxicants shift these activity patterns in one or both photoperiods, which may result in biphasic activity patterns, highlighting the various ways that the complexity of the zebrafish brain and nervous system development can be disrupted. While this methodology is promising, there are some limitations. Underlying sources of variability remain poorly understood for zebrafish locomotor activity, especially because there is not a consistent standardized method utilized among laboratories. Development of such should enhance the reproducibility and replicability of developmental neurotoxicity results where discrepancies remain for classifying neurotoxicants among laboratories [53]. Increasing consistency across laboratories and reducing variability will ultimately increase our success of determining significant chemical responses and decrease our minimum number of larvae required, making our assay not only higher throughput but also aligning it with the 3Rs principles: replace, reduce, and refine [82]. In conclusion, we foresee zebrafish larval behavioral assays as being an important adjunct in any neurotoxicity testing battery.

Acknowledgments

We would like to thank Drs. William Boyes and Aimen Farraj for their insight and review of this manuscript and thank Katy Britton for laboratory support and Donald Holman, Guilermo Orozco, Femi Yerumo, Clark Kridler, Jenelle Dunn, Kimberly Wingate, and Leslie Jarrell for oversight, maintenance, and upkeep of our zebrafish facility. This project was supported, in part, by an appointment to the Research Participation Program at the Office of Research and Development administered by the Oak Ridge Institute for Science and Education through an interagency agreement with the U.S. Environmental Protection Agency. K.D. Coldsnow was supported by the National Science Foundation Graduate Research Fellowship under Grant No. DGE 1744655. Any opinion, findings, and conclusions or recommendations expressed in this material are those of the authors(s) and do not necessarily reflect the views of the National Science Foundation.

This manuscript has been subjected to review by the U.S. EPA Center for Computational Toxicology and Exposure and approved for publication. Approval does not signify that the contents reflect the views of the Agency, nor does the mention of trade names or commercial products constitute endorsement or recommendation for use.

References

1. Weiss B (2000) Vulnerability of children and the developing brain to neurotoxic hazards. Environ Health Perspect 108(3):375–381

2. Needham LL et al (2011) Partition of environmental chemicals between maternal and fetal blood and tissues. Environ Sci Technol 45 (3):1121–1126

3. Braun JM et al (2006) Exposures to environmental toxicants and attention deficit hyperactivity disorder in U.S. children. Environ Health Perspect 114(12):1904–1909

4. Yolton K et al (2014) Exposure to neurotoxicants and the development of attention deficit hyperactivity disorder and its related behaviors in childhood. Neurotoxicol Teratol 44:30–45

5. Kalkbrenner AE, Schmidt RJ, Penlesky AC (2014) Environmental chemical exposures and autism spectrum disorders: a review of the epidemiological evidence. Curr Probl Pediatr Adolesc Health Care 44(10):277–318

6. Cannon JR, Greenamyre JT (2011) The role of environmental exposures in neurodegeneration and neurodegenerative diseases. Toxicol Sci 124(2):225–250

7. Albert SM (2007) Projecting neurologic disease burden. Neurology 68(5):322–323

8. Mundy WR et al (2015) Expanding the test set: chemicals with potential to disrupt mammalian brain development. Neurotoxicol Teratol 52:25–35

9. USEPA (1998) Health effects test guidelines OPPTS 870.6300 developmental neurotoxicity study. Office of Prevention, Washington, DC

10. OECD (2007) Test no. 426: developmental neurotoxicity study, OECD guidelines for the testing of chemicals, section 4. OECD Publishing, Paris

11. Grandjean P, Landrigan PJ (2009) Developmental neurotoxicity of industrial chemicals. Lancet 368(9553):2167–2178

12. Krewski D et al (2010) Toxicity testing in the 21st century: a vision and a strategy. J Toxicol Environ Health B Crit Rev 13(2–4):51–138

13. Frank R (2016) Lautenberg chemical safety for the 21st century act, U.S.C. § 448

14. USEPA (2019) Directive to prioritize efforts to reduce animal testing. https://www.epa.gov/sites/production/files/2019-09/documents/image2019-09-09-231249.pdf

15. USEPA (2020) New approach methods work plan: reducing use of animals in chemical testing. U.S. Environmental Protection Agency, Washington, DC. EPA 615B2001

16. de Souza AC et al (2018) Zebrafish (*Danio rerio*): a valuable tool for predicting the metabolism of xenobiotics in humans? Comp Biochem Physiol C Toxicol Pharmacol 212:34–46

17. Cornet C, Di Donato V, Terriente J (2018) Combining zebrafish and CRISPR/Cas9: toward a more efficient drug discovery pipeline. Front Pharmacol 9:703–703

18. Letrado P et al (2018) Zebrafish: speeding up the cancer drug discovery process. Cancer Res 78(21):6048–6058

19. Horzmann KA, Freeman JL (2018) Making waves: new developments in toxicology with the zebrafish. Toxicol Sci1 63(1):5–12

20. de Esch C et al (2012) Zebrafish as potential model for developmental neurotoxicity testing: a mini review. Neurotoxicol Teratol 34(6):545–553

21. d'Amora M, Giordani S (2018) The utility of zebrafish as a model for screening developmental neurotoxicity. Front Neurosci 12(976):1–6

22. Guo S (2009) Using zebrafish to assess the impact of drugs on neural development and function. Expert Opin Drug Discov 4(7):715–726

23. Navratilova P et al (2009) Systematic human/zebrafish comparative identification of *cis*-regulatory activity around vertebrate developmental transcription factor genes. Dev Biol 327:526–540

24. Flentke GR et al (2014) An evolutionarily conserved mechanism of calcium-dependent neurotoxicity in a zebrafish model of fetal alcohol

spectrum disorders. Alcohol Clin Exp Res 38(5):1255–1265

25. Nishimura Y et al (2015) Zebrafish as a systems toxicology model for developmental neurotoxicity testing. Congenit Anom (Kyoto) 55(1):1–16

26. Kalueff AV, Echevarria DJ, Stewart AM (2014) Gaining translational momentum: more zebrafish models for neuroscience research. Prog Neuro-Psychopharmacol Biol Psychiatry 55(3):1–6

27. Stewart AM et al (2015) Molecular psychiatry of zebrafish. Mol Psychiatry 20(1):2–17

28. Stewart AM et al (2014) Zebrafish models for translational neuroscience research: from tank to bedside. Trends Neurosci 37(5):264–278

29. Fontana BD et al (2018) The developing utility of zebrafish models of neurological and neuropsychiatric disorders: a critical review. Exp Neurol 299:157–171

30. Rihel J, Schier AF (2012) Behavioral screening for neuroactive drugs in zebrafish. Dev Neurobiol 72(3):373–385

31. Kokel D et al (2010) Rapid behavior-based identification of neuroactive small molecules in the zebrafish. Nat Chem Biol 6:231–237

32. Rihel J et al (2010) Zebrafish behavioral profiling links drugs to biological targets and rest/wake regulation. Science 327(5963):348–351

33. Rennekamp AJ, Peterson RT (2015) 15 years of zebrafish chemical screening. Curr Opin Chem Biol 24:58–70

34. Tierney KB (2011) Behavioural assessments of neurotoxic effects and neurodegeneration in zebrafish. Biochim Biophys Acta 1812(3):381–389

35. MacPhail RC et al (2009) Locomotion in larval zebrafish: influence of time of day, lighting and ethanol. Neurotoxicology 30(1):52–58

36. Fernandes Y et al (2014) Embryonic alcohol exposure impairs associative learning performance in adult zebrafish. Behav Brain Res 265:181–187

37. Carvan MJ 3rd et al (2004) Ethanol effects on the developing zebrafish: neurobehavior and skeletal morphogenesis. Neurotoxicol Teratol 26(6):757–768

38. Cowden J et al (2012) Developmental exposure to valproate and ethanol alters locomotor activity and retino-tectal projection area in zebrafish embryos. Reprod Toxicol 33(2):165–173

39. Bellinger DC (2008) Very low lead exposures and children's neurodevelopment. Curr Opin Pediatr 20(2):172–177

40. Chen J et al (2012) Developmental lead acetate exposure induces embryonic toxicity and memory deficit in adult zebrafish. Neurotoxicol Teratol 34(6):581–586

41. Westerfield M (2000) The zebrafish book: a guide for the laboratory use of zebrafish (*Danio rerio*). University of Oregon Press, Eugene

42. Kimmel CB et al (1995) Stages of embryonic development of the zebrafish. Dev Dyn 203 (3):253–310

43. Hamm J et al (2018) Characterizing sources of variability in zebrafish embryo screening protocols. ALTEX 36(1):103–120

44. Henn K, Braunbeck T (2011) Dechorionation as a tool to improve the fish embryo toxicity test (FET) with the zebrafish (*Danio rerio*). Comp Biochem Physiol C Toxicol Pharmacol 153(1):91–98

45. Truong L, Harper SL, Tanguay RL (2011) Evaluation of embryotoxicity using the zebrafish model. Methods Mol Biol 691:271–279

46. Basnet RM et al (2019) Zebrafish larvae as a behavioral model in neuropharmacology. Biomedicine 7(23):1–16

47. Zellner D et al (2011) Rearing conditions differentially affect the locomotor behavior of larval zebrafish, but not their response to valproate-induced developmental neurotoxicity. Neurotoxicol Teratol 33(6):674–679

48. Padilla S et al (2011) Assessing locomotor activity in larval zebrafish: influence of extrinsic and intrinsic variables. Neurotoxicol Teratol 33 (6):624–630

49. Hallare A et al (2006) Comparative embryotoxicity and proteotoxicity of three carrier solvents to zebrafish (*Danio rerio*) embryos. Ecotoxicol Environ Saf 63(3):378–388

50. Maes J et al (2012) Evaluation of 14 organic solvents and carriers for screening applications in zebrafish embryos and larvae. PLoS One 7 (10):e43850

51. Huang Y et al (2018) Unsuitable use of DMSO for assessing behavioral endpoints in aquatic model species. Sci Total Environ 615:107–114

52. Chen TH, Wang YH, Wu YH (2011) Developmental exposures to ethanol or dimethylsulfoxide at low concentrations alter locomotor activity in larval zebrafish: implications for behavioral toxicity bioassays. Aquat Toxicol 102(3):62–166

53. Teixidó E et al (2019) Automated morphological feature assessment for zebrafish embryo developmental toxicity screens. Toxicol Sci 67 (2):438–449

54. Irons TD et al (2013) Acute administration of dopaminergic drugs has differential effects on locomotion in larval zebrafish. Pharmacol Biochem Behav 103(4):792–813

55. Villamizar N et al (2014) Effect of lighting conditions on zebrafish growth and development. Zebrafish 11(2):173–181

56. Saszik S, Bilotta J (1999) Effects of abnormal light-rearing conditions on retinal physiology in larvae zebrafish. Invest Ophthalmol Vis Sci 40(12):3026–3031

57. Saszik S, Bilotta J (2001) Constant dark-rearing effects on visual adaptation of the zebrafish ERG. Int J Dev Neurosci 19(7):611–619

58. Li L, Dowling JE (1998) Zebrafish visual sensitivity is regulated by a circadian clock. Vis Neurosci 15(5):851–857

59. Kopp R, Legler J, Legradi J (2018) Alterations in locomotor activity of feeding zebrafish larvae as a consequence of exposure to different environmental factors. Environ Sci Pollut Res Int 25(5):4085–4093

60. Di Rosa V et al (2015) The light wavelength affects the ontogeny of clock gene expression and activity rhythms in zebrafish larvae. PLoS One 10(7):e0132235

61. Dekens MP et al (2003) Light regulates the cell cycle in zebrafish. Curr Biol 13 (23):2051–2057

62. Chapman GB, Tarboush R, Connaughton VP (2012) The effects of rearing light level and duration differences on the optic nerve, brain, and associated structures in developing zebrafish larvae: a light and transmission electron microscope study. Anat Rec (Hoboken) 295 (3):515–531

63. Bilotta J (2000) Effects of abnormal lighting on the development of zebrafish visual behavior. Behav Brain Res 116(1):81–87

64. Burgess HA, Granato M (2008) The neurogenetic frontier--lessons from misbehaving zebrafish. Brief Funct Genomic Proteomic 7 (6):474–482

65. Hartmann S et al (2018) Zebrafish larvae show negative phototaxis to near-infrared light. PLoS One 13(11):e0207264

66. Kristofco LA et al (2016) Age matters: developmental stage of *Danio rerio* larvae influences photomotor response thresholds to diazinion or diphenhydramine. Aquat Toxicol 170:344–354

67. Legradi J et al (2015) Comparability of behavioural assays using zebrafish larvae to assess neurotoxicity. Environ Sci Pollut Res 22:16277–16289

68. Strähle U et al (2012) Zebrafish embryos as an alternative to animal experiments- a commentary on the definition of the onset of protected life stages in animal welfare regulations. Reprod Toxicol 33:128–132

69. de Esch C (2012) Locomotor activity assay in zebrafish larvae: influence of age, strain and ethanol. Neurotoxicol Teratol 34:425–433

70. Fraser TWK et al (2017) Toxicant induced behavioural aberrations in larval zebrafish are dependent on minor methodological alterations. Toxicol Lett 276:62–68

71. Selderslaghs IW et al (2013) Assessment of the developmental neurotoxicity of compounds by measuring locomotor activity in zebrafish embryos and larvae. Neurotoxicol Teratol 37:44–56

72. Irons TD et al (2010) Acute neuroactive drug exposures alter locomotor activity in larval zebrafish. Neurotoxicol Teratol 32(1):84–90

73. Selderslaghs IW et al (2010) Locomotor activity in zebrafish embryos: a new method to assess developmental neurotoxicity. Neurotoxicol Teratol 32(4):460–471

74. Ogungbemi A et al (2019) Hypo- or hyperactivity of zebrafish embryos provoked by neuroactive substances: a review on how experimental parameters impact the predictability of behavior changes. Environ Sci Eur 31(88):1–26

75. Beker van Woudenberg A et al (2013) A category approach to predicting the developmental (neuro) toxicity of organotin compounds: the value of the zebrafish (*Danio rerio*) embryotoxicity test (ZET). Reprod Toxicol 41:35–44

76. Liu Y et al (2017) Statistical analysis of zebrafish locomotor behaviour by generalized linear mixed models. Sci Rep 7(2937):1–9

77. Fitzgerald JA et al (2019) Emergence of consistent intraindividual locomotor patterns during zebrafish development. Sci Rep 9 (13647):1–14

78. Hsieh JH et al (2019) Application of benchmark concentration (BMC) analysis on zebrafish data: a new perspective for quantifying toxicity in alternative animal models. Toxicol Sci 167(1):92–104

79. Wasserstein RL, Schirm AL, Lazar NA (2009) Moving to a world beyond "$p < 0.05$". Am Statist 73(sup1):1–19

80. Ziliak S, McCloskey D (2008) The cult of statistical significance: how the standard error costs us jobs, justice, and lives. University of Michigan Press, Ann Arbor, MI

81. Hubbard R, Haig BD, Parsa RA (2019) The limited role of formal statistical inference in scientific inference. Am Statist 73

82. Russell WMS, Burch RL (1959) The principles of humane experimental technique. Methuen, London

Chapter 16

A Behavioral Test Battery to Assess Larval and Adult Zebrafish After Developmental Neurotoxic Exposure

Andrew B. Hawkey, Zade Holloway, and Edward D. Levin

Abstract

Behavioral test batteries are valuable methods which allow outcomes with varying characteristics and neurobiological bases to be assessed and compared in the same animals. This allows investigators to construct a profile of impairments produced by a pharmacological or toxicological challenge, and to propose mechanisms for further study based on those findings. This profile is valuable in the assessment of potentially hazardous substances, including environmental toxicants, drugs of abuse, and other neuropharmacologically active agents. Behavioral tests and batteries have been developed for a number of species, including a relatively recent and growing body of work with the zebrafish, *Danio rerio*. This chapter discusses the current zebrafish behavioral battery used in our laboratory, and some of the main factors that drove its development. The principal tests include a motility assay for larval fish (6 days post fertilization, dpf), and a battery intended for adolescent (2–3 months) and adult fish (5+ months), which assay sensorimotor, affective, and cognitive-like functions in these fish. Significant progress has been made in the areas of zebrafish neurobehavioral analysis, although further studies, refinements, and task development efforts will be needed to strengthen this approach in the future.

Key words Zebrafish, Neurotoxicology, Behavior, Sensorimotor response, Emotional function, Cognition, Social behavior

1 Introduction

1.1 Rationale for Using Zebrafish in Neurodevelopmental Behavioral Toxicology

Zebrafish are a useful complementary model for toxicology which carry some of the necessary features provided by higher model species like rodents, while preserving many of the practical and methodological advantages of lower models like invertebrates and in vitro models [1]. As such, they have an important role to play in addressing the backlog of chemicals which need a comprehensive toxicology assessment [2], estimated by the EPA to be in the tens of thousands of chemicals. As vertebrates, zebrafish have substantial genetic and functional homology with mammals including humans, which supports their predictive validity and value in risk assessments and mechanistic studies [3, 4]. Additionally, they have a well-studied genome and can be genetically manipulated for

Jordi Llorens and Marta Barenys (eds.), *Experimental Neurotoxicology Methods*, Neuromethods, vol. 172,
https://doi.org/10.1007/978-1-0716-1637-6_16, © Springer Science+Business Media, LLC, part of Springer Nature 2021

mechanistic analyses of toxicant action and vulnerability, similar to mouse models. Zebrafish have been particularly valued in developmental toxicity studies, where the strengths of this model organism are amplified. As fish embryos are translucent and grow *ex utero*, developmental processes are easily visualized, allowing them to be used for longitudinal studies in early development which cannot be reasonably conducted in mammals. Additionally, their lack of parental dependence and rapid motor development allows neurobehavioral testing to begin within the first day after fertilization and be further conducted at any desired point in their lifespan. As they age, their small size, high housing density, and low material costs allow them to be efficiently maintained for lifespan-length aging studies which are possible, but prohibitively costly, in mammals like rats, mice, and nonhuman primates [5, 6]. This advantage also allows studies to include wider concentration-response ranges and more complex mixtures than could be completed with rodents, based on the strains that numerous animals and treatment groups place on materials and space [7].

It is important to note that as with other lower models, there are limits to the predictive validity of zebrafish relative to mammals. A portion of their genes are not homologous with humans, and some features like sex-determination and reproductive systems are not readily comparable with humans [8, 9]. As ectothermic species with genome duplication (naturally evolved in teleost fish), metabolic regulation in zebrafish does have meaningful differences relative to mammals, and these differences must be considered when interpreting relevant findings. Additionally, the nervous system of fish is considerably less complex than mammals and lacks the structural and functional homology that rodent and primate models can provide. Relevant to toxicant exposures, the route of administration for fish is generally immersion, with chemical transfer across membranes, such as those in the gills, which may have pharmacokinetic differences from traditional mammalian methods such as injection or oral consumption/gavage, and these differences may be exacerbated by any additional differences in metabolism. On the other hand, zebrafish provide a more behaviorally and developmentally relevant view of the vertebrate nervous system than is modeled by invertebrate models and cell or slice cultures. Without a mammalian brain structure, zebrafish can still provide valuable insight into how conserved developmental processes act to build a complete and behaving nervous system, and how networks, cells, neurotransmitters, and ecologically relevant behaviors are affected by toxicant exposures [10, 11].

Along the spectrum of models used in toxicology, zebrafish have a profile of advantages and limitations that allow them to fill a gap between mammals and invertebrate or in vitro models. They have moderate throughput, cost, and complexity, and so may be suited to answering research questions that cannot be adequately addressed using other models. Our perspective on the zebrafish is that it is a complementary model, rather than an alternative or replacement for these other models. For applications to human health, toxicology data from zebrafish may be best used to provide preliminary risk assessments and hypotheses that can drive future mechanism studies and help set priorities for future basic and epidemiological research.

1.2 Zebrafish Husbandry and Exposure Considerations

Zebrafish are relatively simple to maintain in a laboratory setting, although certain considerations must be made for their housing, breeding, and exposures. Our zebrafish are fed three times per day using a combination of live food (*Artemia Salina,* Brine Shrimp Direct, Ogden, UT, USA, at 9 am and 4 pm) and solid food pellets (Gemma Micro at 12 pm) and maintained in flow-through aquatic racks (Aquatic Habitats/Pentair AES, Apopka FL, USA; Tecniplast USA, West Chester PA, USA). System water is a mixture of sea salt (Instant Ocean, 0.5 g/L) and buffer (Seachem Neutral Regulator, 0.3 g/L, Seachem Alkaline Regulator, 1 g/L) in deionized water. The flow-through water system connects all tanks to a central reservoir and allows automatic water exchange, temperature control and filtration, as well as community water chemistry testing and maintenance for all tanks on a given system. Importantly, shared water systems share a microbiome of nitrifying bacteria and algae which maintain water quality and combat the buildup of ammonia and nitrites from uneaten food and other waste. These features make flow-through systems superior to similar approaches with standard benchtop tanks, in that they allow a greater degree of control and consistency for housing conditions across all animals in a study. In some cases, we have conducted adult toxicology studies which required adult fish to be housed in benchtop tanks with individual heaters and bubblers for 4–12 weeks [12], and while this can be practically done, precautions must be taken to ensure the health of the fish. Standalone tanks require daily monitoring of the temperature, aerators, and fish health, as well as weekly (at minimum) complete water changes. Sustainable microbiomes cannot be reasonably established in these benchtop tanks, but can be artificially created by supplementing a portion of the tank water either with colonized water from a flow-through system (e.g., 1/3–1/2 of total volume) or with commercial stocks of nitrifying bacteria (e.g., API Quickstart™).

With respect to egg collection, we use a group breeding method to ensure each cohort contains eggs with a high level of genetic homogeneity. Briefly, tanks of mixed-sex breeders (housed at <5 fish per liter) are fed 4 times (rather than the standard 3) and a removable egg collection trap is placed in the bottom of each at ~4 pm. The trap consists of a plastic tray with tapered sides, narrow slits on the side to allow some water circulation, and a latticed lid with square holes large enough to allow eggs to fall through. To encourage the females to lay eggs on this box, artificial vegetation is attached to the lid. When the collection box is placed near the end of the day (4 pm or later), no eggs will be laid or fertilized until the following morning. If they are placed too early, however, some females may lay eggs the same day, resulting in eggs of mixed ages within the same batch.

On the morning of egg collection, the traps are removed and all eggs are washed on a mesh screen to remove debris, rinsed with a 0.01% bleach solution for 60 s to remove microorganisms from outside of the eggs, and then rinsed twice with fresh system water. After this, the eggs are transferred to a petri dish and incubated at 28 ° C ($\pm 1°$). At 3–4 h post-fertilization (hpf), the embryos are sorted under a dissecting microscope and selected embryos are randomly divided into glass petri dishes (9 cm diameter $\times$ 2 cm depth) at a density of 1 embryo per mL of system water (40 embryos per 40 mL). These petri dishes are then randomly assigned to one of the available treatment conditions for a given study. Our standard embryonic exposure protocol begins with a complete water change of the petri dish at 5 hpf, and then a change of the exposure medium every 24 h until 120 hpf. Most of the compounds we test are lipophilic, so our standard exposure medium is 0.1% dimethyl sulfoxide (DMSO). At each exposure medium change, the embryos are examined for death or dysmorphogenesis, and a log is kept on the type and frequency of any deformities. We note visible deformities including truncated bodies or tails, small or missing eyes, pericardial edema, and lordosis/scoliosis. All embryos with visible deformities are removed prior to the medium change. If a certain concentration leads to a majority of embryos in a dish dying or showing dysmorphogenesis (# dead + # deformed $\geq 50\%$ of fish), that dish is excluded from behavioral testing. If this pattern of disruption is consistent across replicate plates, the concentration is excluded entirely from subsequent behavioral testing. In our protocol, the primary goal of counting deaths and deformities is to establish the concentration-threshold for lethality and/or anatomical dysmorphology, and to investigate concentration ranges that fall below that threshold. At 120 hpf, all larvae are rinsed twice and housed in fresh system water until the time of larval motility testing (144 h).

With respect to exposures, there are a few important considerations. Our embryos are group-reared in petri dishes, while many other labs individually rear fish in microtiter plates with up to 96 wells [13]. Individually housing the embryos allows for greater tracking and measurement of the development of each fish, as the identity of a group-housed embryo or larva cannot be verified from one time point to another. The inability to track the identity of each embryo is not a major concern for our protocol, which does not take detailed measurements of the embryos and is primarily interested in concentrations below the threshold for overt dysmorphology. Another advantage of the well plate approach is that the fish are housed in the same 96-well apparatus where larval motility is usually tested. This does eliminate the need to rehousing and acclimation prior to testing; however, once the larvae can freely swim, their movements are restricted by the size of the well. Padilla and colleagues [14] did a comparison of the locomotor activity of fish raised in microtiter plates with larger or smaller wells and found that fish raised in 48- or 96-well plates (smaller wells) were substantially less active than those raised in 24-well plates (larger wells). This seems to indicate that raising embryos in restricted spaces may affect their swimming behavior. Group rearing in a petri dish (9 cm diameter) bypasses this concern and additionally provides the sensory stimulation of living in a shoal of other larvae.

Follow-up neurobehavioral testing can be performed at any age following the end of a drug or chemical exposure, although certain considerations may need to be made for housing and testing based on age. Embryonic or larval testing typically takes place during the yolk-sac stage (0–6 days post fertilization (DPF), while the animals do not require feeding. Equivalent testing is sometimes performed within a span of several days (e.g., 10 dpf) following the end of this stage without changes to housing, although water quality parameters will need to be monitored to ensure that uneaten powdered food does not result in potentially hazardous ammonia production. We generally transfer fish from petri dishes to our flow-through water system at 6 dpf, where they stay for the remainder of their lifespan. To support growth, we provide water supplementation to larval to juvenile tanks (up to 30 dpf) as a drip to eliminate any current which the fish would need to swim against, and progressively transition the fish to a higher flow as they grow. Beginning at 30 days of age, these juveniles are transitioned from the powdered baby food to live food (brine shrimp) and from the drip to a typical water flow condition. Follow-up testing generally takes place in adolescence, adulthood, or both. Zebrafish reach sexual maturity around 3 months of age, although in our experience they do not typically reach a normal adult size until closer to 5 months of age. Based on these observations, we test for adolescent effects of an embryonic exposure at 2–3 months of age, and adult effects in 6+ month old fish. In the latter portions of the lifespan, we assess effects in late adulthood at 14–15 months of age, based on our observations that fish stocks show increased rates of spontaneous scoliosis and morbidity when they exceed 18 months of age.

1.3 Behavioral Test Battery

It is important that a behavioral test battery evaluate a variety of functions including assessments of sensorimotor, emotional, social, and cognitive function. As the battery develops, the tests can be refined to improve validity, sensitivity, reproducibility, and efficiency. Validity can be determined in several ways. The environmental and drug challenges that affect behavioral responses in zebrafish should correspond to those that affect behavioral responses in mammalian experimental models and humans. The behavioral response will not be exactly the same across species but rather will be appropriate for the ethology of each species. For example, in the novel tank diving test described in detail below, an anxiolytic response would increase the time spent near the bottom of the tank, while a similar response in rats would increase the time spent near the sidewalls in an open field or in the wall-bordered arms in an elevated plus maze. Both fit can be interpreted based on a reduction of predatory risk in ways that are specific to their natural environments. Understanding these differences is necessary to ensure the construct validity of each test and the predictive validity between zebrafish and other animal models.

1.3.1   Video Tracking of Behavioral Response

Whenever possible, it is useful to include automated motion tracking in a behavioral battery for zebrafish. A number of companies have developed behavior analysis software which can detect an animal within a video and collect data on its movements, either in real time or using prerecorded testing sessions. These products have a variety of motion detection settings which must be adjusted to optimize animal identification, eliminate the tracking of nonanimal objects, and maintain consistent tracking of the same animal throughout a session. The most common issue with the tracking of zebrafish is a lack of strong color contrast between the fish and the background, which will result in video frames where no subject is being tracked and no data is collected. Our adopted program, EthoVision XT® (Noldus Information Technology, Wageningen, The Netherlands) provides a data filtering option which will include or remove subjects from analysis based on a calculation called "subject not found." Our image sampling rate is 20 frames per second, and our threshold for inclusion in any dataset is <4% of all frames labeled as "subject not found." Color contrast can be optimized by using white or backlit backgrounds, improving lighting, and by adjusting the brightness and image quality settings on the camera. Basic nonscientific cameras can be sufficient for this purpose in adult testing, while a specialized setup like the DanioVision™ lightbox (Noldus) or other design with high-resolution video and/or magnification is needed for embryos and larvae. In our protocol, the appropriate camera feed is run directly through the Ethovision software, live scored, and saved in autogenerated video files.

Automated motion tracking products also have a variety of computational options for the data and can report movement data as total or average distances moved, velocity, average distance from an object or zone, and the frequency of crossings between quadrants or zones, among others. All of these outcomes can be scored and calculated by hand, but the automation of this testing will relieve some of the drawbacks to animal behavioral testing, while also protecting the validity of the data.

Automation of both data collection and stimulus delivery has been a major focus in the development of the zebrafish battery for a few reasons. First, this ensures the consistency and quality of each method over the course of each study and between studies using the same battery. Second, it gathers very detailed movement data on the sub-second time scale, third, it improves the throughput of the battery substantially and allows comprehensive data files to be generated, exported, and analyzed, all on the same day that the tests are completed. Finally, it eliminates the need for the experimenter to be within sight of the fish, which may lead to startling or stressing of the fish each time the experimenter moves.

2 Materials and Methods

2.1 Larval Neurobehavioral Testing

The earliest zebrafish neurobehavioral assays can be conducted during embryonic or larval development. Motor functions develop quickly over the first several days of life, beginning with spontaneous tail flexion in the unhatched embryos around 17–19 hpf. These tail flexions have allowed some groups to observe neurotoxicity-induced locomotor deficits as early as 24–36 hpf [15, 16]. At these early time points, the immature embryos cannot freely swim, but do flex their tails spontaneously at a low rate and a much higher rate in response to a flash of light. This response goes through rapid inhibition and will fail to be elicited by a second flash of light delivered quickly after. As the fish continue to develop, behavior analysis can become more sophisticated and can include swimming and responsivity to other forms of stimulation. For example, Hahn and colleagues have characterized the circuitry underlying a larval startle response and the potential for neurotoxicity within this circuitry [17]. The approach of our lab and others has been to measure motility of free-swimming larvae.

Larval motility is most often tested between 120 hpf and 144 hpf, when the larval swim bladder is inflated and the yolk is still present. In our protocol [18], we perform larval motility testing at 144 hpf, which is 24 h after the end of a standard exposure. Testing is performed in a 96-well plate, fitted with circular 0.5 mL glass well inserts (*see* 96-well plate in DanioVision™ lightbox, Fig. 1a). Since we group-rear our fish in glass petri dishes, the larvae must be acclimated to the 96-well plate prior to testing. On the morning of testing, fish are individually loaded into wells and acclimated for 1 h in a dark incubator. The testing session is 50 min, consisting of 5 alternating periods of dark and lit conditions (*see* representative data, Fig. 1b). The first 10 min are conducted in the dark and treated as an acclimation or habituation phase. We generally observe elevated levels of locomotor activity

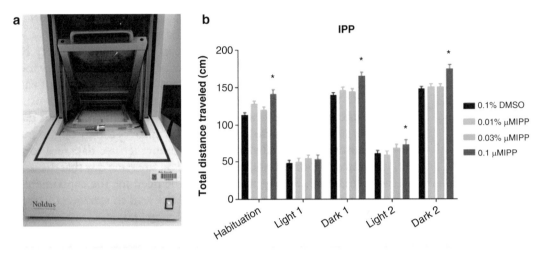

Fig. 1 Light/dark larval motility assay, (**a**) DanioVision Apparatus; (**b**) Larval activity data is shown across 5, 10-min time blocks with alternating dark-lit conditions. Embryonic exposure (5–120 hpf) to the flame retardant IPP (0.1 μM) led to locomotor hyperactivity in 4 of the 5 time blocks (mean ± sem, N = 50+) [18]

during this phase, since this occurs in the dark (0% illumination) and larvae are stimulated under dark conditions. However, the activity in the habituation phase is often noticeably lower than in the later dark phases of testing, likely due to the recent stress of being moved from the incubator to the DanioVision™ lightbox (Noldus). After the 10 min of acclimation, the floor light is turned on (100% illumination, 5000 lux) and left on for 10 min, which decreases locomotor activity for the duration of that lighting phase. Following this, the sequence continues with lights automatically turning off at 20 min, on at 30 min, and off at 40 min. Each lighting change (light-dark, dark-light) is repeated twice in order to detect any interactions between neurotoxic treatments and behavioral adaptations which can occur when stimuli are presented multiple times. The general technique of light and dark motility is widespread in zebrafish toxicology, although a number of variations exist using different numbers of light/dark phases and lengths of time in each phase (e.g., [19–21]).

In terms of data analysis, we primarily rely on the distance moved as the measure of activity. This can be further analyzed in multiple ways, including total distance moved in each min or each lighting phase, average distance per min under lit or dark conditions overall, and the change in activity (either positive or negative) during the transition from dark to light, or light to dark. If there are multiple repetitions of a dark or light phase, as in our protocol, it is statistically necessary to analyze these repetitions as repeated measures factors, prior to any analysis of average-dark and average-light activity score. This is because treatment by replicate interactions could be present (e.g., [18]) and any differences in treatment effects between repetitions must be considered.

2.2 Sensorimotor Response

Acoustic startle assays are available for both larval and adult fish, and demonstrate a sensorimotor reflex triggered by a loud noise, or similar vibration through the water. In zebrafish, startle is measured as a brief spike in locomotion following the onset of the sound stimulus [22]. This is in contrast to rodent startle responses, which are characterized by freezing or a stereotyped contraction of the body without locomotion [23]. An adult zebrafish startle can be simply measured by subtracting the baseline levels of locomotion, measured as the distance moved in the 5 s immediately before the stimulus, from the distance moved in the 5 s immediately after the stimulus [24]. This differential represents the amount of activity that can be attributed to the stimulus. As with other forms of reflexive behavior, not all fish will startle every time a stimulus is delivered, and between-group differences will reflect a combination of the rate of the response being elicited and the relative size of the response when it is elicited.

2.2.1 Tap Startle Testing

Our protocol for tap startle was originally developed by Eddins, Cerutti, and colleagues [24], and uses a custom-built tapping apparatus (Fig. 2a). This apparatus consists of eight clear Plexiglas wells (5.5 cm diameter) arranged in a 2 × 4 setup (40 mL system water, depth 2.5 cm), with short white plastic dividers to prevent the fish from seeing one another. The walls of each well are gently angled to provide a clear line of sight to the mounted camera above and eliminate blind spots. The platform for each well is fitted with a centered hole containing a 24-volt DC push solenoid, which strikes the well floor with a metal pin when triggered. Each session of tap startle consists of 10 tap stimuli, presented 1 min apart. The tap sequence begins 30 s after the start of the trial to provide brief acclimation. Tap startles are defined using the distance moved by the fish in the 5 s prior to the tap (pre-tap), the 5 s following the tap (post-tap), and the differential between them (post-tap – pre-tap).

We typically analyze tap startle data using raw total distance moved scores (*see* representative data, Fig. 2b), treating the 10 replicate taps as a repeated measures variable (e.g., [12]). Sensorimotor habituation and adaptation are measured as main effects of tap on activity, analyzed separately for pre-tap activity, post-tap activity, and the differential (or tap startle magnitude). Pre-tap activity scores tend to show an increasing trend across taps, while post-tap and differential scores tend to show a decreasing trend across taps. The distances moved prior to or after the tap can often be skewed, and may require normalization (e.g., log-transformation) for traditional analyses of variance to be performed. A similar statistical method is using a form of linear regression. Linear mixed effects analyses can generate linear models with intercepts and habituation slopes for each subject, and those parameters can then be compared between groups using simplified ANOVAs (as in [18, 25]). This

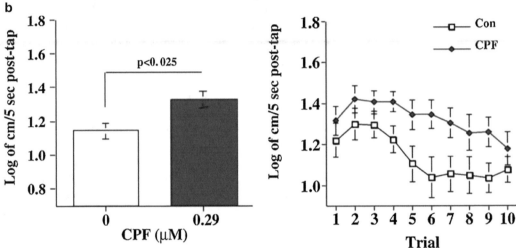

Fig. 2 Tap sensorimotor startle and habituation test, (**a**) Apparatus; (**b**) Post-tap activity (normalized via log-transformation) is reported as an average (left) or for sequential tap stimuli (right). Embryonic chlorpyrifos exposure (0.29 μM, 5–120 hpf) led to elevated post-tap activity and impaired patterns of habituation (mean ± sem, $N = 24$–40) [24]

approach does restrict what post hoc analyses can be performed, and so is not suitable for all hypothesis testing.

2.3 Novel Tank Exploration

One of the most common neurobehavioral tests across model species is exploratory behavior in a novel environment. Common variants for rodents include open field and elevated plus maze, which allow measurement of locomotor activity and spatial preferences while exploring the maze [26]. Rodents show a phenomenon called thigmotaxis, which is an innate aversion to open spaces and preference to stay close to walls. This is understood to be a defense mechanism which removes certain angles of attack and minimizes the chances of predation. Zebrafish have a similar innate aversion to the surface and preference for the floor of a novel tank [27]. The magnitude of this aversion, or any attenuation of it by a treatment, is often interpreted as a change in an anxiety-like function. This task has been behaviorally validated as measuring the diving response when the tank is novel but not when it is familiar and pharmacologically validated with attenuation of the diving response after administration of the anxiolytic drugs diazepam and buspirone [28].

Novel tank behavior can be measured in tanks of various shapes and sizes. Using a tank that is both deep and wide allows for diverse swimming patterns, although it does create some technical problems to overcome. Techniques have been developed to map the location of the fish in 3-dimensional space (e.g., [29, 30]), using two cameras to synchronously track the fish as it moves side to side, front to back, and top to bottom within the tank. Two-dimensional analyses are also quite common and can be reasonably accomplished using a narrow tank viewed from the side.

2.3.1   Novel Tank Dive Test

In our protocol (e.g., [18]), adolescent or adult fish are given 5 min to explore a narrow trapezoidal 1.5 L tank (26 cm × 5.5 cm at the top) filled with 10 cm of system water (1.2 L of system water) (*see* test apparatus and tracking pattern, Fig. 3a). In this design, the fish are allowed to dive to the bottom and continue their stereotypic swimming pattern, which consists of continuous back and forth swimming. Early in the session, the fish will choose to make very few, if any, visits to the surface, but as time passes, they will acclimate to this environment and make progressively more visits to the surface (*see* representative data, Fig. 3b). Similarly, the fish will remain less active early in the session, with some individuals remaining still during the first min(s) of the session, but as time goes on, locomotor activity will tend to increase.

The primary measures we gather represent either locomotor activity or anxiety-like floor preference. We generally represent locomotor activity as the distance moved within each min of the session and across the session, although the behavioral software provides other options, including time spent moving and average swim speed. For attraction to the bottom, this bias can be represented in multiple ways as well. In previous iterations of this test (e.g., [28]), we represented attraction to the floor by drawing a zone around the bottom third of the tank and having the software measure how much time the fish spent in that zone. This is a simple and elegant measurement, although the edges of the lower zone are arbitrary and depth may be more accurately judged as a continuous variable. In our current design (e.g., [18]), we place a thin zone around the floor of the tank and have the software generate an average distance from that zone across each min of the session. This distance from the zone measure reliably increases across time. For interpretability purposes, the distance moved and distance from the bottom should be evaluated side by side in order to allow better interpretation of each measure. Treatments which severely suppress locomotion may reduce distance from the bottom as well, if this

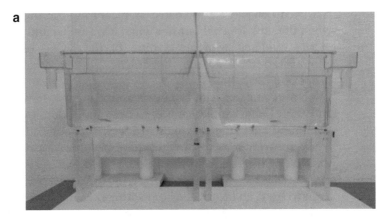

Distance from tank bottom

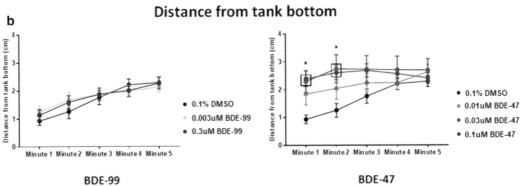

BDE-99   BDE-47

Fig. 3 Novel tank diving test to assess anxiety response vs. risk taking, (**a**) Apparatus; (**b**) Diving responses, represented as distance from the bottom (cm), are shown for two brominated flame retardants (BDE-99, left; BDE-47, right). BDE-47 (0.03–0.1 μM, 5–120 hpf) impaired the diving response during the first two 1-min time blocks of the session (mean ± sem, $N = 21$–28) [25]

results in immobile time, where the fish remain still on the bottom of the tank. In other circumstances, these two measures are fairly independent of one another, and it is common to find a treatment effect on one, but not the other.

2.4 Social Behavior   Zebrafish are social animals and regularly swim in a shoal, weaving around other fish in a loose cluster. This attraction to conspecifics can be incorporated in a formal test of social attraction. The best method for measuring this attraction is a matter of some debate. The most naturalistic method for measuring social behavior is by allowing fish to swim in a group, and then assessing how "cohesive" that shoal is based on the attraction of each fish to the others (e.g., [31]). However, this approach provides some computational and statistical challenges. For example, the software required must be able to reliably track and discriminate between each fish, even when their silhouettes merge and separate repeatedly throughout the test. Additionally, the behavior of each fish is not independent from the other fish in its shoal and treatment effects on social cues and social responsiveness may be conflated. An approach that solves this is using untreated fish as the social cues. This can be effectively done by placing live fish into a visible compartment separated by a divider (e.g., [32–34]), or by playing prerecorded videos of a shoal of zebrafish within view of the test fish (e.g., [18, 35]). In each of these cases, social attraction can be measured using approach behaviors, or relative distance from the shoal when isolated, but in view of the shoal.

2.4.1 Shoaling Assay

Our protocol is adapted from the method developed by Gerlai and coworkers [36] with modifications [18]. Fish are individually housed in narrow 1.5 L tanks for 30 min prior to testing to enhance social attraction. After this deprivation period, the fish are moved to a clear Plexiglas partitioned tank which is separated into lanes (9.25 cm × 28 cm) by black plastic walls, and placed above a light box (Huion Technology, Shenzhen, China) (see apparatus and stimuli, Fig. 4a). The light box serves as a backlight and enhances the video tracking of the fish when recorded from above. Each lane is filled with water to a depth of 10 cm (2.75 L of system water). On either end of the tank are 19.5-inch LCD monitors which display either a video of a zebrafish shoal or control shapes, which are unmoving, size-matched ovals patterned after zebrafish stripes. The test session consists of a 2 min baseline with control shapes on both screens, followed by a 5 min period with a shoaling video played on one of the two screens. Social attraction would then be scored by the relative preference for the end of the tank close to the shoal while the video is playing.

Attraction to the shoal is measured as an average distance from a small zone covering the edge of the tank adjacent to the cue screen (see representative data, Fig. 4b). As zebrafish tend to swim back and forth in a stereotyped fashion, they do not spend all of their time adjacent to the screen even when the video is playing. Instead, they make repeated visits of varying durations to the wall of the enclosure, which leads to an overall preference for the shoal-oriented side of the tank. The relative side preference of a fish is analyzed as average distance from the shoal-adjacent wall during the baseline and video phases, as well as using a differential (baseline – video) to represent the change in preference attributable to the video. An additional endpoint is locomotor activity, which we have found to be considerably higher in the backlit partitioned experimental tank than in the novel tank.

2.4.2 Predator Avoidance

Threat recognition is an important function for zebrafish, as they are prey fish with many natural predators. Vision-based fear-like responses are partly tied to the detection of movement, but zebrafish are also able to discriminate between threatening and non-threatening moving stimuli. For example, zebrafish show fear-like responses to a bird-shaped object moving overhead, videos

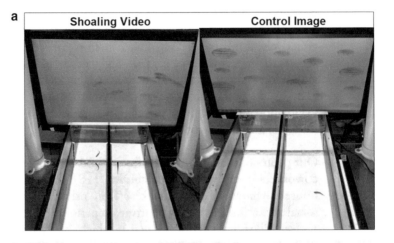

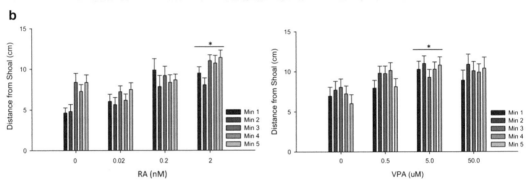

Fig. 4 Shoaling Test for assessing social affiliation, (**a**) Apparatus; (**b**) Social approach is shown as the distance from the screen playing the shoaling video, Embryonic exposure (5–120 hpf) to 2 nM valproic acid (left) or to 5 μM vitamin A (right) led to greater distances from the shoaling stimuli (mean ± sem, $N = 30$) [67]

of Indian leaf fish or needlefish and growing dots, which are 2-dimensional representations of an approaching object [36]. As noted in the previous section, zebrafish can visually identify other zebrafish, and do not show fear-like responses to videos of them. As with other functions, fear-like behaviors can be measured in a variety of ways. Responses to threatening stimuli include an increased preference for the bottom of the tank, erratic movement, jumping or leaping behavior, and freezing or floating [36]. We have further identified spatial aversion to the stimulus (e.g., [37]), which is a directional preference in the opposite direction.

2.4.3 Predator Avoidance Test

Our protocol uses a similar general approach as other fear-response tests [36] with custom stimuli and testing designs [18]. In the current version of this protocol, selected predator stimuli are presented intermittently on one end of the partitioned experimental tank described in the shoaling section above (*see* apparatus and representative stimulus, Fig. 5a). Fish behavior is recorded over a 9-min session containing by four 1-min predator cue presentations

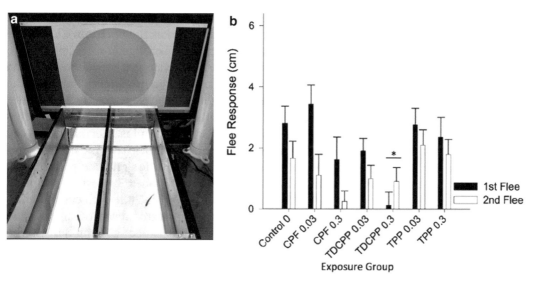

Fig. 5 Predator stimulus escape test for assessing fear response, (**a**) Apparatus; (**b**) Predator avoidance is shown as the magnitude of the fleeing response, or change in position due to the stimulus presentation. Embryonic exposure to 0.3 uM TDCPP (5–120 hpf) led to reduced fleeing responses (mean ± sem, $N = 24$–31) [68]

corresponding with min 2, 4, 6, and 8 of the session. Each predator cue appears on the same screen (left end of the tank) and a blank white screen is shown during the interstimulus interval, and on the control screen on the opposite end of the tank. Each stimulus presentation consists of a looped video of a growing dot which grows from 1 to 23 cm, to simulate a large object approaching that end of the tank. The intensity of the stimulus can be adjusted by modifying the speed of the dot growth. In our protocol, the first two predator cues are slow moving (4-s duration) while the second two are fast moving (1 s duration). The interstimulus intervals are provided to provide a control or comparison period where no predator is present.

We primarily measure the fear-like response of fish in this test as an average distance from a small zone covering the edge of the tank adjacent to the cue screen (*see* representative data, Fig. 5b). Zebrafish will tend not to stay against the wall on the opposite side of the tank, but rather continue to swim back and forth in a stereotyped "pacing" manner. As they do this, they will make some approaches toward the cue-adjacent screen. When the cue is present, it can trigger fleeing or an abrupt turn away from the cue when facing it, supporting a shorter pacing pattern that stays further from the screen-adjacent wall. One recent modification of the statistical method has been the treatment of the interstimulus interval. In many cases, but not all, fish will lose their aversion to the predator-paired screen as soon as the cue is removed and explore both sides of the enclosure equally. Based on this, previous iterations of this method have subtracted the baseline, or pre-stimulus distance from the wall from that distance during the cue presentation. This differential would theoretically represent the change in spatial preference due to the cue. However, some recent datasets have shown that the degree to which a fish returns to "no preference" during the interstimulus interval can differ between treatment groups [12]. In some cases, fish maintain an aversion to the cue-paired screen during one or more of the interstimulus intervals, which can be interpreted as a form of fear-based short-term memory. Therefore, the pre-stimulus spatial preference for one cue presentation may be impacted by carryover effects from the previous stimulus. To allow such effects to be detected, we include each cue presentation and interstimulus phase in the repeated measures analysis and consider all potential transitions between cue phases during interpretation.

2.5   Learning and Memory

In non-verbal species, learning is measured as a change in the likelihood of a response due a particular experience, while memory is a situation where current behavior is under the control of cues or events in which took place in the past. Generally speaking, these concepts are related to one another, but they are assessed somewhat differently (e.g., [38, 39]). Measurement of learning, also termed acquisition, requires repeated measurements to establish a change over time or repeated trials, and the evidence for learning is the magnitude of that change (e.g., pretest – posttest). Memory, also termed retention, can be measured on a single occasion, and the evidence for memory is that the rate or likelihood of the response is different than would be expected if no learning had taken place (e.g., observed – expected). A couple examples of learning in zebrafish have already been discussed in this chapter. Specifically, a reduction in the magnitude of the startle response over successive taps shows learning over successive trials, while the tendency to make more trips to the surface of a novel tank over time reflects learning across an extended experience in a novel environment. These are very basic forms of learning reliably shown in zebrafish, but they are perhaps not sufficient for assessing cognition, as they do not have strong face validity for more complex phenomena that are likely to be impaired by toxicant exposures, such as associative conditioning, learning from consequences, and spatial learning or navigation.

To date, a number of learning and/or memory tasks have been developed for zebrafish, largely modeled after rodent paradigms. In these paradigms, zebrafish can be tested for learned preferences for cues or locations (e.g., [40]), navigation of simple mazes (e.g., [41]), learned avoidance (e.g., [42]), and even learning based on interactions with a touchscreen [43]. Overall, however, adoption of these methods has been rather limited. This may be due to the relatively labor-intensive and low throughput of learning tests relative to other tests, and to certain obstacles that limit the adaptation of rodent designs. In terms of measuring complex behavior, zebrafish lack some basic advantages of rodents. The most important differences are that rodents are much more apt to engage with and manipulate objects and can be easily tail marked to allow identification across multiple sessions and multiple days of testing. With few exceptions, zebrafish express their choices and preferences using their location, rather than by actually touching objects, and as mentioned previously, their location in the environment is somewhat relative given their stereotypic back and forth swimming pattern. A related hazard is that zebrafish show freezing behavior when stressed, a phenomenon which can disrupt normal swimming patterns for up to several min, in our experience. Freezing can make protocols based on repeated handling and testing across trials inefficient and impractical. While a preferable strategy would be to conduct multiple test sessions across multiple days, methods for verifying the identity of a single fish across dates of testing [44] tend to be fairly invasive (e.g., fin clipping, electronic tagging) or cumbersome (e.g., automated color pattern recognition).

In our protocols, we have attempted to develop or adapt learning or memory tests that can provide reliable models of cognitive phenomena, while still remaining practical to run in toxicology studies with large numbers of fish. Below are two paradigms.

2.5.1

Three-Chamber Task

To measure learning from consequences, we developed a three-chamber test which trains the fish to avoid one chamber and to enter another instead. The 3-chamber task was originally developed with a large 40 L tank in which spatial discrimination, color discrimination, and spatial alternation tasks were developed [45, 46]. Subsequently, it was refined to be conducted in a smaller more efficient apparatus (Fig. 6a) [11]. The apparatus consists of a white PVC cylinder with closed ends (split lengthwise to create a long U-shaped channel) and straight sidewalls attached (overall size – 14.5 cm wide × 28 cm long, × 18 cm high), filled to a depth of 6.5 cm (2.5 L of system water), (*see* apparatus, Fig. 6a). The main channel is subdivided into three compartments by two thin plastic walls which can be moved using handles accessible from the outside of the apparatus. The wall placement is 4 cm from the center of the apparatus, leaving a narrow 8 cm center compartment and two 12 cm end compartments. Moving each wall changes of the size of the adjacent end compartment, which allows the experimenter to either confine a fish in that compartment or expand the swimming space available. In the three-chambered task, changes in the size of the compartment can be triggered as a consequence for entering into a particular end compartment. Zebrafish are relatively claustrophobic and will learn to prefer any compartment where the swimming space is expanded upon entry and avoid any compartment where the swimming space is confined. On the inside of the

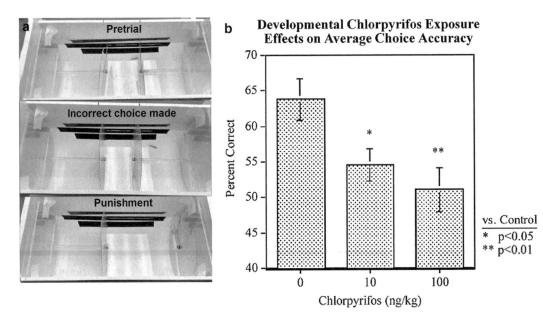

Fig. 6 Three-chamber test of learning and memory, (**a**) Apparatus; (**b**) Memory is shown as percent choice accuracy, representing the number of trials in which the fish chose the correct chamber. Embryonic exposure to chlorpyrifos (10–100 ug/l, 5–120 hpf) led to impaired choice accuracy in the three-chamber test (mean ± sem, *N* = 12–16) [69]

apparatus, three black lines run down one of the two long walls, to provide cues for spatial orientation.

To begin the testing sequence, the fish complete a spatial preference test. Each fish is placed into the central chamber of the testing tank with the circular walls positioned so that no doorways are present to allow access to the other two chambers. After 30 s, the dividing walls are rotated so that each contains a circular opening (6 cm diameter). As soon as the fish enters one of the end chambers, the walls are rotated to remove the opening and trap the fish inside. After noting the choice, the fish is returned to the center chamber and the next trial begins. In order to minimize stress, fish are not handled directly when being moved back to the center compartment. Rather, the border wall is turned so that the opening can pass safely over the fish, and carefully slid to force the fish back into the center compartment. Once the fish is in the center again, the wall can be turned to remove the opening and slid until it reaches the starting position. Spatial preference is established when the fish chooses to enter one chamber on three consecutive trials. If no preference is evident after 10 trials, the fish is excluded from further testing. Beginning with the third consecutive entry into a particular chamber, punishment training begins, whereby any entry into the preferred chamber (now referred to as the "incorrect side") results in the animal being trapped inside and the wall being moved until the chamber is reduced to a width of 1 cm for 60 s. Any entry into the opposite chamber (now referred to as the "correct side") results in no consequence and the fish is allowed to remain there for 60 s. Once the fish is returned to the center chamber, the fish remains there for a 30 s intertrial interval, after which the next trial begins. A standard session consists of 7–10 trials, excluding those that established the preferred chamber. Choice accuracy and response latency are recorded manually using a stopwatch and scoring sheet (*see* representative data, Fig. 6b). Learning can be assessed using the raw accuracy scores of each choice (correct = 100% accuracy, incorrect = 0% accuracy) (e.g., [37]) and/or plotting a linear trend to represent their improvement over trials (e.g., [47]).

2.5.2 Novel Place Recognition

To measure recognition memory, we recently began using a novel place recognition task [12] adapted from a similar Y-maze design for zebrafish developed by Cognato and colleagues [48] and based loosely on a rodent novel object recognition task we regularly use in rat developmental neurotoxicology studies (e.g., [49]). The apparatus for novel place recognition (NPR) (*see* Fig. 7a) is a clear Plexiglas plus maze with a central hub (10 cm × 10 cm) with four rectangular arms (30 cm x 10 cm) extending outward, filled to a depth of 8 cm (10 L of system water). The clear walls of the arms have replaceable covers which can be used to show distinct visuospatial cues. In our protocol, three of the arms have identical visuospatial cues, consisting of a horizontal black and white stripe pattern, while the fourth arm has different cues, consisting of a blue background with lime green dots. Testing consists of two 10-min sessions. The first is an "AA" session (AA, meaning all stimuli are identical) where the fish can explore the three striped arms for 10 min, while the dotted arm is blocked by an opaque white divider which prevents the fish from exploring or seeing it. After this familiarization session, the fish is placed into a 1.5 L holding tank and remains there for a 2 h intertrial interval. After 2 h, the fish is then placed back into the maze for an "AB" session (where one stimulus does not match the others), with the barrier to the dotted arm removed. This target arm is novel and the fish will react differently to it than to the familiar striped arms. Although Cognato and colleagues [48] demonstrated preference for the novel arm in their Y-maze, our fish demonstrate neophobia, or avoidance of the novel arm. In an unpublished pilot study, we found that the tendency to spend very little time in the novel arm could be attenuated by acute pretreatment with an anxiolytic drug, such as nicotine (Fig. 7b), suggesting that avoidance demonstrates an anxiety-like aversion to the novel arm. At sufficient concentrations, this allowed the fish to show recognition through a preference in the opposite direction (Fig. 7c). Recognition memory for the stripes is demonstrated by an altered amount of total time spent

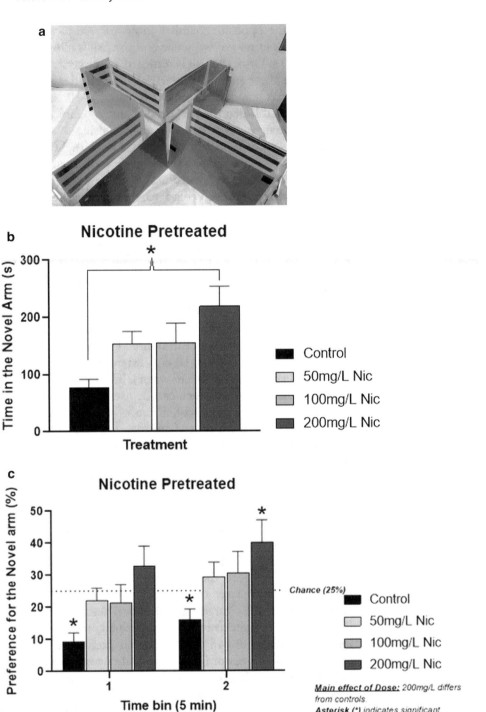

Fig. 7 Position discrimination in the plus maze, (**a**) Apparatus; (**b**) Memory is shown as time spent in the novel arm of the maze. Acute exposure to the anxiolytic drug nicotine (3 min, immediately prior to the familiarization session) dose-dependently increases novel arm time; (**c**) Memory is shown as preference (% of total time) for the novel arm. Acute exposure to the anxiolytic drug nicotine (3 min, immediately prior to the familiarization session) reverses novel arm preference from avoidance to preference (mean ± sem, $N = 20$) [Unpublished]

in the spotted arm in either direction (arm contains 23% of total water volume, so 23% of the time would be the random choice value). As with rodent novel object recognition (Hawkey et al., 2020b), the difference between exploration of the familiar and unfamiliar arms can attenuate over time as the novel cues become more familiar (as in controls in Fig. 7c). Within session-learning can be measured as the change in open arm exploration between the first and second 5-min of the 10-min trial. These time bins can be loaded as a repeated-measures variable in an analysis of variance or used to generate a simple slope which represents the degree of learning across the session.

With respect to interpretation, it is important to note that the time spent in any one arm is partially conflated with the level of activity, as very low numbers of crossings from arm to arm, likely indicating freezing, do not reflect exploration and artificially inflate the preference for the arm the fish happens to be in while immobile. If sessions are video recorded, then automated motion tracking software can be used to measure the distance moved by the fish, the amount of time spent immobile or freezing, and the frequency or duration of visits to each of the arms. This reduces the labor needed for testing, if not the throughput, and eliminates the need for the experimenter to be in the room with the fish. That being said, plus maze exploration can also be manually scored using a stopwatch to record the time spent in the novel arm and a score sheet to tally up the number of times the fish crosses from one arm to another.

3   Discussion

3.1   Interpretation of Results

The use of a battery of tests allows greater clarity on a phenotype than one test can accomplish alone. Not only does it provide a broad screen capable of detecting several different neurotoxic impacts, it also improves the interpretation of each significant difference. A heterogeneous result with some tests being positive and other being negative provides important information about the specificity of neurobehavioral toxicity. Most importantly, comparisons of different tests can indicate whether a finding is due to a general deficit which is not the primary aim of a test, such as general motor, perceptual, or motivational ability, or faithfully demonstrates a change in the target function. For example, in a study of the long-term effects of embryonic exposure to two brominated flame retardants [25], we found that low concentrations of both compounds led to reduced locomotor activity over the course of the predator avoidance test. Given that this effect was observed in this test, but not in either the novel tank or the shoaling test, we can conclude that this reduction in activity is a reaction to the test, and does not indicate a general deficit in arousal or motor ability. This distinction could not be made without complementary tasks that would be similarly affected if very basic and generally necessary functions were impacted.

3.2 Limitations of the Zebrafish Model

One important factor for which zebrafish are likely not a good model are sex differences in neurobehavioral toxicity. Certainly, there are male and female zebrafish, but how they arrive at their sex is not based on an XX vs. XY-like "master switch." Rather, both genetic and environmental factors interact to influence the ratio of males and females within a brood of fish and the development of the juvenile ovary into masculine or feminine gonads (*see* reviews, [50, 51]). Zebrafish do have relatively conserved endocrine systems, including receptors for estrogenic and androgenic signals, and so are also valuable for detecting to endocrine-disrupting chemicals [52]. However, the face validity of zebrafish sex-differences in vulnerability to a toxicant, relative to human health, remains low. Given that zebrafish may still exhibit some behavioral and vulnerability differences between males and females, we load sex as a covariate into our statistics to account for the degree of variability that is due to this factor, but tend not to include sex differences in the interpretation of data.

Another limitation of the zebrafish model is the complexity of the central nervous system. Zebrafish, like rodents and other vertebrates, have complex brains. The caudal portions of their brains are quite similar in structure to mammals. For example, the general structure and duties of the olfactory system, thalamus, amygdala, and habenula appear to be conserved across fish and mammals [53–55]. More rostral portions of the zebrafish brain are differently organized. In mammals, the main overlying brain structure is the neocortex, while with zebrafish it is the optic tectum. Likewise, functions carried out in the mammalian hippocampus and basal ganglia are carried out in less conserved networks in the zebrafish pallium [56, 57]. Given the structural differences, a zebrafish brain cannot carry out all necessary functions in the same way as a mammalian brain does, so structure-function relationships, particularly in cognitive functions, should be interpreted carefully. Zebrafish have higher face validity elsewhere, including in the general organization, synaptic structure and principal neurotransmitters in the brain [58, 59], so these endpoints are a preferable focus for mechanistic investigations of zebrafishneurotoxicology.

3.3 Needs for the Future

To date, a variety of behavioral assessments have been developed for young and adult zebrafish. Further development is still needed in order to improve the breadth, efficiency, and informativeness of existing tests and batteries. The keys to improved testing will be the automation of testing, the enhancement of throughput, improving our knowledge of the neuropharmacology of available tests, and a better understanding of how fish behavior and test sensitivity are affected by developmental age.

Of particular concern, a number of learning and cognitive tests are available, but the slow throughput and labor-intensive nature of most of these tests reduces their effectiveness and desirability. A more rapid throughput cognitive test is needed. Our three-chamber test has been sensitive to the neurobehavioral toxicity of varying compounds, including the organophosphate pesticide chlorpyrifos and the stimulant medication methylphenidate [47, 60]. This test can be completed in a single day with a small number of trials. However, few fish can be run per day and testing is quite intensive for the experimenter. The plus maze NPR task requires no manipulation of the fish or stimuli during testing, and these sessions can be recorded and scored using Ethovision software. This minimizes the labor required for scoring, but again, few fish can be run in a single day. In either case, completing the learning component of the battery for a whole cohort takes as long as the rest of the adult battery, or perhaps longer depending on the number of fish. Improved methods that can be quick and fully automated will be needed in order to best address these functions within an efficient neurobehavioral battery.

An additional concern is our ability to interpret the effect of age on the neurobehavioral effects of toxicant exposures. We typically assess both larval light/dark motility and the adult battery and frequently find discrepancies between the dosimetry and presence of behavioral changes at these two ages (e.g., [18, 25, 61]). Assessments of neurobehavioral toxicity at varying ages across development are needed to better characterize the deficits that are observed, and to explain reportedly discrepant findings. For example, testing at a single time point in early adulthood does not demonstrate whether a finding represents a lifelong deficit or perhaps a temporary deficit due to developmental delays. This is certainly true when considering the radical changes that take place during early life development, but similar issues come up in the study of declines due to aging. With aged fish, multiple testing is needed in order to discriminate between a toxin which exacerbates the processes of aging and one which produces effects that are present early and persist as the animals age (e.g., [12]). Adaptation of tests to novel age ranges is needed, and may have obstacles that will need to be resolved. For example, motion tracking can be problematic with young fish, which are small and light colored or even translucent, so improved test designs and visualization methods are needed. Some standardization can be achieved based on fish size, as testing environments can be made smaller to accommodate for size differences between younger and older fish, although the behavioral relevance of apparatus or stimulus size, water depth, and other parameters will need to be verified as well.

Interpretation will also be aided by the development of an acute pharmacological profile for each behavioral test. Screening each test against drugs with known pharmacological actions is important for two reasons. First, it helps to determine translatable neural mechanisms that underlie the test, and supports the experimenter's ability to suggest neurochemical mechanisms for follow-up investigations based on behavioral toxicology studies. Second, it offers positive controls which can be used to show the validity and sensitivity of the tests, and to support the validity of findings from the experimental groups. Positive controls are common in molecular and tissue-level analyses, but relatively rare in live animal behavioral studies. This is likely due to the additional cost and space required for additional treatment groups, but these concerns are less relevant for zebrafish models. Our typical positive control for embryonic exposure studies is the organophosphate insecticide chlorpyrifos.

An additional area of future research concerns studies being conducted across different labs and institutions. Procedures for husbandry, toxicant administration, and testing may all contribute to the behavior of an animal, and sharing and these techniques vary considerably from lab to lab. Likewise, the selection of a particular strain of zebrafish may influence the outcome (e.g [61, 62].). A lesser appreciated aspect that is becoming more recognized is the role of the microbiome (e.g., [63–65]). Zebrafish have not only an internal microbiome, but also an external biome of microbiota in the tank water. Presently, little is known about the make-up and variability between flow-through water system within a lab or between water systems maintained across labs, institutions, and our field as a whole. Further testing and investigation will be needed to parse out how these issues influence the sensitivity of our fish and our tests to toxicants of interest.

3.4 Summary and Conclusions

Zebrafish offer an outstanding intermediate complementary model that facilitates translation between high throughput in vitro cell-based assays and classic rodent models. Like the cell-based assays, larval zebrafish offer an inexpensive way to assess neurotoxicity of a wide variety of toxicants and mixtures and offer access to the molecular and cellular processes of toxic impact and response. Like classic mammalian models, zebrafish offer an integrated organism [66] with a complex brain and behavior repertoire that is the product of millions of years of evolution. Care must be taken to fashion the behavioral tests around the ethology of the species. For example, the anxiety-like response of zebrafish is to dive to the bottom of the tank rather than the wall hugging seen in rodents. This is just another form of thigmotaxis in the z-dimension that serves the same purpose, to diminish predatory threat. Also, since zebrafish are social species, individual fish are attracted to the sight of a shoal of their conspecifics. This assay likely would not work in a species that was more solitary in habit. Zebrafish show analogous behavioral responses to many toxicant and drug challenges compared with rodents and can provide a critical complementary intermediate model to fill the gap between in vitro and mammalian models.

Acknowledgment

This research and review was sponsored by the Duke University Superfund Center (ES010356).

References

1. Kalueff AV, Stewart AM, Gerlai R (2014) Zebrafish as an emerging model for studying complex brain disorders. Trends Pharmacol Sci 35 (2):63–75

2. Noyes PD, Garcia GR, Tanguay RL (2016) Zebrafish as an in vivo model for sustainable chemical design. Green Chem 18 (24):6410–6430

3. Eimon PM, Rubinstein AL (2009) The use of in vivo zebrafish assays in drug toxicity screening. Expert Opin Drug Metab Toxicol 5 (4):393–401

4. Parng C, Roy NM, Ton C, Lin Y, McGrath P (2007) Neurotoxicity assessment using zebrafish. J Pharmacol Toxicol Methods 55 (1):103–112

5. Gilbert MJ, Zerulla TC, Tierney KB (2014) Zebrafish (Danio rerio) as a model for the study of aging and exercise: physical ability and trainability decrease with age. Exp Gerontol 50:106–113

6. Ruhl T, Jonas A, Seidel NI, Prinz N, Albayram O, Bilkei-Gorzo A, von der Emde G (2016) Oxidation and cognitive impairment in the aging zebrafish. Gerontology 62 (1):47–57

7. Hill AJ, Teraoka H, Heideman W, Peterson RE (2005) Zebrafish as a model vertebrate for investigating chemical toxicity. Toxicol Sci 86 (1):6–19

8. Barbazuk WB, Korf I, Kadavi C, Heyen J, Tate S, Wun E et al (2000) The syntenic relationship of the zebrafish and human genomes. Genome Res 10(9):1351–1358

9. Lieschke GJ, Currie PD (2007) Animal models of human disease: zebrafish swim into view. Nat Rev Genet 8(5):353–367

10. Best JD, Alderton WK (2008) Zebrafish: an in vivo model for the study of neurological diseases. Neuropsychiatr Dis Treat 4 (3):567–576

11. Eddins D, Petro A, Williams P, Cerutti DT, Levin ED (2009) Nicotine effects on learning in zebrafish: the role of dopaminergic systems. Psychopharmacology 202:53–65

12. Hawkey AB, Glazer L, Dean C, Wells CN, Odamah KA, Slotkin TA, Seidler FJ, Levin ED (2020) Adult exposure to insecticides causes persistent behavioral and neurochemical alterations in zebrafish. Neurotoxicol Teratol 78:106853

13. Mandrell D, Truong L, Jephson C, Sarker MR, Moore A, Lang C, Simonich MT, Tanguay RL (2012) Automated zebrafish chorion removal and single embryo placement: optimizing throughput of zebrafish developmental toxicity screens. J Lab Autom 17(1):66–74

14. Padilla S, Hunter DL, Padnos B, Frady S, MacPhail RC (2011) Assessing locomotor activity in larval zebrafish: influence of extrinsic and intrinsic variables. Neurotoxicol Teratol 33 (6):624–630

15. Gauthier PT, Vijayan MM (2018) Nonlinear mixed-modelling discriminates the effect of chemicals and their mixtures on zebrafish behavior. Sci Rep 8(1):1–11

16. Noyes PD, Haggard DE, Gonnerman GD, Tanguay RL (2015) Advanced morphological—behavioral test platform reveals neurodevelopmental defects in embryonic zebrafish exposed to comprehensive suite of halogenated and organophosphate flame retardants. Toxicol Sci 145(1):177–195

17. Panlilio JM, Aluru N, Hahn ME (2019) Domoic acid disruption of neurodevelopment and behavior involves altered myelination in the spinal cord. BioRxiv. https://doi.org/10.1101/842294

18. Glazer L, Hawkey AB, Wells CN, Drastal M, Odamah KA, Behl M, Levin ED (2018) Developmental exposure to low concentrations of organophosphate flame retardants causes life-

long behavioral alterations in zebrafish. Toxicol Sci 165(2):487–498

19. Cowden J, Padnos B, Hunter D, MacPhail R, Jensen K, Padilla S (2012) Developmental exposure to valproate and ethanol alters locomotor activity and retino-tectal projection area in zebrafish embryos. Reprod Toxicol 33 (2):165–173

20. Gauthier PT, Vijayan MM (2020) Municipal wastewater effluent exposure disrupts early development, larval behavior, and stress response in zebrafish. Environ Pollut 259:113757

21. Knecht AL, Truong L, Simonich MT, Tanguay RL (2017) Developmental benzo [a] pyrene (B [a] P) exposure impacts larval behavior and impairs adult learning in zebrafish. Neurotoxicol Teratol 59:27–34

22. Zeddies DG, Fay RR (2005) Development of the acoustically evoked behavioral response in zebrafish to pure tones. J Exp Biol 208 (7):1363–1372

23. Curzon P, Zhang M, Radek RJ, Fox GB (2009) The behavioral assessment of sensorimotor processes in the mouse: acoustic startle, sensory gating, locomotor activity, rotarod, and beam walking.

24. Eddins D, Cerutti D, Williams P, Linney E, Levin ED (2010) Zebrafish provide a sensitive model of persisting neurobehavioral effects of developmental chlorpyrifos exposure: comparison with nicotine and pilocarpine effects and relationship to dopamine deficits. Neurotoxicol Teratol 32(1):99–108

25. Glazer L, Wells CN, Drastal M, Odamah KA, Galat RE, Behl M, Levin ED (2018) Developmental exposure to low concentrations of two brominated flame retardants, BDE-47 and BDE-99, causes life-long behavioral alterations in zebrafish. Neurotoxicology 66:221–232

26. Kraeuter AK, Guest PC, Sarnyai Z (2019) The open field test for measuring locomotor activity and anxiety-like behavior. In: Guest P. (eds) Pre-Clinical Models. Methods in Molecular Biology, vol 1916. Humana Press, New York, NY, pp. 99–103

27. Levin ED, Bencan Z, Cerutti DT (2007) Anxiolytic effects of nicotine in zebrafish. Physiol Behav 90(1):54–58

28. Bencan Z, Sledge D, Levin ED (2009) Buspirone, chlordiazepoxide and diazepam effects in a zebrafish model of anxiety. Pharmacol Biochem Behav 94(1):75–80

29. Al-Jubouri Q, Al-Nuaimy W, Al-Taee MA, Young I (2017) Computer stereovision system for 3D tracking of free-swimming zebrafish

30. Macrì S, Neri D, Ruberto T, Mwaffo V, Butail S, Porfiri M (2017) Three-dimensional scoring of zebrafish behavior unveils biological phenomena hidden by two-dimensional analyses. Sci Rep 7(1):1–10

31. Miller NY, Gerlai R (2008) Oscillations in shoal cohesion in zebrafish (Danio rerio). Behav Brain Res 193(1):148–151

32. Al-Imari L, Gerlai R (2008) Sight of conspecifics as reward in associative learning in zebrafish (Danio rerio). Behav Brain Res 189 (1):216–219

33. Moretz JA, Martins EP, Robison BD (2007) The effects of early and adult social environment on zebrafish (Danio rerio) behavior. Environ Biol Fish 80(1):91–101

34. Peichel CL (2004) Social behavior: how do fish find their shoal mate? Curr Biol 14(13): R503–R504

35. Saverino C, Gerlai R (2008) The social zebrafish: behavioral responses to conspecific, heterospecific, and computer animated fish. Behav Brain Res 191(1):77–87

36. Luca RM, Gerlai R (2012) In search of optimal fear inducing stimuli: differential behavioral responses to computer animated images in zebrafish. Behav Brain Res 226(1):66–76

37. Bailey JM, Oliveri AN, Zhang C, Frazier JM, Mackinnon S, Cole GJ, Levin ED (2015) Long-term behavioral impairment following acute embryonic ethanol exposure in zebrafish. Neurotoxicol Teratol 48:1–8

38. Christian KM, Thompson RF (2003) Neural substrates of eyeblink conditioning: acquisition and retention. Learn Mem 10(6):427–455

39. Whishaw IQ, Mittleman G, Bunch ST, Dunnett SB (1987) Impairments in the acquisition, retention and selection of spatial navigation strategies after medial caudate-putamen lesions in rats. Behav Brain Res 24(2):125–138

40. Mathur P, Lau B, Guo S (2011) Conditioned place preference behavior in zebrafish. Nat Protoc 6(3):338–345

41. Gould GG (2011) Modified associative learning T-maze test for zebrafish (Danio rerio) and other small teleost fish. In A. Kalueff, & J. Cachat (Eds.), Zebrafish Neurobehavioral Protocols, pp. 61–73

42. Wong D, von Keyserlingk MA, Richards JG, Weary DM (2014) Conditioned place avoidance of zebrafish (Danio rerio) to three chemicals used for euthanasia and anaesthesia. PLoS One:9(2)

43. Brock AJ, Sudwarts A, Daggett J, Parker MO, Brennan CH (2017) A fully automated computer based Skinner box for testing learning

and memory in zebrafish. bioRxiv. https://doi.org/10.1101/110478

44. Delcourt J, Ovidio M, Denoël M, Muller M, Pendeville H, Deneubourg JL, Poncin P (2018) Individual identification and marking techniques for zebrafish. Rev Fish Biol Fish 28(4):839–864

45. Arthur D, Levin ED (2001) Spatial and non-spatial visual discrimination learning in zebrafish (Danio rerio). Anim Cogn 4:125–131

46. Levin ED, Chen E (2004) Nicotinic involvement in memory function in zebrafish. Neurotoxicol Teratol 26:731–735

47. Sledge D, Yen J, Morton T, Dishaw L, Petro A, Donerly S, Linney E, Levin ED (2011) Critical duration of exposure for developmental chlorpyrifos-induced neurobehavioral toxicity. Neurotoxicol Teratol 33(6):742–751

48. Cognato GDP, Bortolotto JW, Blazina AR, Christoff RR, Lara DR, Vianna MR, Bonan CD (2012) Y-maze memory task in zebrafish (Danio rerio): the role of glutamatergic and cholinergic systems on the acquisition and consolidation periods. Neurobiol Learn Mem 98 (4):321–328

49. Hawkey AB, Pippen E, White H, Kim J, Greengrove E, Kenou B, Holloway Z, Levin ED (2020) Gestational and perinatal exposure to diazinon causes long-lasting neurobehavioral consequences in the rat. Toxicology 429:152327

50. Liew WC, Orbán L (2014) Zebrafish sex: a complicated affair. Brief Funct Genomics 13:172–187

51. Santos D, Luzio A, Coimbra AM (2017) Zebrafish sex differentiation and gonad development: a review on the impact of environmental factors. Aquat Toxicol 191:141–163

52. Segner H (2009) Zebrafish (Danio rerio) as a model organism for investigating endocrine disruption. Comp Biochem Physiol C 149 (2):187–195

53. Kermen F, Franco LM, Wyatt C, Yaksi E (2013) Neural circuits mediating olfactory-driven behavior in fish. Front Neural Circuits 7:62

54. Cheng RK, Krishnan S, Lin Q, Hildebrand DG, Bianco IH, Kibat C, Jesuthasan S (2016) The thalamus is a gateway for stimulus-evoked activity in the habenula. bioRxiv. https://doi.org/10.1101/047936

55. Perathoner S, Cordero-Maldonado ML, Crawford AD (2016) Potential of zebrafish as a model for exploring the role of the amygdala

in emotional memory and motivational behavior. J Neurosci Res 94(6):445–462

56. Cheng RK, Jesuthasan SJ, Penney TB (2014) Zebrafish forebrain and temporal conditioning. Philos Trans R Soc B 369(1637)

57. Mueller T, Dong Z, Berberoglu MA, Guo S (2011) The dorsal pallium in zebrafish, Danio rerio (Cyprinidae, Teleostei). Brain Res 1381:95–105

58. Panula P, Sallinen V, Sundvik M, Kolehmainen J, Torkko V, Tiittula A, Moshnyakov M, Podlasz P (2006) Modulatory neurotransmitter systems and behavior: towards zebrafish models of neurodegenerative diseases. Zebrafish 3:235–247

59. Santana S, Rico EP, Burgos JS (2012) Can zebrafish be used as animal model to study Alzheimer's disease? Am J Neurodegener Dis 1(1):32–48

60. Levin ED, Sledge D, Roach S, Petro A, Donerly S, Linney E (2011) Persistent behavioral impairment caused by embryonic methylphenidate exposure in zebrafish. Neurotoxicol Teratol 33(6):668–673

61. Crosby EB, Bailey JM, Oliveri AN, Levin ED (2015) Neurobehavioral impairments caused by developmental imidacloprid exposure in zebrafish. Neurotoxicol Teratol 49:81–90

62. Oliveri AN, Levin ED (2019) Dopamine D1 and D2 receptor antagonism during development alters later behavior in zebrafish. Behav Brain Res 356:250–256

63. Breen P, Winters AD, Nag D, Ahmad MM, Theis KR, Withey JH (2019) Internal versus external pressures: effect of housing systems on the zebrafish microbiome. Zebrafish 16 (4):388–400

64. Davis DJ, Bryda EC, Gillespie CH, Ericsson AC (2016) Microbial modulation of behavior and stress responses in zebrafish larvae. Behav Brain Res 311:219–227

65. Phelps D, Brinkman NE, Keely SP, Anneken EM, Catron TR, Betancourt D, Wood CE, Espenschied ST, Rawls JF, Tal T (2017) Microbial colonization is required for normal neurobehavioral development in zebrafish. Sci Rep 7 (1):1–13

66. Russell RW (1992) Interactions among neurotransmitters: their importance to the "integrated organism". In: Neurotransmitter interactions and cognitive function. Levin E. D., Decker M.W., Butcher L.L. (eds). Birkhäuser Boston Birkhäuser, Boston, pp 1–14

67. Bailey JM, Oliveri AN, Karbhari N, Brooks RA, Amberlene J, Janardhan S, Levin ED (2016) Persistent behavioral effects following early life

exposure to retinoic acid or valproic acid in zebrafish. Neurotoxicology 52:23–33

68. Oliveri AN, Bailey JM, Levin ED (2015) Developmental exposure to organophosphate flame retardants causes behavioral effects in larval and adult zebrafish. Neurotoxicol Teratol 52:220–227

69. Levin ED, Chrysanthis E, Yacisin K, Linney E (2003) Chlorpyrifos exposure of developing zebrafish: effects on survival and long-term effects on response latency and spatial discrimination. Neurotoxicol Teratol 25(1):51–57

Chapter 17

Evaluation of Neurotoxic Effects in Zebrafish Embryos by Automatic Measurement of Early Motor Behaviors

Elisabet Teixidó, Nils Klüver, Afolarin O. Ogungbemi, Eberhard Küster, and Stefan Scholz

Abstract

Zebrafish (*Danio rerio*) has rapidly become a popular model species for behavioral studies that may be relevant to drug screening and safety toxicology. Zebrafish embryos show a complex behavioral repertoire already a few hours after fertilization. Particularly, early stage zebrafish show characteristic behavioral features such as spontaneous tail coiling (STC) or induced movements when exposed to a short and bright light flash (called photomotor response—PMR). In this chapter, we provide the methods for assessing STC and PMR in zebrafish embryos and to detect changes provoked by chemicals. One of the protocols uses video analysis suitable for automated high-throughput screening. Moreover, both protocols describe the use of automated video analysis by using an open-source integration platform (KNIME® analytics platform), providing a flexible workflow system that can be adapted to a diversity of video recordings. We also provide a toxicological validation of this assay and show that these protocols can be used to provide an automated, high data-content readout for zebrafish behavioral responses.

Key words Video tracking, KNIME®, High-content screening, behavioral activity, neuroactive, eleutheroembryo, sublethal, developmental neurotoxicity

1 Introduction

The detection of developmental neurotoxicity (DNT) has been recognized as a major challenge by regulatory bodies based on scientific evidence for associations between chemical exposure and learning and memory impairment [1]. The majority of alternatives to animal experimentation and relevant to DNT are focused on key molecular and cellular events in vitro and do not address the potential for chemically induced DNT within an intact whole organism [2]. Bioassays using early life stages of fish offer a promising tool for identification of DNT-inducing chemicals

Supplementary Information The online version of this chapter (https://doi.org/10.1007/978-1-0716-1637-6_17) contains supplementary material, which is available to authorized users.

Jordi Llorens and Marta Barenys (eds.), *Experimental Neurotoxicology Methods*, Neuromethods, vol. 172,
https://doi.org/10.1007/978-1-0716-1637-6_17, © Springer Science+Business Media, LLC, part of Springer Nature 2021

because they capture the complexity of differentiation processes of the nervous system in early embryonic development [3]. Particularly, zebrafish are well suited for neurotoxic and DNT endpoints as they provide key advantages if compared to other models (e.g., they can be obtained in large numbers at a low cost; have a rapid development and they are transparent, accessible to high-throughput assessment and their embryonic behavior represents a promising integrating read-out that is anticipated to capture also DNT). Moreover, the basic processes of neurodevelopment in zebrafish are homologous to those that occur in humans [4].

Zebrafish embryos show a number of typical behavior patterns that are principally suitable to study the impact of chemicals on the development and functions of the nervous system. The early motor activities in zebrafish begin at 17 hpf (hours post fertilization, raised at 28.5 °C) with spontaneous tail coiling (STC). This behavior involves contractions of the trunk (coil) mediated by a discrete group of motoneurons and interneurons in the spinal cord [5]. The frequency of STC peaks between 19 and 24 hpf (depending on the strain and test conditions—e.g., use of non-chorionated or chorionated embryos) and decline progressively over a time course of 6–7 h [6–8]. At the beginning, coiling activity is exclusively driven by periodically depolarizing spinal neurons [9]. Later, during the pharyngula stage, more complex neuronal circuits are integrated, resulting in doubled-sided coiling movements which can be modified by external stimuli such as light and touch [9, 10]. Particularly, between 30 and 42 hpf zebrafish exhibit a stereotypic series of motor behaviors, some of them such as the so-called photomotor response (PMR) provoked by a high-intensity light stimulus [10]. The presentation of a light stimulus elicits vigorous high-frequency body flexions and tail oscillations in zebrafish. After 5–7 s, the excitation ceases abruptly and is followed by a refractory period during which basal activity is suppressed, and animals fail to respond to a second pulse of light. For analysis, the PMR has been divided into 4 broad phases: a pre-stimulus background phase, a latency phase, an excitation phase, and a refractory phase [11]. To quantify effects on this behavior, a motion index is calculated for every phase based on video analysis (changes in pixels).

In this chapter, we describe protocols to measure two embryonic early motor behaviors: STC and PMR. These two early motor behaviors are considered promising endpoints for neurotoxicity assessment and mechanistic studies since several studies have been published analyzing their suitability as endpoints for DNT testing and/or for the screening of neuroactive drugs [7, 10, 12–15].

For both behaviors, automated video assessment procedures are available. However, they require commercial software, cannot be applied to videos from different sources, and/or do not allow assessment of individual embryos. Therefore, we here describe the

STC and PMR measurement using a freely available data analysis software [16] that does not require commercial software licenses and thus is accessible to more scientists including those in academic environments. We also provide examples of typical results showing the effects of diclofenac and salmeterol xinafoate on early behavior.

2 Materials

2.1 Animals

Adult, healthy, and unexposed zebrafish can be obtained from local commercial suppliers or maintained in-house at 26–28 °C in a 14-light, 10-dark cycle according to standard protocols [17, 18]. Spawning of a mixed population of females and males can be triggered by placing spawning trays into the tanks the evening before. The trays are covered with a 2-mm mesh to prevent the fish from eating the eggs. Spawning occurs an hour after the onset of light. Detailed protocols for breeding zebrafish can be found in the zebrafish book [18], which is available online at http://zfin.org/. Eggs may also be provided by a contract lab if they can be made available at the stage that is required for starting the exposure. All zebrafish research should be compliant with local and animal welfare regulations.

2.2 Reagents and Equipment

The following elements are required prior to the execution of the protocol:

- Crystallization dishes (DURAN Group, Mainz, Germany) and watchmaker glasses (DURAN Group, Mainz, Germany). Alternatively, exposure may be conducted in 96-well plates (*see* **Note 1**).

- Embryo medium: Embryos are cultured in embryo medium (also known as dilution water, ISO 7346-3 [19]) from the single-cell stage until analysis. Weigh the salts: 294.0 mg/L of $CaCl_2 \times 2H_2O$, 123.3 mg/L of $MgSO_4 \times 7H_2O$, 5.5 mg/L of KCl, and 63.0 mg/L of $NaHCO_3$. Dissolve the salts in 1 L Milli Q or deionized water by stirring on the magnetic stirrer with the swizzle stick and aerate the ISO water overnight with pressurized air. Store at room temperature (for a maximum of 5 days) and adjust the pH to 7.4 ± 0.1 before the start of test (*see* **Note 2**).

- Endpoint selective controls (optional): Researchers may use a positive and negative control chemicals (which are known from previous experience to reliably affect or not STC count, *see* [8]) to test their effects on behavior. Test solutions are prepared by adding appropriate amounts of a stock solution of the test substance to the embryo medium to give a required concentration. Solvents (e.g., DMSO) can be used to accelerate solubilization of compounds (*see* **Note 3**).

- Computer with installed KNIME® analytics platform software ([16]; available at: https://www.knime.com/knime-analytics-platform) and R (recommended version 3.4.1). Depending on the video source and in case of incompatibility with KNIME® the ffmpeg software might be required to split videos into frames.

2.2.1 Specific Equipment for STC Test

Spontaneous tail coiling in zebrafish embryos are recorded with a video camera (Olympus DP21) mounted in an Olympus SZX7 stereomicroscope ($0.8\times$ magnification, camera with a minimal resolution of 120 frames per minute, *see* **Note 4**) using a black background and dark field transmitted light at the base. Alternatively, videos can be recorded with a high-resolution video camera (4.1 mega pixels, 2048×2048, Viewpoint Zebrabox).

2.2.2 Specific Equipment for PMR Test

For the photomotor response test an automated video recording system is needed (camera with minimal resolution of 30 frame/s, *see* **Note 4**) that analyzes behavior of zebrafish in multiwell plates. We used the ZebraLab from ViewPoint for the execution of this protocol (*see* **Note 5**). Embryos are transferred to squared 96-well plates prior to analysis (with 1 embryo per well and 400μL exposure medium) (Uniplate, Whatman, GE Healthcare, UK) prior to the image recording.

3   Methods

3.1   Zebrafish Embryos

Embryos are collected from the breeding tanks and placed into a petri dish for selection of fertilized embryos. Twenty fertilized eggs are transferred to clean 50 mm glass crystallization dish with 20 mL embryo medium or test solution (*see* Subheading 3.1.1) covered by a watchmaker glass. Incubate the embryos at 28.5 °C on a 14 light-10 dark cycle until ready for testing. Alternatively, exposure and analysis can be conducted in 96-well plates (we recommend squared well plates due to the higher volume per well, e.g., Uniplate, Whatman, GE Healthcare, Little Chalfont, UK). When using well plates care should be taken for potential absorption of test compounds to the plastic well plates which may result in a decline of exposure concentration [20–22].

3.1.1 Testing of Chemicals

Chemical compounds can be directly dissolved in the exposure medium or added using a solvent to accelerate solubilization. The same solvent concentration should be used for each test concentration including controls. The solvent concentration should not affect behavior and embryo survival. In the standard zebrafish embryo toxicity test (zFET) a maximum of 0.01% v/v of solvent concentration is recommended [20] (*see* also **Note 3**). Treatment duration and dosage are subject to variations depending on the chemicals used and may require a preliminary range finding test. In this protocol, a chemical concentration series was made with a constant dilution factor (e.g., factor of 2). Five to seven concentrations were prepared for each tested chemical. To expose the embryos in crystallization dishes, carefully remove all water from the dish where embryos are placed and add 20 mL of exposure medium with test chemicals. Repeat the procedure for each of the tested concentrations. For 96-well plates remove medium and add 400μL of exposure medium.

3.2 Spontaneous Tail Coiling Test (STC)

3.2.1 Time Course Analysis

Firstly, we collected data on the frequency of tail coiling during zebrafish development to identify the time point with peak frequency in our experimental setup (*see* **Note 6**). All subsequent behavioral analyses should be conducted at this time point. The experiment should be performed in a dedicated lab space distant from lab traffic and noise to avoid vibrations that may interfere with video analysis. For our time course analysis, we measured the activity of three dishes containing 20 embryos each from 23 hpf up to 30 hpf in 30 min intervals. Representative results of a time course analysis are shown in supplementary Fig. S1 (*see* **Note 7**).

1. The video camera mounted in the stereomicroscope is switched on and the settings of the video recording system are adjusted. In our experiments we use an ISO speed of 400, time exposure of 1/80, and image size of 400 × 300 pixels. The intensity of light is set up to the maximum using a black background and dark field transmitted light at the base.

2. The dishes are removed from the incubator and allowed to equilibrate at room temperature for a minimum of 15 min (*see* **Note 8**). Dead embryos are removed before the analysis.

3. The first dish is placed under the stereomicroscope and embryos are gently positioned in the middle of the dish, so they are all visible for video recording. Care should be taken to ensure that embryos are not superposed on each other and that dead embryos are removed.

4. The activity is recorded for 1 min.

5. The dish is removed from the stereomicroscope and the next dish is recorded. The dishes are returned to the incubation chamber after recording.

6. The procedure is repeated every half an hour by recording the same dishes (**step 2–4**) until embryos are 30 h old.

7. The STC frequency in each dish and time point is analyzed following Subheading 3.2.3.

In case the 96-well plates are used, the entire analysis may be conducted with only one plate (e.g., 5 exposure concentrations and one control with 16 embryos per treatment), and hence only one video would be recorded (*see* Subheading 3.3.1).

3.2.2 Behavioral Testing of Chemicals

In our protocol we evaluate the STC of embryos exposed to some chemicals by recording the activity (STC) at a specific time window (24–26 hpf) representing the age at which the STC peak in frequency.

3.2.3 Software Requirement and Installation

Mobility of each embryo is evaluated by means of a workflow using the open KNIME® Analytics Platform [16] that can be downloaded from the corresponding webpage (knime.com/installation, *see* **Note 9**). The complete workflow can be found online on Zenodo [23] or KNIME® Hub (https://hub.knime.com/) and a video tutorial is available at https://youtu.be/wgJN71zTvRw . The first time the workflow is opened in KNIME®, the software asks to install some extensions (math formula, quickforms, image Processing, external tool support and interactive R statistics integration). The instructions on the screen should be followed to install the extensions. The workflow uses image analysis of video frames to compute how many times an embryo shows spontaneous tail coiling. The workflow is divided into several sections using metanodes that group several other nodes. Nodes represent visual programming units each representing a specific data or image transformation step (basic description of the KNIME® Analytics Platform can be consulted in [16, 24]). Video input is required as AVI format (*see* **Note 10**). For analysis of petri dishes the workflow iterates over all files in the directory selected. For well plate analysis the selected plate is analyzed. The AVI files are converted to stacks of images using the FFmpeg tool [25] (Fig. 1a). The program can be downloaded from ffmpeg.org and is executed by using the external tool node in KNIME® that should be configured prior to the first analysis (supplementary file Table S1). The workflow also requires the installation of R software (r-project.org) and R package *quantmod* [26]. If it is the first time to use R in KNIME®, *Rserve* package should be installed and the path to R home in KNIME® has to be defined (by going to File→Preferences→Knime→R and browsing to find the R folder).

3.2.4 Image Analysis Steps with the KNIME® Workflow

The analysis of video recordings from petri dishes can be started by loading the directory containing the AVI files into the *List Files* node. Firstly, the *Global Thresholder* node is executed to detect the individual embryos in the image by applying a threshold. For each frame a median filter is applied to smooth the image and the background is subtracted using the *ImageJ macro* node (*Substract-Background* function). Then a threshold is automatically set and applied using a variable node. The threshold is based on the mean

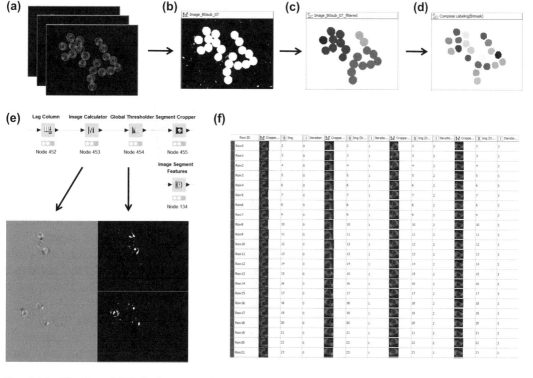

Fig. 1 Identification of STC by image analysis using a KNIME® workflow. (**a**) Video files (AVI format) are converted to image stacks at a 120-frame rate. (**b**) A threshold is applied to identify the location of the embryos in the image, (**c**) the binary image is segmented. (**d**) To separate adjacent touching embryos, we executed a "clump breaking" procedure and some morphological operations (image processing operations that process image based on shapes to improve the identification of each embryo), (**e**) then a lag column is created to identify difference in pixel between consecutive frames indicating movement (using the *Image calculator* node), a binary of the image is created (*thresholder* node) to measure the variance of pixels across all frames (*Image segment features* node). (**f**) The location of each embryo in each frame is binned to identify the identity of the same embryo during the whole sequence of frames. Each column represents the identified embryos and each row shows the embryo in every single frame

pixel intensity of all frames, but it can be adjusted depending on the characteristics of the video files. The thresholded images are inspected by clicking right on to the *Global Thresholder* node and selecting *Threshold Images*. Images with the applied threshold (column name: Image_BGsub_GT) should display embryos completely filled in white and with background in black (Fig. 1b). If this is not the case, threshold should be adjusted in the *Threshold* component node by increasing or decreasing the threshold. In order to separate adjacent touching embryos, a "clump breaking" procedure using the *Wahlby cell clump splitter* node in KNIME® [27] is executed and some morphological image operations (e.g., erosion) are performed. At the end, images are segmented to detect each individual embryo (Fig. 1c, d). The *Interactive Segmentation*

View node is executed to view if images are segmented appropriately to label each embryo independently.

The variance of gray values of two successive images is identified in order to compute the times that an embryo moves during a minute. For that a lag column is created and the difference of pixel across each video frame stamp is calculated using the *image calculator* node. Then a threshold is applied and the variance in pixels is extracted using the *Image segment features* node for each labeled embryo (Fig. 1e, *see* **Note 11**). To identify the same embryo in each frame, the centroid of each labeled embryo is calculated defining the position of the embryo in the image. The centroid is used to bin the labels according to their proximity to each other, so the identity of each embryo could be traced over all the frames (Fig. 1f). Because the whole embryo egg can move between each frame, a distance of 15 pixels was considered for the binning process. This distance depends on the resolution of the video recording and it can be changed under the *Table creator* node named *Pixel difference binner position*. The *Image Viewer* node can be checked to see if eggs are correctly binned.

At the end, an excel file with the same name as the video is automatically saved with the output results. The frequency of movements of each embryo is analyzed by iterating over each label and using as input the pixel variance over all the frames. If the variance increases, it means that there is a tail coiling, so during the time series, peaks are identified using the function "findpeaks" of the R package "quantmod" [26] by means of an *R snippet* node. Figure 2 shows an example graph obtained with the identified peaks for an embryo. The frequency of coils per minute is calculated considering the total duration of the video recordings.

We included a metanode (named *Check for error*) after the *position binner* metanode that serves as an internal control to detect when there is an error during image analysis. For instance, when an individual embryo is not correctly identified. That would require repeating the analysis of the specific video by adjusting the first threshold value (the one that allows identifying each individual embryo) or correcting the distance for the binning process. If there was an error during image analysis, it will be indicated on the first sheet of the output excel file under the column "prediction" as "Error in image analysis."

For analysis of 96-well plate videos recorded with a high-resolution video camera an alternate KNIME® workflow is provided ([23]; detailed instructions are included in the workflow). In brief, one video file has to be selected for each analysis. The video is split into frames (*see* **Note 4**). The user has to mark the upper left and lower right corner of wells A1 and H12. Embryos are not recognized automatically but are labeled by the user by clicking on the center of each embryo. Subsequent frames are subtracted

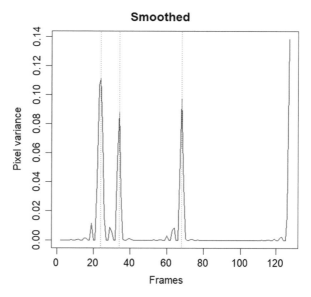

Fig. 2 Example of recorded spontaneous coiling of a control embryo. Representative graph obtained with the STC analysis in KNIME® with the identified peaks (tail coils) for one embryo. X-axis represents the frames (120 frames in total) and y-axis the pixel variance. Vertical blue dotted lines represent the peaks identified using the function "findpeaks" of the R package "quantmod"

from each other and a threshold is used to generate binary images. The threshold needs to be selected by the user and may require initial testing and adjustment for an optimal sensitivity/noise ratio. Based on the embryo centroid a circle label is calculated that covers the area of the embryo. The centroid coordinates are used to assign each embryo with the well number. For each binary subtraction image and embryo label the mean pixel values are calculated. Pixel means are binned for 0.6 s intervals and used to generate time course images for each embryo. An embedded R-script using the library *quantmod* is used to calculate the number of peaks. Initially testing and adjustment of different threshold and smoothing options of the R script may be required to obtain optimal results. An excel file is generated to provide raw data plots, mean plots for each treatment and the entire plate, threshold images for inspection of signal-to-noise ratio and the frequency of peaks (spontaneous tail coiling) for each treatment.

3.2.5 *Data Analysis*   The excel data sheet "Raw data" provides the number of times each embryo moves in a minute (column name: peak.number) and also the mean frequency in that dish (column name: frequency). Concentration-response curves can be generated using the mean frequency from control and treatment dishes with the *drc* package in R [28]. A minimum of two replicates are used. For each replicate, the STC count is normalized to the respective mean control values

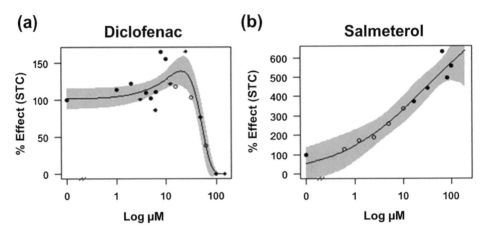

Fig. 3 Effects of (**a**) diclofenac (non-neurotoxic compound) and (**b**) salmeterol xinafoate (neurotoxic, β2-adrenergic receptor agonist) on spontaneous tail coiling. Diclofenac showed a decrease in activity over increasing concentrations while hyperactivity was found after exposing zebrafish embryos to salmeterol. Each data point shows the mean STC of 20 embryos and different symbol types represent different replicates. The solid black line represents the curve fit of the data with the confidence intervals in gray shading

and expressed as a percentage of control (*see* **Note 12**). From the fitted concentration-response curves the concentration at which a 50% change in STC is observed can be calculated.

Figure 3 shows as example the concentration-response curves of zebrafish embryos exposed to two neurotoxic substances. Embryos were exposed to increasing concentrations of the chemicals (diclofenac, CAS 15307-79-6, 99% purity and salmeterol, CAS 94749-08-3, 98% purity) just after fertilization (<2hpf). The results show an activity increase (hyperactivity) compared to control for salmeterol with an EC_{50} of 8.7μM while hypoactivity was found for diclofenac with an EC_{50} of 50.9μM (Fig. 3). Salmeterol is a β2-adrenergic receptor agonist (B2AR) that can provoke a stimulation of the nerve cells by norepinephrine release. Moreover, prenatal exposure to B2AR has been associated with risk of autism spectrum disorders [29]. Diclofenac appears to have some negative effects on both development and differentiation of nerve cells [30, 31] and it has also been linked to alteration of behavioral development in young rats [32].

3.3 Photomotor Response Test (PMR)

For the assessment of the PMR, diverse protocols had been established (e.g., [10, 24]). These protocols were based on using microscope imaging of individual wells and/or more than one embryo per well. Here, we present a protocol that allows the analysis of 96-well plates with one embryo per well. We use the PMR for embryos at 30–32 hpf, which have been incubated at 28 °C. The advantage of this protocol is the compatibility with high-throughput analysis and possibility to combine with automated phenotype assessment and other behavioral assays such as the

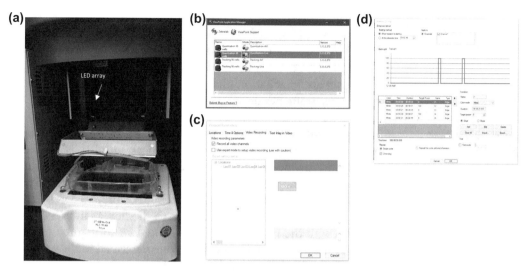

Fig. 4 Automated video recording system (e.g., Viewpoint system with ZebraLab software) that images individual embryos housed in 96-well plate. (**a**) ZebraBox is equipped with an array of white LEDs on top of the well plate to record the light flashes for PMR measurement, (**b**) screenshot showing the zebraLab settings (the Quantization-Live mode is used for recording and measuring the activity on the wells). (**c**) The video is recorded over the full-length of the protocol and (**d**) light flash triggers are set up as 20 s dark (0% target power), 1 s light (100% target power), 9 s dark (0% target power), 1 s (100% target power), and 19 s dark (0% target power)

locomotor response (LMR) [33]. The recording is conducted using the ZebraBox (ViewPoint) but principally the workflow is independent from the system and camera—provided that a video with appropriate frame rates and resolution is generated (*see* **Note 4**). The videos are analyzed with a KNIME® workflow to calculate the mean activity per PMR subphase [12]).

3.3.1 Hardware and Setting up the ZebraLab Program

The ZebraBox is equipped with an array of white LEDs on top of the well plate and the intensity is controllable from 0 up to 40,000 Lux (Fig. 4a). On board is a Full HD camera 4.1mega pixels (2048 × 2048), 1"CMOS image sensor, speed up to 90 fps, additionally a zoom lens is attached to the camera to capture a full 96-well plate. In order to keep the temperature stable, the Zebra-Box is installed in light-protected temperature-controlled incubator. We use 28 °C ± 1 °C for behavior analysis.

In our analysis pipeline we analyze the video recorded with the ZebraLab Software. We use the Quantization-Live mode to set up the light flash triggers and the video is recorded over the full-length of the protocol (Fig. 4b, c). Therefore, the video recording should be enabled with the settings (enable "record all video channels"). Since we are analyzing the PMR video in a KNIME® workflow, no time binning within the time & options settings has to be selected in the ZebraLab. The PMR stimulus setting for the TopLight is 20 s dark (0% target power), 1 s light (100% target power), 9 s dark

(0% target power), 1 s ligth (100% target power), and 19 s dark (0% target power) (Fig. 4d). Thus, the full length is 50 s. Before starting the PMR protocol the embryos at the age of 30–32hpf are placed into squared 96-wells filled with 400µL exposure medium and a minimum acclimatization phase for 5 min in the ZebraBox (in the dark) is recommended.

3.3.2 PMR Analysis in KNIME®

The protocol for PMR assessment is very similar to the STC assessment in 96-well plates. However, the peak number is not computed, and the mean activity is calculated for three different phases of the PMR. The workflow allows the analysis of one plate. Before the analysis a plate layout (excel file) should be prepared as is shown in supplementary file Table S2. Prior to the first analysis, communication with the FFmpeg tool [25] should be configured (supplementary file Table S1). Detailed steps are described in the workflow annotations. Once the plate layout and corresponding video file are uploaded, the center of each embryo is manually labeled using the Interactive annotator node (Fig. 15a). Furthermore, the upper left and lower right corners of the 96-well plate (A1 and H12, Fig. 5b)

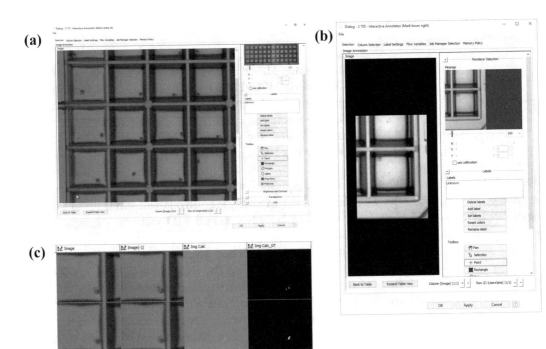

Fig. 5 PMR analysis with the KNIME® analytics platform. (**a**) Each embryo is labeled with a point using an *Interactive annotator* node, (**b**) upper left and lower right corners of the 96-well plate are marked using another *Interactive annotator* node to relate each embryo with a well position. (**c**) The first and second column (*Image* and *Image(−1)*) display subsequent frames that are subtracted from each other (*column 3, ImgCalc*) and a threshold is used to generate binary images (*column 4, ImgCalc_GT*). The threshold may initially require testing and adjustment for an optimal sensitivity/noise ratio

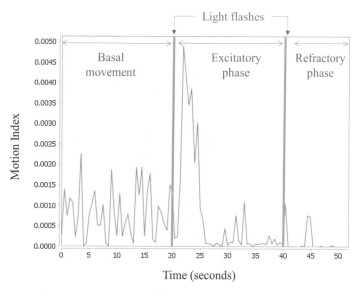

Fig. 6 Activity measurement of a PMR experiment. Representative activity plot of a control group. The x-axis represents time and the y-axis the motion index. A higher motion index represents more movement. Vertical lines at 20 and 40s represent the light pulses. The different phases are indicated in the graph and normalized activities of each PMR phase are calculated for each embryo and the entire plate

are marked to relate each embryo with a well position. A threshold is used to convert subtractive images to binary images. The threshold may initially require testing and adjustment for an optimal sensitivity/noise ratio (Fig. 5c). Finally, the workflow is executed to save results of the analysis as an excel file including normalized activities of each PMR phase, activity plots for each embryo, mean activity plot for each treatment group, and a mean activity plot of the entire plate (Fig. 6).

3.3.3 *Data Analysis*   Concentration-response analysis is typically conducted for normalized activity in the excitatory phase but can be principally applied for any of the phases. In addition, the resulting data tables can be used for any other/user specific analysis.

4   Troubleshooting Section

Problem	Suggestions
Embryos not detected or labels are not separated	Optimize the threshold component node in the KNIME® workflow to the user-specific image qualities. Thresholded images should display

(continued)

Problem	Suggestions
	eggs filled in white and background in black. Try also to change "labeling filter" node settings (minimum and maximum segment area) inside "split label" component node.
Labels are correctly created but still appearing "error in image analysis"	Adjust the distance for the binning process (table creator node) according to the user-specific image qualities. Otherwise, check if embryos change in position (due to their strong movement) and exclude them from the analysis

5 Notes

1. Hydrophobicity of substances should be checked in advance to indicate possible sorption to the exposure well material (check OECD [20] and [21, 22]).

2. Ionizable compounds may be completely charged at the pH in the exposure medium, leading to no or very slow uptake and thus to no or to very high effect concentrations [34, 35].

3. Appropriate control group treated with the vehicle solution (embryo medium) and group treated with a solution containing the solvent used (when applicable) should be used. This is important to detect and rule out undesirable effects of solvent on behavior. Exposure to most solvents can impact embryo development and viability. Therefore, solvent concentrations should be kept as low as possible. Our data suggested that DMSO causes hyperactivity with an EC_{50} of 1.96% (v/v) [8] and no morphological effects after 24 h have been observed up to 2% [36]. Therefore, the recommended solvent concentration of 0.01% (v/v) in the standard zebrafish embryo toxicity test (zFET) are well below those causing effects on embryo survival and development.

4. For the STC measurement, the camera had a resolution of 120 frames/minute with a video image size of 400 × 300 pixels. For the measurement of the PMR and STC in 96-well plate, although 30 frames/s were used, it may be possible to use a lower frame rate.

5. STC and photomotor response videos may be recorded and measured using software made by companies other than View-Point (e.g., Daniovision from Noldus (Wageningen, The Netherlands)).

6. Reduced movement or activity have also been linked to malformations in the zebrafish when exposed to chemicals [15, 36], so it is recommended to annotate which embryos are malformed or to discard them directly from the analysis. That will allow us to be sure that the observed effect in behavior is not due to an indirect effect of abnormal morphology.

7. The background time course on spontaneous movements and PMR should be checked as it may vary depending on genotype, strain, and incubation temperature.

8. The assay is performed under room temperature; dishes are acclimated to room temperature for at least 15 min to ensure a stable temperature during the analysis. There were no significant statistical differences between embryos analyzed between a period of 15–30 min (*see* [8]). Temperature could be monitored over the course of the STC analysis; variations on temperature could principally impact on STC frequencies. We recommend to normalize frequencies to the control in each experiment.

9. The KNIME® platform and ffmpeg tool are also available for MacOs and Linux operating systems. The workflow has been run in Microsoft operating system and it should be possible to run it on MacOs and Linux, albeit it has not been tested in these operating systems.

10. Other video formats can be used, but they should be converted to AVI files or directly to stack of images prior to KNIME® analysis (e.g., using freely available ffmpeg tool [25]). Depending on the video format, frames may be directly analyzed by KNIME® without requirement of the ffmpeg tool.

11. The second threshold in the STC analysis was set by verifying the concordance between the final KNIME® output and the visual count of the STC. The first time you use the workflow, this threshold should be adjusted by comparing the KNIME® STC count and the visual count of a couple of video recordings.

12. Hypo- or hyperactivity effects relative to control can be quantified using hypothesis testing or concentration-response modeling. The normal range of STC count for untreated embryos is from 2 to 5 coils per minute for the used experimental conditions [8]. We recommend estimating the range and mean STC count in your zebrafish strain and experimental conditions to set up a validity criterion of the test.

References

1. Bailey JM, Oliveri AN, Levin ED (2015) Pharmacological analyses of learning and memory in zebrafish (Danio rerio). Pharmacol Biochem Behav 139(Pt B):103–111

2. Crofton KM, Mundy WR, Lein PJ et al (2010) Developmental neurotoxicity testing: recommendations for developing alternative methods for the screening and prioritization of chemicals. ALTEX 28:9–15

3. Coecke S, Goldberg AM, Allen S et al (2007) Workgroup report: incorporating in vitro alternative methods for developmental neurotoxicity into international hazard and risk assessment strategies. Environ Health Perspect 115:924–931

4. Tropepe V, Sive HL (2003) Can zebrafish be used as a model to study the neurodevelopmental causes of autism? Genes Brain Behav 2:268–281

5. Brustein E, Saint-Amant L, Buss RR et al (2003) Steps during the development of the zebrafish locomotor network. J Physiol 97:77–86

6. Saint-Amant L, Drapeau P (1998) Time course of the development of motor behaviors in the zebrafish embryo. J Neurobiol 37:622–632

7. Zindler F, Beedgen F, Braunbeck T (2019) Time-course of coiling activity in zebrafish (Danio rerio) embryos exposed to ethanol as an endpoint for developmental neurotoxicity (DNT) – hidden potential and underestimated challenges. Chemposphere 235:12–20

8. Ogungbemi AO, Teixido E, Massei R et al (2020) Optimization of the spontaneous tail coiling test for fast assessment of neurotoxic effects in the zebrafish embryo using an automated workflow in KNIME®. Neurotoxicol Teratol 81:106918

9. Drapeau P, Saint-Amant L, Buss RR et al (2002) Development of the locomotor network in zebrafish. Prog Neurobiol 68:85–111

10. Kokel D, Bryan J, Laggner C et al (2010) Rapid behavior-based identification of neuroactive small molecules in the zebrafish. Nat Chem Biol 6:231–227

11. Kokel D, Peterson RT (2011) Using the Zebrafish Photomotor response for psychotropic drug screening. Methods Cell Biol 105:517–524

12. Selderslaghs IWT, Hooyberghs J, Blust R et al (2013) Assessment of the developmental neurotoxicity of compounds by measuring locomotor activity in zebrafish embryos and larvae. Neurotoxicol Teratol 37:44–56

13. Vliet SM, Ho TC, Volz DC (2017) Behavioral screening of the LOPAC1280 library in zebrafish embryos. Toxicol Appl Pharmacol 329:241–248

14. Legradi J, el Abdellaoui N, van Pomeren M et al (2015) Comparability of behavioural assays using zebrafish larvae to assess neurotoxicity. Environ Sci Pollut Res Int 22:16277–16289

15. Ogungbemi A, Leuthold D, Scholz S et al (2019) Hypo- or hyperactivity of zebrafish embryos provoked by neuroactive substances: a review on how experimental parameters impact the predictability of behavior changes. Environ Sci Eur 31:88

16. Berthold MR, Cebron N, Dill F et al (2008) KNIME: the Konstanz information miner. In: Preisach C, Burkhardt H, Schmidt-Thieme L et al (eds) Data analysis, machine learning and applications. Springer, Berlin, Heidelberg, pp 319–326

17. Lammer E, Carr GJ, Wendler K et al (2009) Is the fish embryo toxicity test (FET) with the zebrafish (Danio rerio) a potential alternative for the fish acute toxicity test? Comp Biochem Physiol C Toxicol Pharmacol 149:196–209

18. Westerfield M (1995) The zebrafish book: a guide for the laboratory use of zebrafish (Brachydanio rerio). Oregon press, University of

19. ISO International Standards (1996) Water quality – Determination of the acute lethal toxicity of substance to a freshwater fish [Brachydanio rerio Hamilton-Buchanan] ISO 7346-3: Flow-through method. http://www.iso.org

20. OECD (2013) Test no. 236: fish embryo acute toxicity (FET) test. OECD guidelines for the testing of chemicals, subheading 2. Paris, France. 1–22

21. Fischer FC, Cirpka OA, Goss K-U et al (2018) Application of experimental polystyrene partition constants and diffusion coefficients to predict the sorption of neutral organic chemicals to multiwell plates in in vivo and in vitro bioassays. Environ Sci Technol 52:13511–13522

22. Riedl J, Altenburger R (2007) Physicochemical substance properties as indicators for unreliable exposure in microplate-based bioassays. Chemosphere 67:2210–2220

23. Teixidó E, Klüver N, Ogungbemi A, et al (2020) KNIME workflows for the evaluation of neurotoxic effects in zebrafish embryos by automatic measurement of early motor behaviours [Data set]. https://doi.org/10.5281/zenodo.3835947

24. Copmans D, Meinl T, Dietz C et al (2016) A KNIME-based analysis of the Zebrafish photomotor response clusters the phenotypes of 14 classes of neuroactive molecules. J Biomol Screen 21:427–436

25. Developers FFmpeg (2016) FFmpeg tool. http://ffmpeg.org

26. Ryan JA, Ulrich JN (2018) Quantmod: Quantitative Financial Modelling Framework. R package version 0.4-13.

27. Wahlby C, Sintorn I-M, Erlandsson F et al (2004) Combining intensity, edge and shape information for 2D and 3D segmentation of cell nuclei in tissue sections. J Microsc 215:67–76

28. Ritz C, Baty F, Streibig JC et al (2015) Dose-response analysis using R. PLoS One 10: e0146021

29. Croen LA, Connors SL, Matevia M et al (2011) Prenatal exposure to β2-adrenergic receptor agonists and risk of autism spectrum disorders. J Neurodev Disord 3:307–315

30. Kudo C, Kori M, Matsuzaki K et al (2003) Diclofenac inhibits proliferation and differentiation of neural stem cells. Biochem Pharmacol 66:289–295

31. Andreasson KI, Savonenko A, Vidensky S et al (2001) Age-dependent cognitive deficits and neuronal apoptosis in cyclooxygenase-2-transgenic mice. J Neurosci 21:8198–8209

32. Biol TJ, Elibol B, Aritan Oğur B et al (2019) Prenatal exposure of diclofenac sodium alters the behavioral development of young Wistar rats. Turk J Biol 43:305–313

33. Teixidó E, Kießling TR, Krupp E et al (2019) Automated morphological feature assessment for Zebrafish embryo developmental toxicity screens. Toxicol Sci 167:438–449

34. Bittner L, Teixidó E, Keddi I et al (2019) pH-dependent uptake and sublethal effects of antihistamines in zebrafish (Danio rerio) embryos. Aquat Toxicol 38:1012–1022

35. Maes J, Verlooy L, Buenafe OE et al (2012) Evaluation of 14 organic solvents and carriers for screening applications in Zebrafish embryos and larvae. PLoS One 7:e43850

36. Padilla S, Hunter DL, Padnos B et al (2011) Assessing locomotor activity in larval zebrafish: influence of extrinsic and intrinsic variables. Neurotoxicol Teratol 33:624–630

Chapter 18

Application of Fluorescence Microscopy and Behavioral Assays to Demonstrating Neuronal Connectomes and Neurotransmitter Systems in *C. elegans*

Omamuyovwi M. Ijomone, Priscila Gubert, Comfort O. A. Okoh, Alexandre M. Varão, Leandro de O. Amaral, Oritoke M. Aluko, and Michael Aschner

Abstract

The nematode *Caenorhabditis elegans* (*C. elegans*) is a prevailing model commonly utilized in a variety of biomedical research fields, including neuroscience. Due to its transparency and simplicity, it is becoming a choice model organism for conducting imaging and behavioral assessment crucial to understanding the intricacies of the nervous system. Here, the methods required for neuronal characterization using fluorescent proteins and behavioral tasks are described. These are simplified protocols using fluorescent microscopy and behavioral assays to examine neuronal connections and associated neurotransmitter systems involved in normal physiology and aberrant pathology of the nervous system. Here, we aim is to make available to readers some streamlined and replicable procedures using the *C. elegans* models as well as highlighting some of the limitations.

Key words *C. elegans*, Behavior, Fluorescence microscopy, Neurotransmitters, Connectomes

1 Introduction: Strength of *C. elegans* in Neuronal Systems Evaluations

The *Caenorhabditis elegans* (*C. elegans*) model is a soil free-living nematode, whose advantages include small size (~1 mm in adulthood), simple anatomy, short life cycle (~3 days at 20 °C), numerous progeny (300 per worm), simple maintenance, four larval stages (L1, L2, L3, and L4), and finally the adult. Under adverse conditions, the worm may assume a resistant larva stage, referred to as *dauer*. *C. elegans* compact and well-annotated genome presents considerable homologies of 60–80% to mammalian genes [1]. The

The original version of this chapter was revised. The correction to this chapter is available at https://doi.org/10.1007/978-1-0716-1637-6_23

Jordi Llorens and Marta Barenys (eds.), *Experimental Neurotoxicology Methods*, Neuromethods, vol. 172, https://doi.org/10.1007/978-1-0716-1637-6_18, © Springer Science+Business Media, LLC, part of Springer Nature 2021, Corrected Publication 2023

worm can be genetically modified to many target genes through established genetic methods such as mutagenesis and RNA interference (RNAi) knockdown, affording several options for the use of *C. elegans* for studies on gene-mediated processes [2]. In contrast to rodent models, *C. elegans's* transparency allows for easy observations of intra-body structures and the use of labeled (e.g., fluorescence) biomolecules in vivo. Fluorescent proteins to tag specific proteins can elucidate the developmental process, neurotoxicity, fat storage, mitochondria integrity, and characterize protein expression [3–6]. Thus, the construction of transgenics expressing the green fluorescent protein (GFP) in specific neurons enables the evaluation of metal neurotoxicity in *C. elegans* with relative ease [6, 7]. Laboratories interested in *C. elegans* research can quickly start the colony by ordering the strains (wild-type and transgenics) and *Escherichia coli* (*E. coli*) from *Caenorhabditis* Genetics Center (CGC). The wholly mapped *C. elegans* nervous system comprises 302 neurons in adult hermaphrodite worms and 383 in adult male worms resulting in approximately 8000 synaptic connections. Most male-specific neurons are tail-associated and involved in mating behavior [8].

The above features provide quite a number of benefits to the continuous use of *C. elegans* biomedical for research. However, these benefits do not eliminate some of its limitations as an ideal experimental model. First, its small body size and shortened length impedes proper biochemical investigations of oxidative markers and genetic products of nutrition. These have created a relatively limited usage of the nematode in research involving aging and some system-specific markers such as adaptive immunity and tissue-signaling source [9–11]. Furthermore, the simple anatomical plan and biology of *C. elegans* constrains its tissue organization into specialized organ-systems, thereby creating an evolutionary gap when compared to humans [12]. The *C. elegans* model further lacks the appropriate inflammatory cell types, microglial cell, that can equilibrate those expressed in humans during neurodegenerative changes [13, 14]. Other difficulties encountered when adopting this model is in its use to imitate the underlying pathology in neurodegenerative disorders due to mammalian cell interacting complexities which lead to a multifactorial disease nature in humans [15]. Despite the above limitations, the *C. elegans* model is proving vital to unraveling cellular and molecular mechanisms that underlie normal and perturbed processes in the nervous system.

Since changes in ionic gradients continuously activate neurons leading to the transmission of information, any change in homeostatic physiological processes can impact their optimal function triggering neurodegeneration. The *C. elegans* genome encodes conserved neurotransmitter biosynthesis and releasing mechanisms and receptors to mammal neurotransmitters such as acetylcholine, glutamate, γ-aminobutyric acid (GABA), serotonin, and dopamine [16]. Glutamate is an excitatory neurotransmitter in vertebrates and invertebrates; its role and each specific neuron in the physiological orchestration of the worm are still being studied. In

C. elegans, the vesicular glutamate transporter (VGluT/EAT-4) seems to be the only specific marker of glutamatergic neurons, since their synthesis does not occur only in these neurons and the glutaminases are distributed throughout the animal's body. Hermaphrodite worms have 79 glutamatergic neurons and males approximately 98. Glutamatergic neurons are distributed in the head, pharynx, ventral nerve cord, and body and tail [17].

Acetylcholine (ACh) has been reported as the predominant neuromuscular transmitter in *C. elegans*. The *unc-17* gene encodes the worm ACh vesicular transporter (VAChT) which is used to tag the GFP to cholinergic neuron assessment. The number of cholinergic neurons is 160 in hermaphrodites and 193 in males, and are located in the head, ventral nerve cord, and body, pharynx, and tail [17, 18].

The γ-Aminobutyric acid (GABA) is an amino acid neurotransmitter with both excitatory and inhibitory functions in *C. elegans*. The 26 GABA neurons comprised 6 DD (motors), 13 VD (motors), 4 RME (head), RIS (interneuron), AVL, and DVB (enteric muscles). The serotoninergic system is not well established as to the number of neurons in this classification. Hermaphrodites have 11 to 13 serotonergic neurons and males from 20 to 22 [19, 20]. Dopaminergic neurons in adult worms are 8 in hermaphrodites (head: ADE (2), CEP (4); ventral body nerve: PDE (2)) and 16 in males (head: ADE (2), CEP (4); ventral nerve of the body: PDE (2), tail: R5A (2), R7A (2), R9A (2), SPSo (2)). Dopamine signaling has established roles in modulating locomotion behavior and learning [19].

In this regard, the effects of metal exposures, including cadmium (Cd), manganese (Mn) [21], zinc (Zn) [22], mercury (Hg), and chromium (Cr) [23], on neurodegeneration has been evaluated in *C. elegans*. Among the parameters used to define a mature neuron are the position, physical connections (chemical and electrical synapses), electrophysiological properties, molecular means by which chemical signals are propagated and transmitted, and morphology.

Even in comparison to cell culture modeling, studies in *C. elegans* are of low cost and afford the advantage of using a complete organism. *C. elegans* is easy and inexpensive to maintain in laboratory conditions with a diet of *E. coli* OP50. The short hermaphroditic life cycle (~3 days) and a large number (300+) of offspring of *C. elegans* allow large-scale production of animals within a short period. Furthermore, rodent models require brain slices, paralyzed animals, high-technology equipment (e.g., Positron emission tomography (PET) scan), and complicated assays (e.g., electrophysiology) to evaluate neuronal dynamics in vivo [24, 25]. Work with *C. elegans* has led to seminal discoveries in neuroscience, development, signal transduction, cell death, aging, and RNA interference, attracting increased attention, including environmental toxicology. *C. elegans* is predictive of outcomes in higher eukaryotes both at the level of genetic and physiological similarity and at the level of actual toxicity data. Many of the basic physiological processes in higher organisms (e.g., humans) are conserved in *C. elegans*. Twelve out of seventeen known signal transduction pathways are conserved between *C. elegans* and human. Furthermore, the intensively studied genome, complete cell lineage map, knockout mutant libraries, and established genetic methodologies including mutagenesis, transgenesis, and RNAi provide a variety of options to manipulate and study *C. elegans* at the molecular level.

Here, we discuss the main neurotransmitter systems in *C. elegans* and how to perform characterizations of neuronal viability and activity. We also described behavioral assessment as a measure of neuronal functions, and how neuronal connections and neurotransmitter systems can be associated with behavioral outputs using the *C. elegans* model.

1.1 Fluorescent Methods to Characterize the Neurons and Connectomes

1.1.1 Fluorescent Proteins

To map the *C. elegans* nervous system, fluorescent methods are used, which are continually evolving and have been crucial for the development of research in this nematode. *C. elegans* adult hermaphrodite was the first multicellular organism used to express Green Fluorescent Protein (GFP) [3], so most of the reporter genes in use in the *C. elegans*, including those available at the *C. elegans* Genetics Center (CGC), use some form of GFP. However, many fluorescent proteins have been developed and extensively modified in recent years [26]. There are currently several types of fluorescent proteins used in *C. elegans*, some of which are described in Table 1.

The first step in *C. elegans* neurobiology study is to choose the right strain to evaluate neurons by observing whether fluorescent filters are available, and the target studied. It is essential to consider that the optical density quantified through the fluorescence can be influenced by the stability or degradation and synthesis of the fluorescent protein, which is sometimes not directly related to neuronal death. Transgenic strains for neuron-specific proteins tagged with fluorescent proteins are ideally suited for neurotoxic evaluations.

Constructions combining the GFP and the sodium-dependent DA transporter DAT-1 expressed in DAergic neurons in *C. elegans* represent the first choice to observe such neurons [31]. The strains BZ555 [*egIs1* (P$_{dat-1}$::GFP)], BY200 [*vtIs1* (P$_{dat-1}$::GFP) + *pRF4* (*rol-6*(su1006))], and BY250 [*vtIs7* (P$_{dat-1}$::GFP)] are available in *Caenorhabditis* Genetics Center (CGC) [32]. Another option is to use the OH7547 strain [*otIs199* (*cat-2*::GFP + *rgef-1*(F25B3.3):: DsRed) + (*rol-6*(su1006)] which expresses GFP in DAergic neurons and DsRed pan-neuronally [31]. Intrinsic differences specific for each strain in the response to the exposures should be considered [32]. Transgenic constructions with the gene encoding the

Table 1
Fluorescent proteins used in *C. elegans* research

Fluorescent proteins	Use	Reference
YFP, mNeonGreen, mYPet	Yellow fluorescence	[27, 28]
CFP, mCerulean	Cian fluorescence	[27, 29]
GFP, CrGFP, mNeonGreen,	Green fluorescence	[27–29]
BFP, mTagBFP	Blue fluorescence	[29]
RFP, KillerRed, mCherry, tdTomato, TagRFP-T, mRuby2, mKate2	Red fluorescence	[27, 28, 30]

protein alpha-synuclein (α-Syn) are used for PD investigation in *C. elegans*. α-Syn is a key constituent of the Lewy bodies found in PD patients and a genetic risk factor in PD's pathogenesis [33]. The strain UA44 [*baIn11* (P_{dat-1}::α-syn + P_{dat-1}::GFP)] expresses both GFP and human α-Syn in the dopaminergic neurons [34].

Glutamatergic neurons are defined by examining the expression of a fosmid-based *eat-4* reporter construct found in DA1240 [*adIs1240* (*eat-4*::sGFP + *lin-15*(+)) X] and DA1243 [*adIs1240* (*eat-4*::sGFP + *lin-15*(+) X)] strains [35].

The GABAergic neurons can be targeted with *unc-47* [EG1285 strain, *oxIs12* (P_{unc-47}::GFP + *lin-15*(+))], serotoninergic with *tph-1* [GR1366 strain, *mgIs42* (*tph-1*::GFP + *rol-6*(su1006))], and cholinergic neurons with *unc-17* [LX929, *vsIs48* (*unc-17*::GFP)] [36]. Those are just a few examples amid many possible strains constructions already available that can be extensively crossed.

A relatively short incubation or treatment period is needed for examining the mortality or behavioral trials, which may range from a few minutes [37, 38] to hours [7, 39–41]. Even if the exposure was carried out at the L1 larval stage, the neuronal development is only completed at L4 and should be considered under the evaluations. After exposure to a compound or condition, the fluorescent worms can be transferred to the slides for microscopic observations.

1.2 Behavioral Assays as Indicators of Neuronal Damage

The shared homolog genome attribute of the *C. elegans* model with humans indicates close similarity in the cellular and molecular activities between humans and this nematode [42, 43]. This knowledge has proved advantageous over the decades in studying the neuronal activity in various disease states [44]. Corresponding illustrations of the presence of neuronal damage can be monitored by the use of designated behavioral assays. These behavioral reactions are sometimes initiated by the interactive response following exposure of the nematode to environmental stimuli [45, 46]. Some of these behavioral tools include basal slowing response/food sensing behavior, ethanol avoidance/preference, plasticity of chemotaxis, locomotor assay, shrinker and loopy foraging, area restricted searching (ARS) behavior, mechanosensory responses, habituation task/tap withdrawal response, chemotaxis assay, fecundity, *dauer*-dependent behavior, swim to crawl transition and pharyngeal pumping and thrashing behaviors [47–51]. To avoid behavioral biases, certain factors have to be considered which are dependent on the assay to be carried out. These include synchronized age, life cycle, and sex of the nematode among other environmental factors [52]. Here we have described methods for the following behavioral assays.

1.2.1 *Basal Slowing Response*

This is a dopamine-mediated activity exhibited by *C. elegans* in the presence of food. It is adaptive conditioning that allows well-fed wild-type (WT) nematodes to spend more time feeding. The dopaminergic neural circuits are triggered by a sensory response to the mechanical activity of bacteria (mechanosensation of DA circuits). On the other hand, the serotonergic circuits control another mechanism called "enhanced slowing response." This heightened slow response involves the introduction of starved (wild-type; WT) worms to a food source, whereby the serotonergic circuits ignites an incessant need to remain within the region. In this assay, the rate of locomotion (number of bends) is the parameter to be quantified. This is a determinant for the functionality of dopaminergic neural circuits [53]. The *bas-1*, *cat-2*, and *cat-4* mutant strains are defective in the production of these monoamines and can be used as positive controls during the assay [50].

1.2.2 *Ethanol Avoidance/ Ethanol Preference*

This assay demonstrates a form of behavioral plasticity in the *C. elegans* model with emphasis on its tolerance level. Ethanol is a substance of addiction which under normal conditions is to a certain degree initially repulsive to the worms. Depending on the nature of the experiment conducted, *C. elegans* can either adapt/generate a preference to an ethanol contained environment (when preconditioned) or exhibit an avoidance behavior (when dopamine is depleted). Mutant worms like *cat-2* and *tph-1* that are defective in dopaminergic and serotonergic neurons, respectively, can be used during the assay alongside control wild-type strains for comparison. However, in other mutant strains as *ced-10*, the presence of an alternative source may compensate for any dopaminergic loss present [54, 55]. Ethanol preconditioned worms exhibit addictive behavior similar to alcohol intoxication/addiction in humans. Under such environmental condition (also called conditioned place preference), *C. elegans* forms a good model of addiction [56]. This assay attests to the integrity of the abovementioned neurotransmitters, although ethanol preference may not be entirely dependent on serotonin signaling. It could therefore be inferred that dopamine may be the primary neurotransmitter of concern while carrying out this assay. Both neurotransmitters are involved in the regulation of learning, depression, anxiety, and addiction among other functions in humans. Whereas, in the nematode it controls locomotion, egg-laying, pharyngeal pumping, and feeding [57, 58].

1.2.3 Plasticity of Chemotaxis by NaCl

Plasticity is a crucial mechanism that promotes the formation of memories and ignites certain experience-based behavioral responses from the perceived sensory neural transmission. Plasticity by NaCl is a conditioned associated behavior in which starved nematodes learn to withdraw from NaCl-exposed regions after a prolonged time, also called gustatory plasticity. The starved worms under this condition learn to associate food deprivation with the presence of the salt [51]. Although naive worms are typically attracted to NaCl within a concentration of 0.1–200 Mm. However, they act aversively after an extended duration in a concentration of 100 Mm NaCl [59]. This behavior exhibited by C. elegans has been described as associative learning beyond adaptation, as it occurs in the absence of food source and presence of the salt to give a gradient corresponding response [60]. The following neurotransmitters dopamine, serotonin, and glutamate play both active and passive roles in regulating gustatory plasticity [59, 61]. To conduct this assay, four sets of plates are prepared to produce different pre-exposure environments for conditioning the worms. These plates include conditioning plate, mock-conditioning plate, NaCl plug plate, and assay plate [51]. Animals are either pre-exposed to the presence of NaCl or left naive by habituating them in a conditioned environment with or without a particular NaCl concentration. The parameter of focus in this assay is calculating the chemotaxis index of the worms.

1.2.4 Loopy Foraging and Shrinker Behaviors

C. elegans foraging ability involves a number of coordinated neuronal activities that enable the nematode search for available food source then feed sufficiently. The GABAergic system regulates the biomechanical coordination of the worm during foraging (head movement) and body wall movement [62]. This neuronal group forms the basis of examination in this behavioral phenotype. During foraging, the four RME GABAergic motor neurons found in the head control the slight arc-like movement of the nose from one side to the other. However, any defect in these motor neurons will result in a loopy (excess flexion) of the nose during foraging. Furthermore, the 19 D-type GABAergic motor neurons allow for a sinusoidal wave-like movement from head to tail when the nematode is gently touched or prodded. The sinusoidal bend is formed by contraction of muscles on one side of the body wall and relaxation by GABA on the other side [63]. These D-type neurons are found in the neuromuscular junction of the body wall. Lack of these neurons cause a shrinker or twitcher movement which is the pulling inward of the head of the worm, and shorten of body length (hyper contraction) when touched. Such behavioral alteration may negatively affect the survival instincts/defense mechanism of the worms.

1.2.5 Locomotor Assay

Locomotion is an essential part of existence and survival in all living organisms. *C. elegans* move in a sinusoidal (undulatory) motion in search of food and a conducive living environment. This movement is aided by the generation of dorsoventral body bends (controlled by the dorsal and ventral muscles) which spans along its entire body length against the direction of movement [64]. Several factors influence the pattern of motility including presence or absence of food and change in environmental/external stimuli. Such alterations can be seen expressed in worms with a mutated dopaminergic system [47]. The locomotor assay is used to assess abnormal motion in *C. elegans* expressed as either irregular bends or thrashing motion, which depicts motility of nematode in liquid. Here, worms are transferred to unseeded NGM plates and total body bends (rate of locomotion) over a period of time is counted and scored.

1.2.6 Pharyngeal Pumping and Thrashing Behavior

Pharyngeal pumping in *C. elegans* is induced by extrinsic factors due to a lack of muscular pacemakers. These extrinsic factors are sometimes formed from the neuronal release from dopaminergic and cholinergic systems. The pharynx of this nematode regulates feeding via contraction (pumping) pattern of the neuromuscular organ in a rhythmic pattern [65]. There are three main parts of the pharynx which include corpus, isthmus, and terminal bulb. The isthmus connects the anteriorly positioned corpus to the posterior terminal bulb, which contains the grinder. Peristaltic movement within the terminal bulb is more visible than that of the corpus due to the crushing activity of the grinder. Hence, considering that contraction-relaxation cycles during feeding occurs through the entire length of the pharynx, the pumping rate can be counted from monitoring the movements of the grinder [66]. A maximum pumping rate is ~300 pumps per minute in 2 days old adult worms [50]. The thrashing assay is a basic motor performance test for *C. elegans* motility. A thrash is referred to as the movement of the head from the dorsal to the ventral side. It could also be referred to as a change in direction of the midbody as the worm struggles to get off a liquid medium. Adequate measurement of the number of directional changes is done accordingly. Normal worms can perform 90–100 thrashes in 30s in fluid medium [50, 67].

1.3 Behavioral Assay Under Optogenetic Control for Assessing Connectomes

From a broader perspective, connectomes can be described as the science of mapping out the entirety of neural connections within an organism's nervous system. There are a number of techniques being used to detail both the structural and functional connectivity of the neural cells of the brain [68, 69]. Understanding the organization of *C. elegans*connectomes has assisted in expanding knowledge on the molecular and structural basis of sex-based behavioral responses [70]. Earlier researches have made significant efforts to describe the interconnection between the neural circuitry (connectomes), muscular mechanics, and behaviors.

This pioneered the use of advanced techniques called "optogenetics," which is a combined use of light and genetically engineered cells to monitor neuronal activity. In optogenetics, genes are modified to produce specialized membrane-bound proteins called opsins (G-protein coupled receptor also referred to as actuators derived from microbial species), which are light sensitive. The most widely used opsin is the "Channelrhodopsin-2" (ChR-2); this is typically responsive to blue light, though there exists a rarely used subtype which is responsive to green light [71–74].

ChR-2 shares a similar excitation spectrum as Green Florescent Protein; it is used mostly during stimulation/excitatory experiments. Some other opsins which express inhibitory properties include NpHR/Halorhodopsin, Archaerhodopsin-3, and Mac [72, 75]. The optogenetics techniques aid in vivo regulation of neural activity through the use of those specific genetically encoded light-sensitive tools. For example, gated ion channels and calcium indicators are targeted to regulate neural cell excitation and activity, respectively [76]. Therefore, the principal optogenetic tools consist of the actuators (opsins) and sensors. While the actuators manipulate neuronal activity, the sensors (calcium indicators: GECIs, vesicular release and pH sensors: synapto-pHluorins, voltage indicators: GEVIs) act as optical recorders to monitor the manipulated neural activity of actuators [77]. During retinoid metabolism in mammalian cells, exogenous vitamin A is used to produce "All-trans-Retinal," an important cofactor for ChR-2 functions. Unfortunately, it is naturally absent in *C. elegans,* and hence the need of an exogenous source. It is a photosensitizer that serves as a chromophore during polarization neuronal activity [78, 79].

2 Materials

2.1 Preparing *C. elegans* for fluorescent Imaging

Worms are prepared on agarose pads, as they easily get dried if laying down directly on the slide. Furthermore, the pad of agarose is used to keep the same slide thickness, and to standardize analysis across treatment groups. Materials are outlined below:

- *C. elegans* strains.
- Slides.
- Coverslip.
- Plastic Pasteur pipettes.

- Agarose 2% (some protocols have suggested 3% or 4%) (http://wbg.wormbook.org/)).
- Drugs to immobilize the worms: 0.2% tricaine/0.02% tetramisole, 1 mM levamisole, or 10–25 mM sodium Azide (NaN_3)) depending on the behavior to be analyzed.
- Micropipette (2–20 μL).
- Tips.
- Labeling tape.

2.2 Behavioral Assays

Behavioral assays described above require relatively simple tools to perform, and of great value in assessing neurotransmitter systems in *C. elegans*. Most behavioral assays have been performed manually under basic stereomicroscopes; however, the use of video recording system and tracking software is an added advantage that is encourage. Synchronous worm strain populations [51] are mostly used in behavioral assays. Several of the behavioral assays use similar materials that are outlined below:

- Worm strains.
- Petri (culture) plates: 6 cm, 9 cm.
- Standard nematode growth media (NGM).
- Genetically engineered *E. coli* bacteria strains: either OP50 or HB101.
- Reagents: Agar, $CaCl_2$, NaCl, $MgSO_4$, KH_2PO_4, Tween 20.
- Buffer solutions; either M9 buffer (KH_2PO_4, Na_2HPO_4, NaCl, dH_2O, $MgSO_4$) or S-basal buffer (NaCl, K_2HPO_4, KH_2PO_4, cholesterol, dH_2O).
- Glass culture tube.
- Capillary pipette and tip.
- Ethanol and flame source.
- Kimwipe.
- Pasteur Pipette.

2.3 Behavioral Assay Under Control of Optogenetics

- Standard NGM.
- Worm Strains (transgenic strains expressing ChR2 [tph-1p::ChR2::GFP + myo-3p::mCherry] available at Caenorhabditis Genetics Center).
- Bacteria source; OP50 or HB101.
- 100 mM all-trans-retinal (ATR) stock solution (diluted in ethanol and stored in a microcentrifuge tube for light-sensitive samples).
- Petri plates; 6 cm.

- Blue light lens filter.
- Parafilm.
- Pasteur pipette.
- M9 buffer.
- DC power source.
- Multi-Worm Tracker.
- Custom-built LED light apparatus/digital computerized LED lights (recommended for precision).

3 Methods

3.1 Fluorescence Imaging Methods for C. elegans

3.1.1 Mounting the Worms (Fig. 1)

- Prepare the slides with two overlapping pieces of tape. These slides can be saved for the next experiments (1).
- Heat the agarose in the microwave for 30–60 s until start boiling (2). Take care of not heating for a long time because the agarose loses transparency and can implicate changes in the experiment.
- Place the slide side by side and a new slide in the middle of a clean countertop near to the heated agarose. Use a plastic Pasteur pipet to place 4 to 5 drops of agarose in the middle of the slide (3).
- Quickly covering puddle with another slide before agarose solidifies (4). Do not press, release the slide as quickly as possible to avoid the bubbles.
- Carefully recover the pad (1.2–1.6 cm across); work rapidly, as the agarose can dry out quickly when left uncovered (5).
- Cut the exceed pieces until it gets the coverslip shape (6).
- The worms can be transferred by picking (hardest way) or by transfer from a resuspension in M9 buffer. Pipet about 10 μL of worms and 10 μL of the anesthetic (7).
- Cover the pad with a coverslip (8).
- Check the worms in the fluorescence or confocal microscope (9).

3.1.2 Fluorescence Acquisition in C. elegans

The image acquisition can be acquired using confocal or fluorescence microscopy. As neurons are distributed in C. elegans cylindrically, Z-stacking should be considered for representative images. Since C. elegans body presents about 35–40 μm of diameter in L4 larval stage, the Z-stacking images should be taken at different focal distances, e.g., 2.5 μm, resulting in about 16 slices. Depending on the extent of photobleaching, 30–60 time points are collected. Concomitantly, the brightfield should be accessed considering the same set used to acquire the fluorescence. The image format is

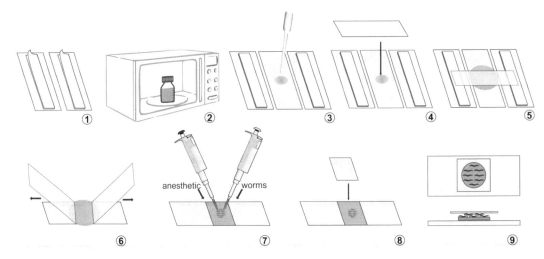

Fig. 1 Procedure to mount *C. elegans* on slides

another crucial information and are generally exported and saved as 16-bit TIFF format. Numerous software can be used to process the images, the most used is ImageJ that is a platform for image processing designed for scientific multidimensional images and inspired by the NIH. To acquire the whole body-length of *C. elegans* in adulthood (>1000 μm), it can be used the 2.5× and 4× objectives.

ImageJ Download and Setting

- Download ImageJ with Java on different operating systems using the link: https://imagej.nih.gov/ij/download.html
- After installing, download the plugging "bioformats_package. jar": https://downloads.openmicroscopy.org/bio-formats/5. 0.3/
- Copy the file and paste into the folder: imageJ > plugins > Jars.
- Open imageJ, click in plugins > install > bioformats_package. jar > save.

Z-Stack Image Sequence Processing in ImageJ (Fig. 2)

- Open ImageJ, click File > Import > image sequence (1).
- Select folder containing images for treatment and set desired sequence options (2).
- Z-stack images are collapsed into a maximum projection image: click image > stacks> Z-Project (3).
- Select "start" slice and "stop" slice and the Projection type (e.g.: sum slice) (4).
- For fluorescence quantification, click "polygon selections" and delimit the total area of the worm (5), then click analyze> measure (6, 7).

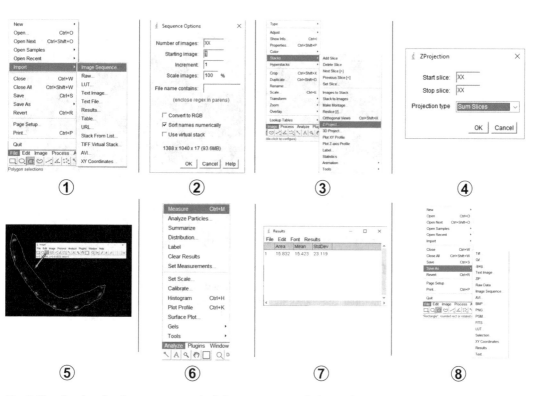

Fig. 2 The step-by-step to process a z-stack image sequence in ImageJ

- To save the image, click file >save as> TIFF> select the destination folder (8).

3.1.3 Neuronal Viability Analysis

The images processed can be analyzed in comparison to a control group. Some endpoints can be examined in vivo as indicatives of neurotoxicity including the neuronal development, punctum, neuronal absence or shrinkage, neuronal gaps, absence of cell bodies, and reduction in intensity of fluorescence [6, 32, 80, 81]. Figure 3, kindly provided by Tao Ke, shows neuronal changes at dopaminergic neurons induced by *Methylmercury* (5 μM MeHg) exposure.

Neurodegeneration can be also classified by range considering some of the endpoints discussed, as used by Schetinger and coworkers [82]. Cholinergic neurodegeneration was ranked from 0 to 3, where 0 meant no degeneration; 1 meant low degeneration; 2 means moderate degeneration, and 3 means high degeneration considering the head, body, and tail of *C. elegans*.

3.1.4 Neuronal Activity Imaging

Regarding neuronal activity, *C. elegans* is an ideal model mainly due to its already mapped nervous system for both hermaphrodites and males [83, 84]. The development of imaging methods enables a fantastic expansion in *C. elegans* neuronal activity field. While in rodents expensive systems must be used to acquire neuronal

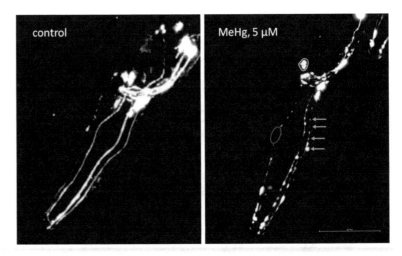

Fig. 3 *C. elegans* dopaminergic neurodegeneration under *Methylmercury* (5 μM MeHg) exposure. Blue arrows indicate the punctum, red delimited area indicates neuronal absence, and green delimited area indicates the shrinking soma. BZ555 strain worms were focused on the head, L4 larval stage, 60× objective. Kindly provided by Tao Ke

function, the fluorescent GCaMP construct allows *C. elegans* in vivo four-dimensional imaging of neuronal calcium implicated in both behavior and neuronal activity patterns [ZM9078 strain; *hpIs587* (*flp-14$_p$*::GCaMP6::wCherry + *lin-15*(+))] [85] and fluorescent voltage reporters can infer the postsynaptic responses [86]. For instance, GCaMP expression is limited to a set of neurons approaches, and the field of neuronal activity imaging is still limited by a lack of tools for robust assignment of all neurons at the same time.

3.2 Behavioral Assays

3.2.1 Basal Slowing Response

Procedure:

- NGM plates can be prepared according to the protocol by [87]. These plates are stored at 20 °C 16–20 h prior to the assay.

- Two (2) sets of assay plates are prepared accordingly: bacteria spread (seeded) and un-spread (unseeded) plates.

- Assay plates to be seeded should be freshly spread with bacteria (OP50 or HB101) depending on the species used for cultivating the worms as follows [51, 53]:

  – Drop approximately a drop of 2 μL bacteria within a circle (bacteria lawn: inner and outer rings of ~1 and 3.5 cm respectively) in the center of the NGM.

  – Gently spread the bacteria within the boundary of the circle with the bottom of a glass culture tube (sterilize with 70% ethanol and flame).

  – Place both sets of plates (spread and un-spread) in an incubator overnight at 37 °C.

- Take assay plates out of the incubator and allow to cool at room temperature before use, dry off lids with a Kimwipe.
- Already cultivated well-fed animals are washed in a buffer (~1.5 mL of M9 or S-basal) after being removed from culture plates before being transferred to assay plates. Washing, to remove bacteria from the worms, is done twice as follows:
  - First, wash worms off plates using the buffer into Eppendorf tubes.
  - Spin in a centrifuge 30–60 s at 1000 rpm or allow worms to settle to the bottom of the tube, then pull off buffer with pipette leaving 100–200 µL.
- Transfer worms (5–10 worms) to the clear zone in the center of the assay plates as follows:
  - In a drop of the buffer, use a capillary pipette or cut a 200 µL tip to transfer ~5 µL of worms. Ensure there is a minimum of 5 worms on the assay plate.
- Gently mop up the drop of the buffer used to transfer the worms from the assay plate with a Kimwipe.
- Allow assay plate sit undisturbed on a bench for about 5 min of adaptation time.
- Body movement count (movement rate) is counted; thus,
  - Count the number of bends per 20 s for each worm, i.e., change in body direction moving either forward or backward of the area behind the pharynx.
  - Frequency of body bends can be recorded manually and digitally for 20–60 s with a data acquisition tracker software [88, 89].
- The analysis is calculated; thus, Basal slowing response =.

    [rate of movement in unseeded plates – the rate of movement in seeded plates] / [rate of movement in seeded plates].

Notes
(a) Controls adapted for use during the assay should be well fed and similar in size relative to the thickness of the bacteria lawn in each plate.
(b) Due caution should be taken on the purity (freshness) of reagents used as this can affect the assay.
(c) Transfer of worms to assay plates should be done one group at a time. This is to allow time to plate the worms and mop up the liquid (buffer) and efficient counting of bends.
(d) To prevent worms from crawling off agar of unseeded assay plates, 100 µL of 4 M fructose can be dropped on the edge of

the plate. Tilt gently to spread the fructose all the way around the late. Allow drying for ~10 min before beginning the assay.

(e) These experiments are done blindly and repeated at least 3 independent times.

(f) Assays can be recorded via digital camera and analyzed later.

3.2.2 Ethanol Avoidance/ Ethanol Preference

Depending on the investigation being carried out, this behavioral procedure can either be tagged an ethanol avoidance or preference assay.

For the ethanol avoidance [90]:

- Synchronized young adult worms (~63 h post-synchronization) to an assay plate which has been divided into four quadrants.

  – Assay plate is prepared thus; Divide 10 cm NGM petri dishes into four quadrants consisting of A and B (ethanol quadrant), C and D (control quadrant), with each pair of quadrants opposite each other. Bore 9 mm holes into quadrants A and B, in which ~50 μL of 100% ice-cold ethanol will be poured. Seal the plate with parafilm to allow ethanol equilibration.

  – Transfer of worms is done by washing them off ethanol plate using S-basal (twice), then once in distilled water.

- About 100–200 animals are transferred into the assay plate which contains a central marked spot (forms the origin). These worms are allowed to freely move for 30 min, after which scoring is done. A video recording device of choice can be employed for use within this duration.

The preference index is calculated thus: [Number of worms in control quadrants - Number of worms in ethanol quadrants] / Total number of worms tested.

For an ethanol preference [58]:

- Worms are pre-exposed to ethanol as follows: synchronized young adult worms (~63 h post-synchronization) are pretreated by incubating for about 4 h in a seeded NGM control or ethanol pre-exposure plate.

  – Ethanol plate is prepared as follows: half of NGM plate is seeded with OP50 and then allowed to dry for ~2 h. The other half is seeded with 300 mM ice-cold ethanol. Plate is sealed with a parafilm so as to allow adequate diffusion of the ethanol within the agar for another 2 h.

- Worms are transferred to assay plate as describe above, and procedure similarly performed as above.

The preference index is calculated thus: [Number of worms in ethanol quadrants - Number of worms in control quadrants] / Total number of worms tested.

Notes

(a) Adequate care should be taken in quantifying the concentration of ethanol used during assay plate preparation.

(b) Avoid overcrowding the assay plate in order to prevent clustering of worms.

(c) Ensure the working bench is horizontal and avoid all external sources of disturbances such as wind, centrifuge, and incubator, in order to eliminate vibrations.

(d) Do not count clumped worms left in the midline (center point) of the assay plate while calculating the chemotaxis index, as these consist of wash-injured animals.

3.2.3 Plasticity of Chemotaxis by NaCl

Procedure:

• Prepare a culture media containing 14 mL of NGM using a 6 cm plate. This volume of NGM is the minimum requirement to get a sufficient spread of bacterial (overnight cultured *E. coli*) lawn. This should be incubated for a day at room temperature or overnight at 37 °C, then stored at room temperature.

• Prepare the four sets of plates as follows [51, 91, 92]:

  – Conditioning Plate: This medium is made up of a combination of 25 g/L agar, 100 mM NaCl, 1 mM $CaCl_2$, 1 mM $MgSO_4$, 5 mM KH_2PO_4(pH 6.0). Autoclave, thereafter pour 8 mL into the 9 cm plate (or 9 mL into 10 cm plate).

  – Mock-conditioning plate: This is prepared with similar constituents as the conditioning plate but with the exemption of NaCl.

  – NaCl plug plate: The plug medium consists of 17 g/L agar, 100 mM NaCl, and 1 mM $CaCl_2$, 1 mM $MgSO_4$, 5 mM KH_2PO_4 (pH 6.0). Autoclave, then pour 6 mL of the medium into a 6 cm plate.

  – Assay plate: This medium is prepared in a similar way as the plug plate but without NaCl. After autoclaving, pour ~3–3.5 mL of the medium into a 6 cm plate.

    Store all plates at 4 °C and use within 2–3 weeks

• Create a NaCl concentration gradient on the assay plate. This can be done in two ways:

  – Diffusion method: Use a Pasteur pipette (remove the tip) to excise a chunk of diameter 5 mm from the NaCl plug plate. Place this excised plug on one side of the assay plate for ~19–23 h. This overlapping plug should be discarded before the commencement of the chemotaxis assay [51, 60].

- Spotting method: This is done by directly applying small drops of the chosen concentration of NaCl to a side of the assay plate at an equidistant point from the control region [91, 93].

- Use a wash buffer to collect well-fed animals from NGM. Wash buffer is made up of 1 mM $CaCl_2$, 1 mM $MgSO_4$, 5 mM KH_2PO_4 (pH 6.0), and 0.05% Tween 20].

  NB-: Wash buffer with or without NaCl can be used for the pre-exposure conditions. Then animals are maintained at 20 °C for 15 min in their respective pre-exposure wash solutions. Thereafter, wash once with buffer without NaCl for 60 s at 450 × g for 60 s; this is to remove the NaCl from the body of the worms [91, 93].

- Wash animals collected from NGM in a centrifuge three times using 1.5 mL tubes for 20 s at ~900 × g.

- Transfer washed animals to the conditioning or mock-conditioning plates, respectively.

- Cover plates with lids, then place in an incubator at 20 °C for 4 h.

- Use wash buffer to collect conditioned and mock-conditioned animals into 1.5 mL tubes.

- Gently place animals in the center of the assay plate (region having the established NaCl concentrated gradient), approximately 50–300 animals. Use a Kimwipe to mop off excess wash solution.

- Allow plate with lids to seat undisturbed on the working / experimental bench for 15 min.

- Put several drops of chloroform on the lid to stop animal movement.

- Count the total number of animals on each fraction of the assay plate.

- Calculation of chemotaxis index is done thus [51];

$$[X - Y]/[X + Y]$$

where
  X = Number of animals on NaCl fraction of the assay plate.
  Y = Number of animals on the other (control) fraction of the assay plate.

Notes
(a) Ensure the working bench is horizontal and avoid all external sources of disturbances such as wind, centrifuge, and incubator, in order to eliminate vibrations.

(b) Do not use old seeded nematode growth media (over 2 weeks).

(c) Do not count clumped worms left in the midline (center point) of the assay plate while calculating the chemotaxis index, as these consist of wash-injured animals.

(d) Presence of food during the assay may alter the results or create biases in the response of worms to the concentration gradient.

(e) Before commencement of the assay procedure, ensure animals are well fed and not from a starved agar chunk.

(f) Ensure room temperature is kept at $25 \pm 1\,°C$ during the assay procedure.

(g) The concentration of NaCl (between 50 mM and 100 mM) used during the pre-exposure/preconditioning stage should be similar to that used for the assay [51, 59].

3.2.4 Loopy Foraging and Shrinker Behaviors

Procedures are adapted from this previously described method [63, 94].

Procedure for Loopy Foraging:
- At young adult stage (~63 h post synchronization), pick 20 worms unto assay plate. Allow plate to seat on bench for 5 min.
  - **Assay Plate***(prepared a day before assay)*:
    Prepare thin seeded NGM plates by dropping ~30 μL OP50 bacterial into 60 mM plates, spread and allow to seat at room temperature for 24 h or overnight at 37 °C.
- Observe and count number of worms with "loopy" foraging per 20 worms.

Procedure for Shrinker Behavior:
- At young adult stage (~63 h post synchronization), pick 20 worms unto assay plate. Allow plate to seat on bench for 5 min.
  - **Assay Plate:**
    Prepare thin seeded NGM plates by dropping ~30 μL OP50 bacterial into 60 mM plates, spread and allow to seat at room temperature for 24 h or overnight at 37 °C.
- Gently prod or touch body of worm with a pick, and observe "shrinking" action (see fig below).
- Count number of "shrinkers" per 20 worms.

For examples of phenotypic pattern of loopy foraging and shrinker behaviors see Jorgenson 2005 [63].

Notes
- CB156 strain [*unc-25*(e156) III] worm strains can be used as appropriate controls. These worms which are unc-25 mutants show both loopy head and body shrinking behaviors.

- Same worms can be used for both assays. First observe and count loopy head worms. Then prod worms for body shrinking.

3.2.5   Locomotion Assay

Procedure is modified [94] from the previously described method in [52].

Procedure:
- Post synchronized young adult worms (~63 h) are collected by washing thrice in S-basal buffer.

- Transfer ~7–10 worms with S-buffer into unseeded NGM assay plates using a Pasteur pipette. The use of unseeded plates is to eliminate the slowing of movements in bacteria lawn, particularly at the edge of the lawn, to assess the general locomotory rate.

- Remove excess buffer with a Kimwipe and allow plate seat for 5 min of acclimatization.

- Count the number of body bends per 3 min. The body bend count is denoted as each time the posterior bulb of the pharynx reaches a maximum turn in the opposing direction from the last count.
    - The assay can be repeated several times and data collected over different experiments.

- Data collected is expressed thus and the average calculated; *Body bends/3 min.*

Notes
(a) Care should be taken not to mistake a worm body reversal in backward motion as a body bend. Such reversal occurs when during forward motion, the worm suddenly reverses thereby curving in the same direction it was previously bent/turned in the backward motion.

(b) Prepare one assay session and finish count before preparing another.

(c) Worms counted should be removed from the assay plate by gently picking up to prevent double counting.

(d) If worms tend to crawl off Agar plate, make fructose (4 M) or glycerol (8 M) rings around unseeded assay plates. Drop 60–100 µL of fructose or glycerol solution on edge of plate, tilt plate to coat the edge all the way around, and allow to dry for about 10 min.

(e) These experiments are done blindly and repeated at least 3 independent times.

(f) Assays can be recorded via digital camera and analyzed later.

3.2.6 Pharyngeal Pumping and Thrashing Behavior

Procedure:

- Transfer at least 10 worms from synchronized population into freshly prepared NGM assay plate.

- Count the number of times worms pump per minute (ppm) in two ways:
  - Using a stereomicroscope, you may choose to manually count pumping rate, 10–20 s per min for 10 min [66] OR.
  - If the pumping rate seems abnormal, one can decide to digitally count the pumping rate with a digital camera (at least 10 frames per sec). It is preferred to transfer a single worm into the NGM plate with a bacteria lawn of 5 mm.
  - Counts could also be done via another technique called the high power differential interference contrast [66].

    Data is calculated thus; the rate of pharyngeal pumping = Total number of pumps/Total time

- For the thrashing assay, plates can be prepared in two ways:
  - Put 10 µL of dH$_2$O in a shallow transparent Petri plate, then transfer worms with a Pasteur pipette into this assay plate [50] OR.
  - Place the worm in the middle of a droplet of the isotonic solution in a standard NGM. Counting is done as the worm tries to reach a solid medium [65].
  - Thrashing assay is monitored via videotaping and assessed using any available recommended assay software [50].

Notes

(a) Make use of well-fed animals, as pharyngeal pumping is decreased in normal animals that are off food.

(b) Replay recorded videos of pumping assays at ½ to $^1/_3$ of the original speed to aid accurate count.

(c) Make use of standard NGM that are well seeded with bacteria to avoid irregular increased motility of hungry worms. Such would make pump count difficult.

(d) L1 worms require a magnification of X100 for accurate viewing.

(e) These experiments are done blindly and repeated at least 3 independent times.

(f) Assays can be recorded via digital camera and analyzed later.

3.3 Behavioral Assay Under Control of Optogenetics

For more detailed description *see* [76, 79].

Procedure:

- Prepare NGM using 12 mL of agar, then store at room temperature for ≥48 h before use.

- Seed NGM with a 50 µL stock mixture of bacteria and ATR in a ratio 49.4 µL:0.6 µL, onto the center of the NGM forming a square (4 × 4 cm area).

- Store seeded plates in the dark (using a foil or kept in a drawer) or under red light to allow bacteria growth (~24 h).

- Age synchronize worms (use 3–4-day old young adult worms). This can be achieved by adapting the use of gravid strains that are allowed a 4 h egg-laying period within the plates. These are later removed to have 40 80 age synchronized worms.

  - To achieve a strain of intense light response, worms expressing extrachromosomal array can be transferred into fresh ATR plates.

- Transfer worms into seeded plates by picking using a Pasteur pipette or pipette with M9 buffer.

- Seal worms with parafilm. Store in a controlled dark environment.

- Attach the blue light filter over the lens of the Multi-Worm Tracker.

- The custom LED light (supplied by the DC power) should then be connected to the stimulus relay of the multi-worm tracker.

- Launch multi-worm tracker software, arrange the light stimulus parameter.

- Begin mounting of each plate on the tracker platform, focus the LED light accordingly at the center of the ring on the plates.

- Analyze data using offline java-command-line Chorography software; use custom scripts to re-arrange output files as desired, investigating neuronal activities and behavioral changes.

Notes

(a) Prepare extra seeded plates in case of contamination or uneven distribution of agar spread.

(b) Ensure to minimize the ATR exposure to light to avoid the photosensitizer losing its gating function on the opsins.

(c) The seeded plates should be used within 72 h to avoid excess bacteria growth.

(d) Reset the worm tracker each time the plate is changed.

4 Conclusions

The set of connections in neural systems called the connectome form the communication in the brain. Detailed understanding of these connections, i.e., mapping high-resolution connectivity, is an essential first step in elucidating how the nervous system processes information and generates behavior [95]. C. elegans is an ideal model for studying neuronal activity due to its small, stereotyped yet relatively complex nervous system [84]. Furthermore, its optical transparency, homologous genomic constitution with humans and availability is an added advantage. Knowledge about nervous system connectivity is critical to understanding how it functions. Cook et al. (2019) described in their studies the concepts of both adult sexual forms of the nematode C. elegans. Serial transmission electron microscopy and some previously related data were used to perform a reconstruction of circuits for the male head, including mainly the nerve ring and retro-vesicular globule [96]. The evaluation of whole-animal connectomes from sensory input to end-organ output across the worm showed a considerable number of connections' differences between the male and hermaphrodites [96]. The connectivity profile can indicate how neurons work, whether by sensory perception or hormone secretion. A thorough understanding of the interconnections of various neurotransmitter systems of this nematode is imperative to understand certain behavioral outputs among species. Although varying protocols could be modified, the use of standard methods is essential to ensure reproducibility of results.

Acknowledgments

OMI acknowledges the 2019 Young IBRO Regions Connecting Awards. MA is supported by National Institute of Health (NIH), USA grants, NIEHS R01 10563, NIEHS R01 07331, and NIEHS R01 020852. We acknowledge Tao Ke of the Albert Einstein College of Medicine for images of C. elegans dopaminergic neurodegeneration under MeHg exposure.

References

1. Consortium CeS (1998) Genome sequence of the nematode C-elegans: a platform for investigating biology. Science 282 (5396):2012–2018. https://doi.org/10.1126/science.282.5396.2012

2. Antoshechkin I, Sternberg PW (2007) The versatile worm: genetic and genomic resources for Caenorhabditis elegans research. Nat Rev Genet 8(7):518–532. https://doi.org/10.1038/nrg2105

3. Chalfie M, Tu Y, Euskirchen G, Ward WW, Prasher DC (1994) Green fluorescent protein as a marker for gene-expression. Science 263 (5148):802–805. https://doi.org/10.1126/science.8303295

4. Boulin T, Pocock R, Hobert O (2006) A novel Eph receptor-interacting IgSF protein provides C. elegans motoneurons with midline guidepost function. Curr Biol 16(19):1871–1883. https://doi.org/10.1016/j.cub.2006.08.056

5. Feinberg EH, VanHoven MK, Bendesky A, Wang G, Fetter RD, Shen K, Bargmannl CI (2008) GFP reconstitution across synaptic partners (GRASP) defines cell contacts and synapses in living nervous systems. Neuron 57 (3):353–363. https://doi.org/10.1016/j.neuron.2007.11.030

6. Benedetto A, Au C, Avila DS, Milatovic D, Aschner M (2010) Extracellular dopamine potentiates Mn-induced oxidative stress, lifespan reduction, and dopaminergic neurodegeneration in a BLI-3-dependent manner in Caenorhabditis elegans. PLoS Genet 6(8): e1001084. https://doi.org/10.1371/journal.pgen.1001084

7. Gubert P, Puntel B, Lehmen T, Fessel JP, Cheng P, Bornhorst J, Trindade LS, Avila DS, Aschner M, Soares FAA (2018) Metabolic effects of manganese in the nematode Caenorhabditis elegans through DAergic pathway and transcription factors activation. Neurotoxicology 67:65–72. https://doi.org/10.1016/j.neuro.2018.04.008

8. DiLoreto EM, Chute CD, Bryce S, Srinivasan J (2019) Novel technological advances in functional connectomics in C. elegans. J Dev Biol 7 (2). https://doi.org/10.3390/jdb7020008

9. Marsh EK, May RC (2012) Caenorhabditis elegans, a model organism for investigating immunity. Appl Environ Microbiol 78 (7):2075–2081

10. Johnson TE (2003) Advantages and disadvantages of Caenorhabditis elegans for aging research. Exp Gerontol 38(11–12):1329–1332

11. Gottschling D-C, Döring F (2019) Is C. elegans a suitable model for nutritional science? Genes Nutr 14(1):1–4

12. Tissenbaum HA (2015) Using C. elegans for aging research. Invertebr Reprod Dev 59 (sup1):59–63

13. Leung MC, Williams PL, Benedetto A, Au C, Helmcke KJ, Aschner M, Meyer JN (2008)

Caenorhabditis elegans: an emerging model in biomedical and environmental toxicology. Toxicol Sci 106(1):5–28

14. Chen X, Barclay JW, Burgoyne RD, Morgan A (2015) Using C. elegans to discover therapeutic compounds for ageing-associated neurodegenerative diseases. Chem Cent J 9(1):65

15. Alexander AG, Marfil V, Li C (2014) Use of Caenorhabditis elegans as a model to study Alzheimer's disease and other neurodegenerative diseases. Front Genet 5:279

16. Hobert O (2013) The neuronal genome of Caenorhabditis elegans. WormBook:1–106. https://doi.org/10.1895/wormbook.1.161.1

17. Loer CMR, Rand JB (2016) The evidence for classical neurotransmitters in C. elegans neurons (updated online review/database in WormAtlas 2.0; original in 2010). WormAtlas

18. Rand JB (2007) Acetylcholine. WormBook:1–21. https://doi.org/10.1895/wormbook.1.131.1

19. Chase DL, Koelle MR (2007) Biogenic amine neurotransmitters in C. elegans. WormBook:1–15. https://doi.org/10.1895/wormbook.1.132.1

20. Vidal-Gadea AG, Davis S, Becker L, Pierce-Shimomura JT (2012) Coordination of behavioral hierarchies during environmental transitions in Caenorhabditis elegans. Worm 1 (1):5–11. https://doi.org/10.4161/worm.19148

21. Tang B, Tong P, Xue KS, Williams PL, Wang JS, Tang L (2019) High-throughput assessment of toxic effects of metal mixtures of cadmium(cd), lead(Pb), and manganese(Mn) in nematode Caenorhabditis elegans. Chemosphere 234:232–241. https://doi.org/10.1016/j.chemosphere.2019.05.271

22. Baesler J, Kopp JF, Pohl G, Aschner M, Haase H, Schwerdtle T, Bornhorst J (2019) Zn homeostasis in genetic models of Parkinson's disease in Caenorhabditis elegans. J Trace Elem Med Biol 55:44–49. https://doi.org/10.1016/j.jtemb.2019.05.005

23. Xing X, Guo Y, Wang D (2009) Using the larvae nematode Caenorhabditis elegans to evaluate neurobehavioral toxicity to metallic salts. Ecotoxicol Environ Saf 72 (7):1819–1823. https://doi.org/10.1016/j.ecoenv.2009.06.006

24. Melo-Thomas L, Engelhardt KA, Thomas U, Hoehl D, Thomas S, Wohr M, Werner B, Bremmer F, Schwarting RKW (2017) A wireless, bidirectional Interface for in vivo recording and stimulation of neural activity in freely

behaving rats. J Vis Exp 129. https://doi.org/10.3791/56299

25. Mann T, Kurth J, Hawlitschka A, Stenzel J, Lindner T, Polei S, Hohn A, Krause BJ, Wree A (2018) [(18)F]fallypride-PET/CT analysis of the dopamine D(2)/D(3) receptor in the hemiparkinsonian rat brain following intrastriatal botulinum neurotoxin a injection. Molecules 23(3). https://doi.org/10.3390/molecules23030587

26. Sands B, Burnaevskiy N, Yun SR, Crane MM, Kaeberlein M, Mendenhall A (2018) A toolkit for DNA assembly, genome engineering and multicolor imaging for C. elegans. Transl Med Aging 2:1–10. https://doi.org/10.1016/j.tma.2018.01.001

27. Green RA, Audhya A, Pozniakovsky A, Dammermann A, Pemble H, Monen J, Portier N, Hyman A, Desai A, Oegema K (2008) Expression and imaging of fluorescent proteins in the C-elegans gonad and early embryo. Method Cell Biol 85:179. https://doi.org/10.1016/S0091-679x(08)85009-1

28. Heppert JK, Dickinson D, Pani AM, Higgins CD, Goldstein B (2016) Comparative assessment of fluorescent proteins for in vivo imaging in an animal model system. Mol Biol Cell 27. <Go to ISI>://WOS:000396046900067

29. Molino JVD, de Carvalho JCM, Mayfield S (2018) Evaluation of secretion reporters to microalgae biotechnology: blue to red fluorescent proteins. Algal Res 31:252–261. https://doi.org/10.1016/j.algal.2018.02.018

30. Kobayashi J, Shidara H, Morisawa Y, Kawakami M, Tanahashi Y, Hotta K, Oka K (2013) A method for selective ablation of neurons in C. elegans using the phototoxic fluorescent protein, KillerRed. Neurosci Lett 548:261–264. https://doi.org/10.1016/j.neulet.2013.05.053

31. Flames N, Hobert O (2009) Gene regulatory logic of dopamine neuron differentiation. Nature 458(7240):885–889. https://doi.org/10.1038/nature07929

32. Nass R, Hall DH, Miller DM 3rd, Blakely RD (2002) Neurotoxin-induced degeneration of dopamine neurons in Caenorhabditis elegans. Proc Natl Acad Sci U S A 99(5):3264–3269. https://doi.org/10.1073/pnas.042497999

33. Dauer W, Przedborski S (2003) Parkinson's disease: mechanisms and models. Neuron 39(6):889–909

34. Harrington AJ, Yacoubian TA, Slone SR, Caldwell KA, Caldwell GA (2012) Functional analysis of VPS41-mediated neuroprotection in Caenorhabditis elegans and mammalian models of Parkinson's disease. J Neurosci 32(6):2142–2153

35. Serrano-Saiz E, Poole RJ, Felton T, Zhang F, De La Cruz ED, Hobert O (2013) Modular control of glutamatergic neuronal identity in C. elegans by distinct homeodomain proteins. Cell 155(3):659–673. https://doi.org/10.1016/j.cell.2013.09.052

36. Lieke T, Steinberg CE, Ju J, Saul N (2015) Natural marine and synthetic xenobiotics get on nematode's nerves: neuro-stimulating and neurotoxic findings in Caenorhabditis elegans. Mar Drugs 13(5):2785–2812. https://doi.org/10.3390/md13052785

37. Soares FA, Fagundez DA, Avila DS (2017) Neurodegeneration induced by metals in Caenorhabditis elegans. Adv Neurobiol 18:355–383. https://doi.org/10.1007/978-3-319-60189-2_18

38. Jacques MT, Bornhorst J, Soares MV, Schwerdtle T, Garcia S, Avila DS (2019) Reprotoxicity of glyphosate-based formulation in Caenorhabditis elegans is not due to the active ingredient only. Environ Pollut 252 (Pt B):1854–1862. https://doi.org/10.1016/j.envpol.2019.06.099

39. Zhang RQ, Hong Y, Xiao JS (2013) Separation and determination of pyrrolidinium ionic liquid cations by ion chromatography with direct conductivity detection. Chin Chem Lett 24(6):503–505

40. Rohn I, Raschke S, Aschner M, Tuck S, Kuehnelt D, Kipp A, Schwerdtle T, Bornhorst J (2019) Treatment of Caenorhabditis elegans with small selenium species enhances antioxidant defense systems. Mol Nutr Food Res 63(9). https://doi.org/10.1002/mnfr.201801304

41. Gubert P, Puntel B, Lehmen T, Bornhorst J, Avila DS, Aschner M, Soares FAA (2016) Reversible reprotoxic effects of manganese through DAF-16 transcription factor activation and vitellogenin downregulation in Caenorhabditis elegans. Life Sci 151:218–223. https://doi.org/10.1016/j.lfs.2016.03.016

42. Kim W, Underwood RS, Greenwald I, Shaye DD (2018) OrthoList 2: a new comparative genomic analysis of human and Caenorhabditis elegans genes. Genetics 210(2):445–461

43. Corsi AK, Wightman B, Chalfie M (2015) A transparent window into biology: a primer on Caenorhabditis elegans. Genetics 200(2):387–407

44. Silverman GA, Luke CJ, Bhatia SR, Long OS, Vetica AC, Perlmutter DH, Pak SC (2009) Modeling molecular and cellular aspects of

human disease using the nematode Caenorhabditis elegans. Pediatr Res 65(1):10

45. Metaxakis A, Petratou D, Tavernarakis N (2018) Multimodal sensory processing in Caenorhabditis elegans. Open Biol 8(6):180049

46. Mujika A, Leškovský P, Álvarez R, Otaduy MA, Epelde G (2017) Modeling behavioral experiment interaction and environmental stimuli for a synthetic C. elegans. Front Neuroinform 11:71

47. Maulik M, Mitra S, Bult-Ito A, Taylor BE, Vayndorf EM (2017) Behavioral phenotyping and pathological indicators of Parkinson's disease in C. elegans models. Frontiers in genetics 8:77

48. Rotroff DM, Joubert BR, Marvel SW, Håberg SE, Wu MC, Nilsen RM, Ueland PM, Nystad W, London SJ, Motsinger-Reif A (2016) Maternal smoking impacts key biological pathways in newborns through epigenetic modification in utero. BMC Genomics 17(1):976

49. Raley-Susman KM, Chou E, Lemoine H (2018) Use of the model organism Caenorhabditis elegans to elucidate neurotoxic and behavioral effects of commercial fungicides. Neurotoxins 37

50. Chen P, Martinez-Finley EJ, Bornhorst J, Chakraborty S, Aschner M (2013) Metal-induced neurodegeneration in C. elegans. Front Aging Neurosc 5:18

51. Ijomone OM, Miah MR, Peres TV, Nwoha PU, Aschner M (2016) Null allele mutants of trt-1, the catalytic subunit of telomerase in Caenorhabditis elegans, are less sensitive to Mn-induced toxicity and DAergic degeneration. Neurotoxicology 57:54–60. https://doi.org/10.1016/j.neuro.2016.08.016

52. Hart AC (2006) Behavior (July 3, 2006), WormBook, ed. the C. elegans research community, WormBook. https://doi.org/10.1895/wormbook.1.87.1

53. Sawin ER, Ranganathan R, Horvitz HR (2000) C. elegans locomotory rate is modulated by the environment through a dopaminergic pathway and by experience through a serotonergic pathway. Neuron 26(3):619–631

54. Aschner M, Chen P, Martinez-Finley EJ, Bornhorst J, Chakraborty S (2013) Metal-induced neurodegeneration in C. elegans. Frontiers in aging neuroscience 5:18

55. Kim H, Calatayud C, Guha S, Fernández-Carasa I, Berkowitz L, Carballo-Carbajal I, Ezquerra M, Fernández-Santiago R, Kapahi P, Raya Á (2018) The small GTPase RAC1/CED-10 is essential in maintaining dopaminergic neuron function and survival against α-Synuclein-induced toxicity. Mol Neurobiol 55(9):7533–7552

56. Engleman EA, Katner SN, Neal-Beliveau BS (2016) Caenorhabditis elegans as a model to study the molecular and genetic mechanisms of drug addiction. In: Progress in molecular biology and translational science, vol 137. Elsevier, pp 229–252

57. Zhang Y, Lu H, Bargmann CI (2005) Pathogenic bacteria induce aversive olfactory learning in Caenorhabditis elegans. Nature 438(7065):179–184

58. Lee J, Jee C, McIntire SL (2009) Ethanol preference in C. elegans. Genes Brain Behav 8(6):578–585

59. Hukema RK, Rademakers S, Jansen G (2008) Gustatory plasticity in C. elegans involves integration of negative cues and NaCl taste mediated by serotonin, dopamine, and glutamate. Learn Mem 15(11):829–836

60. Saeki S, Yamamoto M, Iino Y (2001) Plasticity of chemotaxis revealed by paired presentation of a chemoattractant and starvation in the nematode Caenorhabditis elegans. J Exp Biol 204(10):1757–1764

61. Oda S, Tomioka M, Iino Y (2011) Neuronal plasticity regulated by the insulin-like signaling pathway underlies salt chemotaxis learning in Caenorhabditis elegans. J Neurophysiol 106(1):301–308

62. Zhou X, Bessereau J-L (2019) Molecular architecture of genetically-tractable GABA synapses in C. elegans. Front Mol Neurosci 12:304

63. Jorgensen EM (2005) GABA (August 31, 2005), WormBook, ed. The C. elegans research community. WormBook. https://doi.org/10.1895/wormbook.1.14.1

64. Gjorgjieva J, Biron D, Haspel G (2014) Neurobiology of Caenorhabditis elegans locomotion: where do we stand? Bioscience 64(6):476–486

65. Trojanowski NF, Raizen DM, Fang-Yen C (2016) Pharyngeal pumping in Caenorhabditis elegans depends on tonic and phasic signaling from the nervous system. Sci Rep 6:22940

66. Raizen D, Song B-M, Trojanowski N, You Y-J (2018) Methods for measuring pharyngeal behaviors. In: WormBook: the online review of C. elegans biology [internet]. WormBook

67. Kompoliti K, Verhagen L (2010) Encyclopedia of movement disorders, vol 1. Academic Press

68. Shi Y, Toga AW (2017) Connectome imaging for mapping human brain pathways. Mol Psychiatry 22(9):1230

69. Azulay A, Itskovits E, Zaslaver A (2016) The C. elegans connectome consists of

homogenous circuits with defined functional roles. PLoS Comput Biol 12(9):e1005021

70. Cook SJ, Jarrell TA, Brittin CA, Wang Y, Bloniarz AE, Yakovlev MA, Nguyen KC, Tang LT-H, Bayer EA, Duerr JS (2019) Whole-animal connectomes of both Caenorhabditis elegans sexes. Nature 571(7763):63–71

71. Lim DH, LeDue J (2017) What is optogenetics and how can we use it to discover more about the brain? Front Young Minds 5

72. Fang-Yen C, Alkema MJ (1677) Samuel AD (2015) illuminating neural circuits and behaviour in Caenorhabditis elegans with optogenetics. Philos Trans R Soc B 370:20140212

73. Guru A, Post RJ, Ho Y-Y, Warden MR (2015) Making sense of optogenetics. Int J Neuropsychopharmacol 18(11):pyv079

74. Schild LC, Glauser DA (2015) Dual color neural activation and behavior control with chrimson and CoChR in Caenorhabditis elegans. Genetics 200(4):1029–1034

75. Husson SJ, Gottschalk A, Leifer AM (2013) Optogenetic manipulation of neural activity in C. elegans: from synapse to circuits and behaviour. Biol Cell 105(6):235–250

76. Yu AJ, McDiarmid TA, Ardiel EL, Rankin CH (2019) High-throughput analysis of behavior under the control of Optogenetics in Caenorhabditis elegans. Curr Protoc Neurosci 86(1): e57

77. Rost BR, Schneider-Warme F, Schmitz D, Hegemann P (2017) Optogenetic tools for subcellular applications in neuroscience. Neuron 96(3):572–603

78. Yu J, Chen K, Lucero RV, Ambrosi CM, Entcheva E (2015) Cardiac optogenetics: enhancement by all-trans-retinal. Sci Rep 5:16542

79. Pokala N, Glater EE (2018) Using optogenetics to understand neuronal mechanisms underlying behavior in C. elegans. J Undergrad Neurosci Educ 16(2):A152

80. Martinez-Finley EJ, Avila DS, Chakraborty S, Aschner M (2011) Insights from Caenorhabditis elegans on the role of metals in neurodegenerative diseases. Metallomics 3 (3):271–279. https://doi.org/10.1039/c0mt00064g

81. Qu M, Kong Y, Yuana Y, Wang D (2019) Neuronal damage induced by nanopolystyrene particles in nematode Caenorhabditis elegans. Environ Sci Nano 6:2591–2601. https://doi.org/10.1039/C9EN00473D

82. Schetinger MRC, Peres TV, Arantes LP, Carvalho F, Dressler V, Heidrich G, Bowman AB, Aschner M (2019) Combined exposure to methylmercury and manganese during L1 larval stage causes motor dysfunction, cholinergic and monoaminergic up-regulation and oxidative stress in L4 Caenorhabditis elegans. Toxicology 411:154–162. https://doi.org/10.1016/j.tox.2018.10.006

83. White JG, Southgate E, Thomson JN, Brenner S (1986) The structure of the nervous system of the nematode Caenorhabditis elegans. Philos Trans R Soc Lond Ser B Biol Sci 314 (1165):1–340. https://doi.org/10.1098/rstb.1986.0056

84. Jarrell TA, Wang Y, Bloniarz AE, Brittin CA, Xu M, Thomson JN, Albertson DG, Hall DH, Emmons SW (2012) The connectome of a decision-making neural network. Science 337 (6093):437–444. https://doi.org/10.1126/science.1221762

85. Chen TW, Wardill TJ, Sun Y, Pulver SR, Renninger SL, Baohan A, Schreiter ER, Kerr RA, Orger MB, Jayaraman V, Looger LL, Svoboda K, Kim DS (2013) Ultrasensitive fluorescent proteins for imaging neuronal activity. Nature 499(7458):295–300. https://doi.org/10.1038/nature12354

86. Piatkevich KD, Jung EE, Straub C, Linghu C, Park D, Suk HJ, Hochbaum DR, Goodwin D, Pnevmatikakis E, Pak N, Kawashima T, Yang CT, Rhoades JL, Shemesh O, Asano S, Yoon YG, Freifeld L, Saulnier JL, Riegler C, Engert F, Hughes T, Drobizhev M, Szabo B, Ahrens MB, Flavell SW, Sabatini BL, Boyden ES (2018) A robotic multidimensional directed evolution approach applied to fluorescent voltage reporters. Nat Chem Biol 14(4):352–360. https://doi.org/10.1038/s41589-018-0004-9

87. Chaudhuri J, Parihar M, Pires-daSilva A (2011) An introduction to worm lab: from culturing worms to mutagenesis. J Vis Exp 47:e2293

88. Rivard L, Srinivasan J, Stone A, Ochoa S, Sternberg PW, Loer CM (2010) A comparison of experience-dependent locomotory behaviors and biogenic amine neurons in nematode relatives of Caenorhabditis elegans. BMC Neurosci 11(1):22

89. Swierczek NA, Giles AC, Rankin CH, Kerr RA (2011) High-throughput behavioral analysis in C. elegans. Nat Methods 8(7):592

90. Cooper JF, Dues DJ, Spielbauer KK, Machiela E, Senchuk MM, Van Raamsdonk JM (2015) Delaying aging is neuroprotective in Parkinson's disease: a genetic analysis in C. elegans models. NPJ Parkinson's Dis 1 (1):1–12

91. Matsuura T, Urushihata T (2015) Chronic nicotine exposure augments gustatory plasticity in Caenorhabditis elegans: involvement of

dopamine signaling. Biosci Biotechnol Biochem 79(3):462–469

92. Urushihata T, Wakabayashi T, Osato S, Yamashita T, Matsuura T (2016) Short-term nicotine exposure induces long-lasting modulation of gustatory plasticity in Caenorhabditis elegans. Biochem Biophys Rep 8:41–47

93. Urushihata T, Takuwa H, Higuchi Y, Sakata K, Wakabayashi T, Nishino A, Matsuura T (2016) Inhibitory effects of caffeine on gustatory plasticity in the nematode Caenorhabditis elegans. Biosci Biotechnol Biochem 80 (10):1990–1994

94. Ijomone OM, Miah MR, Akingbade GT, Bucinca H, Aschner M (2020) Nickel-induced developmental neurotoxicity in C. elegans includes cholinergic, dopaminergic and GABAergic degeneration, altered behaviour, and increased SKN-1 activity. Neurotox Res 37:1010–1028. https://doi.org/10.1007/s12640-020-00175-3

95. Bhattacharya A, Aghayeva U, Berghoff EG, Hobert O (2019) Plasticity of the electrical connectome of C. elegans. Cell 176 (5):1174–1189. e1116. https://doi.org/10.1016/j.cell.2018.12.024

96. Cook SJ, Jarrell TA, Brittin CA, Wang Y, Bloniarz AE, Yakovlev MA, Nguyen KCQ, Tang LT, Bayer EA, Duerr JS, Bulow HE, Hobert O, Hall DH, Emmons SW (2019) Whole-animal connectomes of both Caenorhabditis elegans sexes. Nature 571(7763):63–71. https://doi.org/10.1038/s41586-019-1352-7

Part VI

In Vitro Methods

Chapter 19

Effects of Neurotoxic or Pro-regenerative Agents on Motor and Sensory Neurite Outgrowth in Spinal Cord Organotypic Slices and DRG Explants in Culture

Sara Bolívar, Ilary Allodi, Mireia Herrando-Grabulosa, and Esther Udina

Abstract

Classically, primary sensory neuron cultures obtained from the DRG have been used as a model to evaluate neurite growth in vitro. Primary sensory neurons are easily cultured, either dissociated or from explants, from embryonic to adult ages. In contrast, culture of motoneurons is much more complex and limited to the embryonic ones or to postnatal organotypic cultures by using membrane culture inserts. Here we describe a protocol of an easy in vitro assay to culture postnatal rodent spinal cord organotypic slices and DRG explants in 3D collagen matrices that are permissive for neuritogenesis. The main aim of this in vitro assay is to have a similar setting for both types of neurons that allows the measurement and comparison of positive or adverse events on neurite growth of motor and sensory neurons. The matrix can also be modified by adding trophic or tropic factors, cells, or other agents. Immunohistochemistry of the explants and the slices is needed to specifically label myelinated fibers and fairly compare the growth of myelinated primary sensory neurons and motoneurons, as well as neuronal survival.

Key words Motoneuron, Dorsal root ganglia, Spinal cord, Primary sensory neurons, Neurite growth, Collagen matrix, Explant, Organotypic, Postnatal, Mice, Rat

1 Introduction

Spinal motoneurons and primary sensory neurons of the dorsal root ganglia (DRG) are the main contributors of the axons that form peripheral nerves, together with autonomic axons. In contrast to the central nervous system, the peripheral nervous system lacks the blood-brain barrier, which usually protects the brain and the spinal cord from external insults. This fact explains why peripheral axons are more prone to suffer after exposure to neurotoxic drugs. A paradigmatic example is the peripheral neuropathy induced by neurotoxic chemotherapy drugs, extensively used to treat different types of cancers [1]. Although these axonopathic agents would affect both motor and sensory axons, DRG neurons are more commonly affected since their somas are more exposed. Therefore,

Jordi Llorens and Marta Barenys (eds.), *Experimental Neurotoxicology Methods*, Neuromethods, vol. 172,
https://doi.org/10.1007/978-1-0716-1637-6_19, © Springer Science+Business Media, LLC, part of Springer Nature 2021

in vitro settings to study peripheral neurotoxicity have mainly focused on sensory neuronal cultures [2]. Similarly, since rodent primary sensory neurons are easily cultured, from embryonic to adult, in vitro studies regarding neurite outgrowth have also been mainly focused on DRG cultures, either DRG explants or dissociated primary sensory neurons [3].

In contrast, neurotoxicity and neurodegenerative conditions in motoneurons are evaluated using spinal cord organotypic slices [4, 5]. The culture of adult motoneurons is quite complex and few papers reported success [6, 7]. Therefore, classical spinal cord explants are performed in postnatal rats by using permeable membrane culture inserts [8, 9]. These inserts allow long-term survival of spinal cord slices but suppose some constraints regarding neurite growth evaluation.

Since sensory and motor neuron cultures are classically performed separately and using different settings, an in vitro setting that allows a fair comparison between both of them would be highly advantageous. Here we describe an easy in vitro assay to culture postnatal spinal cord organotypic slices and DRG explants in collagen matrices [10] that are permissive for neuritogenesis. The aim of this in vitro model is to have a similar setting for both types of neurons that facilitates the comparative measurement of positive or adverse events on neurite growth of motor and sensory neurons [11].

Although usually this type of culture has been performed using postnatal rats, it can be adapted to postnatal mice. This is of great interest since transgenic mice have become a valuable tool for research.

2   Materials

A table with the reference of the bioreagents and surgical tools used in this protocol is included (Table 1).

2.1   Spinal Cord Organotypic Slices and DRG Explants Cultures

- Light microscope (for dissection). A portable dissection microscope that can be placed into the cell culture hood is recommended for fine cleaning of the tissue, whereas extraction of the primary sample can be performed either in a portable or a fixed one.
- Basic surgical tools (scalp, fine forceps, laminectomy forceps, spring scissors...).
- Gey's balanced salt solution enriched with 6 mg/mL sterile D-(+)-glucose. Make fresh as required and store at 4 °C.

Table 1

Bioreagents and surgical tools used in the protocol, with recommended vendor and reference number

Product	Reference	Vendor	
B27 supplement (50×)	17504044	ThermoFisher (Gibco)	t.3
Gey's balanced salt solution	G9779	Sigma	t.4
L-Glutamine solution	G7513	Sigma	t.5
Minimum essential medium (MEM) 10×	11430030	ThermoFisher (Gibco)	t.6
Neurobasal medium	21103049	ThermoFisher (Gibco)	t.7
Nylon net filter	HNWP04700	Millipore	t.8
Penicillin-streptomycin	P0781	Sigma	t.9
Poly-D-lysine hydrobromide	P6407	Sigma	t.10
Rat tail type I collagen	354236	Corning	t.11
Sodium bicarbonate 7.5%	25080-094	ThermoFisher (Gibco)	t.12
Sterile D-(+)-glucose	G7021	Sigma	t.13
Trizma hydrochloride (HCl)	T5941	Sigma	t.14
Trizma base	T1503	Sigma	t.15
Tris buffer (TB; 0.05 M pH 7'4)	Trizma HCl (6.06 g), Trizma Base (1.39 g), distilled H$_2$O (1 L)		t.16
Tris buffer saline (TBS; 0.05 M pH 7'4)	TB10 × (100 mL), NaCl (8 g), distilled-H$_2$O (900 mL)		t.17
Surgical tools			t.18
Small spring scissors	FST 15003-08	Fine science tools	t.19
Spring scissors	FST 15025-10	Fine science tools	t.20
Dumont 2 (laminectomy)	FST 11223-20	Fine science tools	t.21
Dumont 5 (fine forceps)	FST 11254-20	Fine science tools	t.22

- Ice pack to keep the samples cold. A homemade ice pack can be easily made by using glass Petry dishes of 60 mm diameter filled with gel from commercial ice packs, sealed and kept on the freezer before use. Its flat surface and small size are ideal as a base to manipulate the samples under the dissection microscope.
- Cell culture hood with laminar flux.
- McIlwain Tissue Chopper.
- Petri dishes or 24 multiwell plates.

- Round coverslips.
- Poly-D-lysine (10 μg/mL).
- Rat tail type I collagen solution (3.4 mg/mL); store at 4 °C and keep cold during use.
- 10× Minimum essential medium (MEM).
- 7.5% Sodium bicarbonate solution (can be stored at 4 °C).
- Culture incubator set at 37 °C and 5% CO_2.
- Neurobasal medium (NB), supplemented with B27 (I1×), glutamine (20 mM), glucose (6 mg/mL) and with or without penicillin/streptomycin (1×: 100 U/mL and 0.1 mg/mL, respectively). Make fresh for each culture and store at 4 °C.
- Nylon hydrophilic membrane filter.

2.2 Immuno-histochemistry

- Paraformaldehyde (PFA) 4% pH 7,4, store at 4 °C but warm at 37 °C before use.
- Tris Buffer (TB), Tris-buffered saline (TBS), Tris-buffered saline with 0.3% Triton (TBSL).
- Citrate buffer (10 mM, pH 6.1).
- Methanol (50%, 70%, and 100%).
- Ethanol (70%, 96%, and 100%).
- Mounting media (for example, glycerol supplemented with 10% Mowiol and 0.6% DABCO (Sigma), or Fluoromount (SouthernBiotech)).

2.3 Quantification

- Inverted light microscope (to evaluate fresh cultures).
- Epifluorescence microscope and/or confocal microscope (to evaluate immunocytochemistry processed slices).
- Image J software (NIH, available at http://rsb.info.nih.gov/ij/).

3 Methods

3.1 Matrix Preparation

3.1.1 Preliminary Preparations

The explants are placed on top of a collagen matrix that needs to be prepared before harvesting the samples since its gelation needs around 2 h.

Coating of the coverslips placed on the culture dishes is recommended to facilitate attachment of the collagen matrices, and therefore it is better to prepare it the day before.

119
120
121
122
123
124
125
126
127
128
129
130
131
132
133
134
135
136
137
138
139
140
141
142
143
144
145
146
147
148
149
150
151
152
153
154
155
156
157
158
159
160
161
162
163
164
165
166

3.1.2 Preparation of the Matrices

Coverslips are placed in Petri dishes or multiwell plates (three coverslips in a 60 mm petri dish or single ones in a 24-well plate, *see* **Note1**) and coated with poly-D-lysine (PDL, 10 µg/mL) overnight. Three washes with sterile distilled water are performed afterward. Coverslips have to be completely dry to avoid that water droplets dilute the collagen matrix.

The collagen matrices can be prepared in 1.5 mL Eppendorf (500 µL each) and kept on ice to avoid gelation until it is used. 450 µL of rat tail type I collagen solution (3.4 mg/mL), 50 µL of 10× basal Eagle's medium, and 2 µL of 7.5% sodium bicarbonate solution [10] are mixed with an automatic pipette until obtaining a homogeneous matrix of a yellowish color. The mixture is immediately used to prepare the matrices.

A single drop of 30 µL is deposited on the dried coated coverslips and kept in the incubator at 37 °C and 5% CO_2 for at least 2 h to induce collagen gel formation (*see* **Note 1**).

3.2 Spinal Cord Organotypic Slices and DRG Explants Cultures

3.2.1 Obtention of the Primary Tissue

To increase neuronal survival, the following steps should be swiftly performed, and tissue should be kept cold. Flat ice packs can be used as a base.

Postnatal rats (around p7 would be ideal) are deeply anesthetized with an overdose of pentobarbital (200 mg/kg i.p.) and decapitated. Under a light microscope, spinal cords and DRGs are harvested (*see* **Note 2**). Briefly, a laminectomy is performed using a special forceps to manipulate the bone of mice or postnatal rats (laminectomy forceps). The spinal cord is carefully exposed from the cervical to the sacral levels. DRGs from the desired anatomical localization (usually thoracic and/or lumbar) are harvested by picking the nerve root and cutting the sample by fine spring scissors. After harvesting the DRG, the remaining roots are also cut with the scissors to isolate the spinal cord. The spinal cord can be divided into two segments (cervical-thoracic and lumbar-sacral) to facilitate manipulation and to reduce the risk of damaging the tissue. With a fine forceps and the assistance of the scissors, the spinal cord segment is carefully released from the bone and meningeal tissue before harvesting.

Samples are placed in Gey's balanced salt solution enriched with 6 mg/mL glucose. Three washes using the same solution are performed under the cell culture hood to sterilize the samples.

A portable light microscope is placed into a cell culture hood with laminar flux to clean the samples from blood, meningeal debris, and connective tissue. Samples have to be kept cold during the procedure by using an ice pack as a base.

The connective tissue of the DRG and the meninges of the spinal cords should be thoroughly eliminated while maintaining the integrity of the samples (*see* **Note 3**). To prevent the damage of the spinal cord, carefully remove the meninges from the caudal to the rostral part.

Spinal cords are cut in segments of about 2 cm and carefully placed on top of a round Nylon hydrophilic membrane filter with the dorsal horn facing up, then cut with a McIlwain Tissue Chopper into 350 μm thick slices. The samples are immediately placed in Petri dishes again with Gey's balanced solution supplemented with glucose.

Samples (350 μm spinal cord slices or whole DRG explants) are carefully placed on the gelled collagen droplets and a second drop of 30 μL of the same collagen solution is applied to cover them. DRG explants can be easily picked up with fine forceps whereas spinal cord slices have to be carefully collected with a small spoon. Therefore, DRG can be placed on top of collagen matrices mounted into a 24-multiwell plate, whereas it is better to use a 60 mm Petri dish with three collagen droplets to place the spinal cord slices.

The embedded samples are placed again in the incubator for 45 min to allow gelation of the second droplet. Afterward, Neurobasal medium (NB) supplemented with B27, glutamine, glucose, and penicillin/streptomycin is added to each well or Petri dish. The volume added depends on the size of the recipient (1.5 mL for the 60 mm petri dish and 0.5 mL for a 24 multiwell plate). After 1 day in culture, the medium is replaced by a free-antibiotic medium (same NB-based medium, but without penicillin/streptomycin). It is recommended to culture spinal cords for 4 days to allow some spontaneous growth of motor neurites, usually less than 500 microns length (*see* **Note 4**). In contrast, DRG explants show an extensive growth already at 2 days post culture, with neurites extending about 1000 microns. During the culture, explants and spinal cord slices can be examined under an inverted light microscope. Viability of the cultures can be confirmed when migration of exogenous cells and some neurites can be observed (Fig. 1), but immunohistochemistry is needed in order to corroborate these findings and to visualize both the somas and the neurites (Fig. 2).

3.3 Modification of the Matrix

Different factors and drugs can be added directly to the matrix during its preparation, to test their effects on neuron survival and growth. Examples of agents that can be added to the matrix range from a wide variety of trophic factors like BDNF, NT3, GDNF, NGF, and FGF [11, 12], tropic factors like laminin or fibronectin [13], combination of both types of factors [14], and peptides [15] among others. Briefly, factors can be added to 50 μL of 10× basal Eagle's medium before mixing with collagen. As a reference, addition of 10 ng/mL of classical trophic factors like GDNF or BDNF is enough to promote neurite outgrowth in the samples ([11], *see* also Figs. 2 and 3).

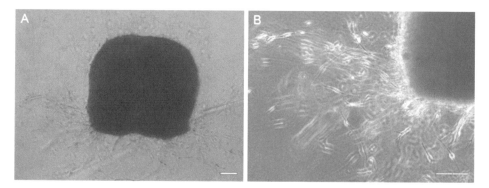

Fig. 1 Representative image of a fresh 4-day spinal cordorganotypic cultures taken from an inverted microscope (**a**) and magnification of the ventral horn (**b**). Neurite growth from the ventral horn and migration of cells from the explant into the matrix can be appreciated, indicative of a viable culture. Bar = 150 μm

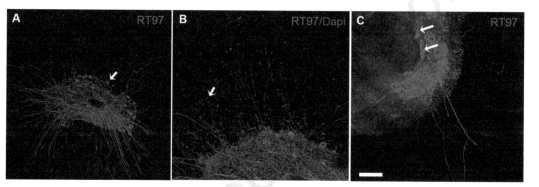

Fig. 2 Representative images of confocal photomicrographs of rat DRG explants and spinal cordorganotypic slices after 2 and 4 days in culture, respectively. Samples have been processed for immunocytochemistry to label myelinated fibres and their somas (by using RT97 antibody (red)). DRG explant with addition of 10 ng/mL GDNF into the matrix, where somas of myelinated sensory neurons (arrow) are labeled, as well as their neurites, that extend into the collagen matrices from all around the explant (**a**). Higher magnification of the explant to observe migration of cells (arrow) from the DRG to the matrix. Nuclei of cells have been labeled with DAPI (blue) (**b**). Spinal cord slice with addition of 10 ng/mL BDNF into the matrix. Motoneurons (arrows), with the typical triangular shape can be appreciated in the ventral horn, with neurites extending into the collagen matrix (**c**). Bar: 50 μm (**a**) and 100 μm (**c**)

Similarly, cells can also be included in the matrix to perform cocultures. Mesenchymal stem cells, fibroblasts, Schwann cells, or olfactory ensheathing glia [11, 16, 17] have been added successfully to the collagen matrix. The specific cells are obtained previously from primary cultures. Then, an adequate amount of cell suspension in the growing medium is carefully mixed in the collagen solution to get the final concentration. As a reference, between 10.000 and 50.000 cells can be mixed to each volume of collagen matrix used to embed the spinal cord slices or DRG explants. This number of cells is sufficient to influence the fate of the cocultures (Figs. 3 and 4, [18]).

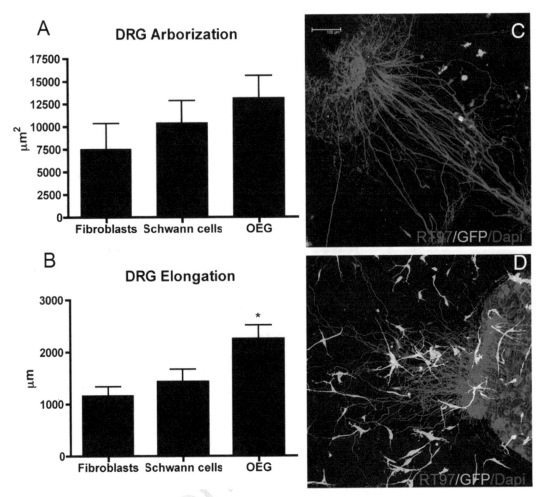

Fig. 3 Evaluation of neurite outgrowth of rat DRG explants cultured into collagen matrices seeded with 10,000 fibroblasts, Schwann cells (SC) or olfactory ensheathing glia (OEG). Column plots of neurite arborization (**a**) and elongation (**b**) when the different cells were added into the matrix. Results are expressed as mean ± SEM. One-way ANOVA with post hoc Bonferroni method was used to statistically analyze neurite outgrowth, *$p < 0.05$ compared to fibroblasts cocultures. Representative images of DRG neurite outgrowth in cocultures with GFP transfected SCs (**c**) and OECs (**d**). In both situations, cells are well integrated in the collagen matrix and no cluster formations can be observed. RT97 in red, GFP in green, and DAPI in blue. Bar = 100 μm. Both OEC and SC promote growth of primary sensory neurons. Adapted from [18]

3.4 Source of Primary Tissues

This protocol is extensively used in our lab using postnatal rats, and it has been adapted recently to postnatal mice. Since survival of postnatal mouse motoneurons can be more challenging than that of rat motoneurons, adding GDNF in control conditions is recommended in other types of spinal cord explants [19]. Addition of GDNF (10 ng/mL) into the matrix promotes considerable neurite growth in our cultures, although non-treated spinal cord slices show some growth as well (Fig. 5a and d vs. b and e). In contrast, survival of primary sensory neurons does not differ between mouse and rat neurons.

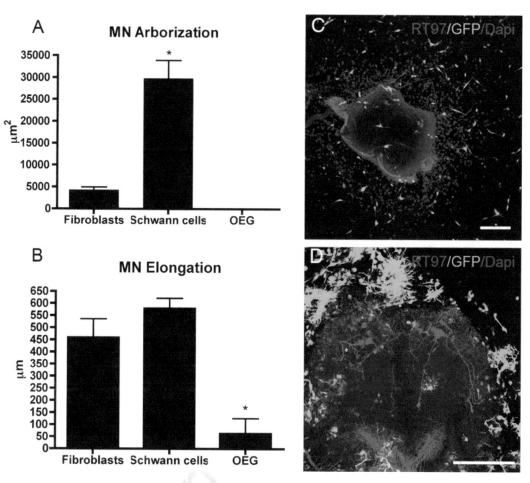

Fig. 4 Evaluation of neurite outgrowth of rat spinal cordorganotypic slices cultured into collagen matrices seeded with 10.000 fibroblasts, Schwann cells (SC) or olfactory ensheathing glia (OEG). Column plots show the mean values for motor neurite arborization (**a**) and elongation (**b**) in cocultures with the three different cell types and in control condition. Results are expressed as mean ± SEM. One-way ANOVA with post hoc Bonferroni method was used to statistically analyze neurite outgrowth, *$p < 0.05$ vs. fibroblast conditions. Confocal pictures showing motoneuronneurite outgrowth from the spinal cord slice cocultured with 10.000 SCs (arrows indicate neurite elongation into the matrix) (**c**) or OECs (**d**). SCs appeared spread in the collagen matrix and some in relation to neurites elongating within the matrix (**c**). In OEC cocultures, clustering of the grafted cells (GFP+, green) can be observed surrounding the spinal cord slice. Motoneurons elongate neurites within the spinal cord slice (arrows) but not into the matrix, thus avoiding contact with the cell clusters. Therefore, OEC are creating a nonpermissive environment for growth into the matrix (**d**). RT97 in red, GFP in green, and DAPI in blue. Bar = 100 μm (**c**) and 200 μm (**d**). Adapted from [18]

3.5 Immuno-
histochemistry

Spinal cord slices and DRGs embedded in the collagen matrix are fixed with PFA 4% pH 7.4 for 30 or 15 min at room temperature (RT), respectively, and then washed with TBSL 3 times (20 min each) at RT on a shaker.

For the spinal cord slices, antigen retrieval is recommended, keeping samples into hot citrate buffer for at least 1 h. Then, a wash with TBSL for 5 min at RT on a shaker is performed. Next, samples

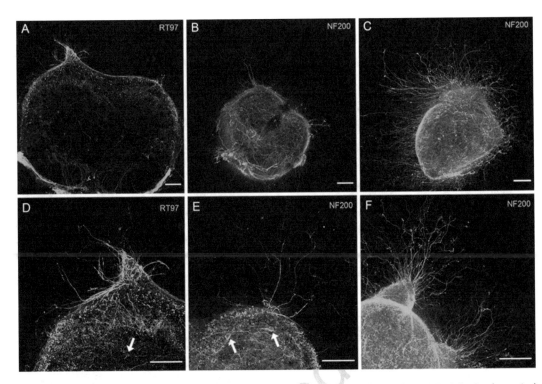

Fig. 5 Representative images of a 4-day spinal cord organotypic culture of an 8-day postnatal mice in control conditions (**a, d**) or treated with GDNF (10 ng/mL; **b, e**). Extensive neurite outgrowth from the ventral horn can be observed in GDNF treated culture, although some neurites and motoneurons are also observed in the control culture (**e** vs. **d**). The samples were labeled with RT97 (**a, d**) or NF200 (**b, c, e, f**). Some motoneurons with the typical triangular shape can be observed (arrow **d, e**), although the soma is not as well labeled as the neurites. Example of a mice DRG explant cultured for 2 days and immunostained with NF200 (**c, f**). Similar to spinal cord slices, there is a fain staining of the somas, being neurites highly contrasted. Bar = 150 μm

are incubated in methanol of increasing concentrations (50%, 70%, and 100% H₂Od) for 20 min and washed two times with TBSL (for 20 min).

Both DRGs and spinal cords are incubated with primary antibody for 48 h at 4 °C in TBSL with 1.5% specific serum (*see* **Note 5**). Three washes with TBSL for 1 h are performed, and then samples are incubated with secondary antibody overnight at 4 °C in TBSL with 1.5% specific serum (the same serum used for the primary incubation). Then, samples are washed with TBSL 2 times, and with TBS and TB once (for 40 min each).

Samples are dehydrated in progressive ethanol concentration (70%, 96%, and 100%, for 1 min). The round coverslips are removed from the dishes with the help of a 26G needle and fine forceps and let dry. Coverslips are finally covered with the desired mounting media (for example, MOWIOL, an aqueous mounting media already prepared with DAPI (1:1000)).

3.6 Quantification

3.6.1 Neuronal Survival Evaluation

To assess the potential contribution of the different conditions tested on neuronal survival, spinal cord slices can be stained with anti-neurofilament antibody to label motoneurons and myelinated sensory neurons, and their growing neurites (*see* **Note 6**). For a fair comparison between spinal cord slices and DRG explants, both myelinated motor and sensory neurons can be labeled (Figs. 2 vs. 3, Fig. 5), although non myelinated neurons can also be stained in the DRG (*see* **Note 7**). Immunofluorescent neurons are counted at least in six samples per condition under a confocal microscope. In the spinal cord slices, only cells with morphology corresponding to motoneurons and a clear visible nucleus have to be counted in each section.

3.6.2 Neurite Growth Measurement

For quantitative analysis of neurite growth, microphotographs are taken at 20× with a digital camera attached to an epifluorescence microscope, then automatically photomerged (when needed) and analyzed with free ImageJ software. The length of the longest neurite is measured for at least 20 samples per condition (*see* **Note 8**). To evaluate the arborization area, micrographs are transformed to 8-bit image into grayscale and quantification is performed after defining a threshold for background correction (*see* panels a and b from Figs. 2 and 3 for examples of quantification of both neurite length and arborization).

4 Conclusions

The use of 3D collagen matrices allows the culture of both organotypic slices and DRG explants in a setting that is permissive for neuritogenesis and enables the measurement and comparison of positive or adverse events on neurite growth of motor and sensory neurons. Moreover, the matrix can be modified by adding trophic or tropic factors, cells, or other agents and, thus, it is a useful tool to evaluate and fairly compare neurotoxic or pro-regenerative effects in both motor and sensory neurons in a short-term culture setting.

5 Notes

1. When placing the spinal cord slices onto the collagen gel, the 24-multiwell plate can be a nuisance. Therefore, placement of the collagen matrices in a small Petri dish would facilitate the task. A 60 mm Petri-dish can contain 3 coverslips and, therefore, 3 matrices. In contrast, DRG explants are easy to manipulate with fine forceps and their placement in a 24-multiwell plate is practical.

2. Some authors prefer to obtain slices from thoracic segments of the spinal cord and DRG explants from the same thoracic level of the cord since these are more uniform. In our hands, slices obtained from lumbar and sacral segments of the spinal cord can also be cultured, with the sacral ones showing higher neuronal survival rate, probably due to smaller size and more favorable oxygenation conditions. It is important to consider the heterogeneity of the samples when comparing different treatments. Similarly, lumbar DRGs are bigger, and therefore the observed growth can be different than the one observed in the more spherical thoracic ones.

3. For the spinal cords, it is important to clean the tissue from the meningeal layers, both the external and the internal one. If some meningeal debris are kept, the slices can get attached to the Nylon membrane filter during the cutting process with the chopper.

4. This type of culture is designed for short-term survival of organotypic slices, less than a week, since oxygenation of the tissue is limited. For longer survival, the use of membrane culture inserts (e.g., Millicells) is needed. Spinal cord slices are placed on these permeable membranes, to avoid direct contact with the culture media, allowing long-term survival of the tissue. However, motor neurites do not have a substrate to grow, and evaluation of neurite growth in this setting is trickier than into a matrix. Moreover, there is no equivalent culture for DRG explants and, thus, there is not a fair comparison between motor and sensory neurite growth when using these membrane inserts to culture spinal cord slices.

5. Incubation times are quite long due to the thickness of the spinal cord slices. DRG explants can be incubated half of the time.

6. Primary antibodies that are good markers for motoneurons and myelinated sensory neurons in rat samples are mouse RT97 (1:200, Developmental Studies Hybridoma Bank), mouse SMI32 (1:2500, Sternberger Monoclonals Inc.), and chicken anti-NF200 (1:1000, Millipore). It is important to note that phosphorylated neurofilament (RT97) labels mainly neurites, and non-phosphorylated one (SMI32) the somas of intact motoneurons. Since the phosphorylate form can also be detected in the soma after injury [11], this form is our preferred choice to label these cultures. Unfortunately, in mice cultures, immunolabeling against R797 in somas is weaker than the one observed in rats (compare Fig. 5 vs. Fig. 2).

7. The pan-neuronal marker PGP is a good antibody to label all the neurons of the DRG, but then neurite growth and survival would also include unmyelinated populations of neurons, whereas motoneurons are exclusively myelinated.

8. Control spinal cord slices are viable but do not always have neurite growth after 4 days. Therefore, it is important to take into account that basal neurite growth can be close to zero in control conditions. When evaluating adverse effects on neurite growth, an optimized control has to be considered (for example, addition of cells or classical trophic factors into the culture).

References

1. Argyriou AA, Briani C, Cavaletti G et al (2013) Advanced age and liability to oxaliplatin-induced peripheral neuropathy: post hoc analysis of a prospective study. Eur J Neurol 20 (5):788–794

2. Lehmann HC, Staff NP, Hoke A (2019) Modeling chemotherapy induced peripheral neuropathy (CIPN) in vitro: prospects and limitations. Exp Neurol 326:113140. https://doi.org/10.1016/j.expneurol

3. Tucker BA, Mearow KM (2008) Peripheral sensory axon growth: from receptor binding to cellular signaling. Can J Neurol Sci 35 (5):551–566

4. Rothstein JD, Jin L, Dykes-Hoberg M, Kuncl RW (1993) Chronic inhibition of glutamate uptake produces a model of slow neurotoxicity. Proc Natl Acad Sci U S A 90(14):6591–6595

5. Kosuge Y, Sekikawa-Nishida K, Negi H et al (2009) Characterization of chronic glutamate-mediated motor neuron toxicity in organotypic spinal cord culture prepared from ALS model mice. Neurosci Lett 454(2):165–169

6. Montoya GJ, Sutachan JJ, Chan WS et al (2009) Muscle-conditioned media and cAMP promote survival and neurite outgrowth of adult spinal cord motor neurons. Exp Neurol 220:303–315

7. Pandamooz S, Salehi MS, Zibaii MI, Safari A, Nabiuni M, Ahmadiani A, Dargahi L (2019) Modeling traumatic injury in organotypic spinal cord slice culture obtained from adult rat. Tissue Cell 56:90–97

8. Guzmán-Lenis MS, Navarro X, Casas C (2009) Drug screening of neuroprotective agents on an organotypic-based model of spinal cord excitotoxic damage. Restor Neurol Neurosci 27(4):335–349

9. Herrando-Grabulosa M, Mulet R, Pujol A et al (2016) Novel neuroprotective multicomponent therapy for amyotrophic lateral sclerosis designed by networked systems. PLoS One 11 (1):e0147626

10. Tucker A, Lumsden A, Guthrie S (1996) Cranial motor axons respond differently to the floor plate and sensory ganglia in collagen gel co-cultures. Eur J Neurosci 8(5):906–916

11. Allodi I, Guzmán-Lenis MS, Hernàndez J et al (2011) In vitro comparison of motor and sensory neuron outgrowth in a 3D collagen matrix. J Neurosci Methods 198:53–61

12. Allodi I, Casals-Díaz L, Santos-Nogueira E et al (2013) FGF-2 low molecular weight selectively promotes neuritogenesis of motor neurons in vitro. Mol Neurobiol 47:770–781

13. Gonzalez-Perez F, Alé A, Santos D et al (2016) Substratum preferences of motor and sensory neurons in postnatal and adult rats. Eur J Neurosci 43:431–442

14. Santos D, González-Pérez F, Giudetti G et al (2017) Preferential enhancement of sensory and motor axon regeneration by combining extracellular matrix components with neurotrophic factors. Int J Mol Sci 18(1):65. https://doi.org/10.3390/ijms18010065

15. Auer M, Allodi I, Barham M et al (2013) C3 exoenzyme lacks effects on peripheral axon regeneration in vivo. J Peripher Nerv Syst 18:30–36. https://doi.org/10.1111/jns5.12004

16. Allodi I, Mecollari V, González-Pérez F et al (2014) Schwann cells transduced with a lentiviral vector encoding Fgf-2 promote motor neuron regeneration following sciatic nerve injury. Glia 62:1736–1746

17. Torres-Espín A, Corona-Quintanilla DL, Forés J et al (2013) Neuroprotection and axonal regeneration after lumbar ventral root avulsion by re-implantation and mesenchymal stem cells transplant combined therapy. Neurotherapeutics 10(2):354–368. https://doi.org/10.1007/s13311-013-0178-5

18. Allodi I (2012) Changing the intrinsic growth capacity of motor and sensory neurons to promote axonal growth after injury. Thesis dissertation. https://www.tdx.cat/handle/10803/96355

19. Rakowicz WP, Staples CS, Milbrandt J et al (2002) Glial cell line-derived neurotrophic factor promotes the survival of early postnatal spinal motor neurons in the lateral and medial motor columns in slice culture. J Neurosci 22 (10):3953–3962

Chapter 20

Integrative In Vitro/Ex Vivo Assessment of Dopaminergic Neurotoxicity in Rodents Using Striatal Synaptosomes and Membrane Preparations

Raúl López-Arnau and David Pubill

Abstract

The striatum is a brain area with a high density of dopaminergic terminals and so it is considerably affected when dopaminergic neurotoxicity occurs. Several methods have been developed to evidence such neurotoxicity, some of which can be performed in vitro, using membranes or synaptosomes obtained from fresh tissues. Concretely, here we detail our methodological experience assessing reactive oxygen species in striatal synaptosomes, as well as measuring brain terminal damage after a neurotoxic treatment leading to decreased [^{3}H]WIN 35428 binding and tyrosine hydroxylase expression measured by Western blot. Also, we have assessed impairment of dopamine transporter after dopaminergic neurotoxicity measuring uptake of [^{3}H]dopamine in striatal synaptosomes. These techniques are complementary and can be useful to assess the dopaminergic neurotoxic potential of drugs such as amphetamine derivatives, among others.

Key words Dopamine uptake, Dopaminergic neurotoxicity, Reactive oxygen species, Striatum, Synaptosomes, Tyrosine hydroxylase, [^{3}H]WIN 35428 binding

---

1 Introduction

It is well known that neurotoxicity occurs when a physical, chemical, or biological agent produces, directly or indirectly, any long- or short-term adverse effect in the central or peripheral nervous system structure or function [1]. In this context, the dopaminergic system plays a key role in the neurotoxic effects observed in the central nervous system (CNS) following abuse of certain drugs (e.g., amphetamines), leading to neurochemical impairments which, to a certain extent, can resemble those of diseases such as Parkinson's (PD).

Synaptosomes are synaptic terminals isolated from a certain brain area which keep the morphological features and most of the chemical properties of the original nerve terminal [2]. They contain numerous small clear synaptic vesicles, sometimes larger dense-core

Jordi Llorens and Marta Barenys (eds.), *Experimental Neurotoxicology Methods*, Neuromethods, vol. 172,
https://doi.org/10.1007/978-1-0716-1637-6_20, © Springer Science+Business Media, LLC, part of Springer Nature 2021

vesicles, and frequently one or more mitochondria. For these reasons, synaptosomes are used to study synaptic transmission in vitro because they contain the molecular machinery necessary for the neuronal uptake, storage, and release of neurotransmitters. They maintain a normal membrane potential, contain presynaptic receptors, translocate metabolites and ions, and when depolarized, release multiple neurotransmitters (including catecholamines) in a calcium-dependent manner.

The striatum is a brain area with predominant dopaminergic innervation, which allows studying effects related with dopamine (DA) (i.e., uptake, storage, and release) but striatal synaptosomes can also be used to assess oxidative stress measuring reactive oxygen species (ROS) induced by substances at the nerve terminal, which is indicative of possible neurotoxicity. The works carried out by Myhre and Fonnum [3] on the neurotoxic effects of hydrocarbons and our studies on the oxidative effects of amphetamine derivatives [4–8] are good examples. Many amphetamine derivatives (e.g., methamphetamine,   3,4-methylenedioxy-methamphetamine (MDMA)) are taken up into the nerve terminals and displace monoamines (DA, serotonin) from their storage vesicles, inducing their release through reverse transport. However, this effect is followed by a blockade of the transporter and free monoamines can undergo cytosolic metabolism producing several types of ROS (i.e., hydrogen peroxide, hydroxyl radical, peroxynitrite, quinones) which react with cell structures and produce neurotoxicity [9–11].

Thus, measuring increased ROS production in nerve synaptosomes is indicative of neurotoxicity. As DA is the predominant neurotransmitter in the striatum, ROS production in this area presumes dopaminergic neurotoxicity. Moreover, the fact of being an in vitro preparation also allows studying the involved mechanisms in an isolated and composition-controlled medium by the addition of specific drugs acting at receptors, inhibiting enzymes, etc.

Some techniques involve functional imaging of dopamine transporter (DAT) in human striatum by positron emission tomography or single photon emission computerized tomography using different radiolabeled DAT ligands (e.g., cocaine or WIN 35428). In fact, DAT imaging is used for detection, for example, of early stage PD [12, 13]. Moreover, neuroimaging studies, which reveal a decrease in DAT levels, have demonstrated the neurotoxic effects in humans induced by some amphetamine derivatives, leading to memory and motor impairments [14]. Changes in DAT expression have been also involved in other human brain disorders, such as dementia with Lewy bodies, Wilson's disease, Machado–Joseph disease, Lesch–Nyhan disease, Gilles de la Tourette's syndrome, and schizophrenia [12, 15, 16]. In addition, and along with some neurodegenerative changes, a reduction in DA uptake and a decrease in tyrosine hydroxylase (TH) levels, the rate-limiting enzyme of the DA biosynthesis, have been also observed in studies with animals [17, 18] (*see* also review [19]) and are broadly used as markers of dopaminergic impairment.

Here we show the protocols we used in our experiments, which were focused on studying the dopaminergic neurotoxicity of amphetamine derivatives such as methamphetamine, MDMA, and mephedrone, among others.

2   Animals, Materials, Buffers, and Reagents

Protocols involving laboratory animals must be compliant with local and national animal welfare regulations and be approved by the corresponding Ethics Committees. Although the techniques explained in this chapter can be adapted to other species, the methods we report have been applied to rats and mice. Rats and mice from the most popular strains can be used. We mostly used Sprague Dawley and Dark Agouty rats, as well as Swiss-CD-1 mice. The age will depend on the study design. For experiments involving in vivo drug treatment and ex vivo determinations, no less than six animals per group should be used. For experiments on preparations from drug-naïve animals, please *see* Subheading 3.1.1.

All the reagents should be of analytical grade and some can be obtained from the most habitual commercial sources.

- Tris-sucrose buffer: 5 mM Tris–HCl and 320 mM sucrose. It can be prepared by diluting an aliquot of a stock 50 mM Tris–HCl (pH 7.4, store at 4 °C) in bidistilled water and adding the sucrose. Dissolve by stirring and adjust pH to 7.4. It is advisable to freshly prepare it, as the presence of sucrose may favor growth of microorganisms.

- Tris-sucrose 1.6 M buffer: prepare as described above but using the corresponding amount of sucrose.

- PBS-sucrose buffer: it consists of sodium phosphate-buffered (100 mM) solution with sucrose 320 mM at pH 7.9.

- HBSS-glucose buffer: it consists of the standard Hank's Balanced Saline Solution (HBSS) plus 20 mM HEPES sodium and 5.5 mM glucose. HBSS composition is (in mM): 140 NaCl, 5.37 KCl, 1.26 $CaCl_2$, 0.44 KH_2PO_4, 0.49 $MgCl_2$, 0.41 $MgSO_4$, 4.17 $NaHCO_3$, 0.34 Na_2HPO_4. This buffer can be purchased from some commercial sources (e.g., Biological Industries, Inc.) and HEPES and glucose should be freshly added to the desired volume. Alternatively, HBSS plus HEPES can be prepared in the laboratory as a concentrated stock (e.g., 2×) and the desired volume can be diluted, and the glucose added the day of the experiment. HBSS should be stored at 4 °C.

– 2′,7′-Dichlorofluorescin diacetate (DCFH-DA). It is the reagent that will produce fluorescence after being taken up by the synaptosomes and react with ROS. Molecular Probes is a habitual source to purchase it. It is a powder which can be previously dissolved and aliquoted as convenient to perform the experiments. Typically, dissolve 10 mg in 2 mL of anhydrous analytical-grade DMSO. This gives a 10 mM stock. Minimize light exposure during this process. Store in aliquots of 100 μL or as convenient in eppendorfs at −20 °C, protected from light. Adding 5 μL of this solution to 1 mL of synaptosome suspension will provide the desired final concentration of 50 μM.

– Radioligands: [^{3}H]DA (DAT substrate for uptake experiments) and [^{3}H]WIN 35428 (DAT ligand, for binding experiments) can be purchased from Perkin Elmer (Boston, MA, USA). [^{3}H]DA stock is stored frozen and at 4 °C once thaw. [^{3}H]WIN 35428 ethanolic solution must be stored at −20 °C. Both compounds must be diluted previously as described in the corresponding Subheading 3.

– Antibodies and reagents for Western blot: mouse monoclonal antibody against TH (Cat. N°: 612300 Transduction Labs, Lexington, KY, USA), peroxidase-conjugated anti-mouse IgG antibody (GE Healthcare, Buckinghamshire, UK), beta-actin (mouse monoclonal antibody, Sigma-Aldrich), and chemiluminescence-based detection kit (Immobilon Western, Millipore, USA) are used for TH detection. Equivalent reagents from other brands should also work.

– Western blotting itself would merit a complete chapter as many variations can be performed. For this reason, please refer to [20–22] for general methodology and apparatus required (e.g., tanks, cassettes, power supply) and we will explain the particularities for TH determination in the corresponding section. All the reagents used for preparing the buffers should be of electrophoresis-reagent quality.

– Other materials: scintillation liquid (e.g., Ultima Gold, Perkin Elmer), Bio-Rad Protein Reagent for total protein determination (Bio-Rad Labs. Inc., Hercules, CA), GF/B glass fiber filters (Whatman, Maidstone, UK), polyvinylidene fluoride (PVDF) sheets (Immobilon-P, Millipore, Billerica, MA, USA), vacuum manifold (or Cell Harvester), scintillation counter (beta), borosilicate glass homogenizing tube, standard glass tubes (≈ 6 mL) (*see* **Note 1**), and scintillation plastic vials.

3 Methods

3.1 Measurement of ROS Production in Striatal Synaptosomes

3.1.1 Obtention of Striatal Synaptosomes for ROS Measurement

Although the P2 fraction, a crude synaptosome preparation whose obtention is described in Subheading 3.3, can be used for many purposes (e.g., uptake experiments), measuring ROS production requires a more purified preparation devoid of free mitochondria which could generate ROS outside the synaptosomes and provide inconsistent effects. Several methods are available to purify synaptosomes, including Percoll gradient centrifugation which is delicate to apply. However, the protocol we use is based on that described by Myhre and Fonnum [3] and allows enough purity of the preparation to perform such determinations (Fig. 1). Basically, the modification of the protocol consists in increasing the sucrose concentration when pelleting the synaptosomes, so that the free mitochondria layer separates from the synaptosomes. The protocol to be used is as follows:

The whole experiment must be carried out within the same day, with fresh samples. Freezing of synaptosomes results in loss of some of their functional properties.

1. Animals: for a typical experiment, two rats (Sprague Dawley weighing 250–325 g, but other strains should work) or 7–8 mice (30–35 g, we used Swiss-CD-1 but others should work) (*see* **Note 2**) provide enough material for a typical experiment. Rats are sacrificed by decapitation under isoflurane anesthesia, while mice can be sacrificed by rapid cervical dislocation followed by decapitation.

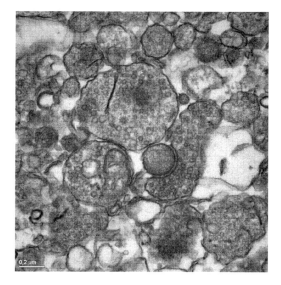

Fig. 1 Representative electron microscopy picture of the synaptosomal fraction obtained by the method described (X 65,000). Reproduced from Escubedo et al., 2009, with permission from Elsevier

2. Dissection of the striatum: immediately after the sacrifice, the skull must be opened, the brain removed and put on a refrigerated surface (hint: a sheet of modeling wax on an icepack provides the appropriate temperature and avoids tissue sticking to the surface). Using thin forceps, separate the cortex following the cerebral longitudinal fissure and cut the white fibers just below (corpus callosum). Separate the two hemispheres and the striata can be seen in the middle of the anterior moiety as two spheroidal corpora. Excise carefully using the forceps and keep in a glass Petri dish or a beaker on ice while processing the rest of animals and until the next step.

3. Weigh the striata and homogenize them in 20 volumes of Tris-sucrose buffer (weight in g x 20 mL/g) using a motor-driven Teflon/glass homogenizer (6 strokes at approximately 1600–1800 rpm). Keep the tube on ice during the homogenization to avoid heating.

4. Filter the homogenate through two layers of surgical gauze in order to retain eventual unbroken tissue. Transfer the homogenate in two equal volumes to centrifuge tubes.

5. Centrifuge at $1000 \times g$ at 4 °C in a refrigerated centrifuge for 10 min.

6. Recover the supernatant with a Pasteur pipette and transfer it to a tube or small beaker (keep always on ice). Then, measure the volume (V) of the supernatant using a micropipette while transferring it to a clean centrifuge tube. Add $V \times 0.6$ mL of cold Tris-sucrose 1.6 M buffer, cap the tube, and mix well by repetitive inversion, so that the final sucrose concentration will be 0.8 M. Split it into two clean centrifuge tubes.

7. Centrifuge at $13,000 \times g$ for 30 min at 4 °C. Then, discard the myelin-rich supernatant. The pellet will appear as a bottom brown-colored layer consisting in mitochondria, covered by a white layer of synaptosomes.

8. Carefully add 1 mL of cold Tris-sucrose buffer sliding along the tube walls and resuspend the white layer by gently shaking or tapping the tube. Avoid resuspending the brown layer. Recover the resuspended synaptosomes with a Pasteur pipette and transfer them to a tube (e.g., a conic 50 mL Falcon-type tube).

9. Dilute the synaptosomes with HBSS-glucose buffer up to a volume around 18 mL. This should give a final protein concentration around 0.1 mg/mL. Keep on ice until starting the incubation steps.

10. This suspension can be distributed in 1 mL aliquots into centrifuge tubes. This is advisable as further centrifugations will be performed after incubation. We prefer using regular centrifuge tubes than eppendorfs, as they permit better shaking of the suspension in the incubation bath.

ROS production is measured by using the fluorochrome 2′,7′-dichlorofluorescin diacetate (DCFH-DA) which passively diffuses through membranes and, after being deacetylated by esterases, is accumulated inside the synaptosomes in the form of 2′,7′-dichlorofluorescine, which is not fluorescent. This compound reacts quantitatively with oxygen species to produce the fluorescent dye 2′,7′-dichlorofluorescein (DCF), whose intensity can be measured to provide an index of oxidative stress.

1. Add 5 µL of DCFH-DA solution to each tube to reach a final concentration of 50 µM. At the same time, modulatory drugs (e.g., antagonists, enzyme inhibitors) are also added to the suspension if needed at a volume of 5–10 µL (*see* **Note 3**). Remember to always perform a control tube of synaptosomes which only will receive the fluorochrome, as it will provide the basal ROS production (*see* **Note 4**). Gently vortex each tube for 2 s. Avoid direct exposure to light which would degrade the fluorochrome, so work under low-light conditions and incubate in the dark or alternatively, with the tube rack covered by aluminum foil.

2. Incubate in a thermostatic shaking bath at 37 °C for 15 min at moderate speed (e.g., 70 shakes/min). During this time, synaptosomes will take up the fluorochrome and the modulatory drugs will equilibrate with its target.

3. Add the oxidative stimulus to test. Use small amounts (e.g., 5–10 µL) to not significantly increase the final volume. A positive control can also be performed in a tube adding 100 µM H_2O_2. Moreover, a nonspecific antioxidant effect of any of the modulatory compounds used can be assessed in tubes containing the compound at the concentration used and 100 µM H_2O_2.

4. Gently vortex and incubate for the final desired time. A preliminary time-course experiment is advisable, with times ranging between 30 min and 2 h. More extended incubation times are not recommended as they may compromise the viability of the synaptosomes.

5. Once the incubation is finished, centrifuge at 13,000 × *g* for 20 min at 4 °C. Discard the supernatant and resuspend the pellet with 1 mL Tris-sucrose buffer by gently vortexing. This step will wash out residual drugs. Repeat the centrifugation and resuspend the final pellets with 0.2 mL HBSS-glucose buffer. Keep the samples on ice and protected from direct light until fluorescence measurements are carried out.

ROS-induced fluorescence can be measured by means of a flow cytometer equipped with an argon laser at an excitation wavelength of 488 nm and detecting emission at 525 nm (*see* **Note 5**). The advantage of using this equipment is that synaptosomes can be

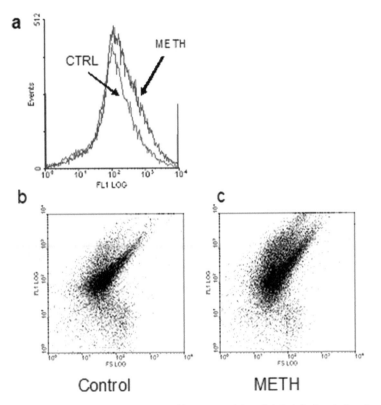

Fig. 2 Representative flow cytometry histograms (**a**) and dot plots (**b**, **c**) showing the change in DCF fluorescence of rat striatal synaptosomes after 2-h incubation at 37 °C alone (Control, CTRL) or with 2 mM methamphetamine (METH). Reproduced from Escubedo et al., 2009, with permission from Elsevier

separated in different populations according to their shape/size and fluorescence, so that different analysis possibilities are available. To perform the measurements, dilutions of the suspension must be made in order to achieve a flow rate of 500–900 synaptosomes/s and each sample should be measured for 1 min. This flow rate is usually achieved by diluting 10 μL of synaptosome suspension in 1 mL of HBSS/glucose buffer. It is advisable to measure triplicates of each sample. Processing of the cytometry data using dedicated software allows generating histograms and dot plots showing the displacement of the fluorescence of the synaptosomal population by effect of the treatment (Fig. 2). Also, the mean fluorescence is given in order to perform comparisons among treatment groups.

Obtain the basal fluorescence by calculating the mean of the control samples. This value will be assigned the reference value of 1 or 100. Normalize the rest of the data with respect to this value for each experiment.

Alternatively, total fluorescence can be assessed transferring the samples to a multiwell plate with black walls and transparent bottom and use a fluorimeter to measure fluorescence using an excitation wavelength of 488 nm and an emission wavelength of 525 nm [3, 23, 24].

3.2 Radioligand Binding and Western Blot Assays

3.2.1 Preparation of Striatal and Cortical Crude Membranes and Cytosolic Fraction from Rodent Brains

Immediately after sacrifice by cervical dislocation and decapitation (mice) or by decapitation under isoflurane anesthesia (rats), the brains are rapidly removed from the skull. Striatum and/or frontal cortex are quickly dissected out on an ice-cold surface as described above, but then they must be frozen on dry ice, and can be stored at $-80\ ^\circ$C until use.

1. The day of the experiment, tissue samples are weighed, thawed, and homogenized with a sonicator homogenizer at cycle mode of 0.95 (for the sonication of heat-sensitive tissue) and amplitude of 95–100 for 20 s (mice) or with a Polytron homogenizer, at around 12,000 rpm, for 15 s (rats) at 4 $^\circ$C in 10 (mice) or 20 (rats) volumes of Tris-sucrose buffer containing protease inhibitors (aprotinin 4.5 μg/μL, 0.1 mM phenylmethylsulfonyl fluoride, and 1 mM sodium orthovanadate) at pH 7.4.

2. The homogenates are centrifuged at 1000 × g for 15 min at 4 $^\circ$C. Aliquots ($\approx$100 μL) of the resulting supernatants can be stored at $-80\ ^\circ$C until use for Western blot experiments. This strategy allows using the tissues from the same animals for both WB and binding assays. If membranes for binding are not going to be used, total tissue lysates in, for example, RIPA buffer [20–22] can also be used for WB experiments.

3. The content of the tube is mixed again by vortexing and centrifuged at 15,000 × g for 30 min at 4 $^\circ$C.

4. The resulting pellets are resuspended (in the same volume used in **step 1.**) in cold Tris-sucrose buffer with protease inhibitors and incubated at 37 $^\circ$C for 5–10 min to remove endogenous neurotransmitters that could interfere the determinations.

5. Samples are centrifuged at 15,000 × g for 30 min at 4 $^\circ$C and resuspended again in the same volume as above. Repeat this centrifugation.

6. Discard the supernatant and resuspend the final pellets in 0.4 or 0.75 mL for mice and 1 or 1.5 mL for rats (*see* **Note 6**) of PBS-sucrose buffer for striatal or cortical membrane preparation, respectively.

7. Reserve an aliquot (20–30 μL) of each sample to determine protein content and store the rest at $-80\ ^\circ$C until use in radioligand binding experiments.

3.2.2 [³H]WIN 35428 Binding Assay in Crude Membrane Preparation

The density of DAT in striatal or frontal cortex membranes from mouse or rat can be measured by [³H]WIN 35428 binding assays, as a marker of dopaminergic nerve terminals (*see* Fig. 3).

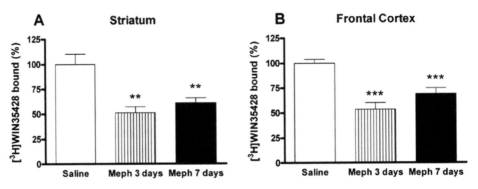

Fig. 3 Effect of a mephedrone (Meph) exposure (3 doses of 25 mg/kg, s.c. at 2 h interval for 2 consecutive days) on DAT density in mouse striatum (**a**) and frontal cortex (**b**). Results are expressed as mean ± SEM from 8 to 10 animals in percentage of control (saline). *$p < 0.05$, **$p < 0.01$, and ***$p < 0.001$ vs. saline. Authors found a relationship between the decrease in the [³H]WIN 35428 specific binding and the decrease in TH expression in the frontal cortex (saline: 100.00 ± 2.46%; mephedrone: 60.04 ± 9.34% $p < 0.01$, 3 days after exposure and 55.30 ± 9.11% $p < 0.001$, 7 days after exposure). Modified from Martínez-Clemente et al., 2014 [27]

1. Crude membrane preparation (*see* Subheading 3.2) must be previously diluted in PBS-sucrose buffer at 4 °C to a concentration of 1 μg/μL. This buffer has been demonstrated to enhance the specific binding of the radioligand [25].

2. The experiment can be performed in borosilicate glass tubes as follows:

   (a) *Total binding tubes,* 50 (striatum) or 100 μL (frontal cortex) of crude membranes preparation previously diluted (50 μg or 100 μg, respectively) and 150 (striatum) or 300 μL (frontal cortex) of PBS-sucrose buffer.

   (b) *Nonspecific binding tubes,* contain the same as total binding tubes but also bupropion (30 μM final concentration) is added so that all the specific binding sites are occupied by this cold ligand and radioligand only will be able to bind nonspecific sites under these conditions.

   Prepare at least duplicates for each condition (sample/total and nonspecific binding tubes).

3. Start the incubation with the addition of 50 (*striatum*) or 100 μL (*frontal cortex*) of [³H]WIN 35428 diluted in PBS-sucrose buffer so that a final radioligand concentration of 5 nM is reached in the binding tubes.

4. Carry out the incubation for 2 h at 4 °C by putting the tubes in a rack on ice. Thereafter, rapidly filter the content of the tubes under vacuum (Vacuum manifold or cell harvester) through GF/B glass fiber filters previously soaked (30 min minimum) in 0.5% polyethyleneimine solution.

5. Rapidly wash tubes and filters three times with 4 mL of ice-cold Tris–HCl 50 mM buffer. Transfer the filters to plastic scintillation vials and add 5–7 mL of scintillation liquid.

6. At least 4 h after adding scintillation liquid, measure radioactivity in the filters by means of a beta scintillation counter.

Calculate the means of each set of replicates and the specific binding by subtracting nonspecific binding from total binding values. Then perform the appropriate statistical analysis between groups.

3.2.3   TH Expression in Cytosolic Fraction

In order to determine the expression of TH in rodents, a general Western blotting and immunodetection protocol can be performed as described elsewhere [26].

1. For each sample (cytosolic fraction obtained from Subheading 3.2.), mix the appropriate amount with sample buffer, boil for 5 min and load 15–20 μL containing 15 μg of protein onto a gel consisting in an upper 5% polyacrylamide stacking part followed by the 10% polyacrylamide separating gel. 1 mm-thick gel is enough.

2. Run the electrophoresis at a constant voltage of 100 V until the protein front crosses the stacking gel and continue electrophoresis at constant intensity of 35 mA until elution of the protein front (blue line).

3. Transfer the proteins to PVDF sheets using a wet transfer tank for 2 h at 0.2 A.

4. Block the PVDF membranes by incubating overnight with 5% defatted milk in TBS-Tween buffer.

5. Then, incubate the membranes for 2 h at room temperature with a primary mouse monoclonal antibody against TH diluted in 5% defatted milk in TBS-Tween buffer (dilution 1:5000). Incubation can be carried out in a small tray or in a capped cylindric tube so that a moderate incubation volume can be used (e.g., 5 mL). Incubation and washing processes must be performed under continuous shaking (e.g., on an orbital shaker or on a tube roller).

6. After washing the membranes with TBS-Tween buffer (3 × 10 min), incubate with a peroxidase-conjugated anti-mouse IgG antibody diluted in TBS-Tween buffer (dilution 1:2500) for 45 min at room temperature.

7. Wash with TBS-Tween (3 × 10 min), add 1–2 mL of reagent from a chemoluminescence-based detection kit following the manufacturer's protocol, and visualize and scan immunoreactive proteins using a luminescence-based gel documentation system (e.g., BioRad ChemiDoc). Scanned blots can then be

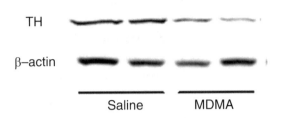

TH

β–actin

Saline   MDMA

Fig. 4 Representative Western blot for TH in striatum from mice treated with saline or MDMA and sacrificed 7 days after the treatment. The corresponding β-actin blots are shown below and were used to normalize the values and correct deviations in total protein loading. MDMA (25 mg/kg, s.c.) was given every 3 h, for a total of three doses. The mice were maintained in an environmental temperature of 26 ± 2 °C and were kept under these conditions until 1 h after the last dose. Adapted from Chipana et al., 2006 [5], with permission from Elsevier

analyzed using dedicated software (e.g., BioRad Image Lab) and dot densities expressed as a percentage of those taken from the controls (*see* Fig. 4).

8. Beta-actin is a cytosolic protein whose expression is not affected by drug treatments, so its immunodetection in the same membrane serves as a control of loading uniformity for each lane and is used to correct differences in TH expression due to protein content. The incubation with the anti-beta-actin antibody (mouse monoclonal, dilution 1:2500) is carried out for 1 h at room temperature. After washing the membranes with TBS-Tween buffer (3 × 10 min), incubate for 30 min at room temperature with a peroxidase-conjugated anti-mouse IgG antibody (dilution 1:2500), wash again (4 × 5 min) and visualize and quantify the chemiluminescence as described above.

9. Normalize each TH value dividing by its corresponding beta-actin. Calculate the means for each treatment group. The values for control group can be set to 100% or 1 and the rest normalized accordingly. Then perform the appropriate statistical analysis.

3.3 Obtention of Striatal Synaptosomes for Uptake Experiments

As mentioned before, crude P2 fraction from synaptosomal preparation can be used for monoamine uptake experiments. The procedure is similar to that described before in Subheading 3.2, with some modifications:

1. Follow the **steps 1–5** from Subheading 3.2.

2. After centrifugation, recover the whole supernatant and centrifuge at $13,000 \times g$ for 30 min at 4 °C.

3. Thereafter, discard the supernatant and dilute the pellet in 3–18 mL (*see* **Note 6**) HBSS-glucose buffer containing pargyline (20 μM) and ascorbic acid (1 mM). Keep on ice until starting the uptake experiment.

3.3.1 Plasmalemmal [³H] DA Uptake Experiments

In order to study the plasmalemmal functionality/density of DAT, [³H]DA uptake assays can be carried out as follows:

1. Synaptosomes from rat striatum are prepared as described above.

2. A thermostatic water bath with shaking is needed. Fill it to a level ensuring that the bottom of the tubes is immersed but they do not float and set it to 37 °C with gentle shaking. Use always an appropriate tube rack and keep all the cautions needed when handling radioactive materials.

3. The experiment should also be performed in borosilicate glass tubes as follows:
    (a) *Total uptake tubes:* 125 μL of HBSS-glucose buffer (containing 20 μM pargyline and 1 mM ascorbic acid) and 100 μL of the corresponding synaptosomal suspension.

    (b) *Nonspecific uptake tubes:* composed by the same as above but the buffer contains cocaine for a final concentration of 300 μM. As these will be incubated at 4 °C, specific uptake should already be blocked, so adding cocaine would be optional. We prefer adding it to ensure the complete blockade of specific uptake, but both options can be found in the literature.

4. Incubation starts with the addition of 25 μL of [³H]DA diluted in HBSS-glucose buffer (containing pargyline and ascorbic acid) for a final concentration of 5 nM to each tube. The three components, buffer, synaptosomal suspension, and [³H]DA stock solution, must be previously warmed separately for 5 min at 37 °C before mixing.

5. The addition of [³H]DA starts the reaction and then the incubation in total uptake tubes is carried out for a further 5 min at 37 °C.

6. For nonspecific uptake, incubate the corresponding tubes at 4 °C, also for 5 min.

7. Then, terminate the uptake reaction by rapid vacuum filtration through Whatman GF/B glass fiber filters (as described in Subheading 3.2.1.) presoaked in 0.5% polyethyleneimine.

8. Rapidly wash tubes and filters three times with 4 mL ice-cold 50 mM Tris–HCl buffer.

9. Transfer the filters into plastic scintillation vials and add 5–7 mL of scintillation liquid.

10. After at least 4 h, measure radioactivity in the filters using a liquid scintillation counter.

Each condition (total and nonspecific uptake tubes) must be performed at least per duplicate. As for binding assays, specific uptake is defined as the difference between the radioactivity measured in the absence (total uptake) and in the presence (non-specific uptake) of an excess of unlabeled DAT blocker (cocaine). Then, calculate the means for each group and perform the appropriate statistical analysis.

4   Notes and Troubleshooting

1. Other tubes or recipients may work (i.e., eppendorfs or plastic tubes) but some radioligands can adhere to plastic, so make sure that this is not your case if plastic is going to be used.

2. Other strains and from different age or weight can be used; however, different total amount of protein (membranes or synaptosomes) must be expected.

3. Some drugs need to be dissolved in DMSO. In this case, prepare stocks so that, after dilution, final concentration of the solvent does not exceed 0.5%.

4. Remember to always perform a control tube to assess basal ROS production.

5. Synaptosomes are not cells, so ask advice to the cytometer technician, who will set up the apparatus properly in order to detect the synaptosomes and will define the appropriate populations to obtain reliable results.

6. Different final volumes can be used in order to increase/decrease protein concentration, depending on how many experiments, conditions, or experimental tubes are required.

References

1. Erinoff L (1995) General considerations in assessing neurotoxicity using neuroanatomical methods. Neurochem Int 26:111–114. https://doi.org/10.1016/0197-0186(94)00105-4

2. Evans GJO (2015) Subcellular fractionation of the brain: preparation of synaptosomes and synaptic vesicles. Cold Spring Harb Protoc 2015:462–466. https://doi.org/10.1101/pdb.prot083469

3. Myhre O, Fonnum F (2001) The effect of aliphatic, naphthenic, and aromatic hydrocarbons on production of reactive oxygen species and reactive nitrogen species in rat brain synaptosome fraction: the involvement of calcium, nitric oxide synthase, mitochondria, and phospholipase A. Biochem Pharmacol 62:119–128. https://doi.org/10.1016/s0006-2952(01)00652-9

4. Escubedo E, Chipana C, Camarasa J (2005) Methyllycaconitine prevents methamphetamine-induced effects in mouse striatum : involvement of α7 nicotinic receptors. J Pharmacol Exp Therap 315:658–667. https://doi.org/10.1124/jpet.105.089748

5. Chipana C, Camarasa J, Pubill D, Escubedo E (2006) Protection against MDMA-induced dopaminergic neurotoxicity in mice by

methyllycaconitine: involvement of nicotinic receptors. Neuropharmacology 51:885–895. https://doi.org/10.1016/j.neuropharm.2006.05.032

6. Chipana C, Torres I, Camarasa J et al (2008) Memantine protects against amphetamine derivatives-induced neurotoxic damage in rodents. Neuropharmacology 54:1254–1263. https://doi.org/10.1016/j.neuropharm.2008.04.003

7. Chipana C, García-Ratés S, Camarasa J et al (2008) Different oxidative profile and nicotinic receptor interaction of amphetamine and 3,4-methylenedioxy-methamphetamine. Neurochem Int 52:401–410. https://doi.org/10.1016/j.neuint.2007.07.016

8. Chipana C, Camarasa J, Pubill D, Escubedo E (2008) Memantine prevents MDMA-induced neurotoxicity. Neurotoxicology 29:179–183. https://doi.org/10.1016/j.neuro.2007.09.005

9. Cadet JL, Krasnova IN (2009) Molecular bases of methamphetamine-induced neurodegeneration. Int Rev Neurobiol 88:101–119. https://doi.org/10.1016/S0074-7742(09)88005-7

10. Song B-J, Moon K-H, V Upreti V et al (2010) Mechanisms of MDMA (ecstasy)-induced oxidative stress, mitochondrial dysfunction, and organ damage. Curr Pharm Biotechnol 11:434–443. https://doi.org/10.2174/138920110791591436

11. Yamamoto BK, Moszczynska A, Gudelsky GA (2010) Amphetamine toxicities: classical and emerging mechanisms. Ann N Y Acad Sci 1187:101–121

12. Marshall V, Grosset DG (2003) Role of dopamine transporter imaging in the diagnosis of atypical tremor disorders. Mov Disord 18:S22–S27. https://doi.org/10.1002/mds.10574

13. Tissingh G, Bergmans P, Booij J et al (1998) Drug-naive patients with Parkinson's disease in Hoehn and Yahr stages I and II show a bilateral decrease in striatal dopamine transporters as revealed by [123I]β-CIT SPECT. J Neurol 245:14–20. https://doi.org/10.1007/s004150050168

14. Volkow ND, Chang L, Wang GJ et al (2001) Association of dopamine transporter reduction with psychomotor impairment in methamphetamine abusers. Am J Psychiatry 158:377–382. https://doi.org/10.1176/appi.ajp.158.3.377

15. Bannon MJ (2005) The dopamine transporter: role in neurotoxicity and human disease. Toxicol Appl Pharmacol 204:355–360. https://doi.org/10.1016/j.taap.2004.08.013

16. Wong DF, Harris JC, Naidu S et al (1996) Dopamine transporters are markedly reduced in Lesch-Nyhan disease in vivo. Proc Natl Acad Sci U S A 93:5539–5543. https://doi.org/10.1073/pnas.93.11.5539

17. Todd G, Noyes C, Flavel SC et al (2013) Illicit stimulant use is associated with abnormal substantia nigra morphology in humans. PLoS One 8:e56438. https://doi.org/10.1371/journal.pone.0056438

18. Ellison G, Eison M, Huberman H, Daniel F (1978) Long-term changes in dopaminergic innervation of caudate nucleus after continuous amphetamine administration. Science (80-) 201:276–278. https://doi.org/10.1126/science.26975

19. Büttner A (2011) Review: the neuropathology of drug abuse. Neuropathol Appl Neurobiol 37:118–134

20. Hirano S (2012) Western blot analysis. Humana Press, Totowa, NJ, pp 87–97

21. Kurien BT, Hal Scofield R (2015) Western blotting: an introduction. In: Western blotting: methods and protocols. Springer, New York, pp 17–30

22. Taylor SC, Posch A (2014) The design of a quantitative western blot experiment. Biomed Res Int 2014

23. Lebel CP, Bondy SC (1990) Sensitive and rapid quantitation of oxygen reactive species formation in rat synaptosomes. Neurochem Int 17:435–440. https://doi.org/10.1016/0197-0186(90)90025-o

24. Mai HN, Sharma N, Shin E-J et al (2018) Exposure to far infrared ray protects methamphetamine-induced behavioral sensitization in glutathione peroxidase-1 knockout mice via attenuating mitochondrial burdens and dopamine D1 receptor activation. Neurochem Res 43:1118–1135. https://doi.org/10.1007/s11064-018-2528-5

25. Coffey LL, Reith MEA (1994) [3H]WIN 35,428 binding to the dopamine uptake carrier. I. Effect of tonicity and buffer composition. J Neurosci Methods 51:23–30. https://doi.org/10.1016/0165-0270(94)90022-1

26. López-Arnau R, Martínez-Clemente J, Abad S et al (2014) Repeated doses of methylone, a new drug of abuse, induce changes in serotonin and dopamine systems in the mouse. Psychopharmacology 231:3119–3129. https://doi.org/10.1007/s00213-014-3493-6

27. Martínez-Clemente J, López-Arnau R, Abad S et al (2014) Dose and time-dependent selective neurotoxicity induced by mephedrone in mice. PLoS One 9:e99002. https://doi.org/10.1371/journal.pone.0099002

Chapter 21

Quantification of Oligodendrocytes and Myelin in Human iPSC-Derived 3D Brain Cell Cultures (BrainSpheres)

David Pamies, Megan Chesnut, Hélène Paschoud, Marie-Gabrielle Zurich, Thomas Hartung, and Helena T. Hogberg

Abstract

Myelination is considered a critical process in the development of the vertebrate brain. This process is susceptible and can be affected by exposure to developmental neurotoxicants or by numerous diseases (e.g., Schizophrenia, bipolar disorder, amyotrophic lateral sclerosis). Studying human myelination has been very difficult due to the lack of in vitro human models capable of reproducing this process. A human 3D iPSC-derived brain model (also called BrainSpheres—BS), developed by Johns Hopkins University, is a multicellular culture that includes different neuronal and glial cell types such as neurons, astrocytes, and oligodendrocytes, and is able to mimic human myelination in vitro. Here, we describe all the methods developed in this model in the last years to quantify oligodendrocytes and myelination. Application of Computer-assisted Evaluation of Myelin (CEM) (developed by Kertman et al.) and other immunochemistry quantification methods are here adapted to a 3D culture BrainSpheres.

Key words Myelin, Oligodendrocytes, IPSC, BrainSpheres, Organotypic model, 3D in vitro culture

1 Introduction

During the development of the nervous system, the processes of glial cells (oligodendrocytes in the central nervous system (CNS) and Schwann cells in the peripheral nervous system (PNS)) wrap membrane around axons to form the myelin sheaths. Myelin is a lipid-rich substance that is produced with the objective to isolate the axons to increase speed and efficiency of the electrical impulse in neurons and is essential for the function of the nervous system [1]. The myelin sheath is discontinuous, and the gaps between the segments of the sheath, called nodes of Ranvier, contain sodium channels [1–4]. Myelination allows fast saltatory conduction of action potentials along these nodes [2, 4] that increases the speed of an action potential along a myelinated axon approximately 10–100 times [5]. This increase reduces the need for ATP-dependent sodium-potassium exchange required for

Jordi Llorens and Marta Barenys (eds.), *Experimental Neurotoxicology Methods*, Neuromethods, vol. 172,
https://doi.org/10.1007/978-1-0716-1637-6_21, © Springer Science+Business Media, LLC, part of Springer Nature 2021

maintenance of the resting membrane potential along axons [6]. Myelin is mostly comprised of lipids (70%) and proteins (30%), and is formed by multiple protein-lipid layers [3, 7, 8]. The main structural proteins of myelin are myelin basic protein (MBP), proteolipid protein (PLP), 2'3'-cyclic-nucleotide 3-'-phospodiesterase (CNP), myelin-associated glycoprotein (MAG), and myelin oligodendrocyte glycoprotein (MOG) [5, 7], and the main structural lipids of myelin are phospholipids, glycolipids, and cholesterol [5, 7–9]. Myelinating oligodendrocytes participate in neuronal signaling networks [10–12] and have a symbiotic relationship with the axon [13], promoting axonal growth, integrity, and survival [2, 6, 13–18]. In addition, a recent study has also demonstrated that oligodendrocytes play a role in inflammation [19]. Myelin is also metabolically active, maintains its own structure [20], regulates the ion composition and fluid volume around axons [21], and releases molecular factors that inhibit axon development and synapse formation, thus contributing to neural plasticity [22].

Areas in the CNS with mainly myelinated axons are termed white matter. Diffusion tensor imaging studies, which allow the mapping of white matter by tractography, have associated its maturation during childhood with improved language, memory [23], and reading performance [24]. Likewise, the failure to form or maintain myelin can disrupt neuronal signal transmission or trigger degradation of axons, and a reduction in the velocity of action potentials can lead to physical or mental disability [4]. Animal studies have implicated white matter deficits during critical periods in development in learning and memory impairments [25, 26] as well as an inability to learn motor skills [27]. Moreover, many human neurologic diseases have been associated with myelin deficits, white matter injury, or dysfunctional oligodendrocytes [4, 28]. These include schizophrenia [29–33], bipolar disorder [32], amyotrophic lateral sclerosis (ALS) [34, 35], depression [36], periventricular leukomalacia (PVL) [37, 38], which can result in cerebral palsy [39], and multiple sclerosis [40], among others.

There are concerns about the increase of developmental diseases, such as autism and attention deficit hyperactivity disorder (ADHD), and their possible link to exposure to environmental chemicals [41–43]. However, only few developmental neurotoxicity (DNT) chemicals have been identified, as the assessment of environmental chemicals for DNT is not done routinely due to the high testing costs resulting from the existing guidelines [44, 45]. In addition, there are concerns regarding the relevance of these tests to human toxicity. Indeed, current DNT guidelines are based on animal testing where the dosage is performed in a different way than in the human exposure scenario. Moreover, the assessment is based on behavior and histology of the animals that are difficult to interpret and translate to human outcomes. In addition, the evaluation of DNT induced by exposure to environmental chemical depends not only on the duration and dose but also on the developmental stage of the brain at the time of exposure [46]. In order to better evaluate DNT, experts have suggested a battery of in vitro human test methods integrated into Integrated Approaches to Testing and Assessment (IATAs) to routinely evaluate compounds [47–49]. This battery must cover the key events (KE) of neurodevelopment at the cellular level (proliferation, migration, apoptosis, differentiation, synaptogenesis, neurite growth, **myelination,**neuronal network formation and function) [49]. European Food Safety Authority (EFSA) has published a detailed report on the evaluation of the currently available in vitro test methods suitable for DNT testing [50] and their readiness for regulatory application and the development of IATAs was recently discussed [51]. A first testing battery of in vitro and non-mammalian assays for DNT has been selected and is currently challenged with approximately 100 chemicals. At the same time, the Organisation for Economic Co-operation and Development (OECD) is in preparation of a guidance document that will describe the testing battery, its usage, and interpretation. However, one KE of neurodevelopment, the myelination process has been a challenge to study in vitro and is currently missing in the testing battery.

Very few in vitro models have shown de novo myelination. A human 3D iPSC-derived brain model (also called BrainSpheres—BS), created at Johns Hopkins University, has shown to produce a multicellular culture that includes different neuronal and glial cell types, such as astrocytes and very importantly oligodendrocytes [52] mimicking the histology of the central nervous system. Several markers for oligodendrocytes (CNP, O1, O4, NOGOA, MOG, MBP) have been evidenced by immunohistochemistry and qRT-PCR [52, 53]. In BrainSpheres up to 40% of the axons are myelinated [52], with a relatively high level of compaction of the myelin sheath, as demonstrated by electron microscopy [52]. Electron microscopy demonstrated that the myelin sheaths are able to wrap axonal structures [52]. Here, we describe a method to assess myelination by quantifying MBP staining in human cells (Fig. 1). The protocol presented is adapted from a recent publication using mouse embryonic stem cells by the Cage group [25] and applied to BrainSpheres [52, 54].

2 Method

2.1 iPSC Generation

iPSCs are cultured on irradiated mouse embryonic fibroblasts (MEFs) in human embryonic stem cell (hESC) medium comprising D-MEM/F12 (Invitrogen), 20% KnockOut™ Serum Replacement (KSR, Invitrogen), 2 mM L-glutamine (Invitrogen), 100μM MEM

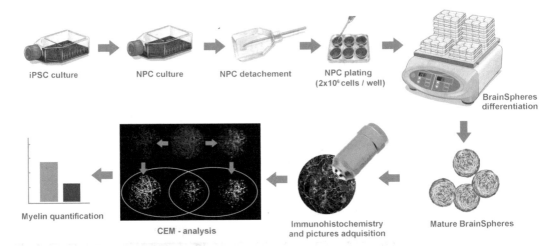

Fig. 1 Diagram of myelin quantification workflow. *CEM* computer-assisted evaluation of myelin formation

NEAA (Invitrogen), 100µM β-mercaptoethanol (Invitrogen), and 10 ng/mL human basic FGF (bFGF, PeproTech). iPSCs colonies are manually picked after 3–6 weeks of expansion for characterization. Only iPSCs with passage ≤20 are used for NPC differentiation. Medium is changed daily and iPSC lines are passaged using collagenase (Invitrogen, 1 mg/mL in D-MEM/F12 for 1 h at 37 °C). Cell lines need to be karyotyped and mycoplasma-free. Feeder-free media can also be used. iPSCs have also been successfully cultured in feeder-free media, using mTeSR™ Plus (STEM CELLS techonologies) on Matrigel (Corning Life Sciences) coated plates.

2.2 NPC Generation

BrainSpheres (BS) are formed from neural progenitor cells (NPCs). NPCs can be generated from pluripotent stem cells (PSC) using different protocols. In our hands, two protocols have been used to generate NPCs; however, other protocols could also be appropriate. NPCs are generated using previously published protocols based on embryonic bodies formation [55], or using the protocol "Induction of Neural Stem Cells from Human Pluripotent Stem Cells Using PSC Neural Induction Medium" (Gibco, Publication Number MAN0008031). Depending on which differentiation protocol is used, two different expansion NPC medias are applied. StemPro® NSC SFM (Life Technologies) is used for NPCs derived from EB and Neural Induction Expansion Medium is used for the NPCs derived from the Gibco protocol. Both protocols have shown to produce quality NPCs and subsequently BrainSpheres. NPCs can be cryopreserved. Specifications can be found in "Cryopreserve NSCs" section in the Gibco protocol.

2.3 BrainSpheres Differentiation	Preparation of the organotypic human BrainSphere model follows a protocol that has been described by Pamies et al. (2017) [52]. Briefly, NPCs thawed from frozen stocks are expanded for 2–3 weeks, mechanically scraped, and dissociated into a single cell suspension. Accutase® can also be used to detach the NPCs and make the single cell solution. Two million cells per well (counted using TrypanBlue (GIBCO) and an automated cell counter (Countess; Invitrogen)) are plated in non-coated 6-well plates in 2 mL of NPC medium and incubated at 37 °C, 5% CO_2 as free-floating cultures under constant gyratory shaking (Kuhner or Thermo Scientific MaxQ 2000 CO2) at 88 rpm, 19 mm orbit. After 2 days, NPCs medium is exchanged to differentiation medium containing Neurobasal® electro medium (GIBCO) supplemented with 1% Penicillin Streptomycin Glutamine (GIBCO), 1% Glutamax (GIBCO), 1% B-27®Electrophysiology supplement (GIBCO), 0.02µg/mL human recombinant glial-cell derived neurotrophic factor (GDNF, Gemini), and 0.02µg/mL human recombinant brain-derived neurotrophic factor (BDNF, Gemini). Cultures are maintained for 8 weeks at 37 °C, 5% CO_2 under constant gyratory shaking at 88 rpm, 19 mm orbit with medium exchange 3 times per week [52]. The treatment of the compound of interest is from week 7 to week 8. No freezing protocol has been used successfully in BrainSpheres yet.
2.4 Chemical Exposure	Cuprizone (CAS number: 370-81-0), a well-known demyelinating compound [56, 57], was used as a reference compound to establish the method. Cuprizone-induced toxicity has been extensively used as a model for demyelinating diseases [56, 57]. Cuprizone oxalic acid bis(cyclohexylidene hydrazide) (SIGMA) produced oligodendrocytes cell death resulting in demyelination [58]. BS are exposed at 7 weeks of the differentiation process. Samples are collected at 8 weeks (after 1 week exposure).
2.5 Immuno-histochemistry and Confocal Microscopy	Confocal microscopy is required to perform myelin quantification according to this method (Fig. 2). For immunohistochemistry, BS are exposed to test compounds from 7 to 8 weeks of differentiation with the compound solution renewed at each media exchanged (3 times per week). After the treatment, aggregates are collected with 1 mL micropipette and collected into 1.5 mL Eppendorf tubes. Media are removed and BS aggregates are washed two times with PBS. PBS used was always without calcium chloride and magnesium chloride (Gibco 14,190-144). Afterwards, BS are fixed for 1 h in 4% paraformaldehyde. After the fixation, BS are washed three times in PBS and incubated for 1 h in blocking solution (5% normal goat serum (NGS) in PBS with 0.4% Triton X-100). After the blocking step, BS are incubated at 4 °C for 48 h with a combination of primary antibodies (NF200, MBP, PLPC1, OLIG1 or O4, Table 1) in 5% NGS, 0.1% Triton X-100 in PBS.

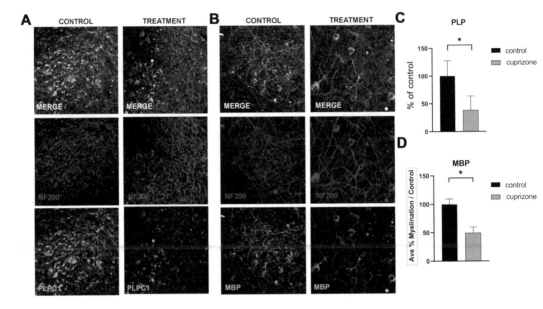

Fig. 2 Myelin quantification on BrainSpheres by CEM. BrainSpheres have been differentiated for 7 weeks. Cuprizone was added to differentiation media to reach a final concentration of 50μM. BrainSpheres were treated for 7 days and collected for immunohistochemistry for (**a**) PLPC1 and NF200, (**b**) MBP and NF200, (**c**) total fluorescence quantification of PLPC1 staining, (**d**) CEM quantification of MBP staining, *Ave* average

Table 1
Primary antibodies

Antibody	Host	Type	Source	Code	Dilution
NF200	Rabbit	Polyclonal	Sigma	N4142	1:500
MBP/SMI99	Mouse	Monoclonal	BioLegend	808,401	1:200
PLPC1	Mouse	Monoclonal	BioRad	MCA839G	1:200
O4	Mouse	Monoclonal	R&D Systems	MAB1326	1:200
OLIG1	Mouse	Monoclonal	Sigma	MAB5540	1:200

Subsequently, BS are washed in PBS three times and incubated with secondary antibody for 1 h in PBS with 5% NGS at room temperature using the proper combination of secondary antibodies (e.g., goat anti-rabbit secondary antibody conjugated with Alexa 594 and goat anti-mouse secondary antibody conjugated with Alexa 488 (Molecular Probes)). For nuclei staining different dyes can be used, including DRAQ5 dye (Cell Signaling; 1:5000 in PBS), Hoechst 33342 (1:10,000 in PBS), or DAPI (1μg/mL, in PBS) for 1 h incubation. After nuclei staining, BS are washed two times in PBS and mounted on slides with coverslips and Prolong Gold antifade reagent (Molecular Probes). Images are taken using a Zeiss UV-LSM 510 confocal microscope or similar. At least

10 pictures per condition are used for the imaging analysis, each photo taken in a different sphere (10 BrainSpheres).

3 Computer-Assisted Evaluation of Myelin (CEM)

For myelin quantification, we use Computer-assisted Evaluation of Myelin (CEM). CEM is an ImageJ plugin developed by Kerman et al., 2015 [59] that can be downloaded for free.[1] In the supplementary material of the Kerman et al., 2015 article a tutorial demonstrates how to use the plugin. In addition, we developed a macro to automatically generate the binary pictures necessary for the analysis (Table 2). Briefly, create a file directory and transfer all images (LSM files) you would like to analyze to one folder. Create a new folder for the binary single channel images that will be generated in this protocol. It is important that neither of these folders is contained within the other. In order to determine the optimal threshold settings for your images, open an image of a control sample and an image from a treated sample (positive control, e.g., Cuprizone) where less myelination is observable. Open the Brightness/Contrast window in ImageJ. Adjust the threshold of the green and red channels for both images by changing the minimum threshold value in the Brightness/Contrast window. Check that background is reduced, but that important features are still observable. Record which values you have selected for the red and green channels. Next, determine which channel numbers correspond to your markers. Record these channel numbers. In the example below, MBP (green) is channel 1, NF (red) is channel 2, and Hoechst 33342 (blue) is channel 3. Edit the macro text in Box 1 to reflect the correct channel numbers (highlighted in the color of the channel, Hoechst, MBP, NF) and replace the words "min" and "max" (highlighted in yellow) with the minimum and maximum threshold values chosen for the green and red channels. If the images from the experiment do not contain a channel with nuclear staining, delete the text for this portion of the macro (red font). Afterward, run the batch binary conversion macro and follow. It is also possible to use total fluorescence for other antibodies related to myelin different than MBP such as PLPC1 (Fig. 3a). After having obtained the files with the binary pictures, follow CEM instructions for myelin quantification. In Fig. 2, we see an example of demyelination after cuprizone treatment for 1 week, having been measured by total fluorescence (Fig. 3a) or CEM (Fig. 3b).

---

[1] http://www.biologists.com/DEV_Movies/DEV116517/DEV116517_Appendix%20S2-CEM%20package.zip

Table 2
Macro for ImageJ to generate binary pictures for CEM

```
macro "Batch binary macro for myelin quantification by MC" {
// chose the folders where the photos are stored and where the binary
folders will be stored //
dir1 = getDirectory("Choose Source Directory");
dir2 = getDirectory("Choose Destination Directory");
list = getFileList(dir1);
// imageJ batch mode //
setBatchMode(true);
for (i=0; i<list.length; i++) {
        showProgress(i+1, list.length);
        open(dir1+list[i]);
// retrieve title of image //
T = getTitle();
// split NF, MBP, and Hoechst channels //
run("Split Channels");
// select Hoechst image and save //
selectWindow("C3-"+T);
saveAs("Tiff", dir2+"Hoechst" + T);
run("Close");
// select MBP image, adjust threshold, convert to binary, and save //
selectWindow("C1-"+T);
//run("Brightness/Contrast...");
setMinAndMax(min,max);
setOption("BlackBackground",true);
run("Make Binary");
saveAs("Tiff", dir2+"Binary_MBP" + T);
run("Close");
// select NF image, adjust threshold, convert to binary, and save //
selectWindow("C2-"+T);
//run("Brightness/Contrast...");
setMinAndMax(min,max);
setOption("BlackBackground",true);
run("Make Binary");
saveAs("Tiff", dir2+"Binary_NF" + T);
run("Close");
setBatchMode(false);
}
run("Close");
```

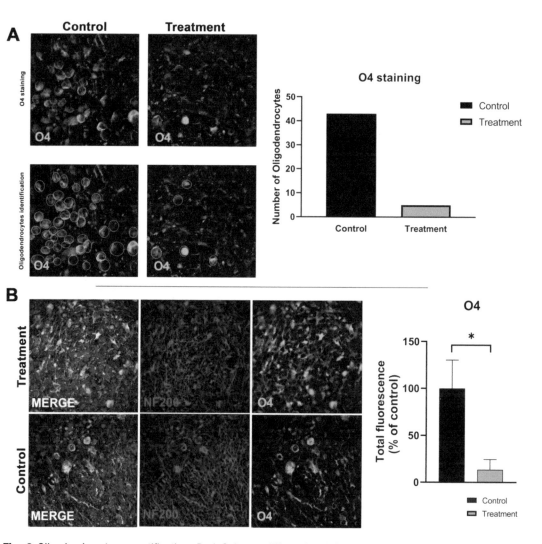

Fig. 3 Oligodendrocytes quantification. BrainSpheres differentiated for 7 weeks. Cuprizone was added to differentiation media to reach a final concentration of 50μM. BrainSpheres were treated for 7 days and collected for immunohistochemistry. (**a**) O4 positive staining identification and manual quantification, (**b**) oligodendrocytes quantification for O4 by total fluorescence imageJ measurement. Total fluorescence is normalized to percentage of control

4 Oligodendrocytes Quantification

For the quantification of oligodendrocytes, BS are differentiated for 8 weeks. Approximately ten randomly selected pictures, from ten BrainSpheres and from at least three experiments, are acquired by confocal microscopy. Oligodendrocytes can be quantified using antibodies for different markers such as CNPase, NOGO1, O4, and O1 [60]. Examples of oligodendrocytes quantification with these antibodies can be found in Pamies et al., 2017 [52] and Zhong et al., 2020 [61]. Quantification should be performed manually by independent investigators working on blinded slides, since oligo-dendrocytes are easy to identify by cell morphology [53] (Fig. 3a) or by using different commercialized software, such as Cellomics' Target Activation image-analysis software package (ThermoFisher Scientific) [52]. Moreover, ImageJ total fluorescence measurement can also be used for quantification (Fig. 3b). More information about oligodendrocytes assessment can be found in Zhong et al., 2020 [53]. In Fig. 3, we see an example of a decrease of O4-positive cells after the exposure to cuprizone by counting the oligodendro-cytes identified (Fig. 3a) and by measuring total fluorescence (Fig. 3b).

5   Conclusions

The test methods presented here enable studying oligodendrocytes and myelin. BS have shown to be reproducible in shape, size, and cell content [52] with a relatively easy differentiation protocol. Moreover, they present unique properties, such as containing func-tional oligodendrocytes and a high degree of myelination for an in vitro system. In addition, the well-known demyelinating com-pound cuprizone decreases myelin integrity, quantified by various techniques described here. However, to assess the predictive per-formance of the test method, additional demyelinating and DNT chemicals are tested in ongoing research.

When the readiness criteria have been met [51], the test method should be incorporated into the in vitro testing battery to assess compounds for DNT. Readiness criteria evaluate whether a method could be potentially acceptable to both regulators and test developers [51]. In order to be able to accept alternative DNT test data, some criteria need to be studied in-depth such as the uncer-tainty due to genetic background, cell type and topography, life-stage, and exposure temporality in dose-response modeling [62, 63]. Readiness criteria have been summarized elsewhere [51]. In addition, such method would be very useful in drug discovery and mechanistic understanding of different myelin-related diseases such as schizophrenia, bipolar disorder, amyo-trophic lateral sclerosis (ALS), depression, periventricular leukoma-lacia (PVL), which can result in cerebral palsy, and multiple sclerosis, among others. Despite decades of research using animal and in vitro models, effective drugs to treat these disorders are still missing. Pharmaceutical companies are seeking new human-relevant alternatives. The BS combined with myelination and oli-godendrocytes quantification as well as other endpoints such as electrical activity, synaptogenesis, and neurite outgrowth has the potential to become such an alternative approach.

Conflict of Interest

TH, HH, and DP are named inventors on a patent by Johns Hopkins University on the production of mini-brains (also called BrainSpheres), which is licensed to AxoSim, New Orleans, LA, USA. They consult AxoSim and TH is shareholder.

References

1. Bunge MB, Pappas GD, Bunge RP (1962) Electron microscopic demonstration of connections between glia and myelin sheaths in developing mammalian central nervous system. J Cell Biol 12(2):448–453

2. Branson HM (2013) Normal myelination: a practical pictorial review. Neuroimaging Clin N Am 23(2):183–195

3. Guleria S, Kelly TG (2014) Myelin, myelination, and corresponding magnetic resonance imaging changes. Radiol Clin N Am 52 (2):227–239

4. Mathews ES, Appel B (2016) Oligodendrocyte differentiation. Zebrafish: cellular and developmental biology. Dev Biol 134:69–96

5. Laule C et al (2007) Magnetic resonance imaging of myelin. Neurotherapeutics 4 (3):460–484

6. Navel KA, Salzer JL (2006) Axonal regulation of myelination by neuregulin 1. Curr Opin Neurobiol 16(5):492–500

7. Podbielska M, Hogan EL (2009) Molecular and immunogenic features of myelin lipids: incitants or modulators of multiple sclerosis? Mult Scler J 15(9):1011–1029

8. van der Knaap MS, Wolf NI, Heine VM (2016) Leukodystrophies five new things. Neurol Clin Pract 6(6):506–514

9. Barkovich AJ (2000) Concepts of myelin and myelination in neuroradiology. Am J Neuroradiol 21(6):1099–1109

10. Bergles DE et al (2000) Glutamatergic synapses on oligodendrocyte precursor cells in the hippocampus. Nature 405(6783):187–191

11. Karadottir R et al (2008) Spiking and nonspiking classes of oligodendrocyte precursor glia in CNS white matter. Nat Neurosci 11 (4):450–456

12. Lin SC, Bergles DE (2004) Synaptic signaling between GABAergic interneurons and oligodendrocyte precursor cells in the hippocampus. Nat Neurosci 7(1):24–32

13. Brady ST et al (1999) Formation of compact myelin is required for maturation of the axonal cytoskeleton. J Neurosci 19(17):7278–7288

14. Funfschilling U et al (2012) Glycolytic oligodendrocytes maintain myelin and long-term axonal integrity. Nature 485(7399):517–521

15. Lee YJ et al (2012) Oligodendroglia metabolically support axons and contribute to neurodegeneration. Nature 487(7408):443–U1502

16. Miller RH (2002) Regulation of oligodendrocyte development in the vertebrate CNS. Prog Neurobiol 67(6):451–467

17. Nave KA, Werner HB (2014) Myelination of the nervous system: mechanisms and functions. Ann Rev Cell Dev Biol 30:503–533

18. Saab AS, Tzvetanova ID, Nave KA (2013) The role of myelin and oligodendrocytes in axonal energy metabolism. Curr Opin Neurobiol 23 (6):1065–1072

19. Peferoen L et al (2014) Oligodendrocyte-microglia cross-talk in the central nervous system. Immunology 141(3):302–313

20. Ledeen RW, Golly F, Haley JE (1992) Axon myelin transfer of phospholipids and phospholipid precursors - labeling of myelin phosphoinositides through axonal-transport (molecular neurobiology, vol 6, pg 177, 1992). Mol Neurobiol 6(4):482

21. Dyer CA (2002) The structure and function of myelin: from inert membrane to perfusion pump. Neurochem Res 27(11):1279–1292

22. Fields RD et al (2017) Cholinergic signaling in myelination. Glia 65(5):687–698

23. Nagy Z, Westerberg H, Klingberg T (2004) Maturation of white matter is associated with the development of cognitive functions during childhood. J Cogn Neurosci 16(7):1227–1233

24. Deutsch GK et al (2005) Children's reading performance is correlated with white matter structure measured by diffusion tensor imaging. Cortex 41(3):354–363

25. Liu J et al (2012) Impaired adult myelination in the prefrontal cortex of socially isolated mice. Nat Neurosci 15(12):1621–1623

26. Makinodan M et al (2012) A critical period for social experience-dependent oligodendrocyte maturation and myelination. Science 337 (6100):1357–1360

27. McKenzie IA et al (2014) Motor skill learning requires active central myelination. Science 346 (6207):318–322

28. Fancy SP et al (2011) Myelin regeneration: a recapitulation of development? Annu Rev Neurosci 34:21–43

29. Hakak Y et al (2001) Genome-wide expression analysis reveals dysregulation of myelination-related genes in chronic schizophrenia. Proc Natl Acad Sci U S A 98(8):4746–4751

30. Najjar S, Pearlman DM (2015) Neuroinflammation and white matter pathology in schizophrenia: systematic review. Schizophr Res 161 (1):102–112

31. Tkachev D et al (2003) Oligodendrocyte dysfunction in schizophrenia and bipolar disorder. Lancet 362(9386):798–805

32. Walterfang M et al (2005) Diseases of white matter and schizophrenia-like psychosis. Aust N Z J Psychiatry 39(9):746–756

33. Yao L et al (2013) White matter deficits in first episode schizophrenia: an activation likelihood estimation meta-analysis. Prog Neuro-Psychopharmacol Biol Psychiatry 45:100–106

34. Kang SH et al (2013) Degeneration and impaired regeneration of gray matter oligodendrocytes in amyotrophic lateral sclerosis. Nat Neurosci 16(5):571–579

35. Philips T, Rothstein JD (2014) Glial cells in amyotrophic lateral sclerosis. Exp Neurol 262:111–120

36. Aston C, Jiang L, Sokolov BP (2005) Transcriptional profiling reveals evidence for signaling and oligodendroglial abnormalities in the temporal cortex from patients with major depressive disorder. Mol Psychiatry 10 (3):309–322

37. Billiards SS et al (2008) Myelin abnormalities without oligodendrocyte loss in periventricular leukomalacia. Brain Pathol 18(2):153–163

38. Volpe JJ et al (2011) The developing oligodendrocyte: key cellular target in brain injury in the premature infant. Int J Dev Neurosci 29 (4):423–440

39. Azzarelli B, Meade P, Muller J (1980) Hypoxic lesions in areas of primary myelination - a distinct pattern in cerebral-palsy. Childs Brain 7 (3):132–145

40. Noseworthy JH et al (2000) Medical progress: multiple sclerosis. N Engl J Med 343 (13):938–952

41. Kuehn BM (2010) Increased risk of ADHD associated with early exposure to pesticides, PCBs. JAMA 304(1):27–28

42. Sagiv SK et al (2010) Prenatal organochlorine exposure and behaviors associated with attention deficit hyperactivity disorder in school-aged children. Am J Epidemiol 171 (5):593–601

43. Grandjean P, Landrigan PJ (2006) Developmental neurotoxicity of industrial chemicals. Lancet 368(9553):2167–2178

44. OECD, Test No. 426: Developmental Neurotoxicity Study (2007) OECD Publishing

45. USEPA, Health effects test guidelines: OPPTS 870.6300, Developmental Neurotoxicity Study [EPA 712–C–98–239]. (1998)

46. Rice D, Barone S Jr (2000) Critical periods of vulnerability for the developing nervous system: evidence from humans and animal models. Environ Health Perspect 108(Suppl 3):511–533

47. Bal-Price A et al (2018) Strategies to improve the regulatory assessment of developmental neurotoxicity (DNT) using in vitro methods. Toxicol Appl Pharmacol 354:7–18

48. EFSA/OECD, Workshop Report on integrated approach for testing and assessment of developmental neurotoxicity (2017) EFSA onlinelibrary

49. Aschner M et al (2017) Reference compounds for alternative test methods to indicate developmental neurotoxicity (DNT) potential of chemicals: example lists and criteria for their selection and use. Altex 34(1):49–74

50. Fritsche E, Alm H, Baumann J, Geerts L, Håkansson H, Masjosthusmann S, Witters H (2015) Literature review on in vitro and alternative developmental neurotoxicity (DNT) testing methods. EFSA Ext Sci Rep 12 (4):778E

51. Bal-Price A et al (2019) Recommendation on test readiness criteria for new approach methods in toxicology: exemplified for developmental neurotoxicity (vol 35, pg 306, 2018). Altex 36(3):506–506

52. Pamies D et al (2017) A human brain microphysiological system derived from induced pluripotent stem cells to study neurological diseases and toxicity. Altex 34(3):362–376

53. Zhong X, Harris G, Smirnova L, Zufferey V, Baldino Russo F, Baleeiro Beltrao Braga PC, Chesnut M, Zurich MG, Hogberg HT, Hartung T, Pamies D (2020) Antidepressant paroxetine exerts developmental neurotoxicity in an iPSC-derived 3D human brain model. Front Cell Neurosci 14:25

54. Hogberg HT et al (2013) Toward a 3D model of human brain development for studying

gene/environment interactions. Stem Cell Res Ther 4:1–7

55. Wen ZX et al (2014) Synaptic dysregulation in a human iPS cell model of mental disorders. Nature 515(7527):414–418

56. Skripuletz T et al (2011) De- and remyelination in the CNS white and grey matter induced by cuprizone: the old, the new, and the unexpected. Histol Histopathol 26(12):1585–1597

57. Matsushima GK, Morell P (2001) The neurotoxicant, cuprizone, as a model to study demyelination and remyelination in the central nervous system. Brain Pathol 11(1):107–116

58. Torkildsen O et al (2008) The cuprizone model for demyelination. Acta Neurol Scand Suppl 188:72–76

59. Kerman BE et al (2015) In vitro myelin formation using embryonic stem cells. Development 142(12):2213–2225

60. Kuhn S et al (2019) Oligodendrocytes in development, myelin generation and beyond. Cell 8 (11):1424

61. Zhong X et al (2020) Antidepressant paroxetine exerts developmental neurotoxicity in an iPSC-derived 3D human brain model. Front Cell Neurosci 14:25

62. Hartung T, Kavlock R, Sturla SJ (2017) Systems toxicology II: a special issue. Chem Res Toxicol 30(4):869

63. Hartung T et al (2017) Systems toxicology: real world applications and opportunities. Chem Res Toxicol 30(4):870–882

Chapter 22

Measurement of Electrical Activity of Differentiated Human iPSC-Derived Neurospheres Recorded by Microelectrode Arrays (MEA)

Kristina Bartmann, Julia Hartmann, Julia Kapr, and Ellen Fritsche

Abstract

Neurotoxicity is caused by a large variety of compound classes and affects all life stages from the developing child to the elderly. Studying for neurotoxicity often involves animal models, which are very resource-intensive and bear the problem of species-differences. Thus, alternative human-based models are needed to overcome these issues. Over the last years, far-reaching advancements in the field of neurotoxicity were made possible by the ability to reprogram human somatic cells into induced pluripotent stem cells (hiPSC). These hiPSCs can be differentiated into neurons and astrocytes, which spontaneously form functional neuronal networks (NN) in vitro. Microelectrode arrays (MEA) are a valuable tool to assess the electrophysiology of such networks. This chapter explains the neural induction of hiPSCs to human neural progenitor cells (hiNPC) in the form of free-floating spheres and their subsequent differentiation into functional neurons on MEAs. The measurement of the electrical network activity, as well as the evaluation of the received data is described.

Key words Neurotoxicology, Neurotoxicity, Human induced pluripotent stem cells (hiPSC), Human induced neural progenitor cells (hiNPC), Microelectrode array (MEA), Neuronal network, Electrical activity

1 Introduction

Adverse effects on the peripheral or central nervous system caused by chemical, biological, or physical agents are referred to as neurotoxicity [1]. Substances such as metals (e.g., lead), industrial chemicals (e.g., acrylamide), pharmaceutical drugs (e.g., doxorubicin), pesticides (e.g., deltamethrin), and natural toxins (e.g., domoic acid) have been shown to cause neurotoxicity [1, 2]. They are often included in industrial, agricultural, and consumer products, which must then be registered and approved by the European Chemical Agency (ECHA) and the European Food Safety

Kristina Bartmann and Julia Hartmann are contributed equally to this chapter.

Jordi Llorens and Marta Barenys (eds.), *Experimental Neurotoxicology Methods*, Neuromethods, vol. 172, https://doi.org/10.1007/978-1-0716-1637-6_22, © Springer Science+Business Media, LLC, part of Springer Nature 2021

Authority (EFSA), prior to entering the European market. Depending on the production volume, different toxicity tests have to be conducted. Current neurotoxicity guideline studies [3, 4] that precede the approval of such products are performed in vivo and are thus highly resource-intensive with regard to time, money, and animals [5]. Additionally, animal and human interspecies variations greatly challenge the translation of the generated data, especially for the nervous system [6–8]. Therefore, we are in urgent need of animal-free alternatives that better mimic human nervous system physiology [9]. The ability to reprogram human somatic cells into induced pluripotent stem cells (hiPSC) [10] has extensively advanced the field of neurotoxicity evaluation [11–14].

The use of human cellular models gave rise to a very important component of hazard identification—the neurophysiological assessment. Cultured in vitro, hiPSCs are able to grow, migrate, and differentiate into functional neuronal networks (NN) [15–19]. Important tools to study the electrophysiology of such NN are microelectrode arrays (MEA). These integrated arrays of electrodes, photoetched into a glass slide or "chip", allow the simultaneous extracellular recording of electrical activity from a large number of individual sites in one tissue [20]. So far, neurotoxicological testing with MEAs has mainly been performed with rodent NN [21–30]. However, the use of human in vitro cultures is preferred, because responses to test compounds might be affected by species differences in toxicodynamics [31–35]. For this reason, human cells are increasingly used to measure neurotoxicity on MEAs [36–40]. In this chapter, we describe the neural induction of hiPSCs to free-floating neural progenitor cells (hiNPC), their differentiation on poly-d-lysin (PDL)/laminin-coated MEAs and the subsequent formation and measurement of functional NN (Fig. 1) [15, 16].

2   Materials

All cell culture procedures need to be performed in a Class II Biological Safety Cabinet under sterile conditions. The cells are cultivated in an incubator at 37 °C and 5% CO_2.

2.1   Generation of Human Induced Neural Progenitor Cells (hiNPC)

2.1.1   Preparation of Neural Induction Medium (NIM)

Mix **DMEM** (high glucose, GlutaMAX™ Supplement, pyruvate, Thermofisher Scientific #31966-021) and **Ham's F12 Nutrient Mix** (GlutaMAX™ Supplement, Thermofisher Scientific #31765-027) 3:1 and add **1% Penicillin-Streptomycin** (10.000 U/mL, Thermofisher Scientific #15140-122), **2% B27™ supplement** (serum-free (50×), Thermofisher Scientific #17504-044), **0.2% Human recombinant epidermal growth factor** (EGF, Thermofisher Scientific #PHG0313), **1% N2 Supplement** (100×, Thermofisher Scientific #17502-048), **20% KnockOut™ Serum**

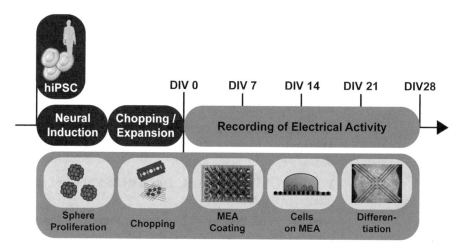

Fig. 1 Experimental setup (adapted from Nimtz et al. [15]). Human iPSCs were neurally induced to hiNPC and cultivated as free-floating 3D neurospheres. Neurospheres proliferate in neural proliferation medium (NPM) supplemented with basic fibroblast growth factor (FGF2) and epidermal growth factor (EGF). By mechanical passaging with a razorblade (chopping), neurospheres are cut into small pieces (100μm) and directly plated onto a PDL-Laminin-coated mwMEA plate. In CINDA the NN differentiate and can be cultured for multiple weeks. The electrical activity can be measured from day 7. *DIV*: days in vitro

Replacement (KSR, Thermofisher Scientific #10828-028), **10μM SB-431542 hydrate** (Sigma #S4317, dissolved at 10 mM in DMSO), and **0.5μM LDN-193189 hydrochloride** (Sigma #SML0559, dissolved at 500μM in ultrapure water).

The medium can be stored at 2–8 °C for at least 3 weeks. Warm it to 37 °C prior to use.

2.1.2 Preparation of Neural Proliferation Medium (NPM)

Mix **DMEM** (high glucose, GlutaMAX™ Supplement, pyruvate, Thermofisher Scientific #31966-021) and **Ham's F12 Nutrient Mix** (GlutaMAX™ Supplement, Thermofisher Scientific #31765-027) 3:1 and add **1% Penicillin-Streptomycin** (10.000 U/mL, Thermofisher Scientific #15140-122), **2% B27™ supplement** (serum-free (50×), Thermofisher Scientific #17504-044), and **0.2% human recombinant epidermal growth factor** (EGF, Thermofisher Scientific #PHG0313). Previously, EGF is dissolved at 10μg/mL in sterile PBS, containing 0.1% bovine serum albumin (BSA, Serva #11920) and 1 mM DL-Dithiothreitol solution (DTT, Sigma #646563-10×). EGF is stored at −20 °C.

The NPM, containing supplements, antibiotics, and growth factors, can be stored up to 2 weeks at 2–8 °C. Prior to use, warm it to 37 °C and add **basic human recombinant fibroblast growth factor** (FGF2, R&D Systems #233-FB) to a final concentration of **20 ng/mL**. FGF2 is dissolved at 10μg/mL in sterile PBS, containing 0.1% bovine serum albumin (BSA, Serva #11920) and 1 mM DL-Dithiothreitol solution (DTT, Sigma #646563-10×). The reconstituted FGF2 can be stored for a maximum of 3 months at −20 °C.

2.1.3 *Preparation of Poly-HEMA Solution*	Dissolve 1.2 g of Poly(2-hydroxyethyl methacrylate) (Poly-HEMA, Sigma-Aldrich #P3932) in 40 mL ethanol (94.8%) using a magnetic stirrer for 5–16 h (*see* **Note 1**).

2.1.4 *Neural Induction*

- iPSC (IMR90) clone (#4), WiCell (*see* **Note 2**).
- mTeSR1, Stemcell Technologies #05850.
- Corning® Matrigel® Growth Factor Reduced (GFR) Basement Membrane Matrix, Phenol Red-free, LDEV-free, Corning #356231.
- Y-27632 (Rock Inhibitor), Tocris #1254, diluted to 10 mM in ultrapure water.
- StemPro™ EZPassage™ Disposable Stem Cell Passaging Tool, Thermofisher Scientific #23181010.
- Corning® cell lifter, Merck #CLS3008.
- McIlwain tissue chopper, Mickle Laboratory Engineering Co. Ltd.
- NIM (*see* Subheading 2.1.1).
- NPM (*see* Subheading 2.1.2).
- Dulbecco's phosphate-buffered saline with $CaCl_2$ and $MgCl_2$ (DPBS, 1×, Gibco).
- Culture dish (Ø10 cm and Ø6 cm).

2.2 Multiwell Microelectrode Arrays (mwMEAs)

2.2.1 *Preparation of Differentiation Medium—CINDA*

Mix **DMEM** (high glucose, GlutaMAX™ Supplement, pyruvate, Thermofisher Scientific #31966-021) and **Ham's F12 Nutrient Mix** (GlutaMAX™ Supplement, Thermofisher Scientific #31765-027) 3:1 and add **1% Penicillin-Streptomycin** (10.000 U/mL, Thermofisher Scientific #15140-122), **2% B27™ supplement** (serum-free (50×), Thermofisher Scientific #17504-044), **1% N2 Supplement** (100×, Thermofisher Scientific #17502-048), **5 mM creatine monohydrate** (Sigma #C3630), **100 U/mL human recombinant interferon-γ** (IFN-γ, Peprotech #300-02), **20 ng/mL human recombinant neurotrophin-3** (Peprotech #450-03) and **20μM L-Ascorbic acid** (Sigma #A5960).

Store the medium at 2–8 °C for up to 2 weeks. Warm the differentiation medium to 37 °C and add **300μM N^6,2'-O-Dibutyryladenosine 3',5'-cyclic monophosphate sodium salt** (Dibutyryl cAMP, Sigma #D0260) prior to use (*see* **Note 3**).

2.2.2 *Differentiation on Poly-D-Lysin/ Laminin-Coated 24-Multiwell MEAs*

- 24-Well plate with PEDOT electrodes on glass (mw-MEA), Multichannelsystems #24W300/30G-288.
- Poly-D-lysine hydrobromide (PDL), Sigma #P0899.
- Laminin from Engelbreth-Holm-Swarm sarcoma basement membrane (Laminin, working solution: $c = 1$ mg/mL), Sigma #L2020 (*see* **Note 4**).

- Autoclaved ultrapure water.
- McIlwain tissue chopper, Mickle Laboratory Engineering Co. LTD.
- Dulbecco's phosphate-buffered saline with $CaCl_2$ and $MgCl_2$ (DPBS, $1\times$, Gibco).
- Counting Chamber Nageotte, Marienfeld #KHY3.1.

2.2.3 Devices and Software for mwMEA Recordings and Analysis

- Multiwell-MEA headstage, Multi Channel Systems MCS GmbH.
- Multiwell-Screen, Version 1.11.6.0, Multi Channel Systems MCS GmbH.
- Multiwell-Analyzer, Version 1.8.6.0, Multi Channel Systems MCS GmbH.

2.3 Lactate-Dehydrogenase Assay

- CytoTox-ONE Homogenous Membrane Integrity Assay Kit (#G7891, Promega; $-20\ °C$).
  - Substrate mix.
  - Assay buffer.
- Triton X-100 (10% in H_2O), Sigma-Aldrich #T8787.
- CINDA (*see* Subheading 2.2.1).
- 96-Well plate compatible with fluorometer.
- Fluorescence plate reader with excitation 530–570 nm and emission 580–620 nm filter pair.

3 Methods

All experiments need to be performed under sterile conditions.

3.1 Neural Induction of hiPSCs into Human Induced Neural Progenitor Cells (hiNPC)

During embryogenesis the human central nervous system develops from the ectoderm, one of the three germ layers that compose the entire body. These complex procedures can now be taken into a dish due to the unique technique of stem cell differentiation. In vitro, pluripotent stem cells, such as hiPSCs, can be directed to the ectodermal lineage by using neural induction media (NIM). The NIM used in this protocol is based on dual SMAD inhibition, and contains the small molecules LDN-193189 and SB431542 [15]. These factors inhibit TGF-β and SMAD signaling pathways and thereby prevent differentiation into non-neural ectodermal directions [41]. The neural induction is performed in Poly-HEMA-coated culture dishes to prevent cell adhesion to the dishes' surface and to ensure neurosphere formation. From day 7, 10 ng/mL FGF2 are added to the NIM. On day 21, the NPCs are transferred to the NPM for further maintenance. With time, the neurospheres grow and need to be reduced in size by cutting into smaller pieces in order to avoid a necrotic sphere core (chopping; [42]). By using this method, neurospheres can be expanded and cultivated of over several months. The number of possible passages depends on the cell line.

1. Add 1.5 mL Poly-HEMA solution to a 6 cm, or 3 mL to a 10 cm, culture dish and distribute the solution evenly across the surface. Let the solution evaporate overnight under sterile conditions or at least for 2 h (*see* **Note 5**).

2. The iPSC colonies are cultivated on Matrigel in a 6-well plate, with mTeSR1 medium. Assess the cell morphology under a microscope. If there are differentiated cells, mark the culture dish at the respective point and scratch off the cells with a pipette (*see* **Note 6**).

3. Add 10μM Y-27632 into the culture medium and swivel the plate in order to distribute it evenly.

4. Incubate the cells for 1 h at 37 °C and 5% CO_2.

5. After the incubation, discard the medium and wash each well with 1 mL pre-warmed DPBS.

6. Add 1 mL NIM with 10μM Y-27632 to the culture and cut the iPSC colonies into pieces by rolling them with StemPro™ EZPassage™ Disposable Stem Cell Passaging Tool once from top to bottom and once from left to right (*see* **Note 7**).

7. Scrape the resulting cell-clusters off the plate using a cell lifter. Use a microscope to make sure that uncut colonies are not lifted from the edge of the well.

8. Transfer the scratched colonies of each two wells to one 6 cm Poly-HEMA-coated culture dish and add 5 mL NIM including 10μM Y-27632 to each dish to reach a final volume of 7 mL.

9. After 2 days, move the culture dish in small circles to collect the formed neurospheres in the middle of the dish and replace 3.5 mL culture medium with the same volume NIM, without Y-27632.

10. Place the dishes into the incubator and carefully move the dish six times horizontally left and right as well as back and forth to distribute the spheres evenly and to avoid clumping.

11. Replace half of the medium every other day by following instructions 9 and 10.

12. Starting from day 7, add 10 ng/mL FGF2 to the NIM.

13. On day 21, collect all neurospheres by gathering them in the middle of the dish and transfer them into a 10 cm Poly-HEMA-coated culture dish. Further cultivate them in 20 mL NPM with 20 ng/mL FGF2.

14. Replace half of the media with fresh NPM every other day, by following instructions 9 and 10.

15. Neurospheres with a diameter of 0.7 mm or above are mechanically passaged with a razor blade (chopped). This is necessary approximately every 7 days.

16. For chopping, use the McIlwan tissue chopper and attach a sterilized razor blade onto the chopping arm. Make sure the razor blade is positioned correctly and screwed on tightly (*see* **Note 8**).

17. Adjust the chop distance to 0.25 mm.

18. Transfer neurospheres from the culture dish to the lid of a sterile 6 cm culture dish, with less medium as possible. Remove the supernatant medium, to prevent moving of neurospheres during the process.

19. Place the lid in the holder of the tissue chopper, move it to the start position, and start the device, by pressing "reset".

20. Stop the tissue chopper when the razor blade reached the end of the dish lid, rotate the lid 90°, and repeat **step 19**.

21. Resuspend the chopped neurospheres in 1 mL NPM by carefully pipetting up and down. Equally distribute the cell suspension into two to three new Poly-HEMA-coated culture dishes (Ø 10 cm), each with 20 mL NPM.

22. Cultivate the cells in an incubator at 37 °C and 5% CO_2.

3.2 Recording Electrical Activity of Neural Networks with mwMEAs

By plating the neurospheres as a monolayer onto an extracellular matrix, here PDL and laminin, and cultivating them in neural differentiation medium, they differentiate into an electrically active NN. Laminin is one of the major integrin interaction factors and, besides facilitating cell adherence, it supports growth, survival, and functional development in iPSC-based neural cultures in vitro [43]. The differentiation medium CINDA contains additional maturation supporting factors (**C**reatine, **I**nterferon-γ, **N**eurotrophin-3, **D**ibutyryl cAMP and **A**scorbic acid) that support neuronal maturation, synapse formation, and spontaneous NN activity [15]. This NN activity can be measured on MEAs, which simultaneously derive the membrane potential of numerous neurons via an electrode recording field. The here used 24-multiwell MEAs have an electrode field of 12 PEDOT-coated gold-electrodes plus four reference electrodes per well. Each electrode has a diameter of 30μm and an interelectrode distance of 300μm. The measured neuronal activity is detected and displayed as spikes and bursts, whereas each spike is derived from one action potential and several consecutive spikes are defined as one burst. Another measurable parameter is network bursts, which occur when the network matures over time and develops a synchronous bursting pattern.

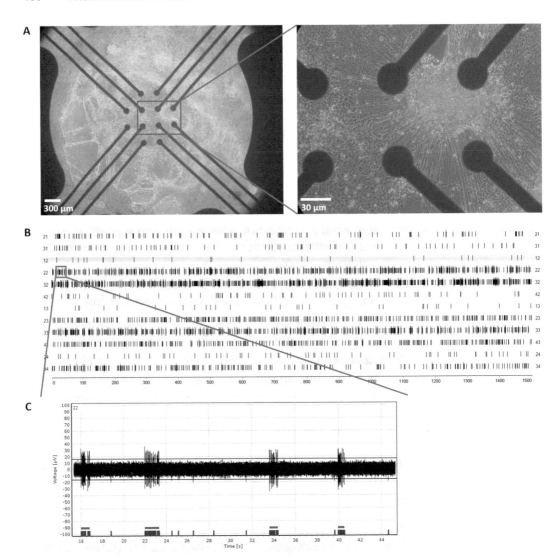

Fig. 2 Measurement of electrical NN activity on mwMEA. (**a**) Representative phase-contrast images of a human NN, differentiated on mwMEA electrodes for 12 days. (**b**) SRP of one well of a 24-mwMEA, after 10 days of differentiation. Each vertical black line represents a spike and each horizontal red bar represents a burst. (**c**) Spike train of the blue marked position in the SRP

During the measurement, the signals are shown as spike raster plots (SRP) and spike trains (Figs. 2 and 3). By setting certain values and thresholds, which have to be adapted for each network, the sensitivity of the measurement can be adjusted and the background noise can be filtered. The measurement data give information about spike count, spike rate [Hz], burst count, mean burst duration [μs], mean burst spike count, percentage of spikes in a burst, mean interburst interval [μs], and number of active electrodes, as well as the same information about network bursts. The combined data set allows a good characterization of the network activity and

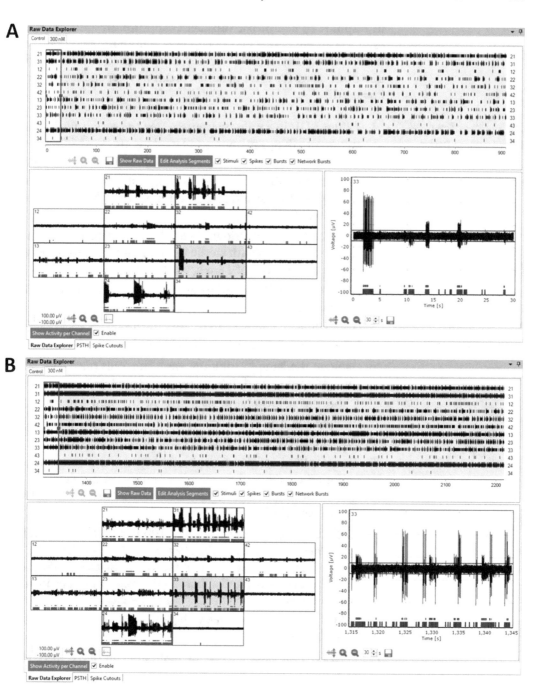

Fig. 3 Neuronal network activity before and during treatment with domoic acid. Domoic acid binds to glutamate receptors with a higher affinity than glutamate itself. This results in an overreaction that can lead to excitotoxicity and finally to cell death. The images show representative software screen shots after 10 DIV of the **a**) baseline activity and **b**) the activity during the treatment with 300 nM domoic acid (after wash-in-phase of 5 min). Each vertical black line represents a spike and each horizontal red line represents a burst. Software: Multiwell-Analyzer, Version 1.8.6.0, Multi Channel Systems MCS GmbH

its response to neurotoxic substance exposure. In order to define a valid experiment, the two parameters minimum spike rate/electrode and number of active electrodes/well need to be monitored. The limit values for these parameters can vary depending on the cell line used.

3.2.1 Differentiation on PDL/Laminin-Coated 24-Multiwell MEAs

1. Dilute PDL in sterile ultrapure water to reach a final concentration of 100µg/mL and store working stocks at $-20\,°C$ until use. Prior to coating, thaw the PDL solution at $37\,°C$.

2. Add 100µL PDL solution (100µg/mL) to each well of a 24-mwMEA plate and incubate at least for 1 h at $37\,°C$ or for 48 h at $4\,°C$.

3. Thaw laminin in the fridge ($4–8\,°C$) to avoid gel formation and dilute the laminin with sterile ultrapure water 1:80 to obtain a final concentration of 12.5 ng/mL.

4. Aspirate the PDL solution and wash wells once with sterile ultrapure water.

5. Add 100µL laminin (12.5µg/mL) solution to each well and incubate at least for 1 h at $37\,°C$ (*see* **Note 9**).

6. Aspirate the laminin solution, wash the wells once with sterile ultrapure water, and directly use the coated plate.

7. Chop the neurospheres to a size of 0.1 mm (for details *see* Subheading 3.1, **step 16–21**).

8. Resuspend the cut spheres in 1 mL prewarmed CINDA.

9. Transfer 80µL of the cut spheres suspension into a Nageotte counting chamber and count the sphere parts. Dilute the cut spheres solution if necessary to reach a maximum of 4000 sphere parts/mL.

10. Collect 200 sphere parts in up to 100µL CINDA and carefully pipet them directly onto the electrodes of a mwMEA well. The solution should form a droplet on top of the electrodes. Allow the spheres to settle and adhere to the well surface, by incubating the droplet for 2 h at $37\,°C$ and 5% CO_2.

11. Add 1 mL CINDA to each well and incubate the plate at $37\,°C$ and 5% CO_2.

12. Feed the cells once a week, by replacing half of the medium in each well.

13. From day 7, measure the electrical activity twice a week as long as the neurons are electrically active (*see* **Note 10**). Preheat the mwMEA headstage to $37\,°C$ and gas the device with carbogen, before adding the mwMEA. For parameter settings *see* Table 1.

14. Let the mwMEA acclimatize for 15 min, before starting the 15 min baseline recording.

Table 1
Parameter settings for mwMEA measurement

Parameter	Setting
Sampling rate	20,000 Hz
Low-pass filter type	Butterworth
Low-pass-filter order	4
Low-pass filter cutoff frequency	3,500 Hz
High-pass filter type	Butterworth
High-pass-filter order	2
High-pass filter cutoff frequency	300 Hz

15. For acute toxicity measurements, stick to the following time schedule:
    - Baseline measurement: 15 min recording.
    - Wash-in-phase: add the substance and equilibrate for 5 min.
    - Treatment: 15 min recording.
    - Cytotoxicity analysis: remove the medium and transfer it into a new 24-well plate for cytotoxicity analysis (*see* Subheading 3.3 for further details).
    - Wash-out-phase: wash each well twice with CINDA to wash out the substance.

16. Further cultivate the cells in fresh CINDA medium at 37 °C and 5% CO_2.

3.2.2 Analysis of mwMEA Recordings

1. Load the .mwr file, which is automatically generated during recording, into the Multiwell-Analyzer Software.

2. Set the "Spike Detector Configuration" to an automatic threshold estimation of 500 ms baseline duration and rising/falling edge of 5× standard deviation (*see* **Note 11**).

3. For the "Burst Detector Configuration" choose the following settings (*see* **Note 12**):
    - Max. Interval to start burst: 100 ms.
    - Max. Interval to end burst: 100 ms.
    - Min. Interval between bursts: 20 ms.
    - Min. Duration of burst: 10 ms.
    - Min. Spike count in burst: 3.

4. To select active wells only, set the "Channel Selection Configuration" to a minimum of 5 spikes/minute (=0.083 Hz) for

each channel and a minimum of 3 active channels for each well [21].

5. Run the analysis.

6. Check each channel for artifacts and defective electrodes (*see* **Note 13**).

7. Export the analysis of all intact electrodes without artifacts as a .csv file.

8. Open the .csv file and copy the data into a software for statistical analysis and data plotting.

3.3 Assessment of Cytotoxicity via the Lactate-Dehydrogenase Assay

To ensure that the adverse effects are due to neurotoxicity and not cytotoxicity, we perform the Lactate-Dehydrogenase (LDH) Assay. The substrate mix contains lactate, NAD^+, and resazurin. If the cell membrane is damaged, the cytosolic lactate dehydrogenase enzyme is released into the cell culture medium and can be quantified by subsequent enzymatic reactions. LDH first catalyzes the conversion of lactate to pyruvate, with a simultaneous reduction of NAD^+ to NADH. By oxidation of NADH to NAD^+, the enzyme diaphorase reduces resazurin to resorufin, which can be measured with a fluorometer (excitation: 540 nm; emission: 590 nm). The amount of produced resorufin is thus proportional to the amount of released LDH. Cells with an intact membrane do not release LDH and therefore no fluorescence is measurable in the culture medium. Two controls are required to perform this assay: a 100% cell lysis control (LC) and a background control (BG; culture medium without cells). By lysing the cells of the LC with Triton X-100, the maximum amount of LDH present is determined.

1. Prepare the CytoTox-ONE Reagent as indicated in the supplier's manual and protect it from light (*see* **Note 14**).

2. For the LC, add 10% (v/v) Triton X-100 solution 1:5 to the desired number of wells (final concentration 2%).

3. Preincubate the LC for 30 min at 37 °C and 5% CO_2.

4. Transfer the complete medium of each well into a new 24-well plate.

5. Transfer 50µL medium of each well of interest of the 24-well plate into a 96-well plate, including the LC and the BG.

6. Add 50µL CytoTox-ONE Reagent to each of the wells.

7. Incubate at room temperature for 2 h, protected from light.

8. Measure the fluorescence of the samples at an excitation wavelength of 540 nm and an emission wavelength of 590 nm in a plate reader (*see* **Note 15**).

9. Calculate the mean of all technical replicate measurements and normalize data by subtracting the mean of the BG from the mean of the different conditions.

10. Calculate the values of each condition as percent of the mean of the LC.

11. The results of at least three independent experiments are pooled, and mean, standard deviation (SD), and standard error of the mean (SEM) are calculated. Data analyses, statistical analyses, and data plotting are performed in GraphPad Prism, using OneWay ANOVA and Bonferroni's post hoc test.

4 Notes

1. The Poly-HEMA solution is stable for up to 2 months at 4 °C.

2. When using a different hiPSC line, the protocol may have to be adapted.

3. Dibutyryl cAMP is sensitive to light and moisture. The CINDA medium should therefore not be exposed to light for more than 1 h.

4. The product concentration depends on the batch number and has to be adjusted to 1 mg/mL with autoclaved ultrapure water.

5. Poly-HEMA-coated dishes can be used for up to 3 months if stored sealed at room temperature and in the dark.

6. iPSCs are small and round in shape and have a high nucleus-to-cytoplasm ratio, with prominent nucleoli. Differentiating cells have a lower nucleus-to-cytoplasm ratio and the morphology differs visibly from the original round shape.

7. We banked our hiPSCs so that we can start each neural induction with a similar passage number. We use the cells starting from passage 3 post-thawing at the earliest and up to passage 10 post-thawing at the latest. This must be adapted for other cell lines.

8. Each side of a razor blade can be used three times.

9. PDL/Laminin-coated plates can be stored in the refrigerator (4–8 °C) for a maximum of 2 weeks prior to use.

10. During cell differentiation, the burst behavior changes and begins to synchronize as the network matures. For this reason, exposure to the same substance in different experiments should always be carried out in the same time frame.

11. This eliminates the detected background signals.

12. These settings need to be adapted to the specific cell signals and vary depending on the cell line.

13. An artifact can be caused by external influences and is visible as an exactly simultaneous signals on all electrodes. Furthermore,

artifacts and defective electrodes can be excluded by observing the spike signal that should resemble a waveform.

14. Aliquot unused CytoTox-ONE reagent and protect the reagent from light. Aliquots should be labeled with preparation date and reagent number and can be stored tightly capped at −20 °C for 6–8 weeks.

15. If the plate reader measures from above, remove the lid of the plate before measurement.

References

1. Costa LG, Giordano G, Guizzetti M, Vitalone A (2008) Neurotoxicity of pesticides: a brief review. Front Biosci 13:1240–1249. https://doi.org/10.2741/2758

2. Massaro EJ (2002) Handbook of neurotoxicology. Humana Press, Totowa, NJ

3. OECD guideline for the testing of Chemicals (1997) Test no. 424: neurotoxicity study in rodents. In: OECD guideline for the testing of chemicals. OECD, Paris, pp 1–15

4. Forum RA (1998) Guidelines for neurotoxicity risk assessment

5. Bal-Price AK, Suñol C, Weiss DG et al (2008) Application of in vitro neurotoxicity testing for regulatory purposes: symposium III summary and research needs. Neurotoxicology 29:520–531. https://doi.org/10.1016/j.neuro.2008.02.008

6. Leist M, Hartung T (2013) Inflammatory findings on species extrapolations: humans are definitely no 70-kg mice. Arch Toxicol 87:563–567

7. Borrell V, Götz M (2014) Role of radial glial cells in cerebral cortex folding. Curr Opin Neurobiol 27:39–46

8. Toutain P-L, Ferran A, Bousquet-Melou A (2010) Species differences in pharmacokinetics and pharmacodynamics. In: Comparative and Veterinary Pharmacology, Handbook of Experimental Pharmacology, Springer, pp 19-48. https://doi.org/10.1007/978-3-642-10324-7_2

9. Masjosthusmann S, Barenys M, El-Gamal M et al (2018) Literature review and appraisal on alternative neurotoxicity testing methods. EFSA Support Publ 15:1–108. https://doi.org/10.2903/sp.efsa.2018.en-1410

10. Takahashi K, Tanabe K, Ohnuki M et al (2007) Induction of pluripotent stem cells from adult human fibroblasts by defined factors. Cell 131:861–872. https://doi.org/10.1016/j.cell.2007.11.019

11. Pei Y, Peng J, Behl M et al (2016) Comparative neurotoxicity screening in human iPSC-derived neural stem cells, neurons and astrocytes. Brain Res 1638:57–73. https://doi.org/10.1016/j.brainres.2015.07.048

12. Barenys M, Fritsche E (2018) A historical perspective on the use of stem/progenitor cell-based in vitro methods for neurodevelopmental toxicity testing. Toxicol Sci 165:10–13. https://doi.org/10.1093/toxsci/kfy170

13. de Groot MWGDM, Westerink RHS, Dingemans MML (2013) Don't judge a neuron only by its cover: neuronal function in in vitro developmental neurotoxicity testing. Toxicol Sci 132:1–7. https://doi.org/10.1093/toxsci/kfs269

14. Tukker AM, De Groot MWGDM, Wijnolts FMJ et al (2016) Research article is the time right for in vitro neurotoxicity testing using human iPSC-derived neurons? ALTEX 33:261–271. https://doi.org/10.14573/altex.1510091

15. Nimtz L, Hartmann J, Tigges J et al (2020) Characterization and application of electrically active neuronal networks established from human induced pluripotent stem cell-derived neural progenitor cells for neurotoxicity evaluation. Stem Cell Res 45:101761. https://doi.org/10.1016/j.scr.2020.101761

16. Hofrichter M, Nimtz L, Tigges J et al (2017) Comparative performance analysis of human iPSC-derived and primary neural progenitor cells (NPC) grown as neurospheres in vitro. Stem Cell Res 25:72–82. https://doi.org/10.1016/j.scr.2017.10.013

17. Izsak J, Seth H, Andersson M et al (2019) Robust generation of person-specific, synchronously active neuronal networks using purely isogenic human iPSC-3D neural aggregate cultures. Front Neurosci 13. https://doi.org/10.3389/fnins.2019.00351

18. Hyvärinen T, Hyysalo A, Kapucu FE et al (2019) Functional characterization of human pluripotent stem cell-derived cortical networks differentiated on laminin-521 substrate: comparison to rat cortical cultures. Sci Rep 9:17125. https://doi.org/10.1038/s41598-019-53647-8

19. Pamies D, Barreras P, Block K et al (2017) A human brain microphysiological system derived from induced pluripotent stem cells to study neurological diseases and toxicity. ALTEX 34:362–376. https://doi.org/10.14573/altex.1609122

20. Johnstone AFM, Gross GW, Weiss DG et al (2010) Microelectrode arrays: a physiologically based neurotoxicity testing platform for the 21st century. Neurotoxicology 31:331–350

21. McConnell ER, McClain MA, Ross J et al (2012) Evaluation of multi-well microelectrode arrays for neurotoxicity screening using a chemical training set. Neurotoxicology 33:1048–1057. https://doi.org/10.1016/j.neuro.2012.05.001

22. Hogberg HT, Sobanski T, Novellino A et al (2011) Application of micro-electrode arrays (MEAs) as an emerging technology for developmental neurotoxicity: evaluation of domoic acid-induced effects in primary cultures of rat cortical neurons. Neurotoxicology 32:158–168. https://doi.org/10.1016/j.neuro.2010.10.007

23. Mack CM, Lin BJ, Turner JD et al (2014) Burst and principal components analyses of MEA data for 16 chemicals describe at least three effects classes. Neurotoxicology 40:75–85

24. Shafer TJ, Brown JP, Lynch B et al (2019) Evaluation of chemical effects on network formation in cortical neurons grown on microelectrode arrays. Toxicol Sci 169:436–455. https://doi.org/10.1093/toxsci/kfz052

25. Alloisio S, Nobile M, Novellino A (2015) Multiparametric characterisation of neuronal network activity for in vitro agrochemical neurotoxicity assessment. Neurotoxicology 48:152–165. https://doi.org/10.1016/j.neuro.2015.03.013

26. Shafer TJ, Rijal SO, Gross GW (2008) Complete inhibition of spontaneous activity in neuronal networks in vitro by deltamethrin and permethrin. Neurotoxicology 29:203–212. https://doi.org/10.1016/j.neuro.2008.01.002

27. Defranchi E, Novellino A, Whelan M et al (2011) Feasibility assessment of microelectrode chip assay as a method of detecting neurotoxicity in vitro. Front Neuroeng 4:1–12. https://doi.org/10.3389/fneng.2011.00006

28. Valdivia P, Martin M, LeFew WR et al (2014) Multi-well microelectrode array recordings detect neuroactivity of ToxCast compounds. Neurotoxicology 44:204–217. https://doi.org/10.1016/j.neuro.2014.06.012

29. Wallace K, Strickland JD, Valdivia P et al (2015) A multiplexed assay for determination of neurotoxicant effects on spontaneous network activity and viability from microelectrode arrays. Neurotoxicology 49:79–85. https://doi.org/10.1016/j.neuro.2015.05.007

30. Novellino A, Scelfo B, Palosaari T et al (2011) Development of micro-electrode array based tests for neurotoxicity: assessment of Interlaboratory reproducibility with neuroactive chemicals. Front Neuroeng 4:1–14. https://doi.org/10.3389/fneng.2011.00004

31. Gassmann K, Abel J, Bothe H et al (2010) Species-specific differential ahr expression protects human neural progenitor cells against developmental neurotoxicity of PAHs. Environ Health Perspect 118:1571–1577. https://doi.org/10.1289/ehp.0901545

32. Dach K, Bendt F, Huebenthal U et al (2017) BDE-99 impairs differentiation of human and mouse NPCs into the oligodendroglial lineage by species-specific modes of action. Sci Rep 7:1–11. https://doi.org/10.1038/srep44861

33. Masjosthusmann S, Becker D, Petzuch B et al (2018) A transcriptome comparison of time-matched developing human, mouse and rat neural progenitor cells reveals human uniqueness. Toxicol Appl Pharmacol 354:40–55. https://doi.org/10.1016/j.taap.2018.05.009

34. Oberheim NA, Takano T, Han X et al (2009) Uniquely hominid features of adult human astrocytes. J Neurosci 29:3276–3287. https://doi.org/10.1523/JNEUROSCI.4707-08.2009

35. Perreault M, Feng G, Will S et al (2013) Activation of TrkB with TAM-163 results in opposite effects on body weight in rodents and non-human primates. PLoS One 8:e62616. https://doi.org/10.1371/journal.pone.0062616

36. Tukker AM, Wijnolts FMJ, de Groot A, Westerink RHS (2018) Human iPSC-derived neuronal models for in vitro neurotoxicity assessment. Neurotoxicology 67:215–225. https://doi.org/10.1016/j.neuro.2018.06.007

37. Tukker AM, Wijnolts FMJ, de Groot A, Westerink RHS (2020) Applicability of hiPSC-derived neuronal co-cultures and rodent primary cortical cultures for in vitro seizure liability assessment. Toxicol Sci. https://doi.org/10.1093/toxsci/kfaa136

38. Tukker AM, Bouwman LMS, van Kleef RGDM et al (2020) Perfluorooctane sulfonate (PFOS) and perfluorooctanoate (PFOA) acutely affect human α1β2γ2L GABAA receptor and spontaneous neuronal network function in vitro. Sci Rep 10. https://doi.org/10.1038/s41598-020-62152-2

39. Ylä-Outinen L, Heikkilä J, Skottman H et al (2010) Human cell-based micro electrode array platform for studying neurotoxicity. Front Neuroeng 3. https://doi.org/10.3389/fneng.2010.00111

40. Odawara A, Matsuda N, Ishibashi Y et al (2018) Toxicological evaluation of convulsant and anticonvulsant drugs in human induced pluripotent stem cell-derived cortical neuronal networks using an MEA system. Sci Rep 8:10416. https://doi.org/10.1038/s41598-018-28835-7

41. Qi Y, Zhang XJ, Renier N et al (2017) Combined small-molecule inhibition accelerates the derivation of functional cortical neurons from human pluripotent stem cells. Nat Biotechnol 35:154–163. https://doi.org/10.1038/nbt.3777

42. Svendsen CN, Ter Borg MG, Armstrong RJE et al (1998) A new method for the rapid and long term growth of human neural precursor cells. J Neurosci Methods 85:141–152. https://doi.org/10.1016/S0165-0270(98)00126-5

43. Farrukh A, Zhao S, del Campo A (2018) Microenvironments designed to support growth and function of neuronal cells. Front Mater 5:1–22. https://doi.org/10.3389/fmats.2018.00062

Correction to: Application of Fluorescence Microscopy and Behavioral Assays to Demonstrating Neuronal Connectomes and Neurotransmitter Systems in *C. elegans*

Omamuyovwi M. Ijomone, Priscila Gubert, Comfort O. A. Okoh, Alexandre M. Varão, Leandro de O. Amaral, Oritoke M. Aluko, and Michael Aschner

Correction to:
Chapter 18 in: Jordi Llorens and Marta Barenys (eds.),
Experimental Neurotoxicology Methods, Neuromethods, vol. 172,
https://doi.org/10.1007/978-1-0716-1637-6_18

Owing to an oversight on the part of Springer, the formula [rate of movement in seeded plates – the rate of movement in unseeded plates] / [rate of movement in unseeded plates] in chapter 18 was published with errors. The correct formula is given here:

> [rate of movement in unseeded plates – the rate of movement in seeded plates] / [rate of movement in seeded plates]

The updated original version of this chapter can be found at
https://doi.org/10.1007/978-1-0716-1637-6_18

Jordi Llorens and Marta Barenys (eds.), *Experimental Neurotoxicology Methods*, Neuromethods, vol. 172,
https://doi.org/10.1007/978-1-0716-1637-6_23, © Springer Science+Business Media, LLC, part of Springer Nature 2023

INDEX

Jordi Llorens and Marta Barenys (eds.), *Experimental Neurotoxicology Methods*, Neuromethods, vol. 172,
https://doi.org/10.1007/978-1-0716-1637-6, © Springer Science+Business Media, LLC, part of Springer Nature 2021

Printed in the United States
by Baker & Taylor Publisher Services